Transport Processes in the Middle Atmosphere

NATO ASI Series

Advanced Science Institutes Series

A series presenting the results of activities sponsored by the NATO Science Committee, which aims at the dissemination of advanced scientific and technological knowledge, with a view to strengthening links between scientific communities.

The series is published by an international board of publishers in conjunction with the NATO Scientific Affairs Division

A	Life Sciences	Plenum Publishing Corporation
B	Physics	London and New York
C	Mathematical and Physical Sciences	D. Reidel Publishing Company Dordrecht, Boston, Lancaster and Tokyo
D	Behavioural and Social Sciences	Martinus Nijhoff Publishers
E	Applied Sciences	Dordrecht, Boston, Lancaster
F	Computer and Systems Sciences	Springer Verlag
G	Ecological Sciences	Berlin, Heidelberg, New York, London,
H	Cell Biology	Paris, and Tokyo

Series C: Mathematical and Physical Sciences Vol. 213

Transport Processes in the Middle Atmosphere

edited by

Guido Visconti

Department of Physics,
University of L'Aquila, Italy

Co-editor

Rolando Garcia

National Center for Atmospheric Research,
Boulder, U.S.A.

Springer-Science+Business Media, B.V.

Published in cooperation with NATO Scientific Affairs Division

Proceedings of the NATO Advanced Research Workshop on
Transport Processes in the Middle Atmosphere
Erice, Italy
November 23-27, 1986

Library of Congress Cataloging in Publication Data

NATO Advanced Research Workshop on Transport Processes in the Middle Atmosphere
(1986 : Erice, Sicily)
   Transport processes in the middle atmosphere / edited by Guido Visconti; co-editor,
Rolando Garcia.
      p.   cm — (NATO ASI series. Series C, Mathematical and physical sciences; vol. 213)
      "Proceedings of the NATO Advanced Research Workshop on Transport Processes in the
Middle Atmosphere, Erice, Italy, November 23–27, 1986"—T.p. verso.
      "Published in cooperation with NATO Scientific Affairs Division."
      Includes index.
      ISBN 978-94-010-8262-4   ISBN 978-94-009-3973-8 (eBook)
      DOI 10.1007/978-94-009-3973-8

      1. Middle atmosphere. 2. Dynamic meteorology. I. Visconti, Guido. II. Garcia,
Rolando Victor, 1919–   . III. Title. IV. Series: NATO ASI series. Series C, Mathematical and physical sciences; no. 213.
QC881.2.M53N38   1986
551.5'153—dc19

                                                                  87–18799
                                                                     CIP

TABLE OF CONTENTS

PREFACE

The NATO Advanced Research Workshop on "Transport Processes in the Middle Atmosphere" was held in Erice, Sicily, from November 23 through November 27, 1986. In addition to NATO, the workshop was supported by the International School of Atmospheric Physics of the Ettore Majorana Center for Scientific Culture, and by the National Research Council of Italy. The Organizing Committee was fortunate to enlist the participation of many of the experts in the field, and this book is an account of their contributions. In order to expedite publication and keep the results "as fresh as possible", it was decided to forego formal review of the papers; instead, the authors were asked to solicit internal reviews from their colleagues. Further, each paper was thoroughly discussed and criticized during the meeting, and those discussions have been taken into account in the preparation of the final version of the manuscripts. Occasional short presentations were made by some of the Workshop participants who wished to provide information complementary to that given in the invited talks. These presentations are not included in this book, which contains only the invited papers.

The book is organized into five chapters corresponding to the different topics covered by the Workshop. The first two chapters contain general reviews of the dynamical climatology of the middle atmosphere and of the growing body of data available on the distribution of chemical constituents. The chapter on dynamical climatology places particular emphasis on gravity waves, and the quasi-biennial and semiannual oscillations (see, e.g., the papers by Hamilton and Fritts). Gravity waves are increasingly recognized as crucially important in the dynamics of the mesosphere and lower thermosphere, while consideration of the QBO and SAO is essential for a meaningful explanation of tracer distributions at low latitudes. The role of planetary scale waves is examined in the paper by O'Neill and Pope, who use Ertel's potential vorticity to study meridional transport by large amplitude waves.

The observations and analyses of chemical constituents presented in the second chapter are based mostly on satellite data. Nevertheless, recent balloon-sonde observations of the variance of ozone and other chemical constituents described by Roth and Ehhalt demonstrate once again the advantages of a combination of measuring techniques for obtaining a complete description of constituent distributions. A possible interpretation of the observed variance is given by Holton in Chapter 3. Problems associated with measuring techniques are reported in the last chapter. At the time of the Workshop, initial results from the U.S. National Ozone Expedition were becoming available, and a discussion of the Antarctic "ozone hole" phenomenon was unavoidable. Although this topic was not part of the original agenda, an entire afternoon was devoted to it and a review by Schoeberl of certain observations relating to ozone and temperature changes is included in Chapter 2.

Interpretation of climatology and other data is the aim of physical theory and numerical modeling. These aspects are the subject of the third and fourth chapters of this volume. Comprehensive reviews on the different aspects of the transport problem by Andrews, McIntyre and Plumb are included, together with recent developments in the interpretation of some specific features. The review papers emphasize the importance of such concepts as planetary wave breaking and the Eliassen-Palm flux divergence for understanding large scale transport in the stratosphere. A parameterization of the EP flux in terms of mean quantities is suggested by Tung as a way to solve in a self-consistent manner the transport problem in two-dimensional models. The papers by Lyjak and Austin show how calculation of parcel trajectories can be applied to estimate eddy diffusivities and to test the accuracy of chemical schemes, respectively. Applications of some of these ideas to two-dimensional modeling are described by Brasseur and Hitchman. Recent advances in three-dimensional dynamical/chemical modeling are presented by Mahlman and Umscheid, and Grose et al.

The fourth chapter also deals with data interpretation and modeling but with special emphasis on radiative processes. The introduction of the Lagrangian and residual Eulerian mean circulations has simplified the description of transport in the meridional plane, although in most two-dimensional models a "diabatic" circulation is used instead. The accuracy and the problems related to this approximation are discussed in the papers by Beagley and Harwood, and Remsberg. Fels offers a very clear and simple summary of the basic concepts developed in the chapter.

The final chapter is devoted to measurement techniques and contains reviews of satellite and ground based methods. The paper by Salby explores the information content and limitations of satellite measurements which are intrinsically asynoptic. Rottger and Chanin and Hauchecorne give general reviews on the results obtained both with MST and lidar techniques that have contributed so much in the last few years to the understanding of the upper atmosphere. One wishes only that more stations using such instrumentation could be built and operated as a network.

The organization of the meeting and the publication of the book required the efforts of many people. G. Fiocco and J. Pyle, members of the Organizing Committee, worked hard to help prepare a comprehensive program. The generosity of the Ettore Majorana Center and of its President, Prof. A. Zichichi, must be acknowledged especially because this Workshop marked the renaissance of the dormant School on Atmospheric Physics. On the part of NATO the kindness and support of L. Da Cunha, director of the Scientific Affairs Division, has been essential. We are especially grateful to R. Cicerone, director of the Atmospheric Chemistry Division, for his hospitality during the stay of G.V. at NCAR in the summer of 1986, and to Donna Sanerib, whose assistance with typing, correspondence and other logistical details was invaluable. In Italy, Gianna Vittorini helped on a part time but enthusiastic basis. Without their help and the support and understanding of our families the Workshop could not have been possible. Finally, one of us (G.V.) wishes to give special thanks to all the Workshop participants who came to Erice even after asking themselves "Who is this guy and what does he want from me?"

Boulder, Colorado   Guido Visconti
August 3, 1987   Rolando R. Garcia

1. DYNAMICAL CLIMATOLOGY OF THE MIDDLE ATMOSPHERE

TROPOSPHERE-STRATOSPHERE GENERAL CIRCULATION STATISTICS

Marvin A. Geller and Mao-Fou Wu
Laboratory for Atmospheres
NASA/Goddard Space Flight Center
Greenbelt, MD   20771

ABSTRACT. Four years of NOAA/NMC temperature data are used to calculate mean zonal winds, planetary wave structures, and eddy heat fluxes. It is found that the interhemispheric differences in planetary wave activity give rise to substantial differences in the annual variation of the mean zonal winds and temperatures. The annual variation in the observed total ozone distribution is also found to be very different in the two hemispheres as a result of the differing ozone transports accompanying the observed planetary wave structures.

1. INTRODUCTION

Recently, there has been much research activity directed toward using newly available stratospheric satellite data to obtain an improved picture of the stratospheric general circulation. Some of the recent papers of this type are those of Hamilton (1982), Smith (1983), Geller et al. (1983, 1984), Hirota et al. (1983), and Hartmann et al. (1984). There have also been recent efforts to derive various representations of the mean meridional circulations from stratospheric satellite temperature data using radiative heating calculations with the thermodynamic equation (e.g., Kiehl and Solomon, 1986, Solomon et al., 1986, and Rosenfield et al., 1987). Recently, there have also been efforts to calculate ozone transports from satellite data such as the works of Wu et al. (1985, 1987). It is the purpose of this paper to discuss some of the main findings of these studies of the stratospheric general circulation. We will do this using the results derived from the four years of NOAA/NMC data that have served as the basis for the studies of Geller et al. (1983, 1984), Rosenfield et al. (1987), and Wu et al. (1987). The results of other papers will be discussed within this context.

The NOAA/NMC data set has been described extensively in Geller et al. (1983) and succeeding papers. Two points should be repeated here however. First, although this data set extends up to the 0.4 mbar level (about 55 km), the temperatures at the very highest levels (2 mbar and above) are not considered reliable since there is very little

3

G. Visconti and R. Garcia (eds.), Transport Processes in the Middle Atmosphere, 3–17.

radiance information at these levels. The publication by Rodgers
(1984) gives more information on this. Another point is that these
four years of data are derived from a succession of instruments with
differing analysis techniques. Temperature corrections have been used
to give consistency throughout the data set, but these corrections
were derived from comparison between analyses of the satellite-derived
temperature fields and rocketsonde data that are almost exclusively
representative of the northern hemisphere low and middle latitudes.
Finally, the method of analysis applied to this data set uses hydrostatic buildup of the geopotential field from 1000 mb with subsequent
application of the geostrophic equations to derive the eddy wind
fields. Thus, derived quantities depend both on the geostrophic
assumption and on the lower boundary condition of the geopotential
height field. Consideration of these points leads us to exercise care
against overinterpreting the derived results at high levels and in the
southern hemisphere. In the following, we will point out results that
we believe are independent of these shortcomings of this data set.
The reasons for our confidence in our presented results stem from
verifications using other data sets as well as these features being
sufficiently large and consistent in their variation.

2. RESULTS

2.1 Temperatures and Mean Zonal Winds

Two of the most basic general circulation parameters are the zonally-
averaged temperatures and the mean zonal winds. These are physically
related by the thermal wind relation between the vertical shear of the
mean zonal wind and the meridional variation of the zonally-averaged
temperature.

Figure 1 shows the zonally-averaged temperatures for both
northern and southern hemispheres from the Earth's surface to 0.4 mbar
for the months of January, April, July, and October. These figures
were derived from 48 months of data beginning with December 1978 and
ending with November 1982. The signature of the seasonally varying
heating pattern is clearly seen in the distinct upward slope of the
isotherms toward the northern hemisphere in January with the opposite
slope in July. The situations in October and April are a bit different than what might have been expected, however, with the April
pattern looking rather symmetric about the equator but with October
largely resembling the January pattern. Clear hemispheric asymmetries
are seen in the minimum polar night temperatures and the summer polar
temperatures. The minimum polar night temperatures are about 198K in
the January lower stratosphere whereas these temperatures are about
185K in July. When one looks at the entire twelve months (not shown
here), one sees that the minimum northern hemisphere lower stratosphere polar temperatures occur in January (198K) whereas in the
southern hemisphere they occur in August (184K). The summer pole is
warmer in the southern hemisphere (in excess of 290K) than in the
northern hemisphere (about 284K) at stratopause levels. A distinct

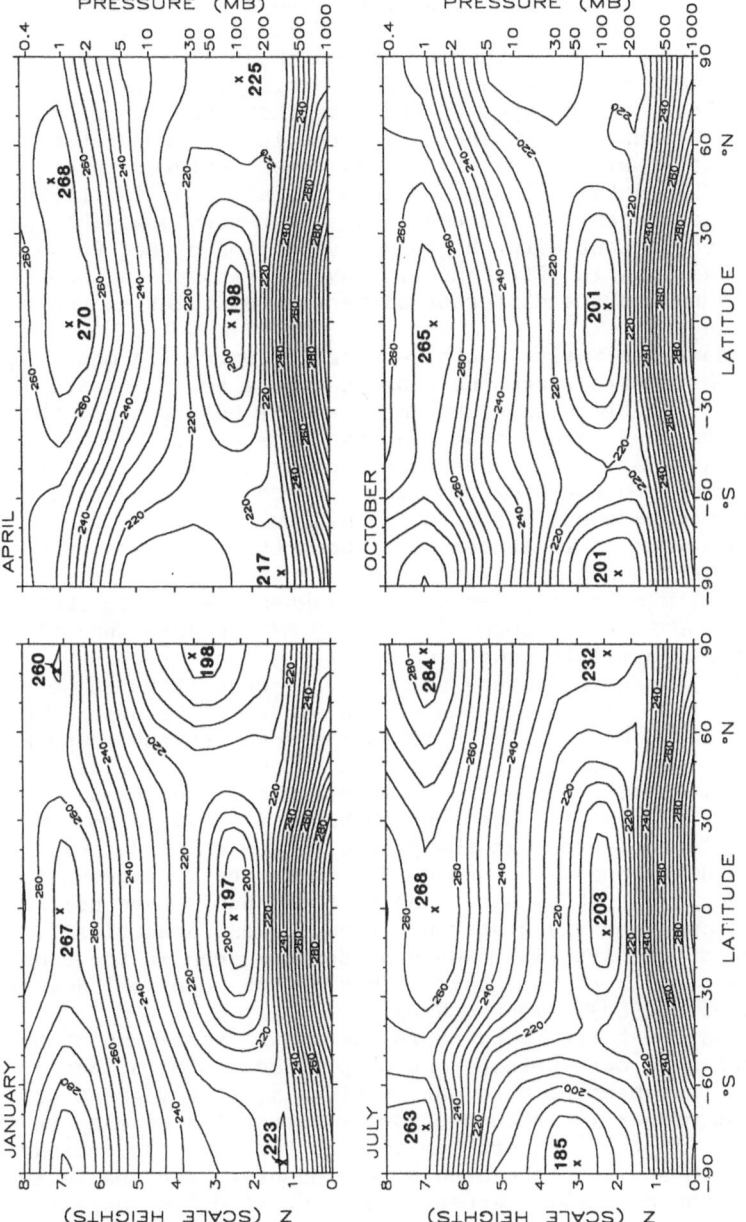

FIG. 1. Monthly mean zonally-averaged temperatures (in degrees Kelvin) for January, April, July, and October derived from the four year data set discussed in the text.

annual cycle is seen in the equatorial tropopause temperature (197K in
January and 203K in July). Finally, we wish to point out the existence of a cold belt in the upper stratosphere which is a weaker
feature in the northern hemisphere in January than what is seen in the
southern hemisphere in July. These features are seen in Figure 1 as a
slight upward bowing of the isotherms between about 1 and 2 mb. near
60N in January with a more marked feature of this type being seen in
July near 60S. Previous analyses of SCR data (Labitzke and Barnett,
1973; Barnett, 1974) have noted these features in the temperature
fields, and Hirota et al. (1983) have noted these features in their
analysis of NOAA/NMC data. It has also been pointed out in previous
papers that the hemispheric differences in summer stratopause temperatures are quite consistent with the annual variation of the Earth-Sun
distance. The hemispheric difference in lower stratosphere temperatures is clearly due to the different wintertime dynamics in the two
hemispheres, however.

Figure 2 shows the twelve monthly-average geostrophic mean zonal
wind distributions derived from the four-year data set. One sees
different annual variations in each hemisphere of both the jet that
exists near the tropopause and that near the stratopause. For
instance, the lower jet stream undergoes a clear annual variation in
the northern hemisphere reaching a maximum of about 45 m/s in February
when it is centered at about 30N with its minimum (about 20 m/s)
occurring in July when it is centered at about 45N. In the southern
hemisphere, however, a semiannual variation is seen with minimum
values of about 27 m/s in November and April when the jet is centered
around 45S with a primary maximum of about 39 m/s in June and July
when it is centered a little equatorwards of 30S and a secondary
maximum of about 31 m/s in January and February when it is centered a
bit polewards of 45S. The evolution of the winds in the upper stratosphere also proceeds quite differently in the northern and southern
hemispheres. The mean zonal winds are much stronger in the southern
hemisphere, particularly in winter. This has also been shown by
earlier authors, for example, Leovy and Webster (1976). Also, these
winds evolve differently through the year with the southern polar
night jet descending from July through November when it appears to
join up with the lower jet system. The northern hemisphere polar
night jet is strong in early winter but then weakens dramatically
showing no sign of the systematic descent that is seen in the southern
hemisphere. This southern hemisphere behavior has been well described
by Hartmann et al. (1984), among others.

Although most of these features in the annual variation of the
mean zonal winds have been pointed out by previous authors, one
feature that has not been pointed out previously, to any great extent,
is the different manner in which the summer easterlies develop in the
two hemispheres. In the northern hemisphere, two easterly regions
start appearing in April in the middle stratosphere, and the subsequent growth of the easterlies appears to occur simultaneously in two
regions, one at low latitudes and the other at high latitudes. In
August, the easterlies appear to diminish throughout the entire
northern hemisphere giving rise to westerlies in September. In the

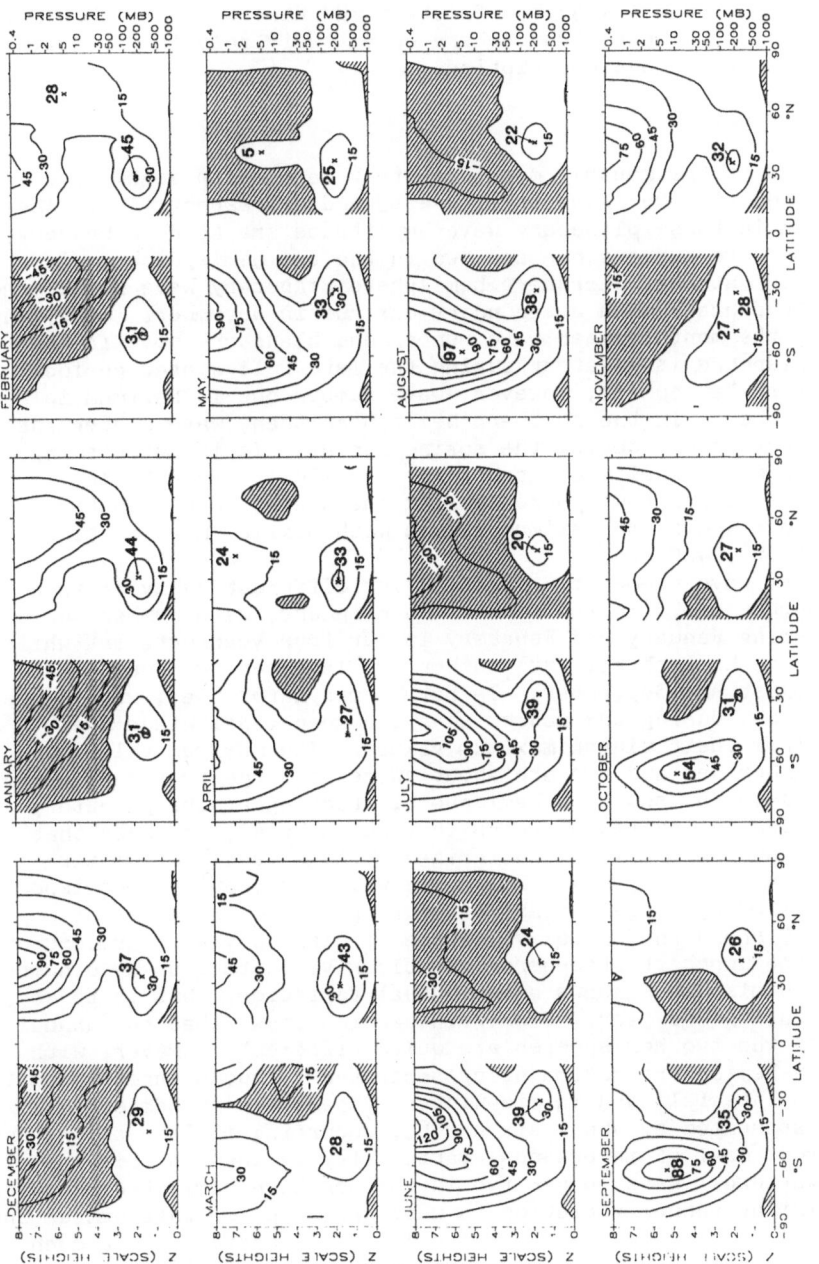

FIG. 2. Monthly mean zonal winds (in m/s) for the twelve months of the year derived from the four year data set discussed in the text.

southern hemisphere, easterlies appear to descend in November. The
easterlies appear to begin ascending in February at high latitudes.
Although Hirota et al. (1983) and Shiotani and Hirota (1985) show mean
zonal winds for the upper stratosphere only, their results are quite
consistent with the above description.

2.2 Planetary Waves

The amplitudes of the monthly mean planetary waves with zonal wave
numbers one and two are shown in Figures 3 and 4, respectively. The
annual cycles in these planetary wave amplitudes are seen to be very
different for wave numbers one and two in the two hemispheres. In
this four year data set, northern hemisphere planetary wave number one
is maximum in January with a secondary maximum in November. Thus, the
amplitude of the monthly mean wave number one planetary wave in the
northern hemisphere is greatest during the late fall-winter period.
In the southern hemisphere, however, wave number one is maximum in
September-October. In the southern hemisphere then, wave number one
has its largest values during the spring months. It is interesting to
note that the mean zonal wind structures are similar in both the
northern and southern hemispheres during the months when planetary
wave number one shows its maximum values with maximum mean zonal winds
in the range of 50-80 m/s.
    Planetary wave number two shows a very different behavior than
does wave number one. In the northern hemisphere, wave number two
maximizes during January and February in our four year data set while
in the southern hemisphere, wave number two is a maximum during the
months of August and September. In both hemispheres then, wave number
two is a maximum during the months when the mean zonal winds start
decreasing from their winter maximum values. The maximum values of
both wave number one and two are about twice as large in the northern
hemisphere as in the southern hemisphere. Looking at the planetary
wave amplitudes in individual years (not shown here), one sees that
the northern hemisphere is characterized by high levels of winter
variability in planetary wave amplitudes while the southern hemisphere
is characterized by large variability during the equinox seasons.
    Figure 5 shows the annual variation of planetary wave numbers one
and two in the upper stratosphere more clearly. Note that wave number
one in both hemispheres shows a semiannual variation, that is to say,
two maxima during the year. Closer inspection shows that the annual
variation in the two hemispheres are quite different, however, with
the northern hemisphere maxima being separated by about three months,
occurring in late fall and in mid-winter. The southern hemisphere
maxima are separated by about six months, occurring in late spring and
fall. A great deal of interannual variability is seen in this annual
variation pattern. Wave number two in the northern hemisphere shows a
somewhat similar annual variation to what was seen for wave number one
in the northern hemisphere with a weak late fall maximum and a much
stronger mid-winter maximum. Wave number two shows quite a different
annual behavior in the southern hemisphere than was seen for wave
number one.

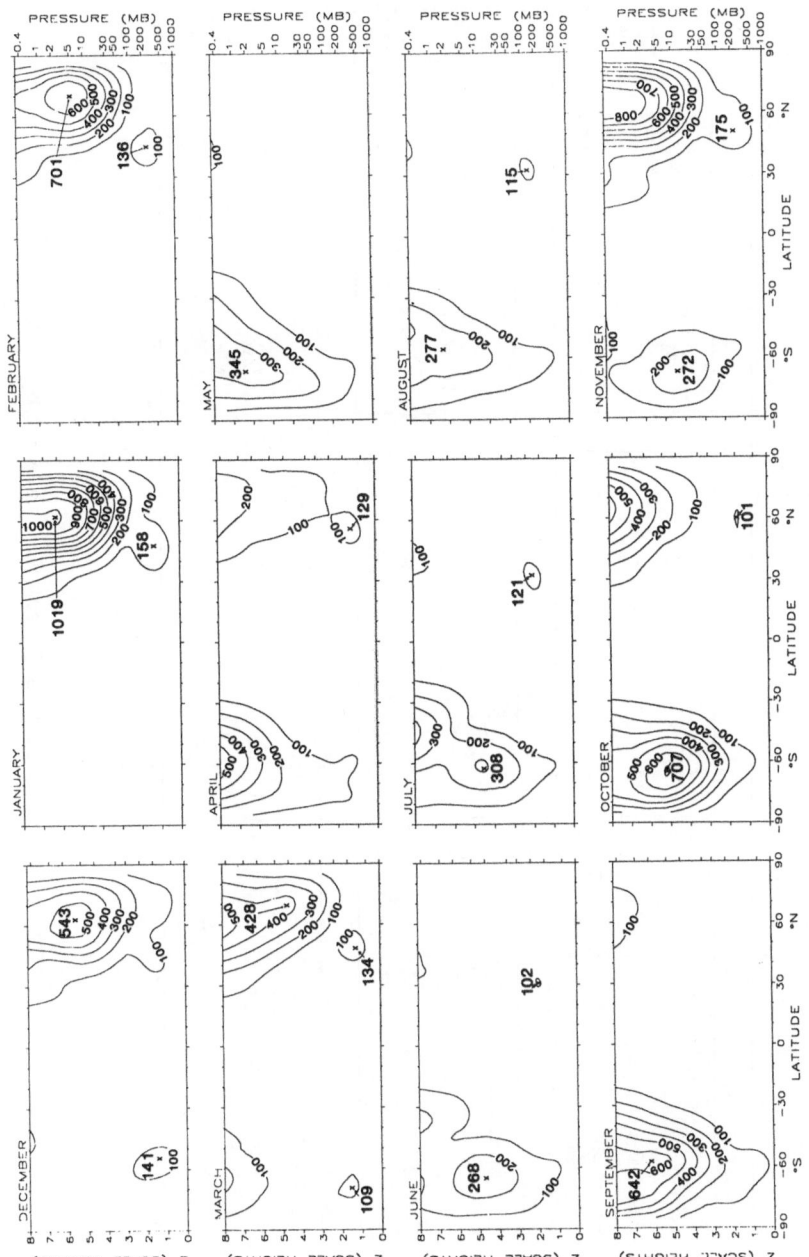

FIG. 3. Amplitudes for the monthly mean wave number one in geopotential height (in meters) for the twelve months of the year derived from the data set discussed in the text.

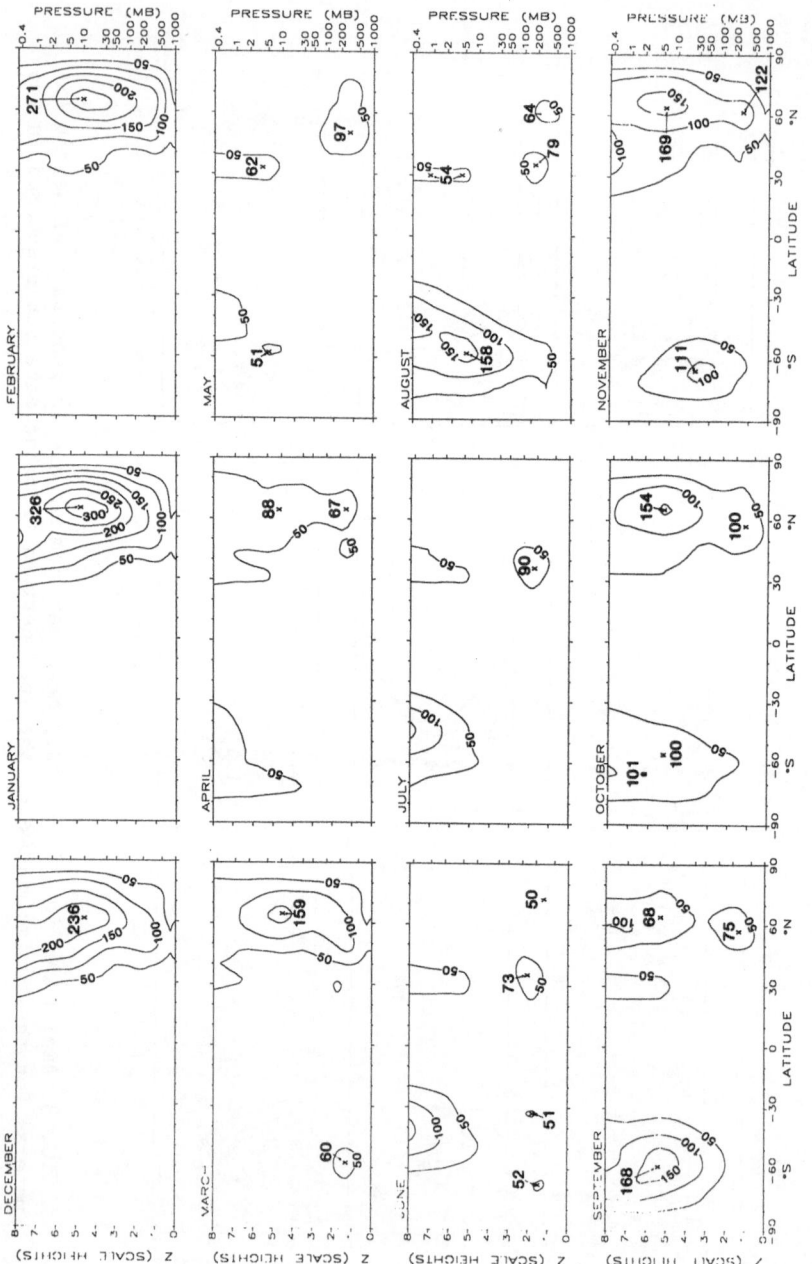

FIG. 4. Same as Figure 3, but for wave number two.

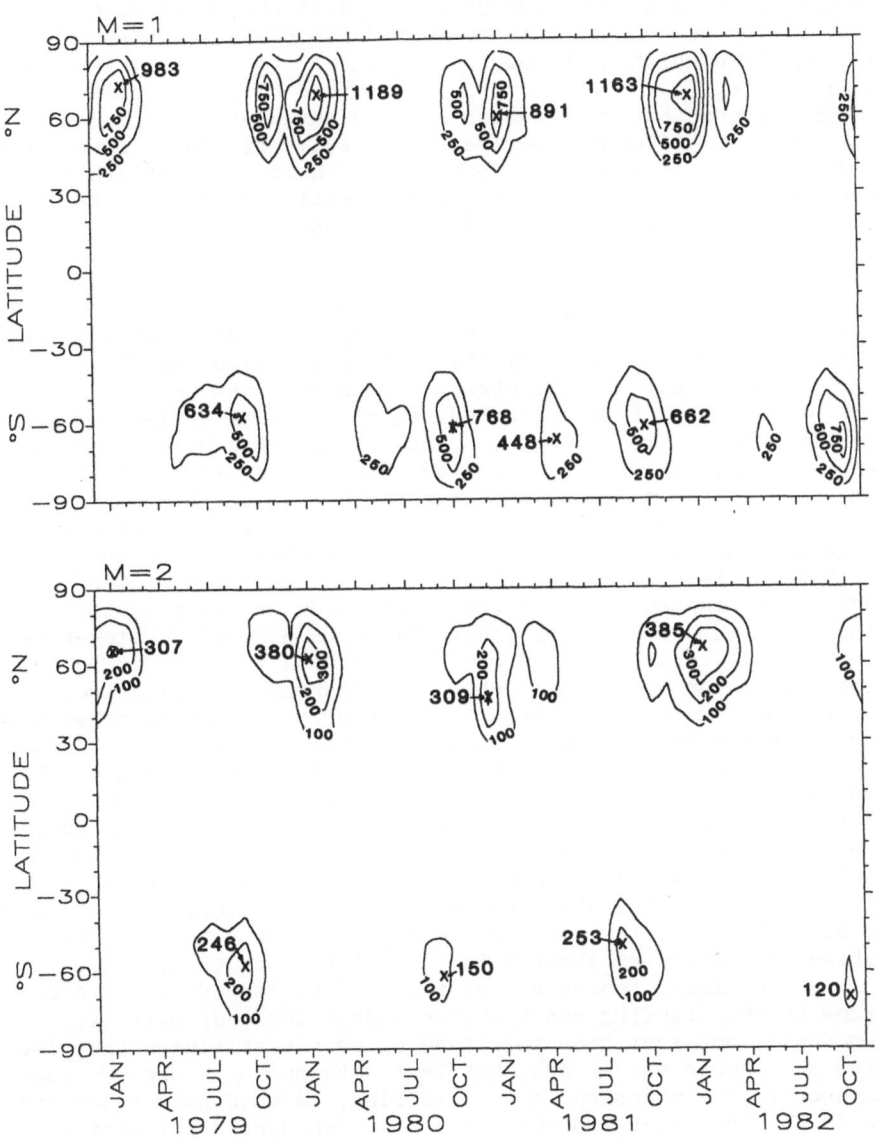

FIG. 5. Variations of the monthly mean wave number one (top) and two (bottom) in geopotential height (in meters) at 5 mb throughout the four years discussed in the text.

It should be pointed out that, given the amount of winter variability that is seen in the northern hemisphere in planetary wave
structure, different four year data sets will show different annual
variation. An example of this is that Smith (1983) found that the
winter month in which stationary wave number one was a minimum was
January for her four year northern hemisphere data set while Geller et
al. (1983) found that wave number one had its winter maximum in
January using a different four year data set. Similar discrepancies
between analyses of planetary wave structure in the southern hemisphere using data sets of a few years' duration might also be
expected, especially during the equinox months.

2.3 Eddy Heat Fluxes

Figure 6 shows a comparison of the monthly mean northward standing
eddy heat transport to that by the transient eddies for the months of
January, April, July, and October. In January, the northern hemisphere heat transport is dominated by the standing eddies (by about a
factor of three). Of course, the January southern hemisphere heat
transport is negligible given the near absence of eddy activity there
at these levels. In April, the southern hemisphere heat transports by
both the standing and transient eddies are of the some order and are
much larger than those in the northern hemisphere during this month.
The situation is similar in July in that the heat fluxes from the
standing and transient eddies contribute about equally in the southern
hemisphere, and northern hemisphere fluxes are small. October is
quite an active month in both hemispheres with the standing eddy contributions to the heat transports dominating the transient eddies in
both hemispheres (by a factor of about two in the northern hemisphere
and by a factor of about five in the southern hemisphere). The months
in which the transient eddy heat fluxes dominate those due to the
standing eddies are February in the northern hemisphere (not shown)
when the transient heat fluxes slightly exceed the standing eddy heat
fluxes and August in the southern hemisphere (not shown) when the
transient eddy heat fluxes exceed those due to the standing eddies by
about a factor of two. Looking at Figure 6, one sees that the largest
eddy heat fluxes in the northern hemisphere exceed those in the
southern hemisphere by about a factor of three to four.
    When one looks separately at the wave number one and two contri-
butions to the standing and transient northward eddy heat fluxes (not
shown here), one sees that the standing eddy heat fluxes are always
dominated by wave number one, but that wave number one shows much less
dominance in the transient eddy heat flux. A wave number two dominance in the transient eddy heat flux is seen during the entire
southern hemisphere winter and is probably attributable to the
eastward traveling wave number two which was first pointed out as
being important in southern hemisphere winter by Harwood (1975).
    Most of the features that we have pointed out with respect to the
heat fluxes are also true for the momentum fluxes.

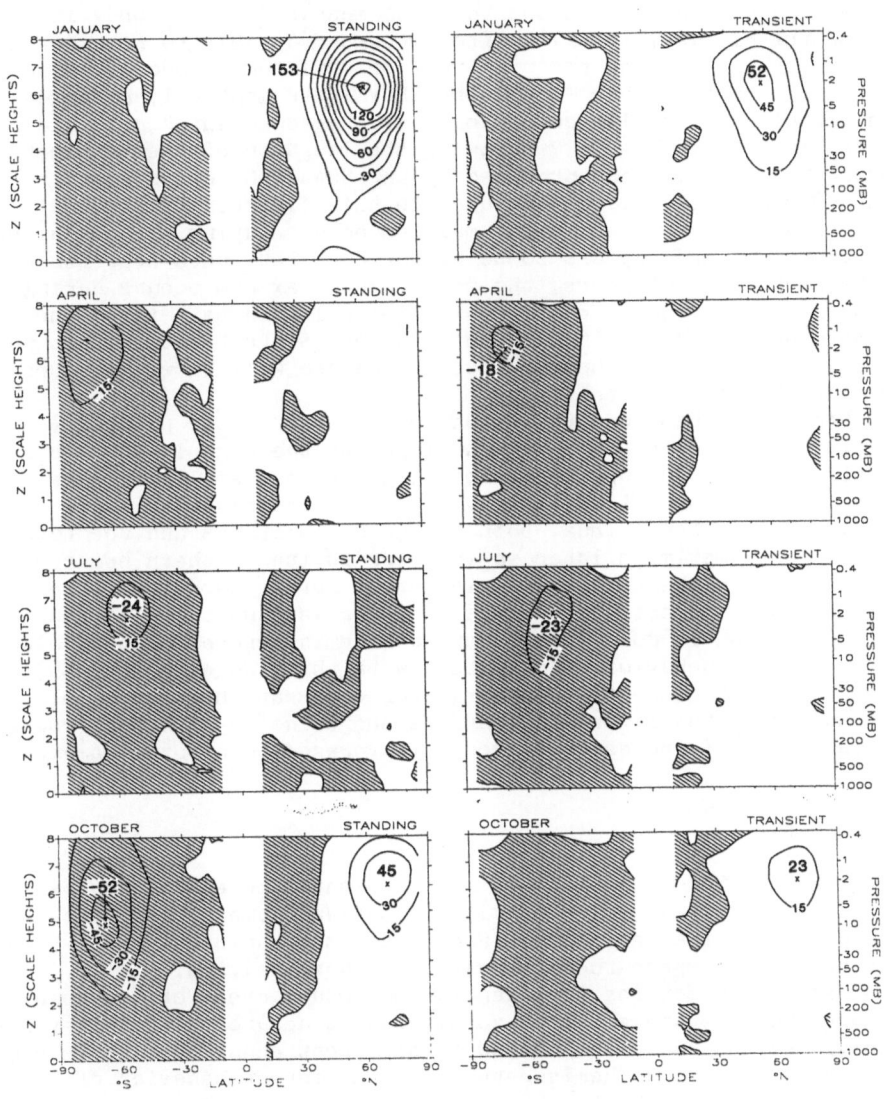

FIG. 6. Monthly mean heat transport (in degrees K m/s) by the standing eddies (left) and by the transient eddies (right) for the months of January, April, July, and October derived from the four year data set discussed in the text.

2.4 Ozone Transport and Planetary Waves

Figure 7 shows the total ozone that was measured by the SBUV instrument on the Nimbus-7 satellite from December 1, 1978 to November 30,
1979, together with the rms amplitude of wave number one at 100 mb
during this same period. Since the SBUV instrument only measures
total ozone under sunlit conditions, all "measurements" poleward of
the dashed line in Figure 7 (showing the location of the terminator)
represent extrapolations produced by the NOAA/NMC analysis of the SBUV
data. One notices several features in this figure. First, the
northern hemisphere total ozone maximum occurs during the months of
February-April and appears to be centered at latitudes near the pole.
In the southern hemisphere, the total ozone maximum occurs during the
month of October and is centered in middle latitudes. Also, the
northern hemisphere maximum in total ozone is greater than that which
occurs in the southern hemisphere. These are well known features of
the distribution of total ozone.
        Comparing the ozone variations with those of the planetary wave
activity, one sees that in both hemispheres the high latitude ozone
amounts increase during the period when the wave amplitudes are the
largest and reach their maximum values when the planetary wave
activity diminishes. Thus, both the wave amplitudes and the total
ozone amounts maximize later in the year in the southern hemisphere
than they do in the northern hemisphere. Furthermore, one also sees
that maximum wave activity occurs at lower latitudes in the southern
hemisphere than it does in the northern hemisphere consistent with the
observed ozone behavior. Consistent with this, Wu et al. (1985, 1987)
have shown that ozone flux convergences occur at the time of planetary
wave pulses in the lower stratosphere and coincide with both the times
and latitudes of increases in zonally-averaged total ozone amounts.

3. SUMMARY

In the preceding discussion, we have presented selected results from
our analysis of four years of global NOAA/NMC temperature data
extending from the earth's surface to the stratopause level. We have
also tried to compare these results to other analyses.
        Marked asymmetries are seen in the annual cycle of the zonally
averaged temperature and mean zonal wind structures. Planetary wave
numbers one and two also show quite different annual cycles in the
northern and southern hemispheres. The different behavior of the
planetary waves in the two hemispheres are reflected in the structure
of the eddy heat fluxes. The different annual evolution of the total
ozone distribution in the two hemispheres is clearly controlled by the
hemispheric differences in the planetary wave behavior.

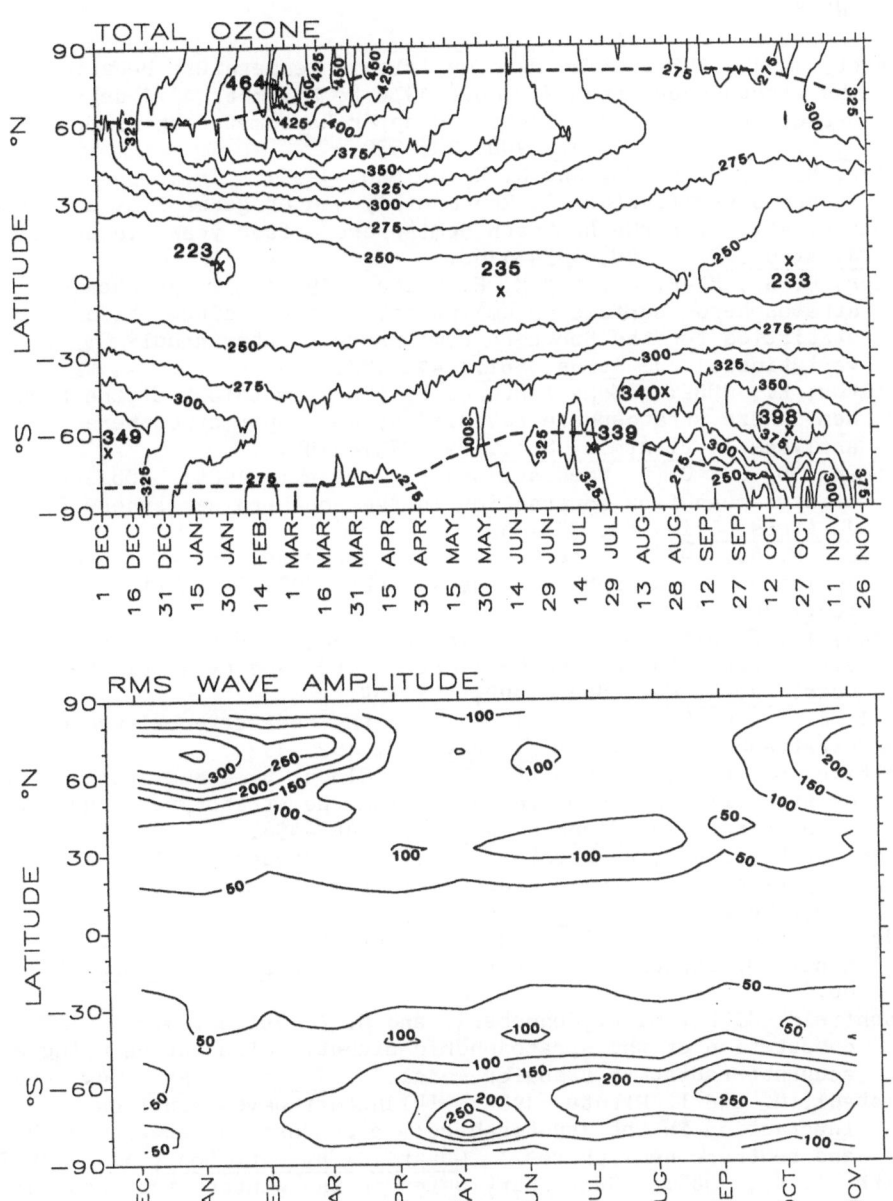

FIG. 7. Top — Variation of SBUV-measured total ozone (in Dobson units) throughout the period December 1, 1978, to November 30, 1979. Bottom — Variation of the wave number one rms wave amplitude (in units of geopotential meters) at 100 mb for the period December 1, 1978 to November 30, 1979.

REFERENCES

Barnett, J. J., 1974:   'The mean meridional temperature behavior of
    the stratosphere from November 1970 to November 1971 derived from
    measurements by the Selective Chopper Radiometer on Nimbus IV.'
    Quart. J. R. Met. Soc., 100, 505-530.
Geller, M. A., M.-F. Wu, and M. E. Gelman, 1983:   'Troposphere-
    stratosphere (surface-55 km) monthly winter general circulation
    statistics for the Northern Hemisphere - four year averages.'
    J. Atmos. Sci., 40, 1334-1352.
Geller, M. A., M.-F. Wu, and M. E. Gelman, 1984:   'Troposphere-
    stratosphere (surface-55 km) monthly general circulation
    statistics for the Northern Hemisphere - interannual
    variations.'   J. Atmos. Sci., 41, 1726-1744.
Hamilton, K., 1982:   'Some features of the climatology of the Northern
    Hemisphere stratosphere revealed by NMC upper atmosphere
    analyses.'   J. Atmos. Sci., 39, 2737-2749.
Hartmann, D. L., C. R. Mechoso, and K. Yamazaki, 1984:   'Observations
    of wave-mean-flow interaction in the southern hemisphere.'
    J. Atmos. Sci., 41, 351-362.
Harwood, R. S., 1975:   'The temperature structure of the southern
    hemisphere stratosphere August-October 1971.'   Quart. J. R. Met.
    Soc., 101, 75-91.
Hirota, I., T. Hirooka, and M. Shiotani, 1983:   'Upper atmosphere
    circulations in the two hemispheres observed by satellites.'
    Quart. J. R. Met. Soc., 109, 443-454.
Kiehl, J. T. and S. Solomon, 1986:   'On the radiative balance of the
    stratosphere.'   J. Atmos. Sci., 43, 1525-1534.
Labitzke, K. and J. J. Barnett, 1973:   'Global time and space changes
    of satellite radiances revealed from the stratosphere and lower
    mesosphere.'   J. Geophys. Res., 78, 483-496.
Leovy, C. B., and P. J. Webster, 1976:   'Stratospheric long waves:
    comparison of thermal structure in the northern and southern
    hemispheres.'   J. Atmos. Sci., 33, 1624-1638.
Rodgers, C. D., 1984:   'Coordinated study of the behavior of the
    middle atmosphere in winter (PMP-1).'   Handbook for MAP, 12, 154
    pp.
Rosenfield, J. E., M. R. Schoeberl, and M. A. Geller, 1987:   'A
    computation of the stratospheric diabatic circulation using an
    accurate radiative transfer model.'   J. Atmos. Sci., 44, 859-876.
Shiotani, M. and I. Hirota, 1985:   'Planetary wave-mean flow
    interaction in the stratosphere: a comparison between northern
    and southern hemispheres.'   Quart. J. R. Met. Soc., 111, 309-334.
Smith, A. K., 1983:   'Stationary waves in the winter stratosphere:
    seasonal and interannual variability.'   J. Atmos. Sci., 40, 245-
    261.
Solomon, S., J. T. Kiehl, R. C. Garcia, and W. Grose, 1986:   'Tracer
    transport by the diabatic circulation deduced from satellite
    observations.'   J. Atmos. Sci., 43, 1603-1617.
Wu, M.-F., M. A. Geller, J. G. Olson, A. J. Miller, and R. M.
    Nagatani, 1985:   'Computations of ozone transport using Nimbus 7

Solar Backscatter Ultraviolet and NOAA/National Meteorological Center data.' J. Geophys. Res., **90**, 5745-5755.

Wu, M.-F., M. A. Geller, J. G. Olson, and E. M. Larson, 1987: 'A study of the global ozone transport and the role of planetary waves using satellite data.' J. Geophys. Res., **92**, 3081-3097.

A REVIEW OF OBSERVATIONS OF THE QUASI-BIENNIAL AND SEMIANNUAL
OSCILLATIONS OF WIND AND TEMPERATURE IN THE TROPICAL MIDDLE
ATMOSPHERE

Kevin Hamilton
McGill University
Department of Meteorology
805 Sherbrooke Street West
Montreal, Canada H3A 2K6

ABSTRACT. A brief review is presented of studies of the quasi-
biennial and semiannual oscillations of the circulation in the
tropical stratosphere and mesosphere. Results from investigations
employing balloon, rocket and satellite observations are discussed in
an effort to elucidate the vertical, meridional and zonal structures
of these oscillations.

1. HISTORICAL BACKGROUND

The first scientific knowledge of the winds in the tropical stratosphere was obtained from observations of the motion of the dust cloud
produced by the eruption of Mount Krakatoa in August 1883. The
optical phenomena caused by the dust (such as twilight glow and
coloured suns and moons) were sufficiently remarkable for their first
appearance to be widely noted. Russell (1888) collected observations
from over 30 locations in the tropics (including ships) and used these
to plot the map reproduced here as Fig. 1. This shows the progression
of the first observations of significant optical phenomena. The
regular westward motion of the dust cloud is quite evident and Russell
computed a mean easterly velocity of something between 70 and 76
miles/hour (about 31 to 34 m/sec). The exact height of the dust cloud
was unknown, but estimates could be made from observations of the
duration of the twilight glow. On this basis Archibald (1888)
concluded that the maximum height of the cloud was about 121,000 feet
(37 km) after the initial eruption in August, and declined to about
64,000 feet (20 km) by January 1884. The 20 km height for the cloud
several months after the eruption is reasonably consistent with modern
observations of volcanic dust clouds (e.g., Dyer and Hicks, 1968).
Present day observations of the winds in the tropical stratosphere
suggest that prevailing easterlies exceed 30 m/sec only at heights
between about 20 and 40 km (and, of course, only near the extreme
easterly phase of the quasi-biennial oscillation).
    The wind in the tropical lower stratosphere was measured
directly with pilot balloons for the first time by Von Berson at two
locations in East Africa (about 2°S and 7°S) during the period

19

G. Visconti and R. Garcia (eds.), Transport Processes in the Middle Atmosphere, 19–29.

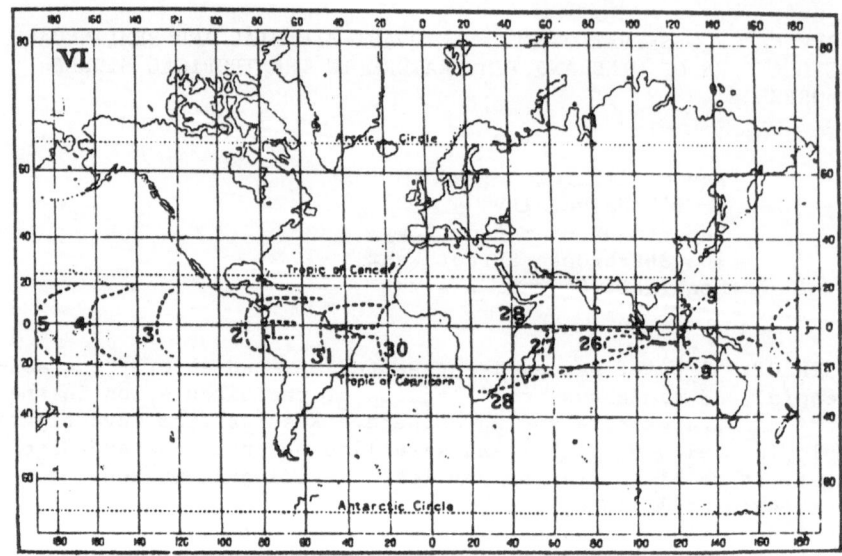

Fig. 1. The spread of the optical phenomena observed after
the eruption of Mount Krakatoa on August 26, 1883. The
dotted lines give the western boundary of the region where
the phenomena had first been observed on August 26, August
27 ... September 9. Reproduced from Russell (1888).

August-October 1909 and by Van Bemmelen at Batavia in Indonesia (6°S)
during 1909-1918. Ebdon (1963) gives the zonal wind Measurements for
each ascent extending above 17 km from both these series of observations. These observations showed the presence of westerlies in the
lower stratosphere at some times. While Ebdon (1963) has pointed out
that the Batavia measurements do hint at the existence of a quasi-
biennial oscillation (QBO) as now understood, at the time the observations were interpreted as reflecting the existence of a narrow
ribbon of westerlies meandering within the predominantly easterly flow
revealed by the Krakatoa dust cloud (e.g., Palmer, 1954).

2. MODERN OBSERVATIONS OF THE QUASI-BIENNIAL OSCILLATION

Extended daily observations of the lower stratospheric winds began at
a number of tropical stations in the late 1940's and early 1950's
(e.g., Palmer, 1954). These observations led to a reassessment of the
classical view of a thread of "Berson westerlies" embedded in the
prevailing "Krakatoa easterlies" (e.g., Graystone, 1959; Ebdon, 1960).
By 1961 the modern view of the variability of tropical lower stratospheric wind was established (Reed et al., 1961; Veryard and Ebdon,
1961). Height-time sections of the balloon measurements of the zonal
wind at various tropical stations have been published over the last
three decades (e.g., Reed and Rogers, 1962; Reed, 1965a; Wallace,

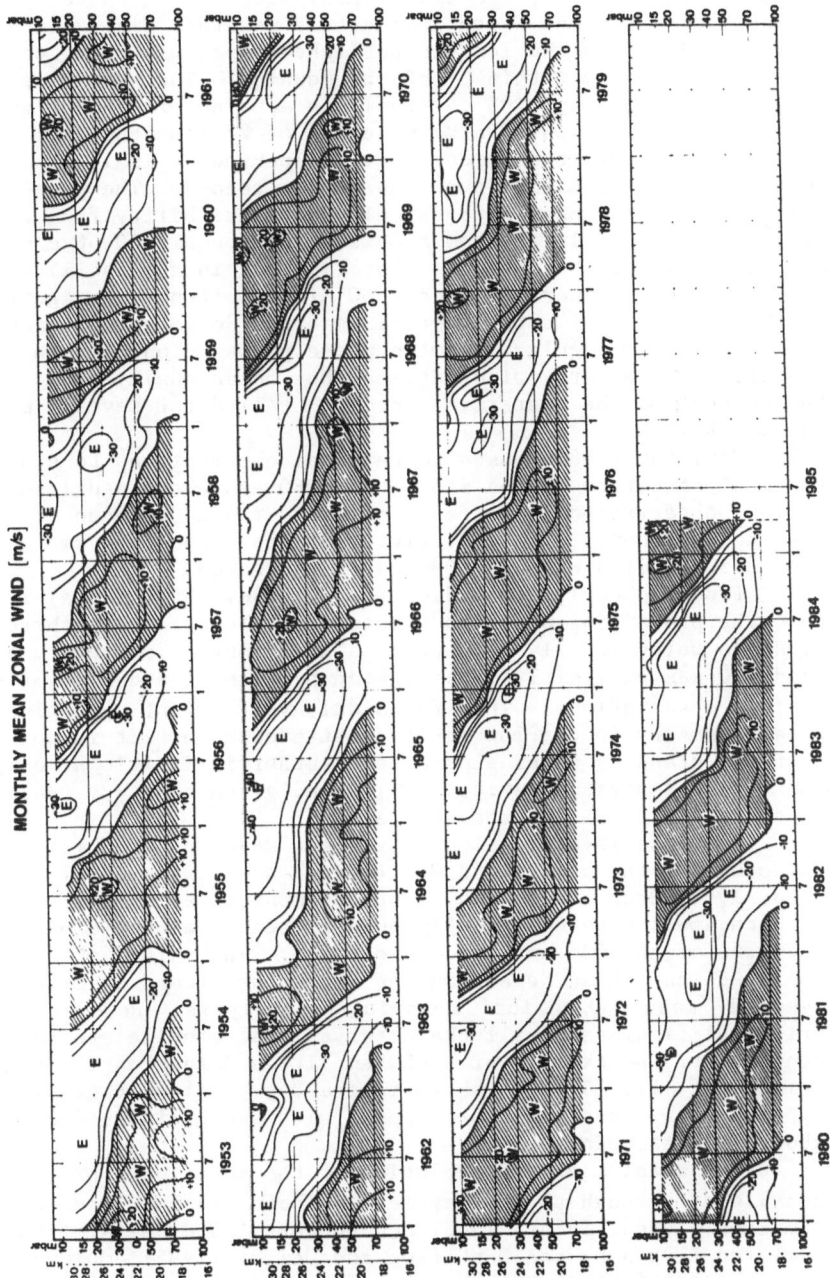

Fig. 2. Time-height section of the monthly mean zonal wind near the equator. Westerlies are shaded and the contour interval is 10 m/sec. Reproduced from Naujokat (1986).

1973; Naukojat, 1986), and all display alternating easterly and westerly wind regimes which propagate downward. Fig. 2 (taken from Naujokat, 1986) is the longest published zonal wind section and is based on data from within 3° of the equator. The general pattern of downward propagating wind reversals is apparent in this figure. Naujokat (1986) computes a mean period of 27.7 months from these observations, but it is clear that the period of individual cycles has considerable variability over the entire record. The peak-to-peak amplitude of the zonal wind QBO also has variability from cycle to cycle, but generally reaches a maximum of roughly 40-50 m/sec near 30 mb. There is also an obvious easterly bias in the oscillation. Easterly maxima generally exceed 30 m/sec (often over significant height ranges), while the westerlies often do not even reach 20 m/sec. There is fairly regular downward phase propagation of both the easterly and westerly wind regimes, but occasionally (1964, 1967, 1978) the descent of the easterlies becomes stalled for several months around the 30 mb level.

Hamilton (1981b) used rocketsonde observations to examine the vertical structure of the zonal wind QBO above the usual ceiling for balloon observations. He found a fairly rapid decrease of amplitude with height above 30 km. At altitudes above about 45 km there is no indication that a distinct QBO can be detected.

From the earliest studies (e.g., Reed, 1964, 1965a) it was clear that the prevailing winds in the tropical lower stratosphere are quite close to being zonally symmetric. The azimuthal structure of the zonal wind QBO was explicitly investigated by Belmont and Dartt (1968) using balloon observations. They found only very slight deviations from zonal symmetry in the low latitude stratosphere. It is not surprising that the standing waves should be a rather insignificant component of the tropical stratospheric circulation, given the very weak prevailing winds in the tropical troposphere.

Numerous investigations have been made into the meridional structure of the zonal wind QBO (Angell and Korshover, 1962, 1970; Reed, 1964; Shah and Godson, 1966; Belmont et al., 1974). The results of such studies have typically been presented as meridional cross-sections of the QBO amplitude and phase. The values obtained for the amplitude and phase depend on the length of record examined and the assumed "mean QBO period" used in the analysis. However, the principal findings of all the observational studies are reasonably consistent. An example of a QBO amplitude determination (again from Belmont et al., 1974) is shown in the top panel of Fig. 3. The QBO appears to be equatorially-trapped, with a meridional decay scale of about 15° latitude.

The details of the evolution of the meridional profile of the zonal wind through the QBO cycle have been examined recently by Hamilton (1984a, 1984b, 1985) and Dunkerton and Delisi (1985). Hamilton (1984a,b) and Dunkerton and Delisi (1985) employed monthly mean observations of the 30 mb and 50 mb zonal winds from a large number of stations within about 25° of the equator. On the basis of these station measurements they tried to infer the meridional profile of the zonally-averaged zonal wind in individual months. The time

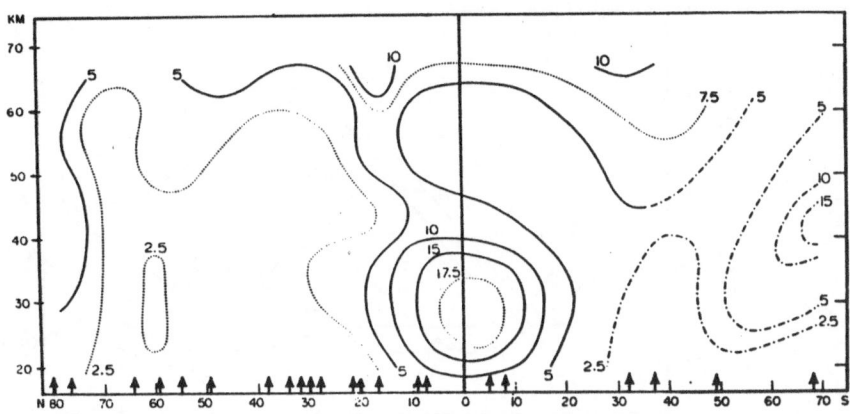

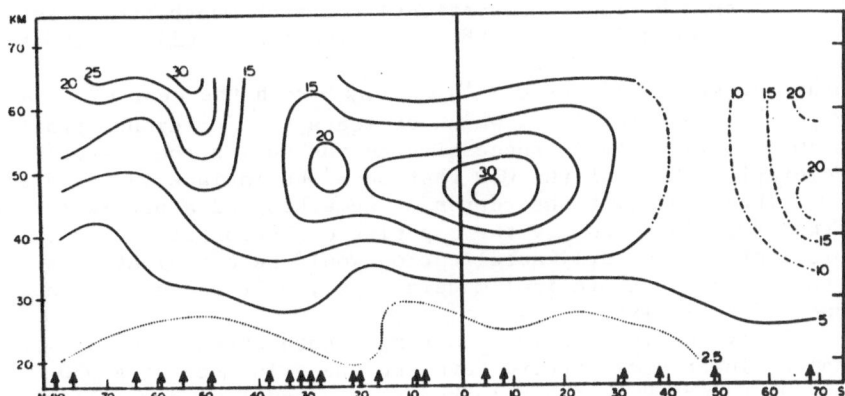

Fig. 3. (Top) The amplitude of the quasi-biennial
oscillation of the zonal wind as a function of height and
latitude determined by Belmont et al. (1974) from rawind-
sonde and meteorological rocket data at 24 stations. The
amplitude was computed by fitting a 29 month harmonic to the
data. Contour labels are in m/sec. The arrows at the bottom
show the station locations. (Bottom) The amplitude of the
semiannual oscillation of the zonal wind. Also reproduced
from Belmont et al. (1974).

series of the resulting profiles display some very interesting features. In particular, a significant asymmetry was noted between the
initial easterly and westerly acceleration phases. While the easterly
accelerations occur over a broad meridional scale throughout the whole
easterly acceleration phase, the initial westerly accelerations are
concentrated in a region within only a few degrees of the equator.
Then the westerly accelerations become broader, and within a few

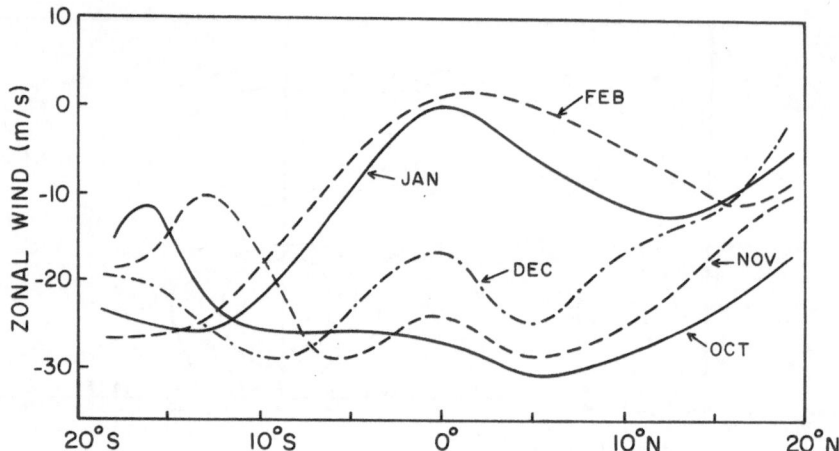

Fig. 4. Profiles of the mean zonal wind at 30 mb (deduced
from balloon observations) for each month from October 1979
through February 1980. Adapted from Hamilton (1984b).

months the flanks of the QBO "catch up" with the centre. A nice
example is shown in Fig. 4 which presents the meridional profiles of
the 30 mb zonal wind in successive months during the westerly
acceleration phase of the QBO that occurred in late 1979 and early
1980. These observations represent something of a puzzle for theories
of the QBO, since exactly the opposite evolution of the meridional
profile of the westerly acceleration would be anticipated on the basis
of the standard theoretical models (e.g., Holton and Lindzen, 1972;
Plumb and Bell, 1982).

Early analyses of in situ temperature observations in the
tropical lower stratosphere revealed the existence of a QBO in the
temperature which is phase locked to the zonal wind oscillation. Reed
and Rogers (1962) showed that, within the precision of the observations,
the monthly mean zonal wind field in the tropical lower stratosphere
is in thermal wind balance with the temperature field, even very close
to the equator. Thus one typically finds warmer (colder) than normal
temperatures on the equator in westerly (easterly) shear zones. The
near-geostrophy of the zonally-averaged zonal wind is consistent with
expectations based on scaling considerations (e.g., Holton, 1975, p.43).

3. OBSERVATIONS OF THE SEMIANNUAL OSCILLATION

Reed (1965b) used rocketsonde observations to examine the zonal wind
variability in the tropical upper stratosphere and lower mesosphere.
The present Fig. 5 is reproduced from Reed's paper and shows both
individual measurements and monthly mean values for the zonal wind at
various levels above Ascension Island in the South Atlantic (8°S). It
is apparent that as the QBO dies out in the tropical upper strato-

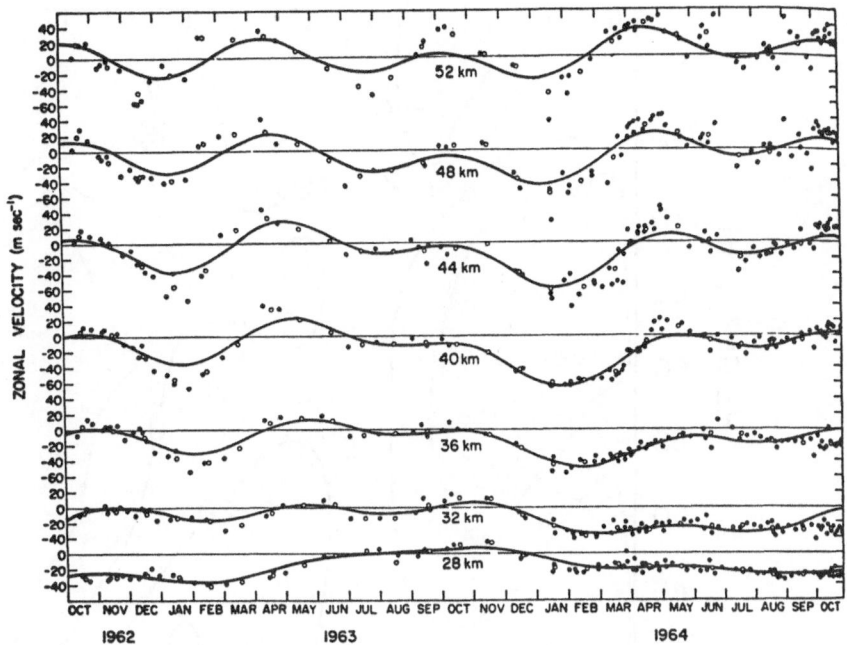

Fig. 5. Zonal wind measurements taken at Ascension Island
(7.9°S) during the period October 1962 through October 1964.
The solid circles show individual measurements. The open
circles are monthly means. The curves are obtained by
summing the biennial, annual and semiannual harmonics
obtained by Fourier analysis of the monthly means.
Reproduced from Reed (1965a).

sphere it is replaced by a strong semiannual oscillation (SAO). In
contrast to the QBO, the SAO appears to be phase locked to the seasonal
cycle. This means that the observations can be composited in a
straightforward manner by simply plotting means for each month. This
has been done in Fig. 6 using seven years of rocketsonde observations
of the zonal wind at Kwajalein in the tropical North Pacific (8.7°N).
In the upper stratosphere (say 45-55 km) the wind attains its maximum
easterly values about a month after the solstices and its maximum
westerly values around the equinoxes. The SAO in this region does
have some downward phase propagation, but (as noted by Reed, 1966) the
easterly phase descends much more rapidly than the westerly phase. At
the altitude of the peak amplitude (near 50 km) the easterly and
westerly phases of the zonal wind SAO have roughly the same strength.
At lower levels there is an easterly bias and at higher levels a
westerly bias. The amplitude of the semiannual component of the wind
variation drops significantly in the lower mesosphere (see Hamilton,
1982a) where the wind evolution becomes dominated by an annual
harmonic. At still higher levels the semiannual harmonic begins to

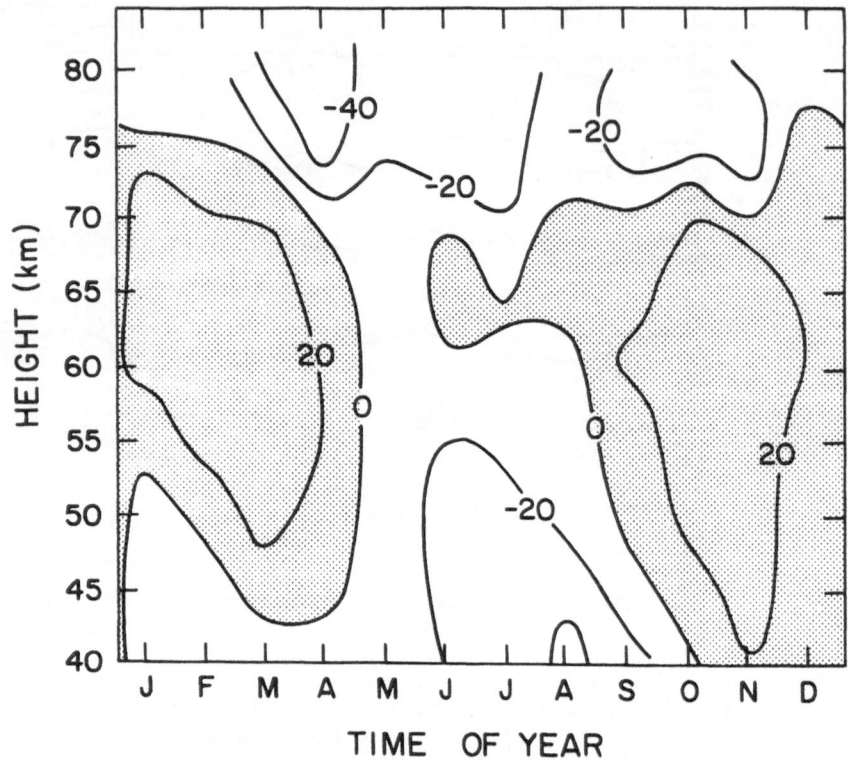

Fig. 6.   Time-height section of the monthly mean zonal wind
averaged over the seven year period 1969-75 at Kwajalein
(8.7°N).   Westerlies are shaded and the contour interval is
20 m/sec.   From Hamilton (1982).

grow in amplitude, and it is now customary to speak of a "mesopause
SAO" peaking near 80-85 km (Hirota, 1978, 1980; Hamilton, 1982). This
oscillation is comparable in strength to the upper stratospheric SAO,
but the two oscillations are roughly 180° out of phase. The easterly
bias in the low latitude upper mesosphere and lower thermosphere is
very pronounced, and may be attributable to the effects of momentum
deposition from the diurnal tide in this region (Miyahara, 1978;
Hamilton, 1981a).

     Various investigators have employed rocketsonde observations to
investigate the meridional structure of the zonal wind SAO (Angell and
Korshover, 1970; Belmont et al., 1974; Hopkins, 1975). The simple
characterization of the zonal wind variation with an amplitude and a
phase is more appropriate for the SAO (which has a fixed period) than
for the QBO. The SAO cross-sections that have been published have
been based on relatively few rocket soundings (particularly in low
latitudes). An example of an amplitude section is shown in the bottom
panel of Fig. 3 (which has been taken from Belmont et al., 1974).

These results suggest that, to first order, the SAO is equatorially-trapped with a meridional e-folding scale of a little over 20° latitude. There is some suggestion that the SAO may be somewhat stronger on the southern side of the equator than on the northern side. Also appearing in Fig. 3 are SAO amplitude maxima in the high latitudes of both hemispheres. There is no particular reason to suppose that these high latitude manifestations of the SAO are connected in any simple way to the tropical SAO. In fact, it may be appropriate to consider the high latitude values in the bottom part of Fig. 3 as simply reflecting the fact that the large annual wind variation in these regions (associated with the alternation between the wintertime polar night westerly jet and the summertime easterly jet) does not take the form of a pure twelve-month harmonic.

The interpretation of some of the details in the SAO meridional cross-sections such as that shown in Fig. 3 is complicated by the paucity of rocket observations. The hint of interhemispheric asymmetry in the zonal wind SAO amplitude in Fig. 3 is supported by rocket observations at only four low latitude stations (Ft. Sherman, 9.3°N, 80°W; Kwajalein, 8.7°N, 168°E; Natal, 5.8°S, 35°W; Ascension Is., 8°S, 14°W). Thus the differences seen could actually reflect either interhemispheric or azimuthal variations in the SAO.

4. CONCLUSION

The main features of both the quasi-biennial oscillation and the stratospheric semiannual oscillation are now well documented. By contrast, the SAO near the tropical mesopause has been observed in only a few dozen rocket soundings at two stations. Further progress in determining the structure of the tropical wind variations in the upper mesosphere will probably await the deployment of MST radars at low latitude sites.

REFERENCES

Angell, J.K., and J. Korshover, 1962: 'The biennial wind and temperature oscillation of the equatorial stratosphere'. Mon. Wea. Rev., 90, 127-132.
Angell, J.K., and J. Korshover, 1970: 'Quasi-biennial, annual, and semiannual wind and temperature harmonic amplitudes and phases'. J. Geophys. Res., 75, 543-550.
Archibald, E.D., 1888: 'Diurnal and secular variation in the duration and brilliancy of the twilight glows of 1883-84'. In The Eruption of Krakatoa and Subsequent Phenomena, Trubner and Co., London, pp. 340-381.
Barnett, J.J., M. Corney and K. Labitzke, 1985: 'Annual and semi-annual cycles based on the middle atmosphere reference model'. Middle Atmosphere Program Handbook, 16, 175-180.
Belmont, A.D., D.G. Dartt, 1968: 'Variation with longitude of the quasi-biennial oscillation'. Mon. Wea. Rev., 96, 767-777.

Belmont, A.D., D.G. Dartt and G.D. Nastrom, 1974: 'Periodic variations in stratospheric zonal wind'. Quart. J. Roy. Met. Soc., 100, 203-211.

Dunkerton, T.J. and D.P. Delisi, 1985: 'Climatology of the equatorial lower stratosphere'. J. Atmos. Sci., 42, 376-396.

Dyer, A.J. and B.B. Hicks, 1968: 'Global spread of volcanic dust from the Bali eruption of 1963'. Quart. J. Roy. Meteor. Soc., 94, 545-554.

Ebdon, R.A., 1960: 'Notes on the wind flow at 50 mb in the tropical and subtropical regions in January 1957 and 1958'. Quart. J. Roy. Meteor. Soc., 86, 540-543.

Ebdon, R.A., 1963: 'The tropical stratospheric wind fluctuation'. Weather, 18, 2-7.

Graystone, P., 1959: 'Meteorological discussion'. Meteor. Mag., 88, 113-119.

Hamilton, K., 1981a: 'Numerical studies of wave-mean flow interaction in the stratosphere, mesosphere and lower thermosphere'. Ph.D. Thesis, Princeton University, 384 pp.

Hamilton, K., 1981b: 'The vertical structure of the quasi-biennial oscillation'. Atmos.-Ocean, 19, 236-250.

Hamilton, K., 1982: 'Rocketsonde observations of the mesospheric semiannual oscillation at Kwajalein'. Atmos.-Ocean, 20, 281-286.

Hamilton, K., 1984a: 'Mean wind evolution through the quasi-biennial cycle in the tropical lower stratosphere'. J. Atmos. Sci., 41, 2113-2125.

Hamilton, K., 1984b: 'Monthly average tropical mean zonal wind profiles for the 30 mb level'. Univ. British Columbia Dept. of Oceanography Report #42, 128 pp.

Hamilton, K., 1985: 'The westerly acceleration phase of the stratospheric quasi-biennial oscillation revealed in FGGE analyses'. Atmos.-Ocean, 23, 188-192.

Hirota, I., 1978: 'Equatorial waves in the upper stratosphere and mesosphere in relation to the semiannual oscillation'. J. Atmos. Sci., 35, 714-722.

Hirota, I., 1980: 'Observational evidence of the semiannual oscillation in the tropical middle atmosphere'. Pure Appl. Geophys., 118, 217-238.

Holton, J.R., 1975: 'The Dynamic Meteorology of the Stratosphere and Mesosphere', A.M.S. Meteor. Monogr., 37, 216 pp.

Holton, J.R., R.S. Lindzen, 1972: 'An updated theory of the quasi-biennial cycle of the tropical stratosphere'. J. Atmos. Sci., 29, 1076-1080.

Hopkins, R.H., 1975: 'Evidence of polar-tropical coupling in upper stratospheric zonal wind anomalies'. J. Atmos. Sci., 32, 712-719.

Naujokat, B., 1986: 'An update of the observed quasi-biennial oscillation of the stratospheric winds over the tropics'. J. Atmos. Sci., 43, 1873-1877.

Palmer, C.E., 1954: 'The general circulation between 200 mb and 10 mb over the equatorial Pacific'. Weather, 9, 341-349.

Plumb, R.A. and R.C. Bell, 1982: 'A model of the quasi-biennial oscillation on an equatorial beta-plane'. Quart. J. Roy. Soc., 108, 335-352.

Reed, R.J., 1964: 'A climatology of wind and temperature in the tropical stratosphere between 100 mb and 10 mb'. U.S. Navy Weather Research Facility Rep. #26-0564-092, 56 pp.

Reed, R.J., 1965a: 'The present status of the 26-month oscillation'. Bull. Amer. Meteor. Soc., 46, 374-386.

Reed, R.J., 1965b: 'The quasi-biennial oscillation of the atmosphere between 30 and 50 km over Ascension Island'. J. Atmos. Sci., 22, 331-333.

Reed, R.J., 1966: 'Zonal wind behavior in the equatorial strato-sphere and lower mesosphere'. J. Geophys. Res., 71, 4223-4233.

Reed, R.J., W.J. Campbell, L.A. Rasmussen and D.G. Rogers, 1961: 'Evidence of a downward-propagating annual wind reversal in the equatorial stratosphere'. J. Geophys. Res., 66, 813-818.

Reed, R.J., and D.G. Rogers, 1962: 'The circulation of the tropical stratosphere in the years 1954-1960'. J. Atmos. Sci., 19, 127-135.

Russell, F.A.R., 1888: 'Spread of the phenomena around the world'. The Eruption of Krakatoa and Subsequent Phenomena, Trubner and Co., London, pp. 334-339.

Veryard, R.G., and R.A. Ebdon, 1961: 'Fluctuations in tropical stratospheric winds'. Meteor. Mag., 90, 125-143.

Wallace, J.M., 1973: 'General circulation of the tropical lower stratosphere'. Rev. Geophys. Space Phys., 11, 191-222.

RECENT PROGRESS IN GRAVITY WAVE SATURATION STUDIES

David C. Fritts
Geophysical Institute and Department of Physics
University of Alaska
Fairbanks, Alaska 99775-0800
U.S.A.

ABSTRACT. This paper will present a brief survey of some of the advances made during the last three years in understanding gravity wave saturation as well as their effects and variability in the lower and middle atmosphere. Our emphasis will be on observational results, though theoretical and modeling studies will be discussed where relevant. We will first present recent evidence of the processes contributing to wave saturation. We will also examine the implications of wave saturation and local turbulence production for wave fluxes of energy and momentum, the turbulent diffusion of heat and constituents, and a saturated spectrum of gravity waves throughout the atmosphere. Finally, some of the recent evidence of geographic and temporal variability of the gravity wave field and of the processes that may contribute to this variability will be reviewed.

1. INTRODUCTION

Following the pioneering work of Hines (1960), internal gravity waves have been recognized increasingly to play several important roles in the dynamics of the lower and middle atmosphere. The saturation of such motions as they propagate upward and attain large amplitudes, for example, results in a net drag on the atmosphere and a vertical diffusion of heat and constituents throughout the region of saturation. These effects are particularly important in the mesosphere, where they determine to a large degree the large-scale circulation and thermal and constituent structures. However, recent studies suggest that gravity wave drag, in particular, may play a significant role in the momentum budget near the tropopause as well. The initial theoretical and observational studies addressing gravity wave saturation in the middle atmosphere were reviewed by Fritts (1984) and will not be summarized here. Our purpose in this contribution is to review briefly some of the recent progress that has been made in understanding gravity wave saturation processes as well as their effects and variability in the lower and middle atmosphere.

We begin with a review of linear saturation theory and of the observational evidence supporting this theory in Section 2. Because of its apparent success, we examine in Section 3 the implications of linear saturation theory for gravity wave drag and induced diffusion as well as for a saturated spectrum of gravity wave motions throughout the atmosphere. Consistent with theoretical expectations, gravity wave drag appears to be dominated by relatively high-frequency wave motions. detailed studies of wave saturation and turbulence production as well as observed thermal and constituent profiles suggest a large Prandtl number associated with gravity wave induced diffusion, and observed vertical wavenumber

G. Visconti and R. Garcia (eds.), Transport Processes in the Middle Atmosphere, 31–46.

spectra support the notion of a saturated spectrum at high vertical wavenumbers. Finally, we review in Section 4 some of the evidence of geographic and temporal variability of the gravity wave spectrum and of the processes that act to impose this variability in the lower and middle atmosphere. The conclusions of this brief review are presented in Section 5.

2.   SATURATION THEORY AND PROCESSES

Linear saturation theory was first applied to internal gravity waves by Hodges (1967, 1969) who assumed convective instability of the wave field and inferred the level of diffusion required to limit wave amplitudes. More recently, Lindzen (1981) recognized that gravity wave saturation could also cause a momentum flux divergence and provide the momentum source necessary to balance the thermal and momentum budgets of the mesosphere. Such a momentum source was found subsequently to account crudely for the observed large-scale circulation and thermal structure in numerical studies (Holton, 1982, 1983; Dunkerton, 1982). In this section we review briefly the linear saturation theory and discuss some observational evidence of its validity.

It was assumed by Hodges (1967, 1969) and Lindzen (1981) that gravity waves would be limited by convective instabilities to that amplitude at which such instabilities just become possible, ie.,

$$\Theta_z = \overline{\Theta}_z + \Theta_z' = 0 \tag{1}$$

or, from the linear equations of motion,

$$u' = c - \overline{u}, \tag{2}$$

where Θ and u are potential temperature and horizontal velocity in the direction of wave propagation, primes and overbars denote perturbation and mean quantities, subscripts denote differentiation, and c is the horizontal phase speed of the wave motion. It was suggested by Fritts (1984) and shown by Dunkerton (1984) and Fritts and Rastogi (1985), however, that wave motions with low intrinsic frequencies may be unstable to a dynamical instability at substantially smaller amplitudes due to the transverse shear in the velocity field of such motions. The amplitude limit obtained assuming a dynamical instability with a minimum Richardson number of $1/4$ is given by (Fritts and Rastogi, 1985),

$$u'/(c - \overline{u}) = \frac{1}{2}(\gamma^2 + 4\gamma)^{1/2} - \gamma/2. \tag{3}$$

where

$$\gamma = 4(\omega^2/f^2 - 1) \tag{4}$$

and ω and f are the intrinsic frequency of the wave motion and the inertial frequency, respectively. This amplitude is consistent with that required for convective instability of the wave field at high frequencies, but falls well below as $\omega \to f$. Thus, it is clear that both the dominant instability and the resulting wave amplitude are likely to vary with wave frequency. Motions with high intrinsic frequencies should favor a convective instability because of its more rapid growth while those with low intrinsic frequencies should lead preferentially to a dynamical instability of the wave field. Additional details of the linear saturation theory can be found in the paper by Lindzen (1981) or the reviews by Fritts (1984) and Fritts and Rastogi (1985).

Evidence of the apparent validity of linear saturation theory has been obtained via both in situ and ground-based techniques. The former include balloon- and rocket-borne instrumentation, the latter include radar and lidar systems. Balloon, rocket, and radar wind measurement techniques have provided evidence of an approximate limit on velocity shears,

$u'_z \sim N$. that is consistent with (1) and (2) for high-frequency. two-dimensional gravity wave motions (Balsley et al., 1983: Fritts et al., 1987). Likewise, temperature measurement systems have revealed wave amplitudes that appear constrained to values marginally larger than required for convective instability of the wave field (Philbrick et al., 1983). Amplitude limits alone. however, do not provide confirmation that wave amplitudes are being constrained by linear saturation processes. It is reasonable to assume, for example, that nonlinear wave processes may also act to limit wave amplitudes under certain conditions. Yet direct simulations of nonlinear and convective adjustment processes suggest that the amplitude limit due to convective instability is more stringent (Fritts, 1985), and detailed observational studies reveal turbulence enhancements consistent with the occurrence and location of linear instabilities (Fritts et al., 1987).

Examples of the velocity and signal-to-noise (S/N) data obtained during the STATE experiment and analyzed by Fritts et al. (1987) are shown in Figures 1 - 3. Instantaneous zonal and meridional velocity profiles measures by rocket are illustrated in Figure 1, a time-height section of the Poker Flat MST radar-derived horizontal velocity is shown in Figure 2, and Figure 3 shows an expanded time-height section of S/N. Inspection of the first of these confirms the point made earlier that velocity shears appear constrained even though characteristic velocities and vertical scales continue to increase with height. Also evident at upper heights in this figure is a wave motion with $\lambda_z \sim 30$ km and approximate quadrature between the zonal and meridional velocity components. Additional evidence of this wave motion is seen in Figure 2 where the horizontal velocity field is observed to rotate in a clockwise sense with increasing height and time.

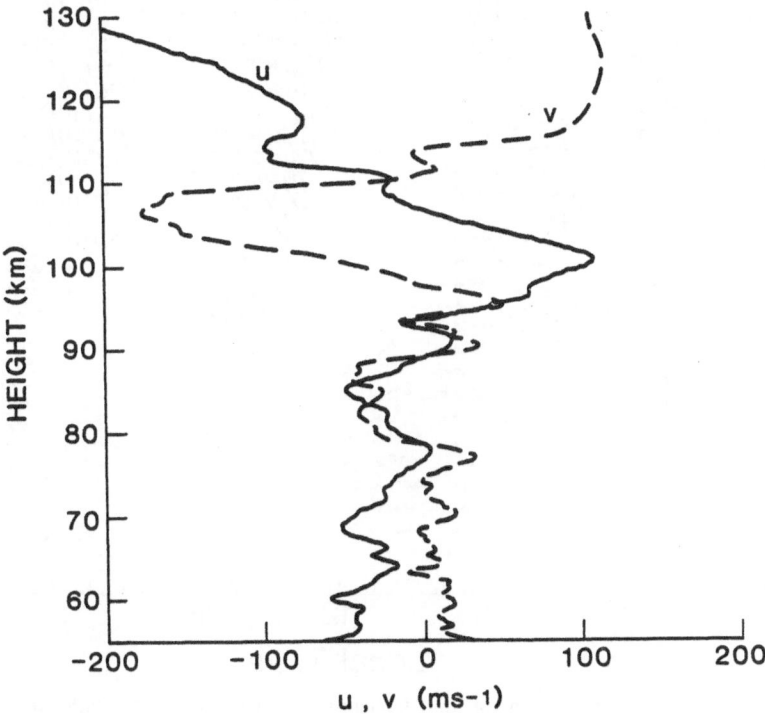

Figure 1. Zonal and meridional velocity profiles derived from rocket-borne accelerometer data obtained during Salvo II of the STATE experiment at Poker Flat, Alaska.

The conjunctive rocket and radar temperature and velocity data collected during this period permitted a fairly complete specification of the dominant wave parameters, including amplitude, vertical scale, direction of propagation, and intrinsic frequency. Thus it was possible to identify that phase of the wave motion that should have been most unstable according to linear theory. The result was good agreement between the most unstable phase based on the observed wave structure (heavy diagonal lines in Figure 3) and the observed variations in radar S/N, which is a qualitative measure of turbulence intensity. This suggests that linear instability processes are indeed responsible in large part both for wave dissipation and for the amplitude limits noted above.

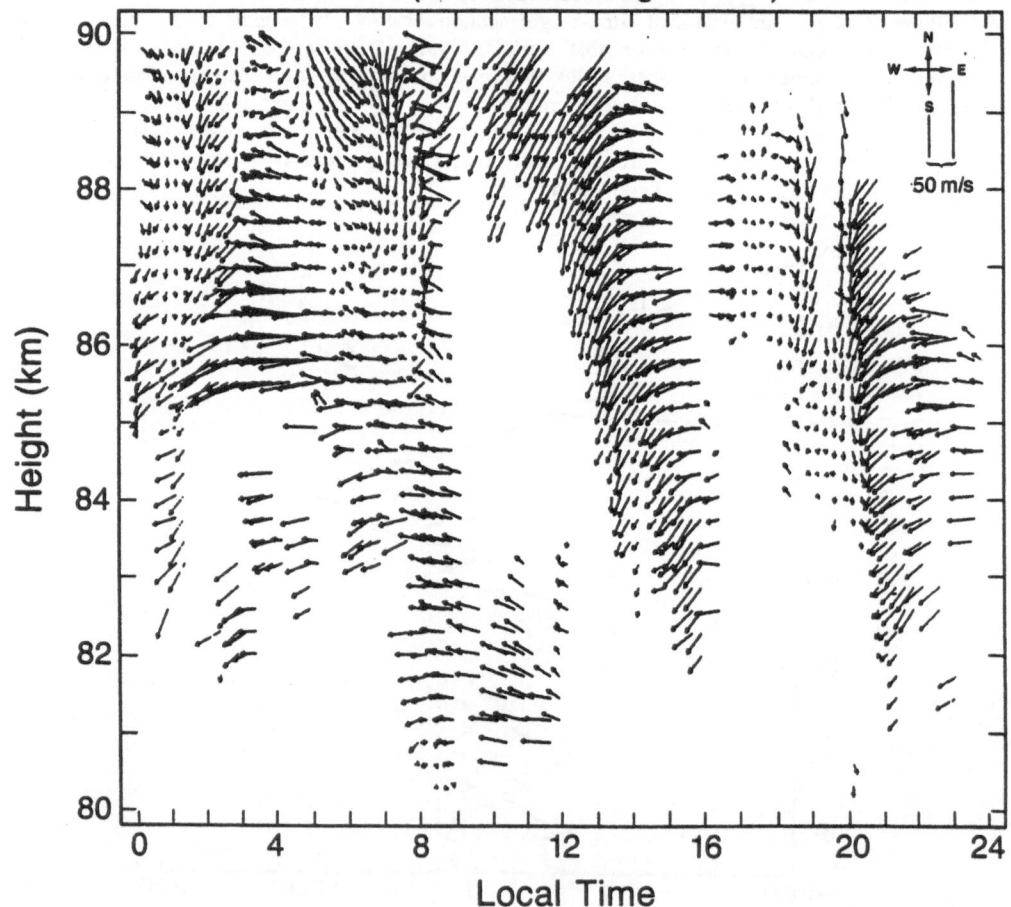

POKER FLAT MST RADAR
15 June 1983
HORIZONTAL WIND VECTORS
(15 Minute Averaged Value)

Figure 2. Time-height cross-section of the horizontal velocity obtained with the Poker Flat MST radar during Salvo II of the STATE experiment. Each vector represents a 15-minute average.

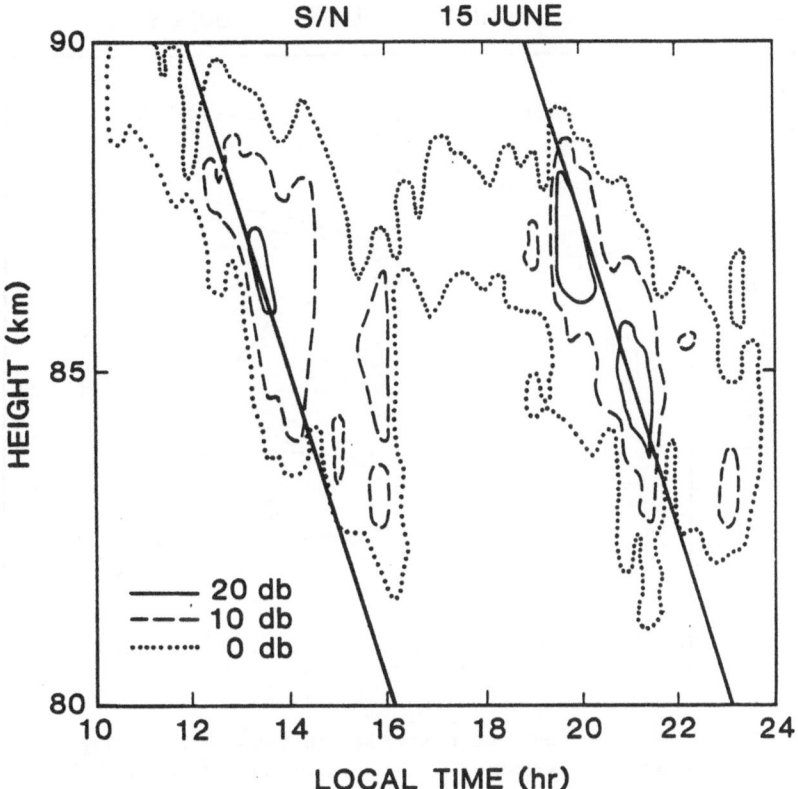

Figure 3. Time-height cross-section of radar S/N during Salvo II of the STATE experiment. Heavy diagonal lines denote inferred unstable phase of dominant wave motion.

3.   IMPLICATIONS OF WAVE SATURATION

3.1   Induced Drag and Diffusion

Observations of gravity wave structure cited in the previous section suggest that wave amplitudes are limited by saturation processes to values generally consistent with (1) and (2). Assuming this applies for all wave motions, we can infer that portion of the wave spectrum that contributes most to the vertical fluxes of momentum and energy. Also assuming, for simplicity, intrinsic frequencies such that $f^2 << \omega^2 << N^2$, where N is the Brunt-Vaisala frequency, the momentum and energy fluxes for a wave of amplitude $u' \sim (c - \bar{u})$ vary as

$$\overline{u'w'} \sim \frac{\omega}{N}(c - \bar{u})^2 \tag{5}$$

and

$$c_{g_z} E \sim \frac{\omega}{N}(c - \bar{u})^3 . \tag{6}$$

Clearly, these fluxes should be expected to be dominated by wave motions with high intrinsic frequencies and large intrinsic phase speeds (or vertical wavelengths).

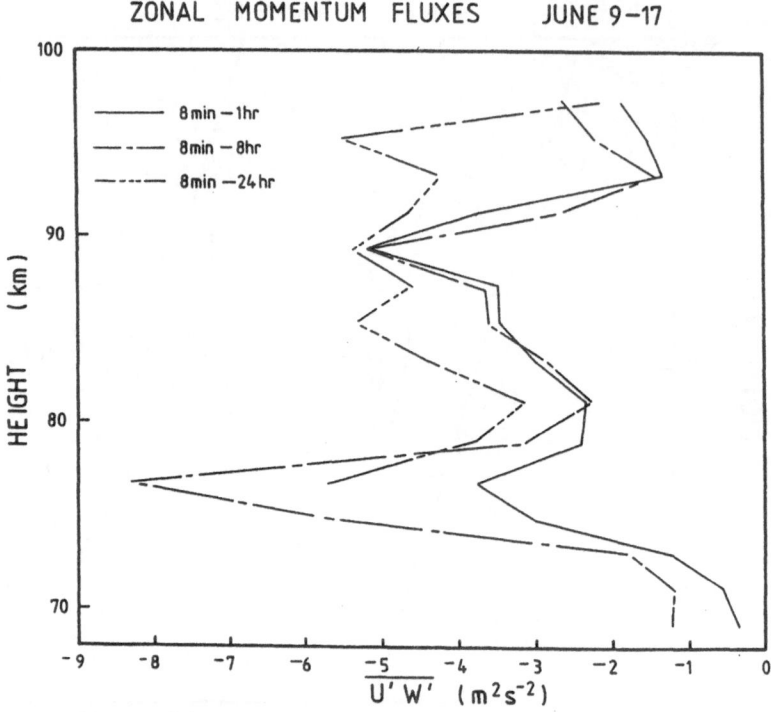

Figure 4. Mean zonal momentum flux profiles for an 8-day period during June 1984. Note that motions with periods < 1 hr account for <font=4>H 70% of the total.

Gravity wave saturation likewise results in the convergence of energy and momentum fluxes, causing a net deceleration of the zonal mean flow and a turbulent heating of the environment. The magnitude of the drag can be inferred from the strength of the observed mean meridional circulation, from differences between the observed mean winds and those calculated from satellite radiance data (Labitzke et al., 1986), or from direct measurements of the momentum flux convergence using radar techniques (Vincent and Reid, 1983; Fritts and Vincent, 1987; Reid and Vincent, 1987a). All of these techniques suggest a zonal drag of ~ 50 m/s/day in the mesosphere and lower thermosphere, consistent with that required to explain the observed circulation and thermal structure of this region. Moreover, the momentum flux determinations of Fritts and Vincent (1987) reveal that this flux and its divergence are dominated by gravity wave motions at high frequencies, with wave motions having periods < 1 hr accounting for ~ 70% of the total flux and divergence. These observations are presented in Figure 4 and are consistent with the arguments presented above and more quantitatively by Fritts (1984).

Another consequence of wave saturation, turbulence generation by wave field instabilities, is responsible both for limiting wave amplitudes, as noted previously, and for the diffusion of heat and constituents. The diffusion efficiency, however, is now believed to be strongly dependent on the distribution of turbulence throughout the wave field. There is clear evidence, such as that presented in Figures 2 and 3, that turbulence intensities are enhanced in the most unstable phase of the dominant wave motion. Chao and Schoeberl (1984) argued that this should lead to a reduced dissipation of the wave in the thermal field. Fritts and Dunkerton (1985) and Coy and Fritts (1987) showed that turbulence localization would result in a substantially reduced diffusion and a large   effective Prandtl number. It is this relative inefficiency of the induced diffusion that allows for strong turbulent dissipation of the wave motions without correspondingly large diffusion rates for heat and constituents (Strobel et al., 1985).

3.2   A Saturated Gravity Wave Spectrum

A potentially significant result of gravity wave saturation in the lower and middle atmosphere is the establishment of a nearly constant, or universal, spectral amplitude of horizontal kinetic energy at high vertical wavenumbers. This appears to arise as a consequence of convective and/or dynamical instabilities within the wave field and yields a saturated vertical wavenumber spectrum that varies as (Dewan and Good, 1986)

$$F_s(m) \simeq b\frac{N^2}{m^2} \tag{7}$$

Here m is vertical wavenumber and b can be obtained by assuming

$$\overline{\theta_z'^2} = \overline{\theta_z}^2/2 \tag{8}$$

for saturation and integrating the contributions to the variance of potential temperature gradient over all m (Smith et al., 1987). For a power spectral density of the form (Desaubies, 1976)

$$F(m) \sim 1/(1 + (m/m_*)^t) \tag{9}$$

with t = 3, together with reasonable choices for the characteristic vertical wavenumber, m∗, and maximum vertical wavenumber, this yields b $\simeq$ 1/6. This amplitude is generally consistent with the spectral observations of Endlich et al. (1969), Dewan et al. (1984), and Smith et al. (1987). However, knowledge of N and more detailed measurements of b would enable estimates of the degree of supersaturation of the high wavenumber portion of the gravity wave spectrum.

At low vertical wavenumbers, the atmospheric gravity wave spectrum departs from the saturation limit and appears to be generally consistent with the form assumed appropriate for the ocean (Garrett and Munk, 1972, 1975; Desaubies, 1976). Confirmation of this is provided by a recent study of gravity wave spectra in the lower stratosphere using Poker Flat radar data, an example of which is shown in Figure 5 (H.-G. Chou, personal communication. 1986). Observations also suggest that those motions at small vertical wavenumbers grow with height until they too achieve saturation amplitudes (Smith et al., 1987). This is consistent with the observed increase in the kinetic energy per unit mass of the gravity wave spectrum of $\sim$ 200 between 8 and 86 km (Balsley and Carter, 1982; Balsley and Garello, 1985). The gradual growth of the dominant vertical scale (decrease in m∗) with increasing height implies in addition a smooth increase with height in both the drag and turbulent diffusion arising from gravity wave saturation and invalidates the notion of a gravity wave breaking level based on monochromatic wave theory.

There are, nevertheless, reasons to believe that the gravity wave spectrum may undergo systematic changes in response to variations in the environment that depart from normal growth with height at low vertical wavenumbers. This can be seen most readily by noting that the saturated momentum flux for an individual wave motion given by (4) depends both on $\overline{u}(z)$ (in the direction of wave motion) and on N. Thus, wave motions that experience a reduction in the intrinsic phase speed, $(c - \overline{u})$, will undergo preferential dissipation while those that experience an increase in $(c - \overline{u})$ will grow with height. This results in a filtering of the gravity wave spectrum (Lindzen, 1981) and may account as well for much of the geographic, seasonal, and short-term variability observed in a number of

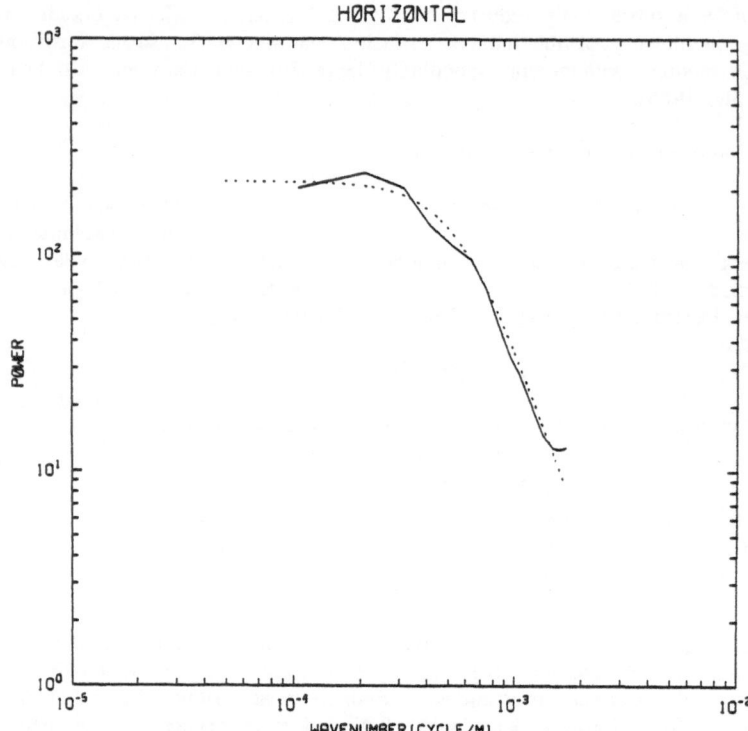

Figure 5. Energy density of 15 minute averaged radial velocity fluctuations (7° beams) in the lower stratosphere for a three-day period during February 1985. The asymptotic slope at high vertical wavenumbers is -3.

observational and numerical studies (Meek et al., 1985a; Vincent and Fritts, 1987; Fritts and Vincent, 1987; Reid and Vincent, 1987a; Miyahara et al., 1986). Likewise, we may expect that wave motions that encounter an increase in N will experience enhanced saturation and a reduction in the vertical energy and momentum fluxes. The same arguments are easily quantified for a saturated spectrum of wave motions and may explain both the apparent need for a wave drag near the tropopause (Palmer et al., 1986) and the enhanced radar echoes observed near the tropopause (T. Tsuda, private communication, 1986) and the high-latitude summer mesopause (Balsley et al., 1983). The saturated spectrum and its variations with height and environment may thus provide a convenient framework within which to develop a simple parameterization of gravity wave effects in the lower and middle atmosphere.

4.   VARIABILITY OF THE GRAVITY WAVE SPECTRUM

It has been known for many years that gravity waves produce fluctuations in temperature (or density), velocity, and constituent concentrations. But it has been only in the last few years that the extent of the geographic and temporal variability of the gravity wave spectrum itself, and of its effects, has begun to be appreciated. The purposes of this section are to exhibit some of the enormous variability and to present evidence pointing to some of the major sources of this variability.

While observations suggest that the atmospheric gravity wave spectrum exhibits considerable universality in a statistical sense (VanZandt, 1982; Dewan et al., 1984; Smith et al., 1987), there are many reasons to expect that the wavenumber and frequency composition and the degree of anisotropy may vary considerably with time and location. This variability may arise in part in response to geographic, seasonal, and short-term variations in the strengths of the important gravity wave source mechanisms. Another contributing factor for which there is increasing evidence is the interaction with and response of the gravity wave spectrum to the variable environment through which it propagates.

An excellent example, both of the geographic variability of a significant source of gravity wave motions (in this case convection) and of the effects of filtering of those motions by the intervening wind profile is the numerical work performed by Miyahara et al. (1986) using the GFDL "SKYHI" model. This model is unique at present in its ability to simulate directly, if crudely, small-scale gravity waves in a high-resolution GCM. In addition to validating the effects of gravity waves expected from previous theoretical, parametric, and observational studies of gravity waves and the middle atmosphere circulation and structure, this work provides clear evidence of the importance of filtering and of geographic variability of source strengths on the resulting wave spectrum at mesospheric heights (see Figure 15 of that paper).

Another example of geographic variability of the (assumed) gravity wave spectrum is provided by a recent observational study using GASP aircraft data (Nastrom et al., 1987). In the present context, the principal finding in this study was a significant enhancement of the amplitude of the motion spectrum over rough terrain (western U.S.) relative to ocean (eastern Pacific) and flat terrain (east-central U.S.). These results suggest that topography

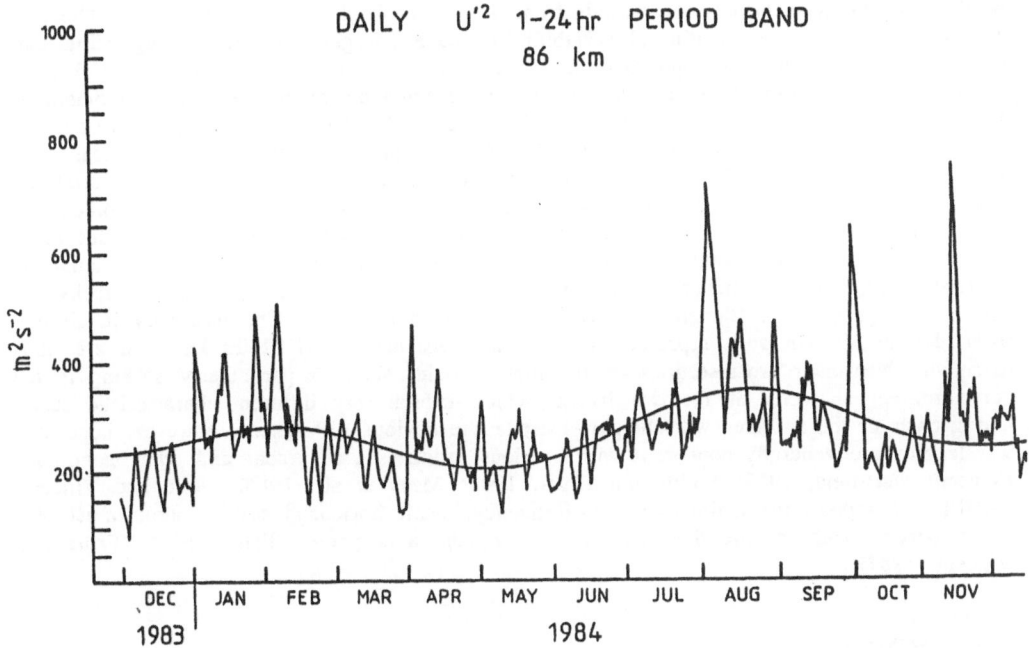

Figure 6. Zonal energy density of 1 - 24 hr period motions at 86 km at Adelaide, Australia computed daily with a 3-day running average applied. Note the large 5 - 10 day fluctuations.

may also contribute substantially to excitation of the gravity wave spectrum and to its variable effects in the lower and middle atmosphere. Other sources, such as wind shear or geostrophic adjustment, are likely to be important and exhibit variability as well, but are even less well quantified than convection and orography at this time. Clearly, considerable work remains in obtaining a detailed knowledge, even statistically, of the dominant gravity wave source mechanisms. Further work will be required to understand the degree to which they vary with location and meteorological conditions.

Temporal variability of the gravity wave spectrum is likewise considerable and is manifested on a wide range of time scales. Seasonal trends are evident in long-term climatologies (Meek et al., 1985a; Vincent and Fritts, 1987), with maximum gravity wave energies occurring during winter and summer and minima near equinoxes. These variations are more pronounced for wave motions with higher frequencies. The activity minima that occur also appear to be well correlated with the transition from summer easterlies to winter westerlies and vice versa, suggestive of a strong modulation of the gravity wave spectrum by filtering and saturation processes. Similar trends have been noted in a long-term study of turbulence intensities by Hocking (1987) and proposed to account for seasonal variations of O_3 at mesospheric heights due to induced variations in the turbulent transports of other chemically active species by Thomas et al. (1984).

Shorter-term variations in gravity wave energies and momentum fluxes have also been observed in a number of studies. The variations in energy of gravity waves with periods of 5 - 10 days are apparent in the mesospheric climatologies that are available (Meek et al., 1985a; Vincent and Fritts, 1987) and appear to correspond closely to the time scales of variations in the observed zonal and meridional winds due to planetary wave activity. The seasonal and planetary wave scale variability in the zonal wave energy observed by Vincent and Fritts (1987) for periods from 1 - 24 hr is illustrated in Figure 6. There is also increasing evidence of diurnal variability of wave energies and fluxes, suggesting that gravity waves may also respond to other factors, such as tides, that act to control their environment. An example of this is the strong modulation of high-frequency momentum fluxes in apparent response to a large-amplitude diurnal tidal motion noted by Fritts and Vincent (1987) and shown in Figure 7. This result implies that filtering by the tidal wind field, like the mean and planetary wave wind fields, may contribute significantly to variability of the middle atmosphere gravity wave spectrum and its transports of energy and momentum.

A final example of the variability of the gravity wave spectrum at mesospheric heights is provided by recently available space shuttle re-entry data. The density fluctuations inferred during two re-entries along virtually identical low-latitude ($\sim$ 20°N) tracks are shown in Figure 8 (R. Blanchard, private communication, 1986). The data vary in altitude from 80 to 60 km and represent a horizontal distance of H 2500 km and are thus essentially horizontal cross-sections of the motion field. What is particularly striking is the very different character of the two tracks, which exhibit very different characteristic scales and growth of those scales with height. Despite the obvious differences, however, horizontal wavelengths are generally consistent with previous estimates using radar and other techniques (Vincent and Reid, 1983; Smith and Fritts, 1983; Meek et al., 1985b; Reid and Vincent, 1987b) and support the claim that high-frequency, small horizontal scale motions must play an important role in the dynamics of the middle atmosphere (Fritts, 1984; Fritts and Vincent, 1987).

5.   SUMMARY

We have presented a brief survey of some of the recent results in studies of gravity wave saturation processes and their variability in the middle atmosphere. This is obviously a field

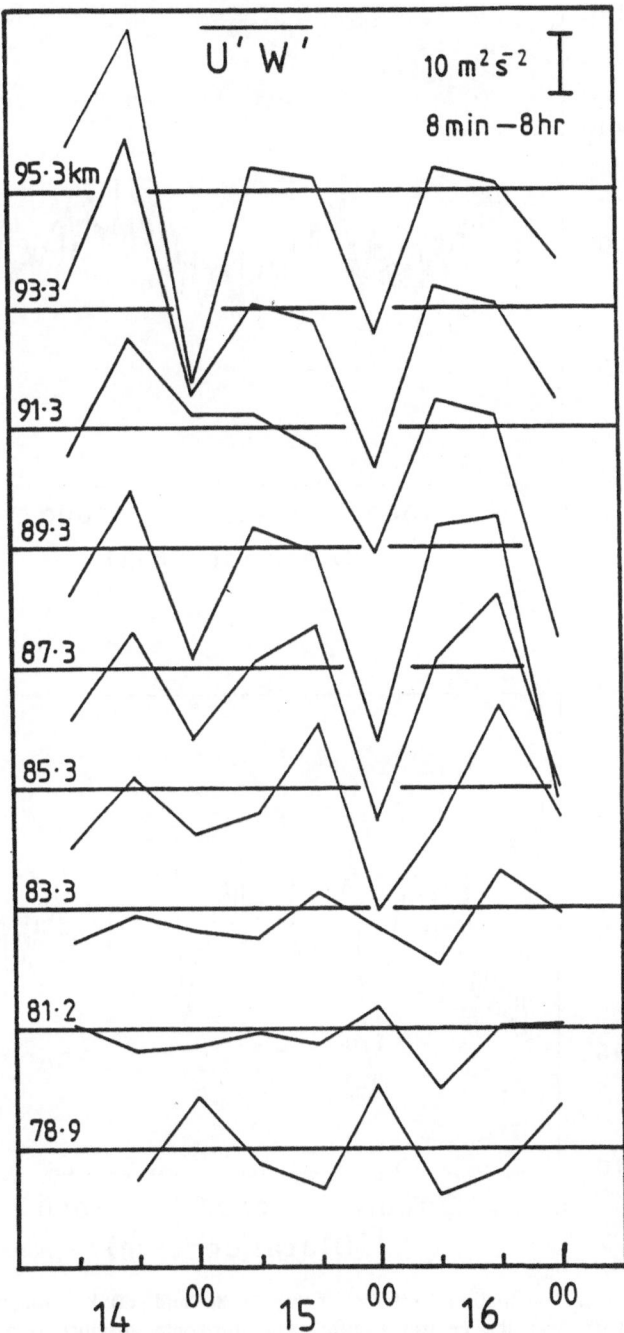

Figure 7. Vertical flux of zonal momentum computed in 8-hr blocks for several heights at Adelaide, Australia for a 3-day period during June 1984. Note the large diurnal modulation at upper levels.

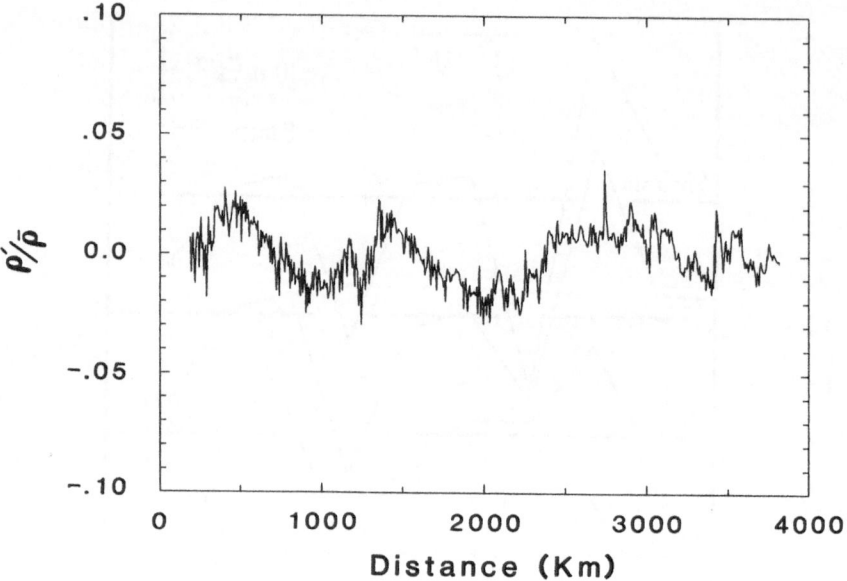

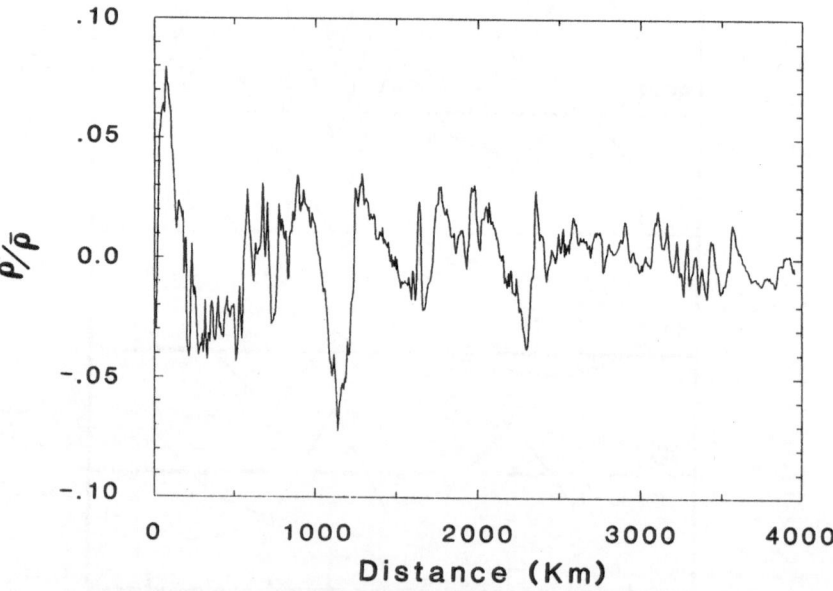

Figure 8. Density fluctuations inferred between 80 and 60 km for two space shuttle re-
entries over the central Pacific. The horizontal distance is ∼ 4000 km. Note the
dramatic differences, both in the amplitudes and the dominant horizontal scales of
the motions encountered.

of very rapid development. and we will no doubt see equally rapid advances made in the future as the importance of gravity wave motions and their middle atmosphere effects becomes more widely appreciated.

During the last few years. progress was made in understanding the nature of the wave saturation processes and of the constraints these processes seem to place on the wave spectrum. Evidence suggests that wave saturation is essentially the consequence of a linear instability of the wave field, with wave dissipation occurring in response to the production of intense local turbulence. This turbulence acts to limit wave amplitudes and causes the diffusion of heat and constituents. Because of the dependence of turbulence intensities on the phase of the wave motion, however, there is a tendency for the turbulence to be substantially less effective in producing vertical diffusion than in limiting wave amplitudes.

Wave saturation and the relative energies of waves with different frequencies suggest that the vertical fluxes of energy and momentum should be dominated by motions with high intrinsic frequencies and large vertical scales. This appears to be confirmed by observations. Another result of saturation is the creation of a quasi-universal vertical wavenumber spectrum that evolves with height in response to the environment through which it propagates. Such a spectrum is both consistent with observations and may provide a useful means of parameterizing gravity wave effects in large-scale models.

Finally, the gravity wave spectrum and its middle atmosphere effects exhibit an enormous degree of variability, both geographically and temporally, as a result of a number of excitation and filtering processes that are only beginning to be understood in detail at this time. Thus, this field will continue to provide a very fertile area for research for some time to come.

Acknowledgments
This review was prepared under support from the Office of Naval Research under contract N00014-86-K-0661.

6. REFERENCES

Balsley, B.B., and D.A. Carter, 1982: The spectrum of atmospheric velocity fluctuations at 8 and 86 km. Geophys. Res. Lett., 9, 465-468.

Balsley B.B., and R. Garello, 1985: The kinetic energy density in the troposphere, stratosphere and mesosphere: A preliminary study using the Poker Flat radar in Alaska. Radio Sci., 20, 1355-1362

Balsley, B.B., W.L. Ecklund, and D.C. Fritts, 1983: VHF echoes from the high-latitude mesosphere and lower thermosphere: Observations and interpretations. J. Atmos. Sci., 40. 2451-2466.

Chao, W.C., and M.R. Schoeberl, 1984: A note on the linear approximation of gravity wave saturation in the mesosphere. J. Atmos. Sci., 41, 1893-1898.

Coy, L., and D.C. Fritts. 1987: Gravity wave heat fluxes: A Lagrangian approach. J. Atmos. Sci., to appear.

Desaubies, Y.J.F., 1976: Analytical representation of internal wave spectra. J. Phys. Oceanogr., 6. 976-981.

Dewan, E.M., and R.E. Good. 1986: Saturation and the "universal" spectrum for vertical profiles of horizontal scalar winds in the atmosphere. J. Geophys. Res.. 91, 2742-2748.

Dewan, E.M., N. Grossbard. A.F. Quesada. and R.E. Good. 1984: Spectral analysis of 10m resolution scalar velocity profiles in the stratosphere. Geophys. Res. Lett.. 11. 80-83. and Correction. Geophys. Res. Lett., 11, 624.

Dunkerton. T.J., 1982: Stochastic parameterization of gravity wave stresses. J. Atmos. Sci., 39, 1711-1725

Dunkerton, T.J., 1984: Inertia-gravity waves in the stratosphere. J. Atmos. Sci., 41, 3396-3404.

Endlich, R.M., R.C. Singleton, and J.W. Kaufman, 1969: Spectral analysis of detailed vertical wind speed profiles. J. Atmos. Sci., 26, 1030-1041.

Fritts, D.C.,1984: Gravity wave saturation in the middle atmosphere: A review of theory and observations. Rev. Geophys. Space Phys., 22, 275-308.

Fritts, D.C., 1985: A numerical study of gravity wave saturation: Nonlinear and multiple wave effects. J. Atmos. Sci., 42, 2043-2058.

Fritts, D.C., and T.J. Dunkerton, 1985: Fluxes of heat and constituents due to convectively unstable gravity waves. J. Atmos. Sci., 42, 549-556.

Fritts, D.C., and P.K. Rastogi, 1985: Convective and dynamical instabilities due to gravity wave motions in the lower and middle atmosphere: Theory and observations. Radio Sci., 20, 1247-1277.

Fritts, D.C., S.A. Smith, B.B. Balsley, and C.R. Philbrick, 1987: Evidence of gravity wave saturation and local turbulence production in the summer mesosphere and lower thermosphere during the STATE experiment. submitted to J. Geophys. Res.

Fritts, D.C., and R.A. Vincent, 1987: Mesospheric momentum flux studies at Adelaide, Australia: Observations and a gravity wave/tidal interaction model. J. Atmos. Sci., 44, 605-619.

Garrett, C.J.R., and W.H. Munk, 1972: Space-time scales of internal waves. Geophys. Astrophys. Fluid Dyn., 3, 225-235.

Garrett, C.J.R., and W.H. Munk, 1975: Space-time scales of internal waves: A progress report. J. Geophys. Res., 80, 291-297.

Hines, C.O., 1960: Internal gravity waves at ionospheric heights. Can. J. Phys., 38, 1441-1481.

Hocking, W.K.. 1987: Radar studies of small scale structure in the middle atmosphere and lower thermosphere. Adv. Space Res.. in press.

Hodges. R.R.. Jr.. 1967: Generation of turbulence in the upper atmosphere by internal gravity waves. J. Geophys. Res.. 72, 3455-3458.

Hodges, R.R.. Jr., 1969: Eddy diffusion coefficients due to instabilities in internal gravity waves. J. Geophys. Res., 74, 4087-4090.

Holton. J.R., 1982: The role of gravity wave-induced drag and diffusion in the momentum budget of the mesosphere. J. Atmos. Sci.. 39, 791-799.

Holton, J.R., 1983: The influence of gravity wave breaking on the general circulation of the middle atmosphere, J. Atmos. Sci.. 40, 2497-2507.

Labitzke, K., A.H. Manson, J.J. Barnett, and M. Corney, 1987: Comparison of geostrophic and observed winds in the upper mesosphere over Saskatoon, Canada. Physica Scripta, in press.

Lindzen, R.S., 1981: Turbulence and stress owing to gravity wave and tidal breakdown. J. Geophys. Res., 86, 9707-9714.

Meek, C.E., I.M. Reid and A.H. Manson, 1985a: Observations of mesospheric wind velocities. II. Cross sections of power spectral density for 48-8h, 8-1h, 1h-10 min over 60-110 km for 1981. Radio Sci., 20, 1383-1402.

Meek, C.E., I.M. Reid and A.H. Manson, 1985b: Observations of mesospheric wind velocities. I. Gravity wave horizontal scales and phase velocities determined from spaced wind observations. Radio Sci., 20, 1383-1402.

Miyahara, S.. Y. Hayashi, and J.D. Mahlman, 1986: Interactions between gravity waves and the planetary scale flow simulated by the GFDL "SKYHI" general circulation model. J. Atmos. Sci., 43, 1844-1861.

Nastrom, G.D., D.C. Fritts, and K.S. Gage, 1987: Topographic effects on mesoscale variability near the tropopause: Waves or turbulence? submitted to J. Geophys. Res.

Palmer, T.N., G.J. Shutts, and R. Swinbank, 1986: Alleviation of a systematic westerly bias in general circulation and numerical weather prediction models through an orographic gravity wave drag parameterization. Quart. J. Roy. Met. Soc., 86, 1001-1040.

Philbrick, C.R., K.U. Grossman. R. Hennig, G. Lange. D. Krankowsky, D. Offermann, F.J. Schmidlin, and U. von Zahn, 1983: Vertical density and temperature structure over Northern Europe. Adv. Space Res.. 2, 121-124.

Reid, I.M., and R.A. Vincent, 1987a: Measurements of mesospheric gravity wave momentum fluxes and mean flow accelerations at Adelaide, Australia. submitted to J. Atmos. Terres. Phys.

Reid, I.M., and R.A. Vincent, 1987b: Measurements of the horizontal scales and phase velocities of short period mesospheric gravity waves at Adelaide, Australia. submitted to J. Atmos. Terres. Phys.

Smith, S.A., and D.C. Fritts, 1983: Estimations of gravity wave motions, momentum fluxes and induced mean flow accelerations in the winter mesosphere over Poker Flat, Alaska. Proceedings of the 21st Conference on Radar Meteorology, Edmonton, 104-110.

Smith, S.A., D.C. Fritts, and T.E. VanZandt, 1987: Evidence of a saturation spectrum of atmospheric gravity waves. J. Atmos. Sci., 44, 1404-1410.

Strobel, D.F., J.P. Apruzese, and M.R. Schoeberl, 1985: Energy balance constraints on gravity wave induced eddy diffusion in the mesosphere and lower thermosphere. J. Geophys. Res., 90, 13,067-13,072.

Thomas, R.J., C.A. Barth, and S. Solomon, 1984: Seasonal variations of ozone in the upper mesosphere and gravity waves. Geophys. Res. Lett., 11, 673-676.

VanZandt, T.E., 1982: A universal spectrum of buoyancy waves in the atmosphere. Geophys. Res. Lett., 9, 575-578.

Vincent, R.A., and D.C. Fritts, 1987: A climatology of gravity wave motions in the mesospause region at Adelaide, Australia. J. Atmos. Sci., 44, 748-760.

Vincent, R.A., and I.M. Reid, 1983: HF Doppler measurements of mesospheric momentum fluxes. J. Atmos. Sci., 40, 1321-1333.

RADAR OBSERVATIONS OF GRAVITY WAVES IN THE MESOSPHERE

R. A. Vincent
Physics Department
University Of Adelaide
Adelaide
Australia 5001

ABSTRACT. Continuous ground-based radar studies of mesospheric gravity waves show that there is a semi-annual variation in wave energy especially at heights below 80 km. The minimum in wave energy coincides with the time of reversal in the zonal winds in the middle atmosphere. The wave motions also appear to be anisotropic with the NS wind perturbations stronger than the EW. The associated non-zero $\overline{u'v'}$ momentum fluxes indicate seasonal variations in the direction of wave propagation. Measurements of $\overline{u'w'}$ and $\overline{v'w'}$ fluxes also show modulation of the waves by planetary and tidal winds. Measurements of horizontal scales and phase velocities are more difficult but radar estimates are in reasonable accord with scales and velocities made by other techniques. The mean horizontal scale appears to increase with the observed wave period.

1. INTRODUCTION

The important role played by gravity waves in the middle atmosphere has long been recognized (e.g. see the review by Fritts (1984) and references therein). However, while there have been a number of theoretical advances in recent years, progress in achieving a better understanding of the effects of gravity waves in the middle atmosphere requires more extensive measurements of fundamental wave parameters. The situation has improved with the commencement at a few sites of ground based radar measurements of winds on a continuous basis, so that it has been practicable to assemble gravity wave climatologies for the mesosphere (Meek et al., 1985a; Vincent and Fritts, 1987) and to study the variations of wave energy with height through much of the middle atmosphere (Balsley and Garello, 1985). With the advent of wave climatologies it is now possible to examine in better detail seasonal and shorter term variations in important parameters such as the energy densities. Recent advances in radar techniques have also enabled measurements of other vital quantities to be made, such as the vertical flux of horizontal momentum (Vincent and Reid, 1983; Fritts and Vincent, 1987; Reid and Vincent, 1987a) and horizontal wavelength and phase speed

47

G. Visconti and R. Garcia (eds.), Transport Processes in the Middle Atmosphere, 47–56.
© *1987 by D. Reidel Publishing Company.*

(Vincent and Reid, 1983; Meek et al., 1985b; Reid and Vincent, 1987b).
        Although wave measurements are now starting to be made in the
troposphere and lower stratosphere by radars operating at both VHF and
UHF most radar observations still pertain to the upper part of the
middle atmosphere between 60 and 100 km. The purpose of this paper is to
summarize some of the recent observations at these heights, with
particular reference to temporal variations in wave fluxes.

2. GRAVITY WAVE CLIMATOLOGIES.

The winds data used to produce the wave climatologies have been obtained
with either the spaced antenna (SA) technique at MF/HF frequencies
(Vincent, 1984) and reported by Meek et al. (1985a) for Saskatoon (52N,
107W) and by Vincent and Fritts (1987) for Adelaide (35S, 138E) or the
Doppler method at VHF by Balsley and Garello (1985) for Poker Flat (65N,
147W). All measurements have some experimental limitations and these
should be recognized. The Poker Flat VHF measurements are limited by a
seasonal variation in echo strengths to heights below 80 km in winter
and above 80 km in summer. The SA observations at lower frequencies do
not have this limitation and consequently are able to measure the wave
motions at heights between 60 and 110 km in all seasons (although at
Adelaide the very low levels of nocturnal ionization at heights below
about 78 km restricts observations to above this height at night). The
vertical resolution that is achievable at frequencies near 2 MHz is
typically 3-4 km which limits the observations to waves with vertical
wavelengths (λ_z) of greater than 6-8 km. From the gravity wave
dispersion relation

$$m = \frac{N}{|u-c|}$$

where m is the vertical wavenumber, N is the Brunt–Vaisala frequency and
$|u-c|$ is the intrinsic phase speed, it is seen that the observations are
sensitive to waves with intrinsic phase speeds greater than about 20-25
ms^{-1}. This limitation imposed by the height resolution tends to place a
lower limit on the wave fluxes which can be measured.
        The observations by the various techniques show good agreement
on the variations in wave amplitude with height. For ease of
presentation and comparison, it has been usual to divide the spectrum
into a number of period ranges, for example for periods less than 1 hr,
1-8 hr and 8-24 hr. If the total mean square amplitude $\overline{V'^2} = \overline{u'^2+v'^2}$,
where u' and v' are the respective deviations from the appropriately
defined mean values of the zonal and meridional wind components, then in
most cases the energy density per unit volume given by $\rho_0\overline{V'^2}$ decreases
exponentially with altitude, where ρ_0 is the neutral density. This
behaviour is often taken as indirect evidence of wave saturation where
the amplitude of the wave motions become comparable to the intrinsic
phase speed i.e. $u' \approx |u-c|$ (Fritts, 1984). It is noteworthy that the
Poker Flat measurements show that $\rho_0\overline{V'^2}$ decreases throughout the middle
atmosphere, from the troposphere upward (Balsley and Garello, 1985)

which suggests that wave dissipation is taking place continuously rather than being confined to altitudes above discrete breaking levels in the mesosphere.

The Adelaide and Saskatoon climatologies reveal similar seasonal behaviour in the mesospheric wave field despite their wide geographic separation. Figure 1 shows a cross-section of $\overline{v'^2}$ in the 1–8 hr period range at Adelaide in 1985–86. The contours are plotted with a 10–20 day time resolution and it is apparent that there are significant short term variations in the fluxes but the overall behaviour is very similar to that reported by Vincent and Fritts (1987) for Adelaide in 1983–84 and by Meek et al. (1985a) for Saskatoon. The hatched regions in fig. 1 correspond to times when the mean zonal circulation was westward (easterly) so the wave amplitudes are smallest at the about the times when the zonal flow is reversing direction, giving rise to a significant semi-annual variation in wave activity which is especially pronounced at heights below 80 km in both hemispheres.

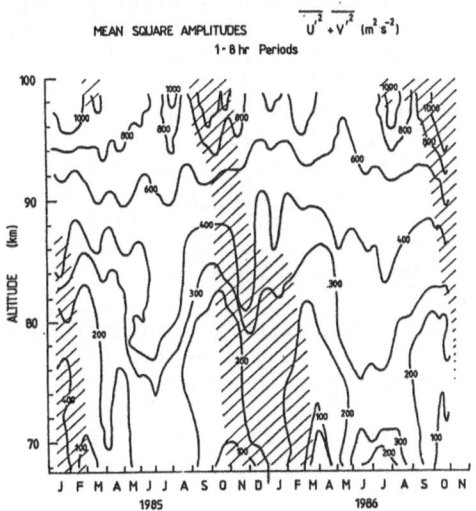

Figure 1. Time–height distribution of the total mean square amplitudes for motions in the 1–8 hr period range at Adelaide.

A significant feature of the Adelaide observations is that meridional amplitudes are usually larger than the corresponding values for u'. Vincent and Fritts (1987) chose to emphasize this property by plotting contours of the ratio v'^2/u'^2. Broadly speaking, the u' and v' amplitudes are comparable only at heights below 75 km and generally the $\overline{v'^2}$ values are 20 to 40% larger than $\overline{u'^2}$, although ratios as high as 2:1 are evident in winter. There is a tendency for the ratio to increase with increasing altitude in all seasons. Again, similar features are found for waves in the 8–24 hr period range and are also evident in subsequent years, although the details may vary from one year to the next. Ebel et al. (1987) report a similar tendency for a preferred north–south orientation of wind fluctuations at Saskatoon.

        The larger values of $\overline{v'^2}$ than $\overline{u'^2}$ show that the gravity wave
motions are anisotropic. The reasons for this anisotropy are not fully
clear but because gravity wave motions tend to be oriented along the
direction of propagation, it suggests that near the mesopause there is a
preponderance of waves propagating in the meridional directions. An
anisotropic angular spectrum is not unexpected on theoretical grounds,
being caused by azimuthal filtering of the gravity waves as they
propagate upward through the strong zonal winds of the stratosphere.
Those waves propagating in the direction of the mean flow will be
removed at critical levels so that the mesospheric gravity wave field
will be deficient in zonally directed waves (e.g. Lindzen, 1981).
Temporal changes in the background wind field, as well as in the source
regions, could then produce the semi-annual variations in wave
amplitude.

        Another potentially important wave parameter which can also be
determined from the wind measurements is the meridional flux of zonal
momentum, $\overline{u'v'}$. While the $\overline{u'v'}$ fluxes are smaller in magnitude than
either $\overline{v'^2}$ or $\overline{u'^2}$, (mean values of about 30 m^2s^{-2}) they are nevertheless
non-negligible and show an interesting seasonal pattern. Overall,
negative fluxes predominate, especially in summer, but with more
variability in winter. As noted above, the horizontal motions are
polarized along the horizontal direction of propagation; e.g., a wave
propagating towards the NE will cause perturbation motions in the NE/SW
quadrants so that the u' and v' motions would be positively correlated
($\overline{u'v'}>0$). The observed negative values of $\overline{u'v'}$ imply a net negative
correlation between u' and v' which in turn means that on average the
wave motions are aligned in the NW/SE quadrants.

        The observations that $\overline{v'^2}/\overline{u'^2}$ ratios are usually greater than
unity and that the $\overline{u'v'}$ fluxes are negative suggest that a certain
fraction of the gravity waves has a relatively well defined direction of
travel which changes with season, i.e., the wave field is "partially
polarized". In an attempt to provide a statistical measure of the degree
of polarization, Vincent and Fritts (1987) devised a model of a
partially polarized gravity wave field in analogy to a partially
polarized spectrum of electromagnetic waves. They found that that the
observations could be explained if, on the average, the degree of
polarization was between 0.10 and 0.30 on a seasonal basis but could be
as high as 0.50 over shorter time intervals. By making assumptions about
the sign of the associated vertical fluxes of zonal momentum ($\overline{u'w'}$) it
is possible to determine the mean direction of propagation of the
polarized fraction of the wave field (Vincent and Fritts, 1987).

        The degree of polarization is usually larger in summer (mean
value ≈ 0.25) than in winter (mean value ≈ 0.15) which suggests that
there is a narrower angular spectrum of waves in summer. While the
details may change from year to year it is usually found that the
polarized waves propagate toward the SE in summer and toward the NW in
winter, directions inferred from the negative sign of the $\overline{u'v'}$ fluxes
and by the choice of $\overline{u'w'}<0$ in winter and $\overline{u'w'}>0$ in summer. However, in
the lower mesosphere near 75 km the waves often have a strong zonal
component but near 95 km the polarized waves are more strongly aligned
toward the north or south which indicates that there has been a removal

of waves propagating zonally. Such a reduction in the zonal flux as the waves propagate up from the lower mesosphere to the lower thermosphere is consistent with the hypothesis of a zonal wave drag exerted by breaking gravity waves; the pronounced meridional component of motion suggests however, that there must also be a significant meridional body force. More direct estimates of wave drag are made from measurements of the $\overline{u'w'}$ and $\overline{v'w'}$ fluxes and their variations with height, which we now consider.

3. MOMENTUM FLUXES

Breaking or dissipating gravity waves exert their drag by the vertical convergence of the momentum fluxes which, in the case of the zonal component, leads to a zonal body force per unit mass or acceleration of

$$F_u = -\frac{1}{\rho_o}\frac{d}{dz}(\rho_o\overline{u'w'})$$

A method to measure $\overline{u'w'}$ using Doppler radar techniques was described by Vincent and Reid (1983). The technique involves measurement of the Doppler shift of backscattered signals received in two beams displaced symmetrically at angles θ from the zenith in the east and west directions. The method is sometimes referred to as the dual complementary beam technique. If the mean square perturbation velocities measured along the east and west pointing beams are V_E^2 and V_W^2 respectively, then it can be shown that

$$\overline{u'w'} = (\overline{V_E^2} - \overline{V_W^2})/2\sin(2\theta)$$

Similar measurements using a pair of beams in the NS plane can be used to determine $\overline{v'w'}$. Since the determination of $\overline{u'w'}$ and $\overline{v'w'}$ depends on taking the difference of two quantities which are similar in magnitude, it is necessary to average the data over several hours to ensure reliable estimates. Further details of the experiment and the sources of error may be found in Vincent and Reid (1983), and Reid and Vincent (1987a).
        Estimates of $\overline{u'w'}$ fluxes measured in early winter periods in each of three different years are shown in fig. 2; the observations were taken over periods ranging from 3 to 8 days. At most heights the fluxes are negative with magnitudes ranging between about 1 and 5 m^2s^{-2}. These values, which are typical of all the values measured at Adelaide, are an order of magnitude smaller than the $\overline{u'v'}$ fluxes but, despite this, the vertical convergence of $\rho_o\overline{u'w'}$ ($F_u \approx \overline{u'w'}/H$, where H is the scale height) is sufficient to cause a significant wave drag, as may be seen in Fig. 3. The accelerations, in $ms^{-1}day^{-1}$, are compared with the Coriolis torques inferred from the measured $\overline{v}$, which are usually poleward in direction in winter with magnitudes of up to 10 to 15 ms^{-1}. Here it is apparent that the wave drag and the torque derived from the locally measured v are in approximate balance, but this is not always the situation.

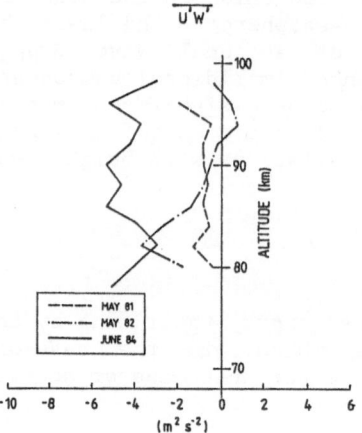

Figure 2. Vertical profiles of $\overline{u'w'}$ fluxes measured at Adelaide in winter.

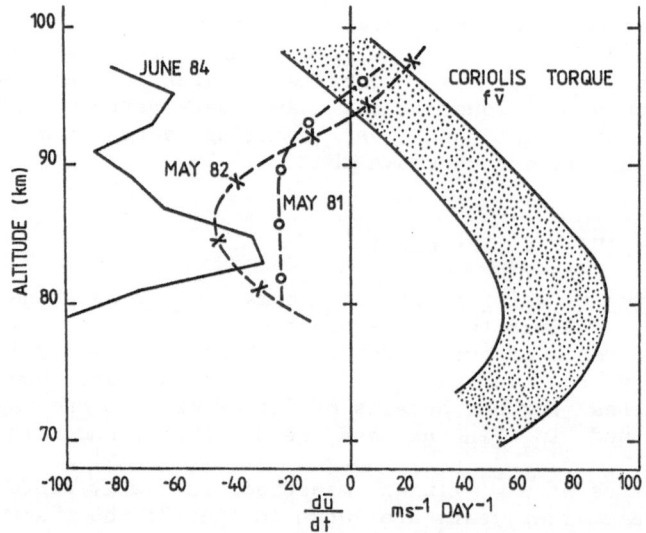

Figure 3. Vertical profiles of the zonal acceleration derived from the u'w' profiles shown in figure 2 compared with the inferred Coriolis torques.

      Reid and Vincent (1987a) found that, while the local zonal acceleration determined from $\overline{u'w'}$ data averaged over several days is often in the correct sense and of the right magnitude to produce balance, significant temporal variability in $\overline{u'w'}$ is also observed so that local balance may not always occur. Reid and Vincent also reported preliminary measurements of $\overline{v'w'}$ fluxes which were found to be comparable in magnitude to the $\overline{u'w'}$ fluxes. In one of the longest sequences of $\overline{u'w'}$ and $\overline{v'w'}$ measurements made at Adelaide they found that the zonal (F_u) and meridional (F_v) components of the total body force were of comparable magnitude so that the total mean flow accelerations were not aligned in the zonal direction. Between 6 and 16 July 1982 the total body force (F_T) was observed to vary considerably as a function of both height and time. These variations coincided with considerable changes in the mean flow at lower mesospheric heights associated with planetary wave activity. The fluctuations in the gravity wave $\overline{u'w'}$ and $\overline{v'w'}$ fluxes appeared to be caused by directional filtering of the waves as they propagated upwards through the changing background winds in the middle atmosphere.

The momentum fluxes can also exhibit fluctuations on time scales of the order of hours. While these variations often seem to be random in nature there are occasions when systematic changes have been observed. A notable example of diurnally modulated $\overline{u'w'}$ fluxes in June 1984 was reported by Fritts and Vincent (1987). The maximum values of $\overline{u'w'}$ were found to occur in the 8 hr period centered on midnight local time. The amplitude of the diurnal tide was quite large (20 ms^{-1}), it had a relatively short vertical wavelength (30–40 km) and the phase was such that during this 8 hr interval the mean zonal wind shear was reversed from the negative values normally observed in winter. To explain their observations Fritts and Vincent (1987) proposed a gravity wave/tidal interaction model based on linear saturation theory in which $\overline{u'w'}$ was predicted to grow in an environment where $|u-c|$ increases with height. At other times of the day, the tidal motions reinforced the negative shear of the mean zonal wind and westward propagating gravity waves (which are believed to dominate the wave spectrum at these heights in winter) would experience a decrease in amplitude, as observed. Other instances of tidally modulated $\overline{u'w'}$ and also $\overline{v'w'}$ fluxes have been observed at Adelaide (D. Murphy, Private Communication).

One interesting aspect of the momentum flux measurements at Adelaide is the dominance of short period (in a ground based reference frame) motions. Most studies to date indicate that waves with periods less than 1 hr carry up to 70% of the upward flux of horizontal momentum (Vincent and Reid, 1983; Fritts and Vincent, 1987; Reid and Vincent, 1987a) although in some recent observations Murphy (Private Communication) finds that fluxes carried by waves in the 1–8 hr range appear predominant. Fritts (1984) has discussed reasons why the mean flow accelerations might be more efficiently accomplished by the higher-frequency motions.

4. HORIZONTAL WAVELENGTHS AND PHASE SPEEDS

The horizontal wavelength (λ_h) and phase speed are amongst the more difficult quantities to measure but a variety of radar methods have been developed in an attempt to determine these wave parameters, for example using spaced antennas (Meek et al., 1985b), Doppler radar (Vincent and Reid, 1983) and meteor techniques (Yamamoto et al., 1986). All techniques measure the time shifts of wave perturbations which propagate

between two or more locations which are horizontally separated by
distances which may range from about 10 km up to a few hundred km. Since
the wave motions may have scales shorter than the separation distance
between observing regions it is possible for ambiguities to occur, and
the techniques also depend on the occurrence of quasi-monochromatic wave
motions which retain their coherence between the observing sites.
Because the wave field in the mesosphere often appears to be broad-band
or random in nature, the characteristics of monochromatic waves may not
be representative of the overall wave field. Nevertheless, the
observations to date are in reasonable accord both with each other and
with waves observed by more direct techniques, such as the photography
of wave perturbations in airglow (e.g. Armstrong, 1982).

Using the dual complementary beam technique, Vincent and Reid
(1983) were able to measure λ_x, the horizontal wavelength in the zonal
direction, while in a later, more comprehensive set of observations at
Adelaide, Reid and Vincent (1987b) reported measurements of the
meridional component as well. These observations were restricted to
motions with periods less than about 100 min and with scales less than a
about 200 km, a limitation imposed by the relatively small beam
separation of about 20 km. Averaged over a number of observing intervals
covering all seasons, Reid and Vincent (1987b) found the mean λ_x ranged
from about 30 km to about 78 km at wave periods of 10 and 100 min
respectively, with a mean value in the whole period range of 55 km. The
corresponding phase speeds were 52 and 13 ms^{-1}. Since these are trace
values, they form upper limits to the actual horizontal wavelengths λ_h
but assuming that the wavefield is isotropic in azimuth then the values
at 10 and 100 min are 22 km and 55 km with associated values of phase
speed of 37 and 10 ms^{-1}.

The increase in λ_h with increasing period found at Adelaide is
typical of other studies (e.g. Meek et al., 1985b). Reid and Vincent
(1987b) in a summary of a large number of wavelength determinations made
with both radar and optical techniques, showed that the scales increase
monotonically out to periods in excess of several hours; at the long
periods the wavelengths range between 1000 and 5000 km. Very
approximately, the horizontal wavelength varies with ground based period
T as $\lambda_h \propto T$ with the constant of proportionality being such that the
mean $\lambda \approx 60$ km for $T \approx 30$ min and $c \approx 30\text{--}35$ ms^{-1}.

5. SUMMARY

The results discussed above show that continuous radar studies of
atmospheric motions are now starting to provide sufficient data on
mesospheric gravity waves that climatologies of important wave
parameters are emerging. Table 1 summarizes some of these results as
they have been observed at one site (Adelaide) at heights near 85 km.
These values must be considered approximate, especially the wavelengths
and phase velocities. Although they only pertain to one site there are
indications that many of the parameters are 'universal' in the sense
that similar values are found at other locations. For example, Vincent
(1984b), who compared observations made by three radars widely separated

Table 1. Gravity wave parameters at 85 km, Adelaide, Australia.

$\overline{u'^2}$	250 m^2s^{-2}
$\overline{v'^2}$	350 m^2s^{-2}
$\overline{V'^2}$	600 m^2s^{-2}
V'	25 ms^{-1}
$|\overline{u'v'}|$	$10-50$ m^2s^{-2}
$|\overline{u'w'}|\approx|\overline{v'w'}|$	$1-5$ m^2s^{-2}
λ_h (10<T<100 min)	$20-55$ km
c	$30-35$ ms^{-1}

in latitude, found that the rms amplitudes were similar at all three sites. The seasonal behavior of wave activity also shows very similar behavior at both Adelaide and Saskatoon. The greater geographical coverage accorded by the recent establishment of new radars in the Antarctic and the imminent installation of radars on the equator will lead to further improvements in our knowledge of gravity waves and of the roles they play in the middle atmosphere.

REFERENCES

Armstrong, E. B., 1982: The association of visible airglow features with a gravity wave. J. Atmos. Terr. Phys., 44, 325-336.

Balsley, B. B. and R. Garello, 1985: The kinetic energy density in the troposphere, stratosphere, and mesosphere: A preliminary study using the Poker Flat MST radar in Alaska. Rad. Sci. 20, 1355-1361.

Ebel, A., A. H. Manson, A. H., and C. E. Meek, 1987: Short-period fluctuations of the horizontal wind measured in the upper middle atmosphere and possible relationships to internal gravity waves. J. Atmos. Terr. Phys., 49, 385-400.

Fritts, D. C. 1984: Gravity wave saturation in the middle atmosphere: A review of theory and observations. Rev. Geophys. and Space Phys. 22, 275-308.

Fritts, D. C. and R. A. Vincent, 1987: Mesospheric momentum flux studies at Adelaide, Australia: Observations and a gravity wave/tidal interaction model. J. Atmos. Sci., 44, 605-619.

Lindzen, R. S. 1981: Turbulence and stress due to gravity wave and tidal breakdown. J. Geophys. Res. 86, 9707-9714.

Meek, C. E., I. M. Reid, and A. H. Manson, 1985a: Observations of mesospheric wind velocities. II. Cross sections of power spectral density for 48-8h, 8-1h, 1h-10min over 60-110 km for 1981. Rad. Sci. 20, 1383-1402.

Meek, C. E., I. M. Reid, and A. H. Manson, 1985b: Observations of mesospheric wind velocities. I. Gravity wave horizontal scales and phase velocities determined from spaced wind observations. Rad. Sci. 20, 1363-1382.

Reid, I. M. and R. A. Vincent, 1987a:Measurements of mesospheric gravity wave momentum fluxes and mean flow accelerations at Adelaide, Australia. J. Atmos. Terr. Phys. (In Press)

Reid, I. M. and R. A. Vincent, 1987b: Measurements of the horizontal scales and phase velocities of short period mesospheric gravity waves at Adelaide, Australia. J. Atmos. Terr. Phys. (In Press)

Vincent, R. A. 1984a: MF/HF radar measurements of the dynamics of the mesopause region - A review. J. Atmos. Terr. Phys. 46, 961-974.

Vincent, R. A. 1984b: Gravity wave motions in the mesosphere. J. Atmos. Terr. Phys., 46, 119-128.

Vincent, R. A., and I. M. Reid, 1983: HF Doppler measurements of mesospheric gravity wave momentum fluxes. J. Atmos. Sci. 40, 1321-1333.

Vincent, R. A., and D. C. Fritts, 1987: A climatology of gravity wave motions in the mesopause region at Adelaide, Australia. J. Amos. Sci., 44, 748-760.

Yamamoto, M, T. Tsuda, and S. Kato, 1986: Gravity waves observed by the Kyoto meteor radar in 1983-1985. J. Atmos. Terr. Phys. 48, 597-603.

THE SEASONAL EVOLUTION OF THE STRATOSPHERE
IN THE NORTHERN HEMISPHERE

A. O'Neill and V. D. Pope
Meteorological Office
London Road
Bracknell
U.K.

ABSTRACT. Isentropic maps of Ertel's potential vorticity are used to
describe the seasonal evolution of the mid stratosphere in the northern
hemisphere during the year 1980/81. Systematic changes in the
circulation are noted and their causes considered. The meridional
transport of air due to planetary-scale disturbances is outlined.

1.   INTRODUCTION

Our aim in this paper is to give a brief description of the seasonal
evolution of the circulation in the mid stratosphere of the Northern
Hemisphere, noting in particular periods when large-scale disturbances
to the flow bring about material exchange between latitudes.
    The movement of air, or transport, and photochemistry change the
distribution of trace gases in the stratosphere. Simplified numerical
models of these processes are often based on a zonal average of the
governing equations (see Brasseur and Solomon, 1984, for a review).
Such averaging may be inappropriate when the flow is strongly
disturbed, i.e. when it is far from being symmetric about the pole. One
reason is that when the intense cyclonic vortex of the winter
stratosphere is displaced from the pole by large-scale disturbances,
the basic physical property that its inner portion retains for a few
weeks the same body of air is obscured by a zonal average.
    We shall infer meridional transport associated with systematic
changes in the stratospheric circulation by using synoptic maps of
Ertel's potential vorticity, Q. For the large-scale flow of the
stratosphere, Q can be expressed approximately as

$$Q \approx - g(\zeta+f)\partial\theta/\partial p \qquad\qquad (1)$$

where g is the acceleration due to gravity, ζ is the vertical component
of relative vorticity, f is the Coriolis parameter, θ is potential
temperature and p is pressure. Q is conserved for adiabatic,
frictionless flow, as is θ. Then, contours of Q plotted on a

G. Visconti and R. Garcia (eds.), Transport Processes in the Middle Atmosphere, 57–69.

quasi-horizontal surface of constant θ are material lines which can be
used to trace the movement of air. Q is also a dynamically active
quantity in that it modifies the flow as it is advected. In principle,
winds and other meteorological fields can be derived from the
three-dimensional distribution of Q, provided the flow is "in balance"
(e.g. in geostrophic balance) and boundary conditions are available.
Hoskins et al. (1985) elaborate on these properties and give examples
of their utility.

Our discussion is based on maps of Q for the 850K isentropic
surface, which lies near 10 mb in the mid stratosphere where the flow
can be taken as conservative for a week or so (the radiative timescale
is about 20 days). This level can be used to illustrate transport
mechanisms operating in the stratosphere as a whole, but accurate
calculations would need to treat variations in height of dynamical and
radiative processes. We cannot infer directly from maps of Q the
transport of a trace gas over a season because Q is not conserved for
this long. We can, however, use such maps to study transport over
shorter, dynamically active periods (lasting a week, say) and gain some
understanding of what the cumulative effects might be if disturbances
persist.

The circulation in the stratosphere in mid winter is dominated by
a huge cyclonic vortex covering most of a hemisphere. The vortex is
created by radiative cooling, and is characterised by strong winds and
a large meridional gradient of Q at mid latitudes (as we show later).
From the dynamical theory of large-scale Rossby waves, it is known that
a gradient of Q acts as a restoring force on meridional displacements
of strings of air parcels (the argument given by Holton, 1979, page
166, must be adapted if winds vary in height, longitude and latitude).
This gradient is therefore a barrier to horizontal transport into the
centre of the vortex by large-scale disturbances. While the vortex is
strong and large, horizontal transport is confined to low latitudes
where winds and the gradient of Q are weak (Plumb and Mahlman, 1987)
though air parcels may drift slowly into the vortex as a result of
radiation. For a trace gas created photochemically at low latitudes to
spread more rapidly over the winter hemisphere, the vortex must be
broken down. We shall outline how this happens and suggest an
explanation of why it does.

The paper is laid out as follows: Section 2 contains a summary of
the data used in this study and of our method of analysis. We describe
the synoptic evolution in the stratosphere during the northern winter
of 1980/81 in section 3, where we focus on the systematic changes in
the circulation, suggest possible causes for them, and infer broadly
how transport varies through the year. Concluding remarks are in
section 4.

2.   DATA AND METHOD OF ANALYSIS

The data used in this study were obtained mainly from a stratospheric
sounding unit (SSU) on board the satellite NOAA-6. Pick and Brownscombe
(1981) describe the instrument. It is a nadir sounding radiometer which

scans across the orbital path giving almost global coverage. The measuring channels of the SSU have weighting functions centred in the stratosphere near 15, 5 and 1.5 mb. The vertical resolution is about 10 km; after fields are smoothed the horizontal resolution is limited to about wave-number 12 in the meridional (pole-to-pole) and zonal directions.

Details of the analysis and of the calculation of isentropic maps of Q are given by Clough et al. (1985). Briefly, thicknesses are derived from radiances by statistical regression. Fields of geopotential height are produced by adding the thicknesses to a global analysis of geopotential height at 100 mb provided by the National Meteorological Center (Washington). Winds and temperatures are calculated using the geostrophic and hydrostatic approximations, and Q is calculated by differentiating and interpolating these fields.

3.   SEASONAL EVOLUTION IN THE NORTHERN HEMISPHERE, 1980/81

3.1 Description of synoptic maps

We now show a selection of maps of Q and horizontal winds on the 850K isentropic surface for the Northern Hemisphere during the period 1980 to 1981. We describe changes in the distribution of Q and associated changes in the circulation that occurred during a sequence of minor and major warmings, and comment on the transport taking place.

The first strong warming of the winter is of the type known as a Canadian warming because warm air is advected over Canada. Its evolution is similar to the one in December 1981 discussed by Clough et al. (1985). At the beginning of November 1980, a westerly vortex lies roughly symmetrically over the North Pole. Q is a maximum near its centre and rapidly decreases towards low latitudes. As the ensuing disturbance develops, low Q is advected north eastwards from low latitudes. The distribution of Q and winds on 23 November is shown in Fig. 1. An island of low Q is cut off from similar values of Q at low latitudes by a closed anticyclonic circulation that forms around it. (Because material lines do not break, there will be filaments of low Q, too thin to be resolved, connecting the two regions unless dissipation has wiped them out.) The westerly vortex is elongated and displaced from the pole, and a tongue of high Q stretches out from the vortex around the anticyclone.

The quasi-horizontal advection of Q gives a good indication of transport in this case. The disturbance develops in about a week and changes in temperature are modest compared with those found during major warmings, so the non-conservative effect of radiation does not stop the contours in Fig. 1 from being regarded as material lines. Also, the pattern shown extends through a deep layer of the stratosphere and therefore the vertical resolution of the measurements is not a limitation.

The meridional displacements of air parcels during the Canadian warming are, for the most part, only temporary. In late November, the island of low Q shown in Fig. 1 is drawn across the polar cap back into low latitudes, slicing through the tongue of high Q in the process. This tongue is eventually reabsorbed by the main vortex which moves back over the pole and becomes roughly symmetric about it.

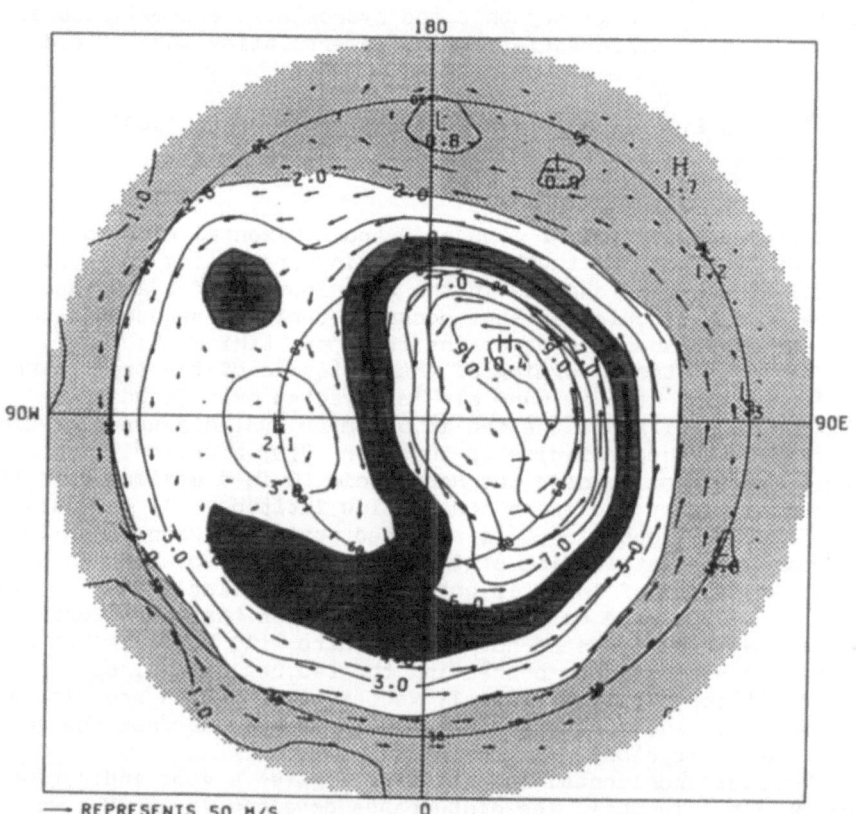

Figure 1.   Polar-stereographic map of Ertel's potential vorticity, Q, and winds on the 850 K isentropic surface for 23 November 1980. The units are 10^{-4} K m^2 kg^{-1} s^{-1}. The light shading is for values less than 2 and heavy shading is for values between 4 and 6 in the above units. The frame surrounding the figure is tangent to latitude 20°N.

During December, the westerly vortex is disturbed only weakly, being roughly symmetric about the pole as shown in Fig. 2 for 2 January 1981. It continues to strengthen owing to radiative cooling in the polar night. This quiescent interlude is ended by the build-up of strong disturbances in January, a prelude to a major stratospheric warming in early February (described by Labitzke et al., 1981). An anticyclone (the so-called Aleutian High) builds in the stratosphere and, unlike its behaviour during the Canadian warming, it persists near the dateline. Systematic changes ensue in the structure of the circulation; in particular the westerly vortex shrinks rapidly. Fig. 3 shows the distribution of Q and winds on 1 February 1981. Broadly, in

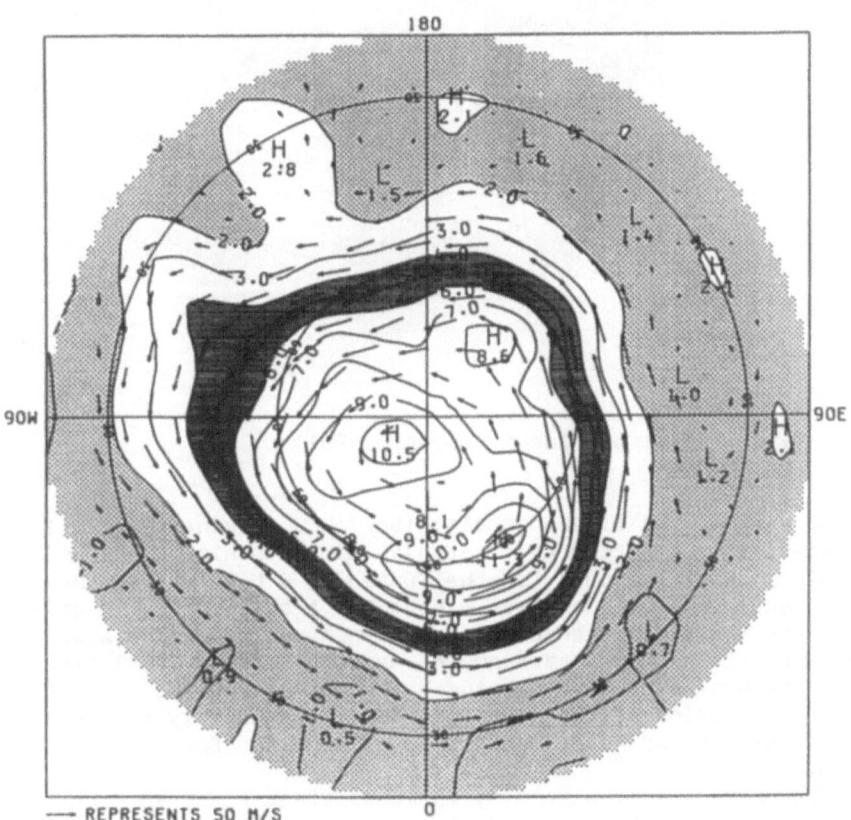

— REPRESENTS 50 M/S

Figure 2.   As Fig. 1 but for 2 January 1981.

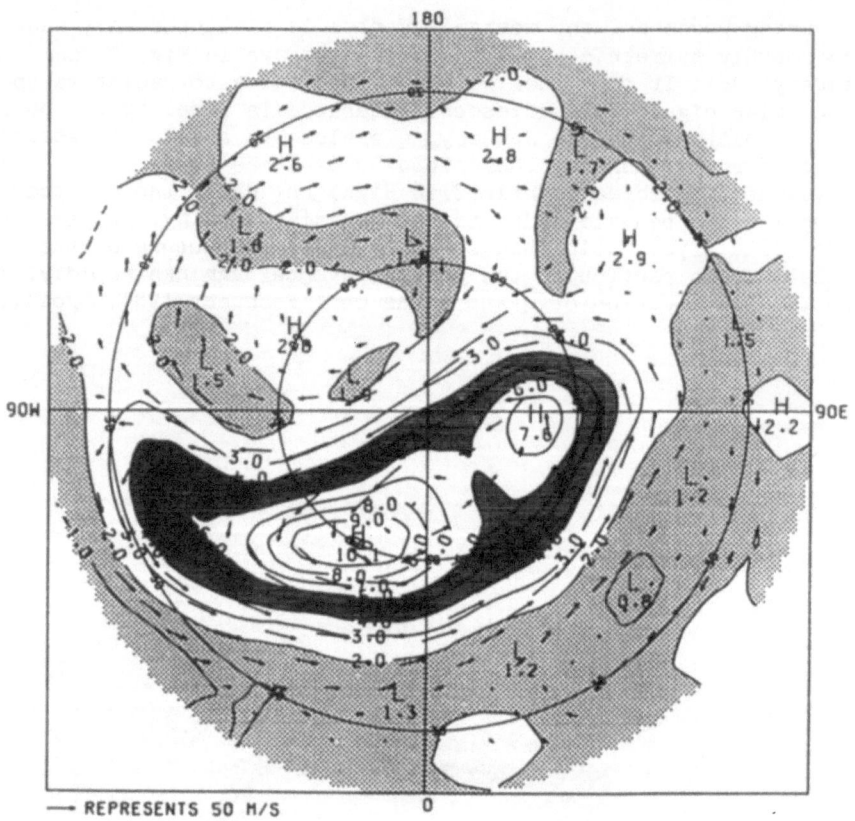

Figure 3. As Fig. 1 but for 1 February 1981.

the top half of the figure is air with low Q originally from low
latitudes; in the bottom half is polar air with high Q. Errors in
analysis probably account for the jumble of Q contours in the
anticyclone; gradients are weak there and a small error in calculating
Q can make a big difference to a contour's position. The Aleutian High
draws air with high Q away from the cyclone towards low latitudes -
notice the tongue of high Q being drawn out near 90°W, 40°N. The
limited resolution of the measurements allows us only a blurred view of
the process, and so we are unable to witness the fate of the tongue of
high Q.
        The capturing of high Q by the persistent Aleutian High seems to
be an important mechanism (radiation is another) for breaking down the
large cyclonic vortex that develops in the stratosphere in early
winter. McIntyre and Palmer (1983) noted this possibility and discussed
some of the possible dynamical implications.

The area covered by the cyclone is much smaller in Fig. 4, the synoptic map for 4 March 1981, than it is in Fig. 3, and values of low Q at the centre of the anticyclone have increased. By mid March, the flow is almost zonally symmetric apart from a weak cyclone displaced from the pole, as is shown by Fig. 5 for 19 March 1981. Q is nearly uniform at high latitudes and winds are correspondingly weak. At lower latitudes, a westerly jet coincides with a zone where the gradient of Q is larger. The residue of the cyclone slowly disappears, and easterly winds eventually replace westerlies in the mid and upper stratosphere. This region is cut off from large-scale, tropospheric disturbances by the easterlies (Charney and Drazin, 1961), and the transport in the hemisphere is limited to a slow meridional circulation driven by radiation.

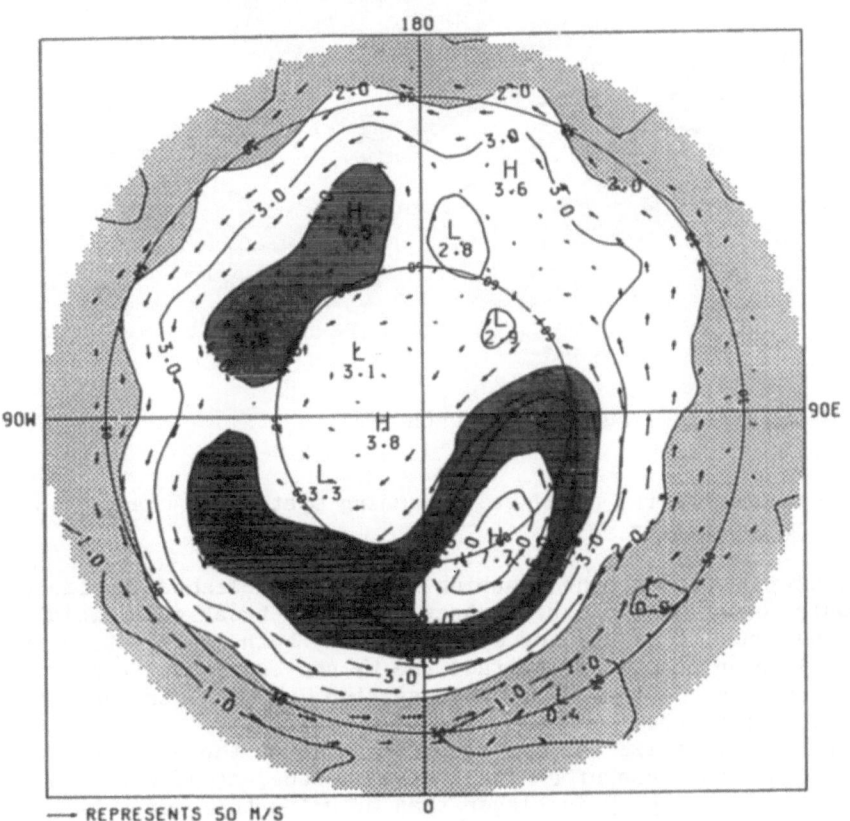

Figure 4. As Fig. 1 but for 4 March 1981

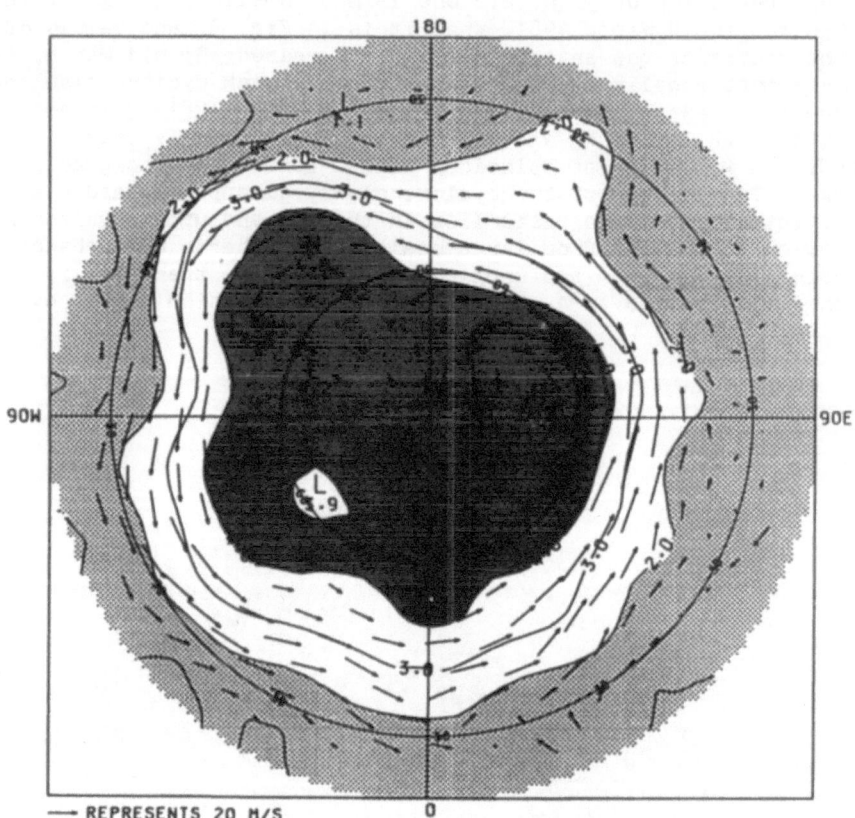

— REPRESENTS 20 M/S

Figure 5. As Fig. 1 but for 19 March 1981.

3.2 Non-conservative processes governing systematic changes in the
     stratosphere

As we have seen, once the Aleutian High is established the large
cyclonic vortex in the stratosphere starts to break down rapidly. The
distribution of Q is changed systematically both by advection of
material on an isentropic surface and by non-conservative processes
which create or destroy Q. In this section, we consider what these
processes might be.
    We may summarize the systematic changes in the distribution of Q
by computing the area of an isentropic surface, A(Q), where the value
of potential vorticity is bigger than Q. Doing this for a number of
values of Q, we are able to estimate the size of the cyclone by
following the prescription given by McIntyre and Palmer (1984): in
brief, find the value of Q at which there is a sharp change in dA/dQ,
then the area surrounded by the contour corresponding to that value is

roughly the area covered by the cyclonic circulation. The method relies on the observed property that gradients of Q in the stratosphere are strong in the cyclone and weak in an anticyclone. For instance, the contour labelled '3.0' in Fig. 3 roughly circumscribes the cyclone, and the difference between A(4) and A(3) is much smaller than that between A(3) and A(2). A diagnostic based on averaging around latitude circles does not reveal systematic changes as clearly as A(Q) does, for such a quantity would vary as the cyclone moved about the polar cap even if the cyclone's structure remained the same.

The evolution of A(Q) may also be used to deduce on average where and when non-conservative processes are important. Our discussion here will be qualitative but it is supported by calculations that we have done (to be reported), and also by the work of Butchart and Remsberg (1986) and of Butchart (1987).

A(Q) is constant if the following hold: (1) radiation and friction can be neglected; (2) the flow is fully resolved; (3) on average, the large-scale flow is non divergent. The reason is that a contour of Q is a material line if (1) and (2) hold, and the area enclosed by a material line is constant in a non-divergent flow. We find (in work to be reported) that assumption (3) is acceptable in a qualitative discussion (but not when accuracy is required). So assumption (1) or (2) or both must be invalid since A(Q) changes with time, as is evident from the synoptic maps introduced earlier.

Radiation and friction change A(Q) irreversibly, i.e. in a time-oriented way. Changes in A(Q) through loss of resolution may be reversible or irreversible depending on the dynamics involved. McIntyre and Palmer (1983, 1984) propose that extreme buckling of Q contours, such as that shown in Fig. 1, will irreversibly mix Q on an isentropic surface to small, unresolved scales in the horizontal. As a result of mixing, A(Q) will decrease for high Q and increase for low Q, perhaps accounting for regions in the stratosphere where the gradients of Q are weak (e.g. near the Aleutian High in Fig. 3). The effects of such mixing have yet to be quantified directly. Using a reasonably accurate radiation scheme (Shine, 1987), we have estimated the rate of change of A(Q) due to radiation at different times of the year (paper in preparation). We shall summarize our findings here, but we stress that the calculations are liable to very large errors owing to the limited resolution of the observations on which they are based.

Fig. 6 shows, for various values of Q, the evolution of A(Q) for the 850 K isentropic surface in the Northern Hemisphere from June 1980 to June 1981. Also plotted is one curve (dashed) for an annual cycle in the Southern Hemisphere. This curve is displaced along the time axis by six months so that corresponding seasons in the two hemispheres may be compared. There is a marked seasonal variation in A(Q) due to the seasonal variation in solar heating. The build up of the cyclone in autumn is reflected by the increase in A(Q) for high values of Q. The cyclone continues to build and areas continue to increase over the course of the Canadian warming. The flow is most disturbed at the end of November, and inadequate resolution accounts for rapid changes in A(4). (Values of Q are given in the units used to plot contours on the synoptic maps.) The disturbance, though strong, does not lead to much

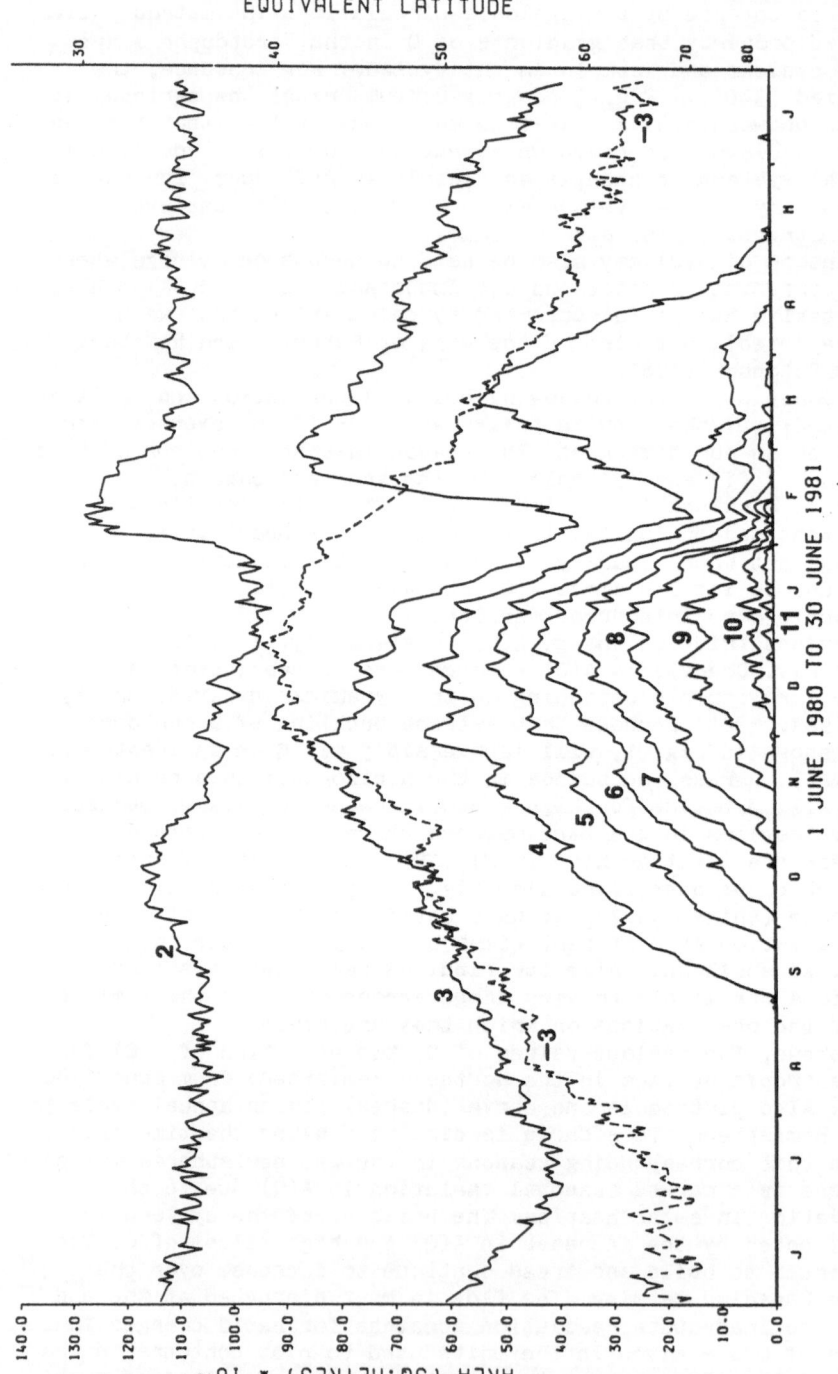

Figure 6. Graphs of areas A(Q) (km² x 10⁶) as described in the text. Each curve is labelled with the corresponding value of Q in units of 10^{-4} K m² kg^{-1} s^{-1}. The continuous curves are for the Northern Hemisphere from 1 June 1980 to 30 June 1981 and the dashed curve is for the Southern Hemisphere from 1 December 1980 to 30 December 1981 (displaced by six months on the time axis). The equivalent latitude, marked on the right hand axis, is calculated as the latitude circle that encloses the area marked on the left hand axis.

erosion of the cyclone because an area of high Q that becomes separated from the cyclone eventually recombines with it. Radiation is apparently more than enough to offset any effect on Q of mixing during the Canadian warming. Clough et al. (1985) found the same for the Canadian warming of 1981.

The vortex is largest at the beginning of January; in Fig. 6, most of the A(Q) reach their maximum values then. As the Aleutian High intensifies, areas decrease rapidly. During January, A(3) is a rough measure of the size of the cyclone (based on the argument given above), and Fig. 6 shows that it almost halves its size in the month. For the less disturbed Southern Hemisphere, A(-3) is about twice A(3) a month after mid winter.

Persistent disturbances take the stratosphere away from a state near to radiative equilibrium and lead to systematic changes in A(Q). Our calculations indicate that radiation reduces the high Q in the cyclone after mid winter (by reducing the stability of the air), but it cannot fully account for the rapid decreases in areas in January. We ascribe the remainder to errors in these calculations, and to high Q being lost to view, e.g. by being drawn into narrow tongues around the southern flank of the Aleutian High. McIntyre and Palmer (1983) noted that Q could apparently disappear in this way. Such a tongue is shown at inception in Fig. 3. Our observations have inadequate resolution to show its elongation and narrowing beyond this stage during January, but a simulation that we made using a numerical model with much better resolution does.

The increases in areas during February and March, on the other hand, can be accounted for by radiation; if our calculations are accurate, the process of mixing, mentioned above, does not need to be invoked. Values of Q at the centre of the anticyclone increase (compare Fig. 4 with Fig. 3) as warm air cools and sinks through isentropic surfaces bringing down higher values of Q. By diminishing the local anomaly in Q, radiation inhibits the growth of the anticyclone, restraining its erosion of the cyclone. The jet at low latitudes that we noted in Fig. 5 is presumably generated by Coriolis forces acting on a poleward and downward circulation driven by the cooling at high latitudes.

We conclude that the formation of weak gradients of Q in the stratosphere - as manifested in the big difference between A(2) and A(3) that develops after mid winter - is a consequence of (1) radiation acting on a highly asymmetric flow, and (2) failure to resolve Q fully e.g. in narrow tongues of high Q.

4.   CONCLUSIONS

Transport of material in the stratosphere is rapid when large-scale disturbances develop in winter. Air from low latitudes is drawn polewards and air from high latitudes is drawn equatorwards as a growing anticyclone displaces the westerly vortex from the pole. An example of this exchange occurs during the strong Canadian warming of early winter 1980/81. But the displaced air mostly returns to its

original latitude as this disturbance decays. There is little net
effect on the distribution of potential vorticity, Q, and hence on the
circulation. Afterwards, the cyclone continues to intensify as a result
of radiative cooling. Strong gradients of Q develop which inhibit the
large-scale displacement of air across the gradient, isolating the air
in the inner portion of the cyclone.

Permanent exchange of air between latitudes occurs once the
so-called Aleutian High becomes a persistent feature, just after mid
winter in 1980/81. The air circulating within the cyclone remains
displaced from the pole, and air towards its edge is drawn into low
latitudes by the anticyclone. Associated with this advection is a loss
of high Q by the cyclone which consequently shrinks, making way for air
originally from low latitudes to encroach into polar latitudes.

The persistence of the Aleutian High leads to systematic changes
in the distribution of Q in the stratosphere, as is evident from the
evolution of the areas on an isentropic surface where Q exceeds
specified values. There are also systematic changes in the circulation.
After mid winter, the areas decrease for high values of Q. We suggest
that this might be accounted for by radiation, and failure to resolve
Q, e.g. in tongues of high Q which extend around the Aleutian High.
Increases in areas for low values of Q are apparently caused by
radiation. Strong cooling over the polar cap after a major warming
results in air moving along and sinking through isentropic surfaces
which would redistribute a trace gas whose concentration varied in
space. The cyclone stays off the pole and radiation gradually weakens
it during spring. Once easterly winds replace westerlies, large-scale
disturbances are trapped in the troposphere, and transport occurs in a
slow meridional circulation driven by radiation.

REFERENCES

Brasseur, G. and S. Solomon, 1984: Aeronomy of the Middle
        Atmosphere: Chemistry and Physics in the Stratosphere and
        Mesosphere, 441 pp. D Reidel, Dordrecht.

Butchart, N., 1987: Evidence of planetary wavebreaking from
        observational data. The relative roles of diabatic effects and
        irreversible mixing. This volume.

Butchart, N. and E. E. Remsberg, 1986: The area of the stratospheric
        polar vortex as a diagnostic for tracer transport on an
        isentropic surface. J Atmos. Sci., 43, 1319-1339.

Charney, J. G. and P. G. Drazin, 1961: Propagation of planetary-scale
        disturbances from the lower into the upper atmosphere. J.
        Geophys. Res., 66, 83-109.

Clough, S. A., N. S. Grahame and A. O'Neill, 1985: Potential vorticity
        in the stratosphere derived using data from satellites. Quart.
        J.R. Met. Soc., 111, 335-358.

Holton, J. R., 1979: An Introduction to Dynamic Meteorology, Second Edition, 391 pp., Academic Press, New York.

Hoskins, B. J., M. E. McIntyre and A. W. Robertson, 1985: On the use and significance of isentropic potential vorticity maps. Quart. J.R. Met. Soc., 111, 877-946.

Labitzke, K., R. Lenschow and B. Naujokat, 1981: The third winter of PMP-1: 1980/81. Beilage zur Berliner Wetterkarte, 16.7.

McIntyre, M. E. and T. N. Palmer, 1983: Breaking planetary waves in the stratosphere. Nature, 303, 593-600.

——— and ———, 1984: The 'surf zone' in the stratosphere. J. Atmos. Terr. Phys., 46, 825-850.

Pick, D. R. and J. L. Brownscombe, 1981: Early results based on the stratospheric channels of TOVS on the TIROS-N series of operational satellites. Adv. in Space Res., 1, 247-260.

Plumb, R. A. and J. D. Mahlman, 1987: The zonally-averaged transport characteristics of the GFDL general circulation/tracer model. J.Atmos. Sci., 44, 298-327.

Shine, K., 1987: The middle atmosphere in the absence of dynamical heat fluxes. Quart. J. Roy. Met. Soc., 113, to appear.

2. OBSERVATIONS AND ANALYSIS OF CHEMICAL CONSTITUENTS

DISTRIBUTIONS OF OZONE AND NITRIC ACID MEASURED BY THE LIMB INFRARED MONITOR OF THE STRATOSPHERE (LIMS)

John C. Gille
National Center for Atmospheric Research
P. O. Box 3000
Boulder, CO   80307

ABSTRACT. The Limb Infrared Monitor of the Stratosphere (LIMS) experiment on the Nimbus 7 spacecraft provided seven months of data on the distribution of ozone and nitric acid. The evaluations of the accuracy, precision and resolution of these data is described before a description of the major features of their distributions is presented. The zonal mean for ozone shows a stratospheric layer, with largest mixing ratios in the tropics and maximum ozone mixing ratios at a given latitude occurring at 10 mb or higher. Nitric acid is distributed in a stratospheric layer that peaks near 30 mb, with the largest mixing ratios occurring in polar regions. Longitudinal variations for a month may be expressed as the sum of stationary and transient zonal harmonics. The amplitudes of the first harmonics can be up to 10-20% of the zonal mean in the winter, but only a few percent in the autumn, and very low in the spring and summer. The second harmonic is generally smaller. These features also appear in the monthly average maps. Time height cross-sections for both display seasonal variations that probably result from dynamical and chemical processes, and rapid dynamically induced fluctuations.

These and related data are making major contributions toward improving the understanding of the mechanisms required to maintain the observed distributions and their annual changes.

1.   INTRODUCTION

Infrared limb scanning is a powerful technique for observing the stratosphere and mesosphere. In this technique, thermal radiation emitted by gases in the earth's atmosphere is measured as a function of angle above the horizon. Some of the advantages were mentioned by Gille and House (1971); many are apparent from the geometry. The long paths through the atmosphere result in the highest sensitivity to small amounts of trace gases, and the ability to observe them to higher altitudes than when looking downward. The largest amounts are often at the lowest altitude along the ray path, which results in narrow weighting functions and high vertical resolution, if a narrow field of view is used. Additionally, because emission is measured, observations can be obtained at all local times, and the dark background behind the emitting atmosphere simplifies the data reduction. The geometry also has

73

G. Visconti and R. Garcia (eds.), Transport Processes in the Middle Atmosphere, 73–85.

limitations; the horizontal resolution is generally no better than about 300 km along the line of sight, and the presence of clouds in the troposphere limits its general applicability to the atmosphere above the tropopause.

The Limb Infrared Monitor of the Stratosphere (LIMS) instrument is a six channel limb scanning infrared radiometer on the Nimbus 7 spacecraft. Results for water vapor and nitrogen dioxide are presented in a companion paper (Remsberg and Russell, ths volume).

The instrument operated extremely well for its planned 7 month lifetime from launch (25 October, 1978) until the depletion of the solid CH_4 cryogen on 28 May, 1979. During this period over 1000 profiles were derived each day, spread from 64 S to 84 N. The experiment has been described by Gille and Russell (1984) with earlier treatments in Russell and Gille (1978) and Gille et al. (1980). The vertical resolution of the individual profiles is not quite as high as that for CO_2 shown by Bailey and Gille (1986).

This paper will discuss ozone and nitric acid results that were derived from the measured radiances. Because of the large amount of data and the richness of phenomena they show, only a few features can be discussed here.

2.   THE OZONE DISTRIBUTION

2.1.   Accuracy and Precision of the Ozone Observations

The quality of the LIMS ozone observations were evaluated by Remsberg et al. (1984). They determined or estimated the several sources of random and systematic error, assessed their individual effects on the results, and combined them for an overall estimate of accuracy.

These estimates of the random error were checked against observations in order to confirm their reliability. The standard deviation of a group of profiles in a region in which the atmosphere is uniform provided an estimate of the retrieval precision. These were $\sim$ 0.10 ppm, in good agreement with the predicted values and $\leq$ 2% from 100 mb into the lower mesosphere.

The absolute accuracy is more difficult to establish, because of the lack of accepted standards. The estimated uncertainty of the accuracy is shown in Table I. These estimates are quite conservative. Comparison with a large number of rocket and balloon comparison observations lead to the differences indicated in Table I. Considering the balloon and rocket uncertainties, these results indicate very good agreement.

Two additional factors should be noted. Observations at the same latitudes on the ascending (mainly daylight) and descending (mainly dark) parts of the orbit show systematic diurnal differences. These are only a few percent at and below 1 mb. Classical photochemistry suggests that there should be a significant (up to a factor of 2) ozone increase from day to night in the mesosphere. In the results presented below, these differences have been ignored, and day and night data are combined to give more detailed coverage.

Solomon et al. (1986) have recently suggested that chemical excitation may lead to a non-Boltzman population of the ν_3 vibrational levels – i.e., the levels may not be in local thermodynamic equilibrium (LTE). This non-LTE emission would lead to an over-estimate of daytime ozone values. Their suggested corrections are only appreciable above 1 mb. No correction for this effect is included in the data presented here.

Table I
LIMS Ozone and Nitric Acid Errors (%)

	Ozone		Nitric Acid	
Pressure Level (mb)	Estimated Systematic Errors	Differences from Correlative Measurements	Estimated Systematic Errors	Differences from Correlative Measurements
100	40	35.2 ± 6.2		
80 (70)			42	(-18.8 ± 24.4)
30	35	3.8 ± 2.4	33	9.4 ± 7.1
10	23	-1.6 ± 2.8	29	26.6 ± 10.7
3	16	2.6 ± 3.1†	65	
1	15	3.9 ± 3.4†		
0.3	26	-10.8 ± 12.9†		

Differences compared to balloon-borne instruments except as noted.
† Compared to rocket measurements.

Some further confirmation of the accuracy of the LIMS data may be obtained from a comparison of LIMS daytime observations with the simultaneous results obtained by the Solar Backscatter UltraViolet (SBUV) instrument on Nimbus 7 (WMO, 1986; Gille and Lyjak, private communication). These show agreement within a few percent everywhere above 10 mb, although the agreement has temporal variations which are not now understood.

Before looking at the spatial and temporal variations of the ozone distribution, it is useful to remind ourselves of some fundamental considerations. The first is that the photochemical lifetime of ozone, or rather the odd-oxygen species (O + O_3) has a strong variation with altitude. In the lower stratosphere it is longer than typical dynamical time scales (e.g., Brasseur and Solomon, 1984), indicating that the distribution will be controlled by atmospheric motions. In the upper stratosphere and mesosphere (where they are illuminated), photochemical times are shorter than dynamical times, and the distribution will tend toward the state given by photochemical equilibrium.

Photochemical theory suggests that the ozone mixing ratio will have a small variation with the solar UV that causes O_2 photolysis (Gille et al. 1984; Hood 1984) and a larger inverse variation with temperature. Evidence of this can be seen in some of the fields, although it can also be produced by vertical motions in regions where the ozone mixing ratio decreases with altitude. Rood and Douglas (1985) have pointed out that more general dynamical effects can also produce these out of phase variations.

2.2.   Zonal Mean Cross-Sections

Latitude-height cross sections of monthly averaged zonal mean LIMS ozone mixing ratios are shown for December and May in Figs. 1 and 2. Here, the LIMS data have been extended with the aid of SBUV data and some extrapolation, as described by Gille and Lyjak (1986). The December distribution (Fig. 1) illustrates some of the common features. To a first approximation, the distribution is horizontally layered, with maximum mixing ratios near 10 mb at low latitudes, but with the maximum tending to move up to higher altitudes toward the poles. The mixing ratio also varies with latitude; at 10 mb and above, the mixing ratio decreases toward the poles, while in the lower stratosphere the mixing ratio increases toward the poles. The gradient in the mesosphere is small.

In late October there were large values in the Southern Hemisphere (SH) lower stratosphere, the result of dynamical effects at the end of the southern winter that resulted in poleward ozone transports when the winter circulation broke down and made the transition to the summer regime. In the northern polar region the ozone mixing ratios were low.

In December, under solstice conditions, the tropical ozone maximum is shifted toward the summer hemisphere and the subsolar location, but there is more in the lower and middle northern polar stratosphere, which must be due to dynamical effects. By March (not shown), at the end of the Northern winter, there are large mixing ratios in the NH polar lower stratosphere as a result of the of the seasonal transition in the wind system. The highest mixing ratio at the pole is near 40 km altitude. By May (Fig. 2), the seasonal reversal is largely complete, and the distribution looks like the latitudinal reversal of the November distribution.

The standard deviation of these zonally averaged monthly averages is less than 2% everywhere in December and May, indicating that these zonal mean values are quite stable, and that even small month-to-month differences are significant.

The monthly averaged zonal average does not give any information on the ozone variation with longitude. One convenient way to see this is to look at the average amplitudes of the stationary and transient waves in the ozone distribution. In order to allow these results to be used as a check on models, in which absolute ozone concentrations may not be correct, wave amplitudes are presented here as a percent of the zonal mean. The amplitude of stationary wave 1 (SW1) for January (Fig. 3) may reach 22% of the zonal mean in the polar stratosphere, and slightly larger in the mesosphere, the transient wave 1 (TW1) is comparable, and larger in the upper stratosphere. In January SW2 is less than 4%, while TW2 reaches 10%. In May the wave amplitudes are smaller, with SW1 and TW1 being less than 6% and 4%, respectively. Wave 2 is anomalous, with SW2 being less than 2% in the SH, but more than twice that in the NH lower stratosphere, but the largest is TW2, which reaches 8% in the SH. The presence of large travelling wave 2 motions in the SH is well known (e.g., Harwood, 1975). Clearly, geographical and temporal variations due to planetary waves can be significant.

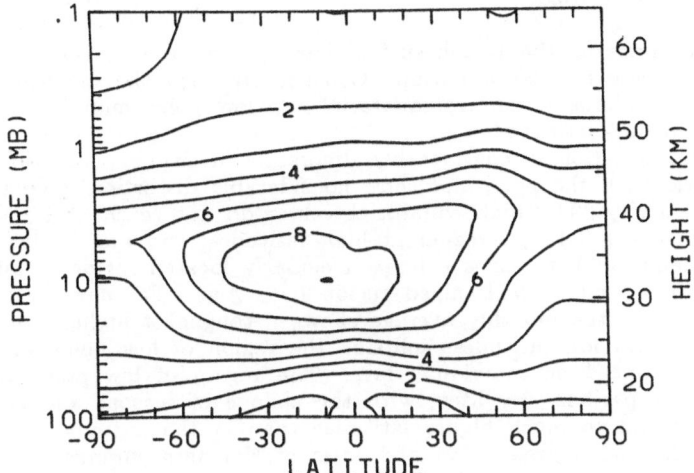

Figure 1. Average zonal mean ozone for December, 1978. Data have been extended to the S. Pole with the aid of SBUV observations, as described by Gille and Lyjak (1986). Contour interval is 1 ppm.

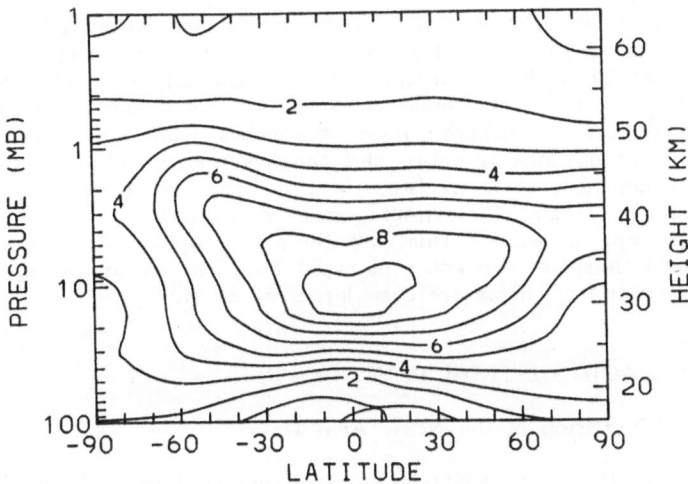

Figure 2. Average zonal mean ozone for May, 1978.

2.3.   Maps on Pressure Surfaces

October, NH ozone maps on the 10 mb surface have nearly circular isolines, indicating wave amplitudes are low for this situation. Consistently with the cross-section, the distribution has high (about 10 ppmv) equatorial and low polar mixing ratios, with a rather steep drop in mid latitudes.

In winter the situation is considerably different. In January the isolines are no longer concentric with the pole, and show considerable distortion, giving evidence of vigorous wave activity. The high equatorial values do not reach as far poleward, and the low values cover a larger region at high latitudes, and are displaced in the direction of Greenland. In between is a large irregularly located region of intermediate values. There is an extensive well mixed region over Asia. February (Fig. 4) is interesting because even the monthly average shows a tongue of higher values coming in toward the pole from low latitudes, splitting the region of low polar values. This tongue is very pronounced on individual days. The region of low polar values is larger than in earlier periods. By May, with the change of seasons and circulation, equatorial values extend to much higher latitudes than in the autumn, and the low values cover a much smaller area before the onset of Northern summer.

The latitudinal variations are rather different at 1 mb (not shown). An irregular collar of high values around the pole in mid latitudes. In May the equatorial values are significantly higher, and rise to an asymetric ridge of still higher values before decreasing toward polar values, which are similar to the January ones.

2.4.   Time-Height Plots

Time-height plots of the ozone distribution are extremely interesting, because they show variations on two distinct time scales. At the equator there is a clear semi-annual variation, with the maximum at the equinox, when the sun is directly overhead, and minimum at the December solstice, when the sun is furthest away. Of course, other factors, such as variations in the residual mean circulation, may be operating.

By contrast, at 32 S the maximum occurs at the solstice, when the sun is relatively close to being overhead, while at 60 S there is a slow increase of the altitude of the maximum as southern winter approaches, while the vertical thickness of the region of maximum mixing ratio, and the lower stratosphere mixing ratios both decrease. By contrast, at 60 N (Fig. 5) there is a minimum amount at the winter solstice, but the region of maximum moves downward, and the mixing ratios increase as summer approaches. In this case there are rapid vertical motions seen during January that illustrate the effect of rapid transient dynamics on the ozone distribution. The dynamical effects are even larger at 80 N.

3.   THE NITRIC ACID DISTRIBUTION

3.1.   Accuracy and Precision of the Nitric Acid Data

The characteristics of the LIMS HNO_3 data were discussed by Gille et al. (1984b). The vertical range is again set by the region of adequate signal to noise ratio, and, at the bottom, by the frequent occurrence of clouds. For the HNO_3 signal, the upper limit occurred at about the 2 mb pressure level, or around 45 km altitude. Clouds usually impose a lower limit at or above the 100 mb pressure level in the tropics.

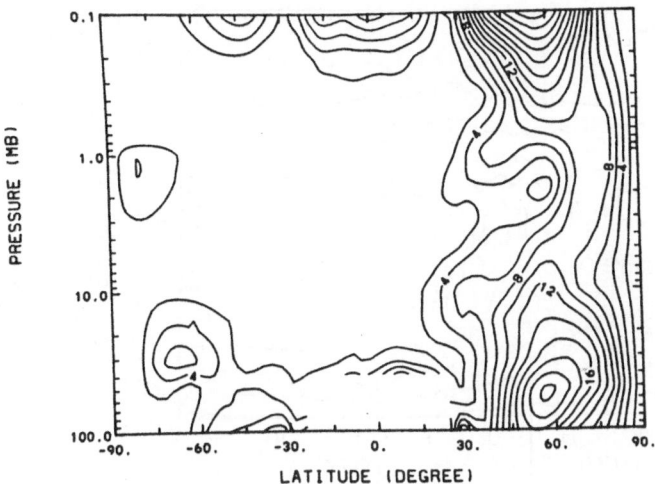

· Figure 3. Amplitude of stationary wave 1 for ozone (as a percent of the local zonal mean), for January, 1979. Contour interval is 2%.

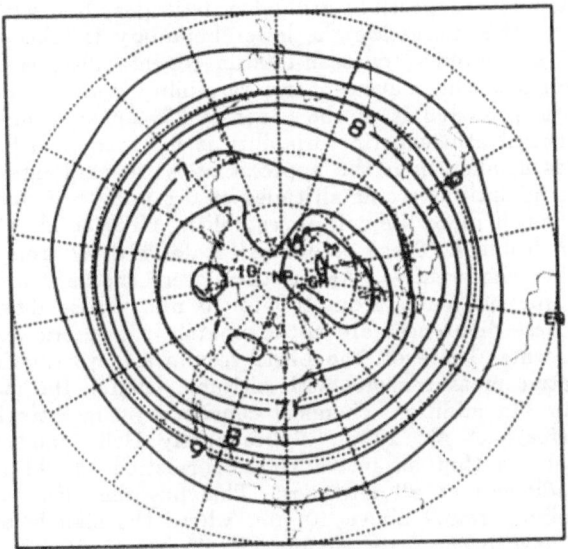

Figure 4. Monthly averaged ozone distribution on the 10 mb surface for February, 1979. Contour interval is 0.5 ppm. The lowest values are displaced from the pole by motions associated with the intrusion of higher values from lower latitudes. The region of small gradients indicates a well-mixed region.

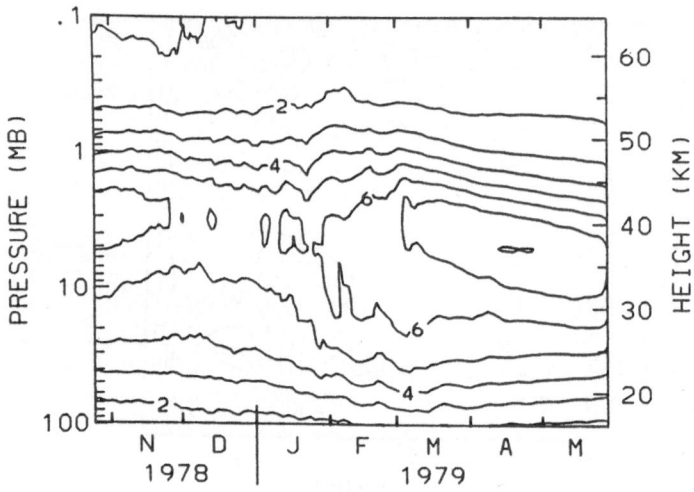

Figure 5. Time-height cross-section of ozone at 60°N.

Retrievals to lower altitude are possible at higher latitudes, but with rather small signal to noise ratios. In this discussion the lower boundary is taken to be 100 mb.

The precision of the profiles, or scan-to-scan repeatability, is about 0.05-0.1 ppbv in undisturbed regions where atmospheric variability does not contribute to the variations. This intrinsic precision is of the order of 2% up to 7 mb, rising to only 5% at 4 mb. When natural atmospheric variability is included, which may incorporate real variations on scales smaller than the approximately 100 km inter-scan spacing, a repeatability at almost all latitudes and altitudes of 0.1 ppbv is found.

The accuracy is much more difficult to establish. Gille et al. (1984b) estimated the errors presented in Table I. These estimates, at least away from the top levels, are thought to be rather conservative. Again, these were checked through comparison with 15 balloon-borne measurements from 100 to 10 mb. These differences are also collected in Table I. These differences are approximately the errors associated with the balloon-borne measurements. However, the LIMS results are systematically increasingly larger than the correlative measurements with altitude, leading the authors to suggest that they were in error. In addition, chemical consistency suggests that the original values are too large (Kaye et al., 1985). Subsequently Bailey and Gille (private communication) have shown that an instrumental correction should be applied that slightly reduces the radiances at all altitudes. This has the effect of significantly reducing the HNO_3 mixing ratios above 10 mb, where the signals are small. The results presented here have been corrected for this effect.

Nitric acid is formed by the reaction of NO_2 + OH, and destroyed by photolysis and to a lesser extent by the reaction HNO_3 + OH. The time scales are several days (Brasseur and Solomon, 1984) at the levels for which LIMS data are available, again indicating that the distribution will be strongly influenced by atmospheric motions.

3.2.   The Zonal Mean Distribution of Nitric Acid

The zonal mean nitric acid distribution for December and May are presented in Figs. 6 and 7. The standard deviation of these means, are less than 1% except for high altitudes and high latitudes in winter, where they are still less than 6%, so the random uncertainties associated with these cross-sections are also quite small. The general features of the nitric acid distribution are illustrated by the December data. There is a broad saddle in the tropics, centered near 20 mb and characterized by values of 2-3 ppbv. Mixing ratios decrease slowly above and below this level, indicating profiles characterized by low and relatively constant value. Maximum values increase toward both poles, with the altitude of the maximum decreasing to the 30 mb level at the highest latitudes. In the NH, the maximum of 3 ppbv at 20 mb for 10 N progresses to a maximum of 10 ppbv at 30 mb for 84 N. The variations are similar in the SH as far as they can be seen. Note also that the isolines are relatively flat on the upper side of the layer, but have fairly steep slopes on the lower side. Finally, there is an indication of an increase at high northern latitudes and high altitudes.

There are significant but regular changes with season of the monthly mean cross-sections. The NH maximum value at 30 mb and 84 N decreases from over 12 ppbv in the fall to about 7 ppbv in May (Fig. 7). At the same time, the values in the SH are increasing. Above 10 mb, the slope of the isolines reverses from October to May, along with the indication of higher concentrations at the winter pole.

In the lower stratosphere in January the amplitudes of SW1 and TW1 are very similar, reaching 16% near 60N at 10 mb, and growing to 40% near 3 mb, where the concentration is quite small. SW2 is much smaller in January, reaching a maximum of only 6% at 60N, 10 mb, but with indications of 10% amplitudes in the lower stratosphere. TW2 is larger, with generally larger amplitudes in the winter hemisphere, reaching 14% in the high latitude upper stratosphere.

Most of the activity in May is in the SH. There, TW1 is slightly larger than SW1, but both are about 5%. TW2 and SW2 are also small, with typical winter hemisphere magnitudes of 5% and 2%, respectively. In both months, all these components show a peak magnitude over the equator, suggesting some modulation associated with equatorial waves.

3.3.   Monthly Mean Maps of Nitric Acid

Monthly mean maps of HNO_3 mixing ratio on the 30 mb surface, which is near the level of the high latitude maximum, show the influence of the circulation. The isolines from low equatorial to high polar values are quite non-circular, even at the end of October, a time of relatively low dynamical activity, and the region of high values is rather small. By December, the area of high polar values is larger, the region of lower equatorial values is smaller, and they are separated by a broad region of intermediate values that suggest horizontal mixing (a surf zone). The polar region of high values is much larger, with a fairly constant latitudinal gradient. The big increase in high latitudes began in February (Fig. 8), after two major disturbances which essentially brought about a change from a winter to a summer circulation. In May there is still a large region of high polar values, but the peak values are lower, and the isolines are nearly circular, reflecting the weak penetration of planetary waves into the stratosphere at this time.

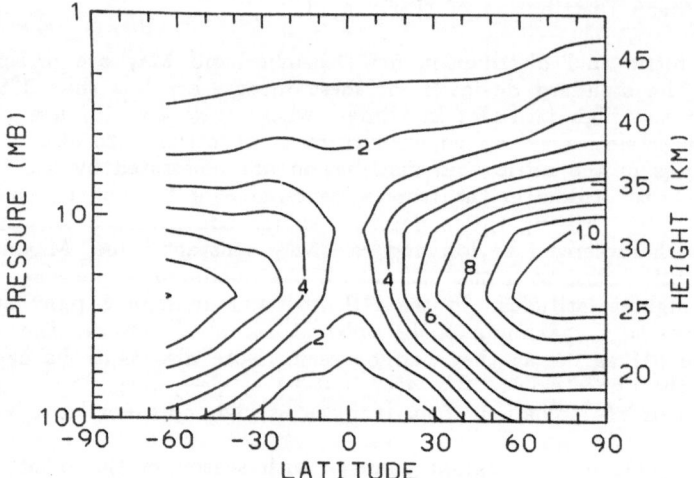

Figure 6. Average zonal mean cross-section of nitric acid for December, 1978. Contour interval is 1 ppb.

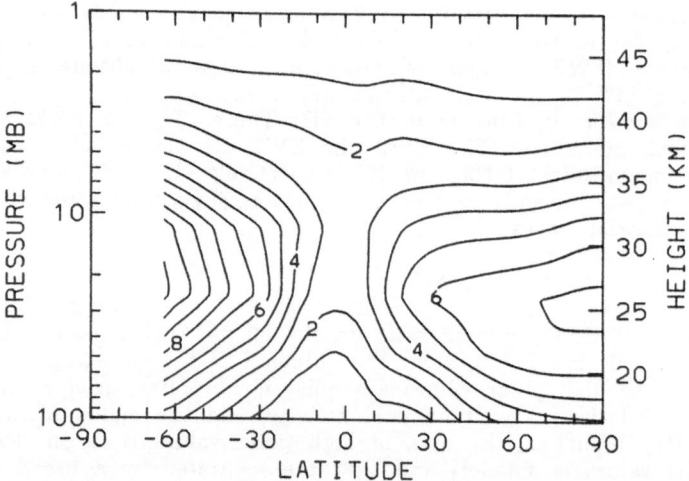

Figure 7. Average zonal mean cross-section of nitric acid for May, 1979.

3.4.   Time-Height Cross Sections

The temporal variation of the zonal mean is shown more explicitly by time-height cross sections. There is a semi-annual oscillation in the tropics, where the HNO_3 maxima are reached at the beginning of December and (probably) June at 30 mb, but about a month later at 16 and 10 mb. The minimum at each level is close to the mid-point between the maxima. This is consistent with semi-annual vertical motions, having maximum strength in March, with minimum motions in December and June.

The patterns are similar at 30 N and S, but the NH maxima and SH minima occur in February at 50 and 30 mb, which also shows the higher SH values in late autumn than in the NH. Again, there is a suggestion that the NH maximum occurs earlier than the SH minimum.

At 60 N and S there is an annual variation, which is out of phase between the two hemispheres, with the NH maximum and SH minimum occurring at the beginning of January at 50 mb, or late December at 30 mb. The patterns are similar above 30 mb, but the spring values in the SH are larger than the NH maxima in February, suggesting a hemispheric asymmetry in the HNO_3 amounts.

At 80 N (Fig. 9) (for which a SH counterpart does not exist), the maximum occurs in November. At 30 mb the maximum of over 12.5 ppbv is reached in mid November, after which the concentration decreases irregularly through the winter and more smoothly in the spring to a value just under 7 ppbv.

These plots also show the long term (seasonal) changes, probably due to photochemical effects, and short period variations, especially during the winter, that are related to dynamical effects. There are marked decreases during the times of major disturbances in the stratosphere, which are to be expected when the downward motions which lead to stratospheric warming through adiabatic compression bring down air which is poorer in HNO_3.

It is clear that the SH maxima and perhaps the minima have larger mixing ratios than those in the NH, indicating an asymmetry between the hemispheres in nitrogen compounds. This has also been seen in NO_2, and estimates of the total odd nitrogen.

4.   CONCLUSIONS

Probably no area has benefitted more from satellite observations than middle atmospheric studies. They have provided a vast amount of information on the distributions of temperature and the trace species, including their spatial and temporal variations. These in turn are leading to an improved understanding of the chemical and dynamical processes that create and modify these distributions, as well as providing more stringent tests of models.

Acknowledgments. I thank Paul Bailey for providing the nitric acid correction factors used here, and Cheryl Craig and Lawrence Lyjak for their careful calculation and plotting of the data. This work was supported by the National Aeronautics and Space Administration under Interagency Agreement W-16215.

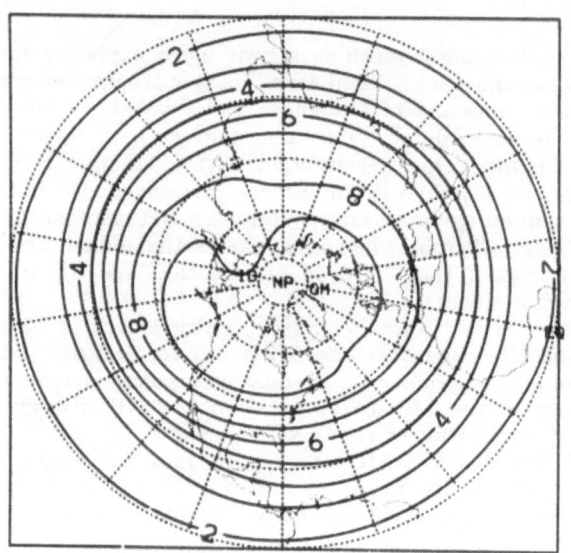

Figure 8. Average nitric acid distribution on the 30 mb surface for February, 1979. Contour interval is 1 ppb.

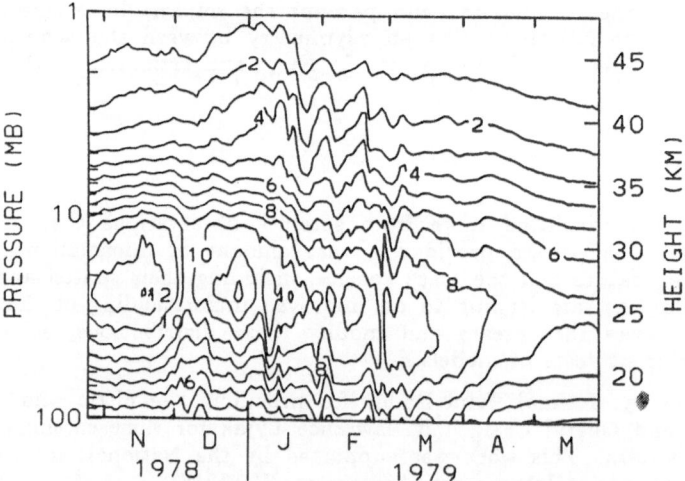

Figure 9. Time height cross-section of nitric acid at 80°N.

REFERENCES

Bailey, P. L., and J. C. Gille, **1986**: Inversion of limb radiance measurements: An operational algorithm, *J. Geophys. Res., 91*, 2757-2774.

Brasseur, G., and S. Solomon, **1984**: Aeronomy of the middle atmosphere. D. Reidel, Dordrecht, 441 pp.

Gille, J. C., and F. B. House, **1971**: On the inversion of limb radiance measurements. I. Temperature and thickness. *J. Atmos. Sci., 28*, 1427-1442.

Gille, J. C., and J. M. Russell III, **1984**: The Limb infrared monitor of the stratosphere: Experiment description, performance, and results. *J. Geophys. Res., 89*, 5125-5140.

Gille, J. C., P. L. Bailey, and J. M. Russell III, **1980**: Temperature and composition measurements from the LRIR and LIMS experiments on Nimbus 6 and 7. *Phil. Trans. R. Soc. Lond. Ser. A., 296*, 205-218.

Gille, J. C., J. M. Russell III, P. L. Bailey, E. E. Remsberg, L. L. Gordley, W.F.J. Evans, H. Fischer, B. W. Gandrud, A. Girard, J. E. Harries, and S. A. Beck, **1984**: Accuracy and precision of the nitric acid concentrations determined by the Limb Infrared Monitor of the Stratosphere experiment on NIMBUS 7. *J. Geophys. Res.*, 5179-5190.

Gille, J. C., C. M. Smythe, and D. F. Heath, **1984**: Observed ozone response to variations in solar ultraviolet radiation. *Science, 225*, 315-317.

Harwood, R. S., **1975**: The temperature structure of the Southern Hemisphere stratosphere: August-October, 1971. *Quart. J. Roy. Meteorol. Soc., 102*, 757-770.

Hood, L. L., **1984**: The temporal behavior of upper stratospheric ozone at low latitudes: Evidence from Nimbus 4 BUV data for short-term responses to solar ultraviolet variability. *J. Geophys. Res., 89*, 9557-9568.

Jackman, C. H., J. A. Kaye, and P. D. Guthrie, **1985**: LIMS HNO_3 data above 5 mbar: Corrections based on simultaneous observations of other species. *J. Geophys. Res., 90*, 7923-7930. Also correction, *Ibid*, 13094.

Remsberg, E. E., J. M. Russell III, J. C. Gille, L. L. Gordley, P. L. Bailey, W. G. Planet, and J. E. Harries **1984**: The validation of NIMBUS 7 LIMS measurements of ozone. *Geophys. Res., 89*, 5161-5178.

Rood, Richard B., and Anne R. Douglass, **1985**: Interpretation of ozone temperature correlations, 1, Theory. *J. Geophys. Res., 90*, 5733-5743.

Russell, J. M. III, and J. C. Gille, **1978**: The Limb Infrared Monitor of the Stratosphere (LIMS) experiment. In *The Nimbus 7 Users Guide*, C. R. Madrid, Ed., NASA Goddard Space Flight Center, Greenbelt, Maryland, 71-103.

Solomon, S., J. T. Kiehl, B. J. Kerridge, E. E. Remsberg, and J. M. Russell III, **1986**: Evidence for nonlocal thermodynamic equilibrium in the ν_3 mode of mesospheric ozone. *J. Geophys. Res., 91*, 9865-9876.

WMO, **1986**: Atmospheric Ozone 1985. World Meteorological Organization Global Ozone Research and Monitoring Project Report No. 16, 1075 pp.

THE NEAR GLOBAL DISTRIBUTIONS OF MIDDLE ATMOSPHERIC H_2O AND NO_2
MEASURED BY THE NIMBUS 7 LIMS EXPERIMENT

Ellis E. Remsberg and James M. Russell III
Atmospheric Sciences Division
NASA Langley Research Center
Hampton, Virginia 23665-5225

ABSTRACT. Two of the six channels of the Nimbus 7 Limb Infrared
Monitor of the Stratosphere (LIMS) experiment were located at 6.9 and
6.2 micrometers for the purpose of sounding the vertical profiles of
atmospheric H_2O and NO_2, respectively. The results from these
channels have provided new information on the global distributions,
variability, and budget of these important molecules. The zonal-mean
cross sections for water vapor show the presence of a hygropause in
the tropical lower stratosphere, a poleward-directed gradient in the
mid and lower stratosphere, and equatorward gradient above. Longitudinal variability is small and temporal changes are slow, occurring
over periods of weeks to seasons. Nitrogen dioxide cross section data
show the presence of an NO_2 "cliff" in winter, higher mixing ratio
levels in the Southern Hemisphere than in the north, pronounced
diurnal variability, and very high mixing ratios in the polar night
mesosphere. The sum of LIMS nighttime NO_2 plus HNO_3 should be a
reasonable approximation of the profiles of total odd nitrogen outside
the polar night regions during 1978-1979.

1. INTRODUCTION

    The Nimbus 7 LIMS experiment operated for its planned lifetime of
7-1/2 months from October 24, 1978, to May 28, 1979 (Russell and
Gille, 1978). It measured vertical profiles of radiance across the
atmospheric limb of the Earth with near global coverage each day.
Those profiles were processed later to infer middle atmosphere
temperature profiles and the concentrations of key compounds believed
to be important in the stratospheric ozone photochemistry (Gille and
Russell, 1984, and references therein). Two of those compounds, H_2O
and NO_2, were retrieved from radiances measured at 6.9 and 6.2
micrometers, respectively, and their distributions are reviewed in
this paper. Results for O_3 and HNO_3 are discussed in a companion
paper (Gille, this volume). Some of the important scientific findings
to date for all the LIMS species were summarized in WMO (1986), but
further studies are anticipated in the areas of both chemistry and
transport. More examples of the monthly, zonal-mean distributions of
H_2O and NO_2 appear in Russell et al. (1986) along with monthly
averaged polar stereographic maps at selected pressure levels.

87

G. Visconti and R. Garcia (eds.), Transport Processes in the Middle Atmosphere, 87–102.

Table 1.   Vertical Resolution of LIMS Retrieved Profiles

Parameter	Amplitude of Atmospheric Feature	Vertical Resolution	
		IPAT (km)	MAT (km)
Temperature	2K	2.5	3.4
O_3 and HNO_3	25%	2.9	3.8
H_2O and NO_2	20%	5.0	6.5

An extensive program to characterize and validate the LIMS H_2O and NO_2 was undertaken and reported in Russell et al. (1984a, b). In general, the LIMS data set possesses two qualities that are particularly valuable for studies of stratospheric dynamics—high precision and excellent vertical resolution.   Precision estimates are of the order of 5 percent for both H_2O and NO_2 in the mid stratosphere.   This means that even relatively small amplitude, large-scale wave activity can be detected in those regions where significant meridional gradients of H_2O and NO_2 occur (e.g. Miles and Grose, 1986).   The vertical resolution of the LIMS retrieved profiles has been determined by simulations.   Resolution in this case refers to the minimum extent or vertical half-wavelength of features in the species profile that are detectable in the corresponding radiance profile with a signal-to-noise (S/N) of 2.   Table 1 summarizes these estimates for the LIMS parameters on both the Inverted Profile Archive Tapes (IPAT) (see Gille and Russell, 1984) and the MAP Archive Tapes (MAT) (see Haggard et al., 1986).   The amplitude of the atmospheric features that are just detectable in the IPAT is also given in Table 1.   Some features were effectively smoothed when the IPAT data were processed to the MAT product as indicated in Miles et al. (1987) for temperature.   Because of the high resolution of the LIMS data, Hitchman and Leovy (1986) and Gray and Pyle (1986) have been able to see effects of tropical waves and the associated semiannual oscillation in the LIMS temperature and constituent distributions.

The accuracy of single profiles of water vapor and nitrogen dioxide has also been reported in Russell et al. (1984a, b).   They noted that the daytime water vapor in the upper stratosphere exceeded that at night by as much as 1 to 2 ppmv.   That variation is attributed to a possible solar induced bias in the LIMS daytime radiance or to non-local thermodynamic equilibrium (NLTE) effects, which still need to be quantified.   Furthermore, Jones et al. (1986) showed that estimates of profiles of total hydrogen are consistent with the sum of the methane profile data from the Stratospheric and Mesospheric Sounder (SAMS) experiment on Nimbus 7 and the descending mode ($\approx$ nighttime) LIMS water vapor.   Thus, the nighttime LIMS water vapor results are considered reliable.   Table 2 contains estimates of uncertainties in zonal-mean nighttime water vapor. In the upper stratosphere, the LIMS NO_2 results vary from day to night by more than a factor of 2 due to photochemical activity. Consequently, Table 2 also contains zonal-mean error estimates for an NO_2 profile, whose magnitude is about midway between that seen for day and night. The cutoff level for accurate NO_2 results in the lower stratosphere is about 40 mb. The accuracies in Table 2 are adequate for addressing some points in the theories about stratospheric photochemistry and the budgets of trace species.

Table 2. Uncertainty in LIMS Zonal Mean H_2O and NO_2

Pressure-Altitude (mb)	H_2O (%)	NO_2 (%)
50	20	-
30	17	43
10	17	18
5	17	16
3	17	16
1	27	26

2. WATER VAPOR DISTRIBUTIONS

Satellite observations of stratospheric water vapor have provided new knowledge of the global distribution of this molecule and also more information about its vertical profile (Russell et al., 1984a; Remsberg et al., 1984; and Jones et al., 1986). Prior to the launch of Nimbus 7, only balloon and airborne in situ and remote observations had been made. Detailed reviews of these measurements were published by Harries (1976) and Ellsaesser (1983). The compilation by these authors gives a picture of stratospheric water vapor which shows a wide range of variability (1-16 ppmv), several profile shapes, and either conflicting representations of altitude trends or altitude increases that were not in accord with methane oxidation theory. Holton (1984) summarized the water vapor distributions in the lower stratosphere derived from in situ data and discussed the implications for studies of the large-scale transport and the water vapor budget. These pictures improved with the LIMS experiment, which provided a near global view of water vapor, and since observations were made by the same instrument over a long time period, large-scale changes were noted with high precision.

Monthly zonal mean cross sections of water vapor from LIMS are shown in Figs. 1 and 2 for December and May. These two figures show the main features of the stratospheric water vapor distribution. It is characterized by a minimum in the Tropics just above the tropical tropopause, a poleward mixing ratio gradient at pressures up to the 4-mb level, and then an equatorward gradient from that point to the stratopause. The distribution also has double minima in mixing ratio at about 6 mb which, at times during the LIMS mission from October to May, merge into just one feature. Gray and Pyle (1986) have suggested that this behavior is due in part to the equatorial semiannual oscillation.

Another characteristic of the distributions is an implied net circulation. The general picture suggested is a "reservoir" of dry air in the Tropics, part of which is being carried upward and poleward with the strongest circulation being toward the winter pole in the upper stratosphere. There is also a seasonal shift in the water vapor in the lower stratosphere, with relatively dry air extending more toward high latitudes of the summer hemisphere at 70 mb. The distribution in Fig. 1 is compared with diabatic circulation calculations for 1 month earlier (November) as obtained using LIMS temperature, ozone, and water vapor data (Callis et al., 1987). Figure 3 shows mass weighted diabatic stream functions in units of kg/km^2-day versus latitude and potential temperature, where 1000 K is near 10 mb and 2000 K is near 1 mb. Bold arrows indicate the sense of that circulation. While it is difficult to use Fig. 3 to compare directly with the distributions 1 month later in Fig. 1 because of the approximations in the circulation (see Remsberg, 1987), it can be seen that the sense of the circulation implied by the stream functions agrees qualitatively with the observed water vapor and that one would expect a seasonal change in the circulation asymmetry by May (Fig. 2). There are also seasonal changes in the tropical and subtropical water vapor in the 1- to 3-mb range, which may be due to changes in transport in the lower mesosphere and upper stratosphere (Hitchman and Leovy, 1986).

Monthly zonal mean vertical water vapor profiles averaged over four latitude bands from 56 N - 84 N, 32 N - 56 N, 28 S - 28 N, and 32 S - 56 S are shown in Fig. 4 to provide a view of changes every 2 months. The high northern latitude profiles (Fig. 4a) are characterized by very little change with altitude, but with some variation near 100 mb. A gradual variation in the lower stratosphere also occurs in the Southern Hemisphere in Fig. 4d. Such changes may be reflective of variations in the net meridional circulation there, and may possibly be due to a weak exchange with the mid latitude troposphere in winter. In the northern mid latitude range (Fig. 4b), there is a slight trend of increasing mixing ratio with altitude. The tropical profiles in Fig. 4c are characterized by only small changes, a very pronounced minimum that occurs above the tropical tropopause, and a clear increase with altitude which is consistent with methane oxidation theory in the upper stratosphere (Remsberg et al., 1984). The minimum just above the tropopause was first observed in balloon

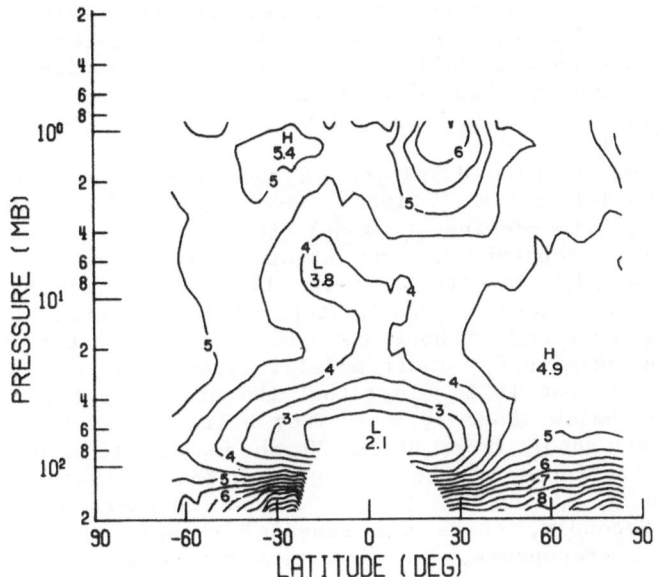

Figure 1. LIMS zonal mean descending mode H₂O cross section for December 1978. Contour interval is 0.5 ppmv.

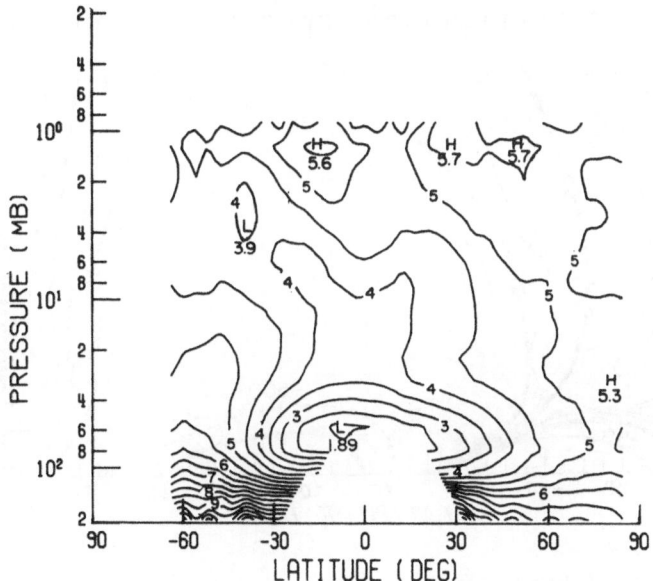

Figure 2. Same as Fig. 1, except for May 1979.

soundings by Kley et al. (1979); this minimum is often referred to as a "hygropause." Remsberg et al. (1984) have determined from simulations that the average LIMS H_2O minimum for the latitude span in Fig. 4c can be too dry by about 0.15 ppmv at 50 mb and 0.3 ppmv at 70 mb. They also pointed out that the position of the LIMS hygropause is too high by 1 to 2 km.

The hygropause feature displayed in the LIMS data is not fully understood. Stordal et al. (1985) developed a two-dimensional, zonally averaged, time-dependent model to describe transport and chemistry of both tropospheric and stratospheric gases, and they compared the model H_2O results with the LIMS data. Figure 5 from their paper shows LIMS data at the Equator and 32N compared with model results obtained with and without the production of water vapor by methane oxidation chemistry. Their model imposes a hygropause at the tropical tropopause, but it is clear that the chemistry has no effect on the hygropause below about 20 mb. They believe that the observed increase in water vapor from 70 to 20 mb is due to two effects. First, vertical gradients in the strength of the diabatic transport and the meridional mixing in the lower stratosphere affects the distribution. Secondly, there are seasonal changes in temperature near the tropical tropopause, and that temperature was a minimum in early 1979. Due to the relatively long residence times of the stratospheric air, the appearance of dryer air in a region containing larger

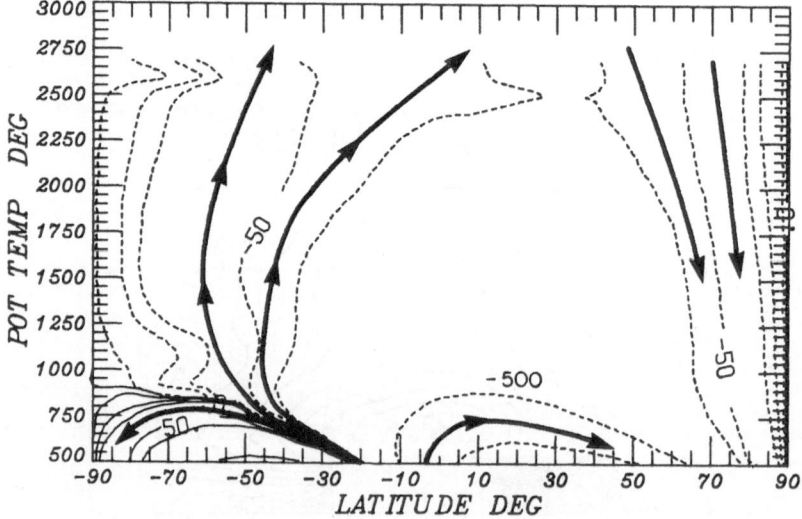

Figure 3.   Mass-weighted diabatic streamfunctions (kg/km²-day) for November 1978. Dashed lines are clockwise, solid lines counterclockwise circulations (see bold arrows). Contour intervals are 0, 1, 5, 10, 50, 100, 500, 1000.

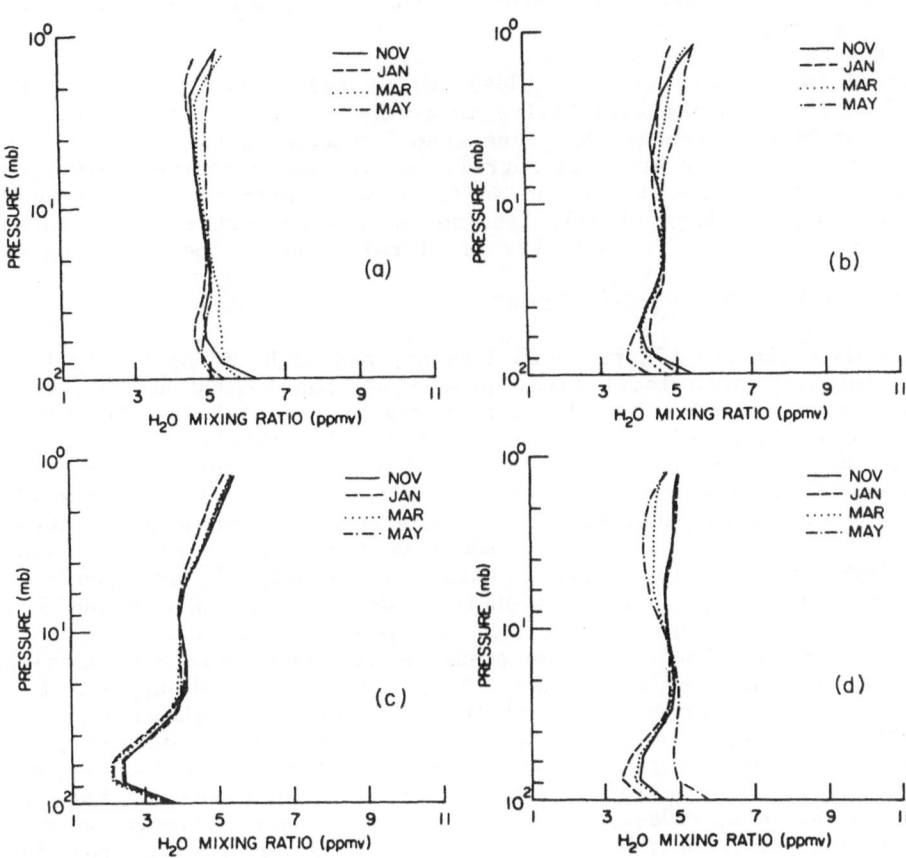

Figure 4. LIMS monthly zonal mean descending mode H₂O profiles averaged over the latitude bands (a) 56 N – 84 N, (b) 32 N – 56 N, (c) 28 S – 28 N, and (d) 32 S – 56 S.

water vapor values from prior seasons, when the minimum temperatures
were warmer, can lead to a steepened vertical gradient of water
vapor. Detailed mechanisms have been proposed for the formation of
the hygropause (see e.g., Danielsen, 1982), but at present, there is
no theory or agreed upon process that adequately describes all
available data.

Evidence of wave activity in the LIMS water vapor fields has been
reported in Remsberg et al. (1984), Miles and Grose (1986), and WMO
(1986). Longitudinal variability is of the order of 1 to 1.5 ppmv in
winter at high latitudes, but less than 0.5 ppmv in the Tropics. The
variations in winter are well correlated with the concurrent distributions of potential vorticity on an isentropic surface, as reported by
Butchart and Remsberg (1986), lending further support to the argument
for using water vapor as a tracer of circulations in the stratosphere.

3. NITROGEN DIOXIDE DISTRIBUTIONS

LIMS collected NO_2 radiance data day and night at nominally 10:30
p.m. and 1:30 p.m. local time, providing about 7,000 horizon scans
each 24-hour period. Descriptions of the near global distribution of
NO_2 have been provided in Russell (1984), WMO (1986), and Russell et
al. (1986). The main features of the LIMS day and night NO_2 distribution can be seen generally in the zonal mean results for January 1979
(Figs. 6 and 7, respectively). Figure 6 reveals a broad layer structure for the daytime NO_2 with peak mixing ratios occurring at about
the 7-mb level. The highest values of nearly 7 ppbv occur in
the Tropics and Southern Hemisphere. Shading in Figs. 6 and 7 is
included to point out the region of the night terminator at about 68 N
and the region below the 40 mb pressure altitude where a climatology
was used in the LIMS retrieval. This climatology was applied to a
progressively larger degree as altitude decreased. A climatology "tie
on" was developed because of the steady loss of NO_2 signal and growth
of the molecular oxygen continuum interference signal in the lower
stratosphere. All of the archived NO_2 was retrieved in this way
(Russell et al., 1984b), but this is the only instance where a
climatology was considered in the retrieval of any of the LIMS
parameters. Data in the shaded regions, therefore, should be
interpreted carefully in any science investigations.

The nighttime NO_2 distribution in Fig. 7 has a character that is
similar to that for daytime, but the variations are enhanced. The
maximum mixing ratio occurs farther to the south and a hemispherically
asymmetric behavior exists. Zonal mean NO_2 distributions for other
periods tend to be very similar in form to the January cross
sections. As time progresses for November to May, the region of the
maximum in the Southern Hemisphere moves equatorward until May when
the nighttime distribution of the NO_2 maximum is more nearly symmetric
about the Equator (see Fig. 8). There is still a southward displacement of the maximum, however. This persistent hemispheric asymmetry
may be an indication of differences in the net circulation of the two

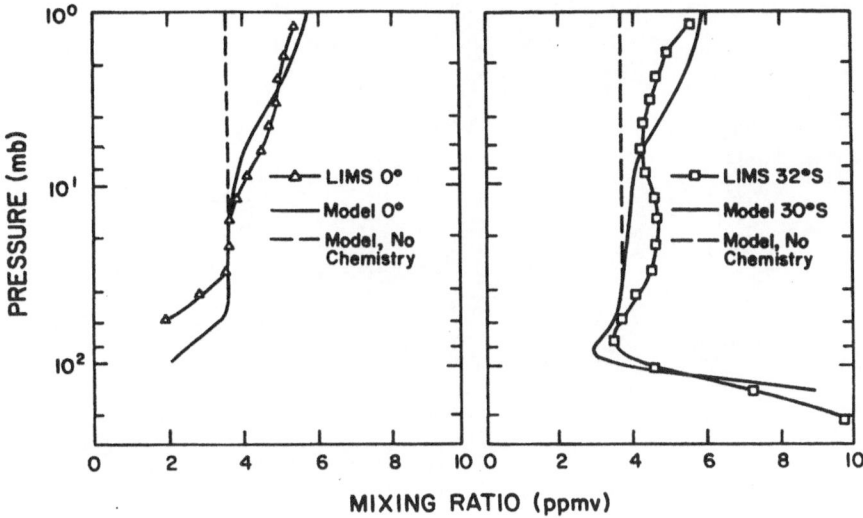

Figure 5.   Comparison of the LIMS monthly zonal mean H_2O profile for May 1979 with 2-D model resuls (from Stordal et al., 1985).

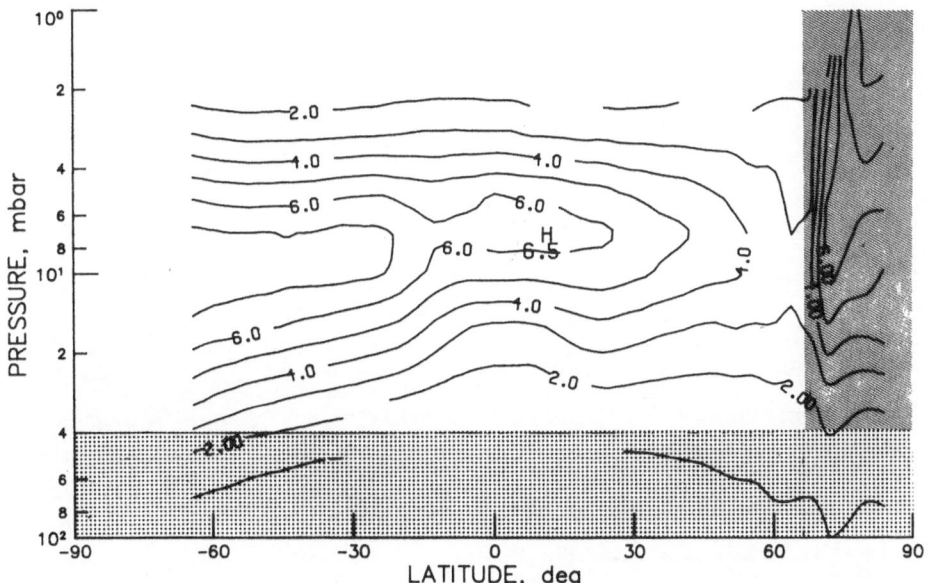

Figure 6.   LIMS daytime monthly zonal mean NO_2 cross section for January 1979.   Contour interval is 1 ppbv.

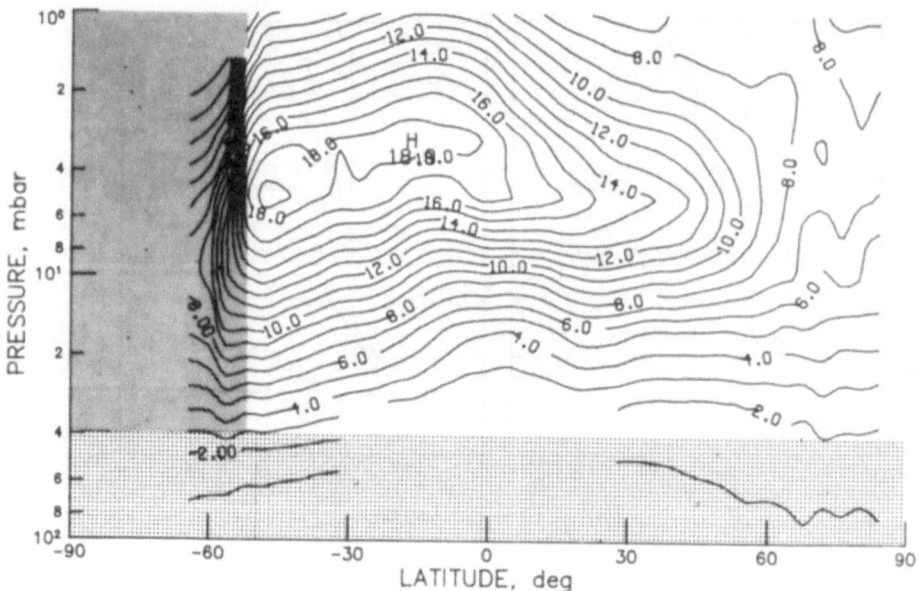

Figure 7.   As in Fig. 6, except for nighttime.

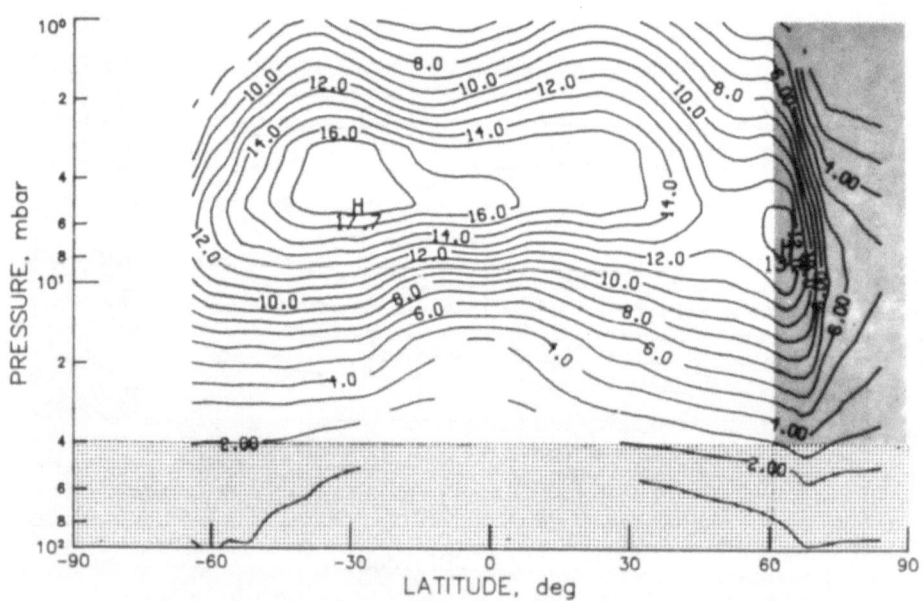

Figure 8.   LIMS nighttime monthly zonal mean NO$_2$ cross section for May 1979.   Contour interval is 1 ppbv.

hemispheres.   Differences in the strength of both sources and sinks of
total odd nitrogen must be considered, too.

A striking feature of the January nighttime cross section (Fig.
7) is the large decrease in mixing ratio with increasing north
latitude which, in the 5 mb region, decreases from a maximum of about
19 ppbv in the Tropics to 7 ppbv at 84N.   A polar stereographic
projection (not shown) for the same time period shows that the
decrease occurs over a broad latitude/longitude region which coincides
with the position of the north polar vortex.   This phenomenon is a
manifestation of the NO$_2$ behavior first reported by Noxon (1979) and
commonly referred to as the "Noxon cliff."   Its occurrence is believed
to be due to a steady conversion of NO$_2$ to N$_2$O$_5$ during the polar night
as air makes circuits around the vortex going into and out of the
night for varying time periods depending on latitude (Solomon and
Garcia, 1983a).   Further work by Solomon and Garcia (1983b) comparing
model calculations to NO$_2$ amounts measured from the ground and by
Callis et al. (1983) comparing model results to LIMS observations
lends further support to this explanation.

Another way to view the NO$_2$ cliff phenomenon is in terms of
column amount above the 30-mb level.   This quantity is similar to the
measurement made by Noxon (1979).   LIMS nighttime and daytime column
data are shown for a 1-week period in January in Fig. 9a for the 240E-
300E longitude sector and for a time period when the NO$_2$ decrease with
latitude was greatest in the Northern Hemisphere.   The retrieval of
these data required the use of radiance-averaging methods (Russell et
al., 1984c) to obtain sufficient signal to do a retrieval in the
region of the cold vortex.   The satellite results show a large (factor
of 6), rather precipitous decrease (within 25 degrees of latitude)
poleward of 45 N.   Figure 9b displays a similar variation in nighttime
NO$_2$ for 1-week periods in November and May.   The NO$_2$ amount at 60 N is
not as low in November as for January, most likely because the southward progression of the polar night region had just begun in November.   The probable nighttime conversion of NO$_2$ to N$_2$O$_5$ was more extensive by January.   Large NO$_2$ decreases at high southern latitudes in
November and January and at high northern latitudes in May are due to
a night-to-day terminator crossing and are not caused by a dynamics/
chemistry interaction effect like the Noxon cliff.   Hemispheric levels
of column NO$_2$ also change with season, i.e., there is more NO$_2$ in the
south in November, but more in the north in May, in Fig. 9b.

In the process of using the radiance averaging method to study
the NO$_2$ cliff, the NO$_2$ retrievals were extended to a higher altitude.
In fact, in polar night, they extend to near the 70-km level and show
large mixing ratio values which are highly variable longitudinally
(Russell et al., 1984c) and which reach magnitudes of 175 ppbv
locally.   As time increases during polar night, mesospheric NO$_2$
continues to increase and is subsequently transported downward into
the stratospheric polar vortex.   The implied downward velocity of 0.8
km/day at the 1 mb level is in accord with estimates from diabatic
circulation models.   Zonal means obtained in late January and early
February 1979 indicate that the high mesospheric, polar night NO$_2$

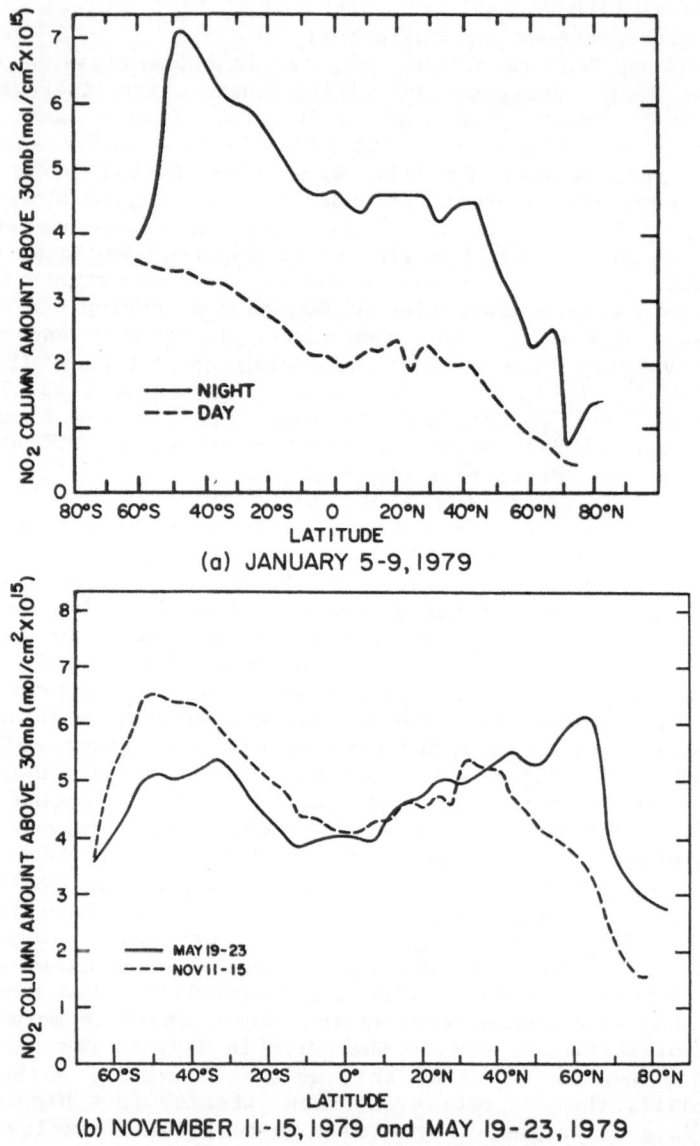

Figure 9. LIMS NO_2 column amount above the 30-mb pressure level averaged over the 240 E to 300 E longitude sector: (a) day versus night for January 5-9, 1979, and (b) nighttime result for periods in November and May.

levels (northward of 60 N) dissipate fairly quickly after polar night ends. An important question that has not been fully studied regards how the stratospheric NO_2 disperses from the vortex region and how long it takes. A reasonable body of data now exists to perform initial studies of the NO_2 budget in the stratosphere, but more data are needed simultaneously in both hemispheres on NO_2 as well as on NO and HNO_3 in order to do a more complete study.

The observed day/night variations agree fairly well with theoretical calculations of diurnal changes with latitude and season (WMO, 1986). However, Solomon et al. (1986) conducted a more detailed study of diurnal variations by using Northern Hemisphere data in spring. NO_2 observations occurred during the afternoon and evening at solar zenith angles ranging from about 35° to 110° during May. Variations across this terminator region were compared with results from a time-dependent, chemical model calculation, and the agreement was good.

Callis et al. (1985) obtained zonal mean cross sections of LIMS nighttime NO_2 plus HNO_3 in order to approximate the total odd nitrogen in the upper stratosphere. By considering just that observed sum, they found levels that were substantially higher than most model estimates of total odd nitrogen, at least for the 1978-79 period. It appears likely that there are interannual variations in total odd nitrogen, and, if so, one must know the changes in its sources and sinks accurately in order to use it as a tracer quantity.

4. CONCLUDING REMARKS

The LIMS experiment has provided the most extensive stratospheric H_2O and NO_2 data set to date. Accuracy and precision of the data are adequate for comparisons with theories of both chemical and transport processes in the middle atmosphere. The observed large-scale variability of both species should be an important diagnostic in analyses of dynamical models of the stratosphere. It may also be possible to consider, once again, budget studies of both total odd nitrogen and water vapor using these data, even though their stratospheric residence times are believed to be much longer than 7-1/2 months. Longer term measurements will be needed in order to evaluate the full seasonal and interannual variability of these species.

5. REFERENCES

Butchart, N., and E. E. Remsberg, 1986: 'The area of the stratospheric polar vortex as a diagnostic for tracer transport on an isentropic surface.' J. Atmos. Sci., **43**, 1319-1339.

Callis, L. B., J. M. Russell III, K. V. Haggard, and M. Natarajan, 1983: 'Examination of wintertime latitudinal gradients in stratospheric NO_2 using theory and LIMS observations.' Geophys. Res. Lett., **10**, 945-948.

Callis, L. B., M. Natarajan, and J. M. Russell III, 1985: 'Estimates of the stratospheric distribution of odd nitrogen from the LIMS data.' Geophys. Res. Lett., **12**, 259-262.

Callis, L. B., R. E. Boughner, and J. D. Lambeth, 1987: 'The stratosphere: Climatologies of the radiative heating and cooling rates and the diabatically diagnosed net circulation fields.' J. Geophys. Res., **92**, 5585-5608.

Danielsen, E. F., 1982: 'A dehydration mechanism for the stratosphere.' Geophys. Res. Lett., **9**, 605-608.

Ellsaesser, H. W., 1983: 'Stratospheric water vapor.' J. Geophys. Res., **88**, 3897-3906.

Gille, J. C., and J. M. Russell III, 1984: 'The Limb Infrared Monitor of the Stratosphere: Experiment description, performance, and results.' J. Geophys. Res., **89**, 5125-5140.

Gray, L., and J. Pyle, 1986: 'The semiannual oscillation and equatorial tracer distributions.' Quart. J. Roy. Meteor. Soc., **112**, 387-407.

Haggard, K. V., E. E. Remsberg, W. L. Grose, J. M. Russell III, B. T. Marshall, and G. Lingenfelser, 1986: 'Description of data on the Nimbus 7 LIMS Map Archive Tape--temperature and geopotential height.' NASA Technical Paper 2553, 53 pp. (available through NTIS, #N86-25924, Springfield, VA).

Harries, J. E., 1976: 'The distribution of water vapour in the stratosphere.' Rev. Geophys. Space Phys., **14**, 565-575.

Hitchman, M. H., and C. B. Leovy, 1986: 'Evolution of the zonal mean state in the equatorial middle atmosphere during October 1978 - May 1979.' J. Atmos. Sci., 43, 3159-3176.

Holton, J. R., 1984: 'Troposphere-stratosphere exchange of trace constituents: The water vapor puzzle.' In Dynamics of the Middle Atmosphere, Terrapub, 369-385.

Jones, R. L., J. A. Pyle, J. E. Harries, A. M. Zavody, J. M. Russell III, and J. C. Gille, 1986: 'The water vapor budget of the stratosphere studied using LIMS and SAMS satellite data.' Quart. J. Roy. Meteor. Soc., **112**, 1127-1143.

Kley, D., E. J. Stone, W. R. Henderson, J. W. Drummond, W. J. Harrop, A. T. Schmeltekopf, T. L. Thompson, and R. H. Winkler, 1979: 'In situ measurements of the mixing ratio of water vapor in the stratosphere.' J. Atmos. Sci., **36**, 2513-2524.

Miles, T., and W. L. Grose, 1986: 'Transient medium-scale wave activity in the summer stratosphere.' Bull. Am. Meteor. Soc., **67**, 674-686.

Miles, T., W. L. Grose, J. M. Russell III, and E. E. Remsberg, 1987: 'Comparison of southern hemisphere radiosonde and LIMS temperatures at 100 mb.' Quart. J. Roy. Meteor. Soc., in press.

Noxon, J. F., 1979: 'Stratospheric NO₂, 2, global behavior.' J. Geophys. Res., **84**, 5067-5076.

Remsberg, E. E., 1987: 'Analysis of the mean meridional circulation using satellite data.' (this volume)

Remsberg, E. E., J. M. Russell III, L. L. Gordley, J. C. Gille, and P. L. Bailey, 1984: 'Implications of the stratospheric water vapor distribution as determined from the Nimbus 7 LIMS experiment.' J. Atmos. Sci., **41**, 2934-2945.

Russell, J. M. III, 1984: 'The global distribution and variability of stratospheric constituents measured by LIMS.' Adv. Space Res., **4**, 107-116.

Russell, J. M. III, and J. C. Gille, 1978: 'The Limb Infrared Monitor of the Stratosphere (LIMS) experiment.' In The Nimbus 7 Users Guide, edited by C. Madrid, pp. 71-103, Goddard Space Flight Center, Greenbelt, MD.

Russell, J. M. III, J. C. Gille, E. E. Remsberg, L. L. Gordley, P. L. Bailey, H. Fischer, A. Girard, S. R. Drayson, W. F. J. Evans, and J. E. Harries, 1984a: 'Validation of water vapor results measured by the Limb Infrared Monitor of the Stratosphere experiment on Nimbus 7.' J. Geophys. Res., **89**, 5115-5124.

Russell, J. M. III, J. C. Gille, E. E. Remsberg, L. L. Gordley, P. L. Bailey, S. R. Drayson, H. Fischer, A. Girard, J. E. Harries, and W. F. J. Evans, 1984b: 'Validation of nitrogen dioxide results measured by the Limb Infrared Monitor of the Stratosphere (LIMS) experiment on Nimbus 7.' J. Geophys. Res., **89**, 5099-5107.

Russell, J. M. III, S. Solomon, L. L. Gordley, E. E. Remsberg, and L. B. Callis, 1984c: 'The variability of stratospheric and mesospheric NO_2 in the polar winter night observed by LIMS.' J. Geophys. Res., 89, 7267-7275.

Russell, J. M. III, S. Solomon, M. P. McCormick, A. J. Miller, J. J. Barnett, R. L. Jones, and D. W. Rusch, 1986: 'Middle atmosphere composition revealed by satellite observations.' MAP Handbook, 22, University of Illinois, Urbana, IL, 302 pp.

Solomon, S., and R. R. Garcia, 1983a: 'On the distribution of nitrogen species in the stratosphere.' J. Geophys. Res., 88, 5229-5239.

Solomon, S., and R. R. Garcia, 1983b: 'Simulations of NO_x partitioning along isobaric parcel trajectories.' J. Geophys. Res., 88, 5497-5501.

Solomon, S., J. M. Russell III, and L. L. Gordley, 1986: 'Observations of the diurnal variation of nitrogen dioxide in the stratosphere.' J. Geophys. Res., 91, 5455-5464.

Stordal, F., I. S. A. Isaksen, and K. Horntveth, 1985: 'A diabatic circulation two-dimensional model with photochemistry: Simulation of ozone and long-lived tracers with surface sources.' J. Geophys. Res., 90, 5757-5776.

World Meteorological Organization, 1986: 'Atmospheric ozone 1985: Global ozone research and monitoring project.' Report no. 16.

SATELLITE MEASUREMENTS OF STRATOSPHERIC AEROSOLS

M. P. McCormick[1] and Pi-Huan Wang[2]
[1]Atmospheric Sciences Division
NASA Langley Research Center
Hampton, Virginia 23665-5225
[2]Science and Technology Corporation
101 Research Drive
Hampton, Virginia 23666

ABSTRACT. This paper describes a global stratospheric aerosol climatology developing from the satellite experiments SAM II, SAGE I, and SAGE II. SAM II, operational since October 1978, has been making measurements of the polar regions. Its data show the effects of volcanic eruptions, which govern the overall trend in stratospheric aerosols, the existence and duration of polar stratospheric clouds, and polar dynamics as manifested in aerosol changes. SAGE I and II data also show these polar region variations, large-scale transport from the equatorial regions in the local winter, the latitudinal distribution of aerosols versus season, and the transport and dispersal of volcanic effluents. Finally, the variations occurring during austral winter and spring and the ozone hole will be discussed.

1. INTRODUCTION

Since the discovery of the stratospheric aerosol layer nearly 30 years ago by Junge et al. (1961), attention has been focused on understanding its behavior and effects on atmospheric chemistry, radiative transfer, and the climate of the Earth. In order to understand this layer on a global basis, NASA has launched three satellite instruments since October 1978: the Stratospheric Aerosol Measurement (SAM II) on Nimbus 7, and the Stratospheric Aerosol and Gas Experiments I and II (SAGE I and II) on the Application Explorer Mission 2 satellite and Earth Radiation Budget Satellite, respectively.

These satellite instruments all utilize the solar occultation technique, which makes them self-calibrating, to measure vertical profiles of limb attenuated solar intensity with a 1 km vertical resolution at desired wavelengths during each sunrise and sunset experienced by the satellite. Table 1 summarizes the launching date, measured species, and latitudinal coverage of these three satellite instruments. The SAM II instrument is a one-channel sunphotometer measuring aerosol extinction at 1.0 μm. The SAGE I instrument is a four-channel sunphotometer which measures aerosol extinction at 1.0 μm and 0.45 μm.

103

G. Visconti and R. Garcia (eds.), Transport Processes in the Middle Atmosphere, 103–120.
© *1987 by D. Reidel Publishing Company.*

Table 1. Satellite Limb Extinction Measurements

Experiment	Satellite	Launch	Channels (Species)	Latitude Coverage
SAM II	NIMBUS-7	Oct. 1978*	1 (Aerosols)	64° – 80°N 64° – 80°S
SAGE I	AEM-2	Feb. 1979**	4 (Aerosols, NO_2, O_3)	79°N – 79°S
SAGE II	ERBS	Oct. 1984*	7 (Aerosols, NO_2, O_3, H_2O)	80°N – 80°S

*Presently still operational.
**Obtained data through November 1981.

In addition, it provides simultaneous observations of stratospheric O_3 and NO_2 at 0.60 and 0.45-μm respectively. The detailed aspects of the SAM II and SAGE I systems have been described by McCormick et al. (1979). The SAGE II satellite instrument is an advanced version of SAGE I. It has three additional channels centered at 0.448, 0.525, and 0.94-μm which provide a differential NO_2 measurement, additional aerosol extinction data, and a H_2O vapor concentration channel (McMaster, 1986). Thus, the SAGE II satellite instrument measures aerosol extinction at four different wavelengths. The simultaneously determined stratospheric H_2O is of particular importance in understanding the aerosol microphysical processes as well as their composition. The purpose of this paper is to describe the global stratospheric aerosol climatology, including polar stratospheric clouds, effects of volcanic eruptions, and the transport processes as manifested in aerosol changes based mainly on SAM II and SAGE I satellite measurements. It is understood that the SAGE II data set is in its preliminary stage and is currently undergoing validation by its science team. The SAM II and SAGE I measurements have been compared extensively with correlative measurements. Under conditions when the sampling locations of both SAM II and SAGE I are close to each other, a comparison between SAM II and SAGE I profiles can be made. Figure 1 shows the result of such a comparison from observations on November 22, 1979 (Yue et al., 1984). Figure 2 shows a typical vertical profile of aerosol extinction obtained by the SAGE II instrument on September 30, 1985. Distinct differences in the aerosol extinction between Figs. 1 and 2 in the altitude range between approximately 11 to 20 km are obvious. As will be shown later, the enhanced aerosol extinction observed by SAGE II on September 30, 1985, is largely due to the El Chichon volcanic eruption which occurred in late March-early April 1982.

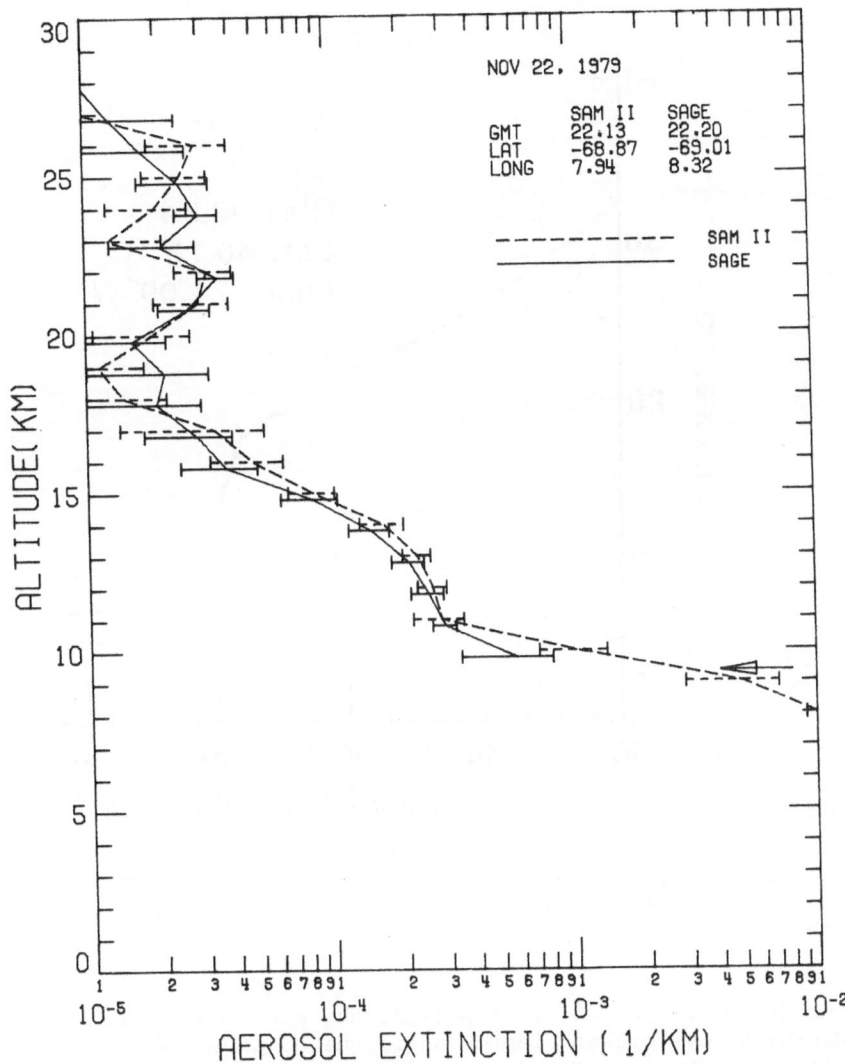

Figure 1. Aerosol extinction profiles at 1 µm wavelength from SAM II and SAGE I during an occasion when both instruments are sampling at locations close to each other. The arrow indicates the approximate tropopause height (Yue and McCormick, 1984).

2. STRATOSPHERIC AEROSOL CLIMATOLOGY

As mentioned in the Introduction, SAGE I provided measurements with a nearly global coverage for a period of 34 months (from February 1979 to

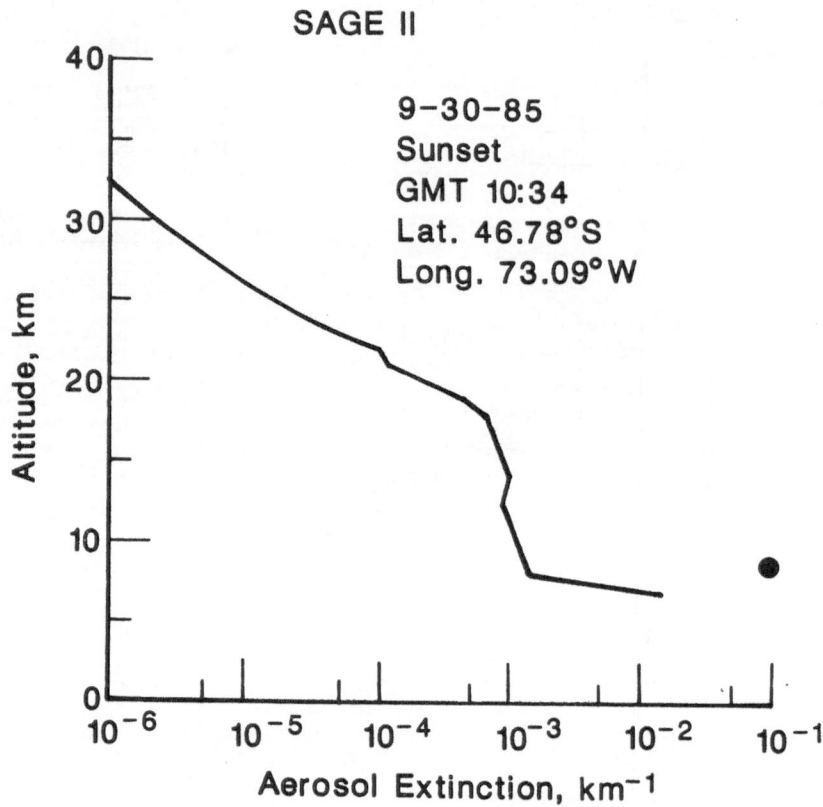

Figure 2. Aerosol extinction profile at 1 μm wavelength from the
SAGE II satellite instrument. The closed dot indicates the tropopause
height.

November 1981). During this period, the stratosphere was relatively
undisturbed by volcanic impact until the eruption of Mount St. Helens
on May 18, 1980. This eruption sent volcanic material up to an altitude of about 23 km (Kent and McCormick, 1984). For this reason, and
from the results of several other measurements by other techniques, one
may consider the observed distribution obtained by SAGE I prior to May
18, 1980, to be characteristic of a background stratospheric aerosol
layer. Figures 3a to d show the derived seasonal mean meridional distributions of the aerosol extinction ratio at 1 μm wavelength during
this undisturbed period for the seasons March-April-May (1979), June-
July-August (1979), September-October-November (1979), and December
(1979)-January-February (1980). The aerosol extinction ratio is
defined as the ratio of the sum of aerosol and molecular extinction to
molecular extinction. Distinct features in these zonal means include a

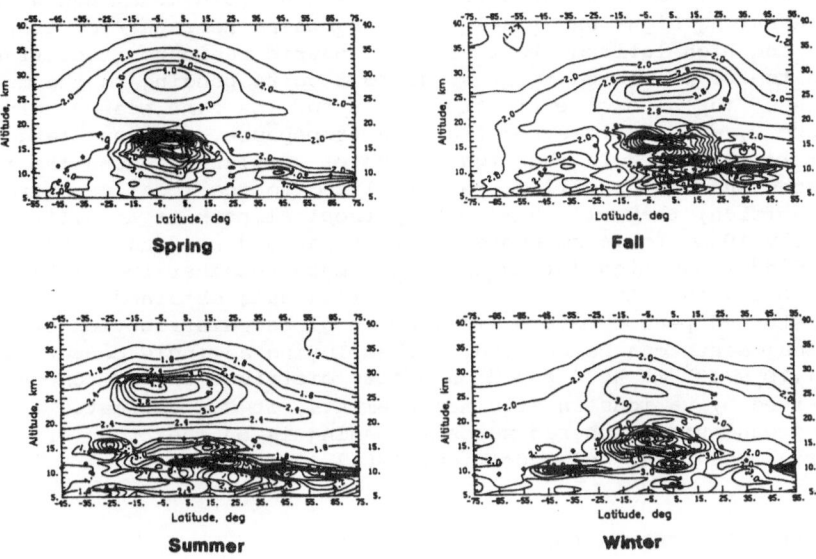

Spring **Fall** **Summer** **Winter**

Figure 3. Seasonal mean meridional distributions of aerosol extinction ratio [(aerosol extinction + molecular extinction)/molecular extinction] derived from SAGE I measurements. The four seasons are Spring (March, April, May 1979), Summer (June, July, August, 1979), Fall (September, October, November 1979), and Winter (December 1979 and January, February 1980). The solid diamonds are the mean tropopause altitudes for these data.

local maximum of aerosol extinction ratio located near an altitude of 25-30 km in the tropics which is indicative of a source region for stratospheric aerosols and a sloping of these ratios with latitude following more or less a constant altitude above the local tropopause.

Perhaps, the most important chemical reaction in the formation of the background aerosol droplets is the oxidation of SO_2 in the stratosphere by

$$SO_2 + OH + M \rightarrow HSO_3 + M$$

and subsequent formation of H_2SO_4 molecules. In general, the aerosol droplets are believed to be produced through heteromolecular condensation processes. Advection, sedimentation and coagulation are important to the distribution of the stratospheric aerosols. To further investigate the latitudinal variation of the SAGE extinction ratio obtained during this unperturbed period, we have derived the mean profile for five latitudinal bins on a seasonal basis. The five latitudinal bins

chosen are 75°S-40°S, 40°S-20°S, 20°S-20°N, 20°N-40°N, and 40°N-75°N.
representing high- and mid-latitude Southern Hemispheric, tropic, and
mid- and high-latitude Northern Hemispheric regions, respectively. The
results are presented in Figs. 4a to d corresponding to the mean profiles of the aerosol extinction ratio for the same seasons given in
Figure 3, and are measured from the tropopause. The outstanding
feature of Fig. 4 is the high degree of similarity among the four
seasons when the profiles are normalized to the local mean tropopause.
In addition, they all show a sharp tropical peak at an altitude approximately 10 km to 14 km above the tropopause depending on the season.
The middle and high latitudes show a much smoother profile.

Using the SAGE I aerosol extinction data obtained during the
undisturbed period before mid-1980, a global distribution of aerosol
optical depth can be derived. By employing a single conversion factor
of 1.10×10^3 m^2 kg^{-1} based on a size distribution and composition
obtained by various in situ measurements, it was estimated that the
background stratospheric aerosol loading is on the order of 0.5×10^6
metric tons (Kent and McCormick, 1984).

3. AEROSOLS IN THE POLAR STRATOSPHERE

As mentioned in the Introduction, the SAM II satellite instrument is
specially designed to provide continuous measurements of aerosol
extinction at 1.0 μm wavelength in the polar regions. More than 8
years of data have been produced since its launch in October 1978, and
the instrument is still operating and providing measurements. The
multiyear SAM II data set can be used for studying the long-term variations of the aerosol layer in the polar regions. Figure 5 presents the
8 years of results of weekly-averaged optical depth (calculated from
the tropopause + 2 km upwards through 30 km). Annotated 3 are times of
various volcanic eruptions which occurred during this period.First,
Fig. 5 shows a value of approximately 1.4×10^{-3} for the background
optical depth in early 1979. The large changes in response to volcanic
eruptions are clearly evident. The most outstanding variation is the
rapid increase in the optical depth after the eruption of El Chichon on
April 4, 1982. A conservative maximum value of about 5×10^{-2} occurred
in March 1983 as the volcanic material peaked in the Arctic region. It
is conservative because the heavy loading of aerosol at this time prevented SAM II observations from reaching down to the tropopause plus 2
km. See McCormick and Trepte (1987) for a complete description of this
effect. This value is approximately 40 times that of the background
optical depth of early 1979. Figure 5 further indicates that a significant amount of material still remained in the stratosphere in October
1986. Note that the time for the El Chichon volcanic material to reach
the high latitude regions is different in the two hemispheres. The
material appeared in the Arctic region shortly after the eruption, but
it took almost 7 to 8 months before it was first observed in the
Antarctic region. During the first 4 years of SAM II Antarctic measurements, periods of enhancements of optical depth occurred in each

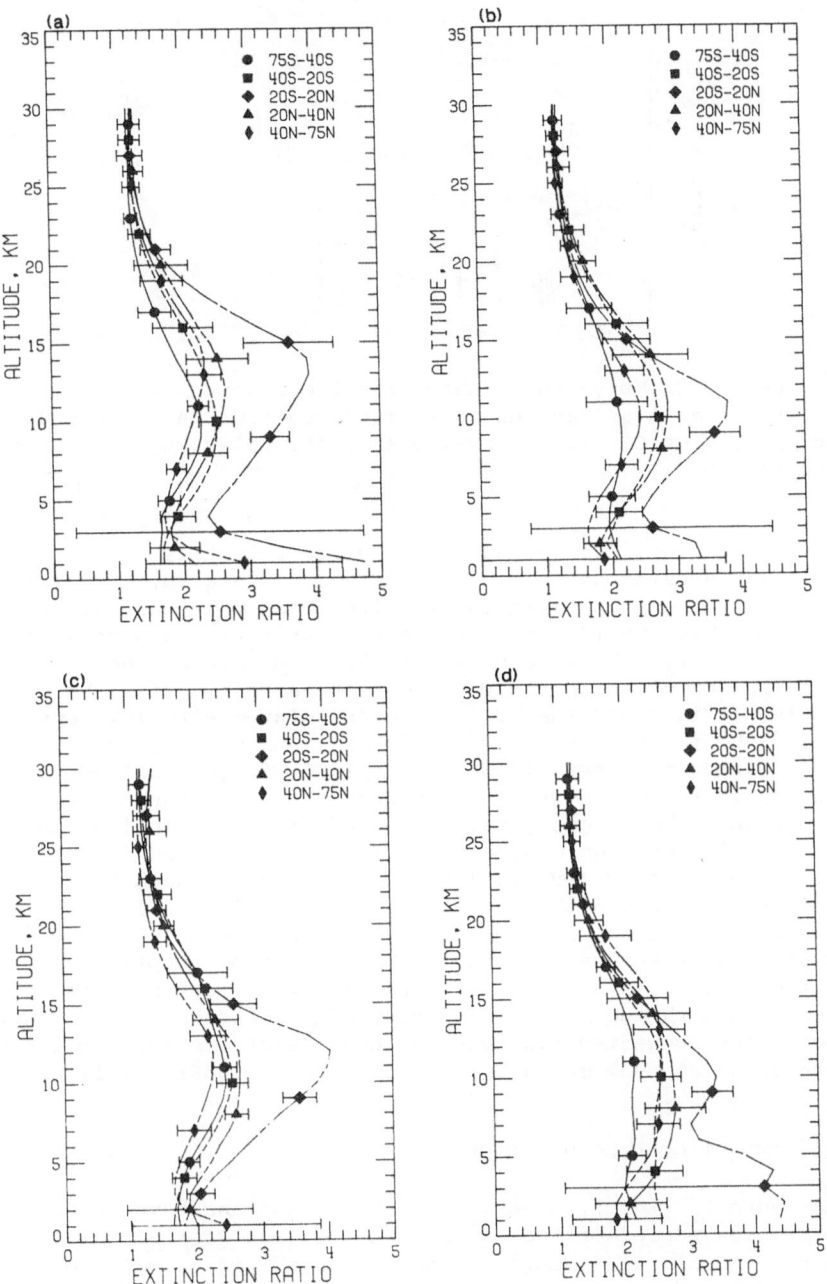

Figure 4. Seasonal mean profiles of aerosol extinction ratio at 1 μm wavelength for tropics, and mid- and high-latitudes in both hemispheres derived from SAGE I measurements; (a) to (d) correspond to the four seasons indicated in Fig. 3.

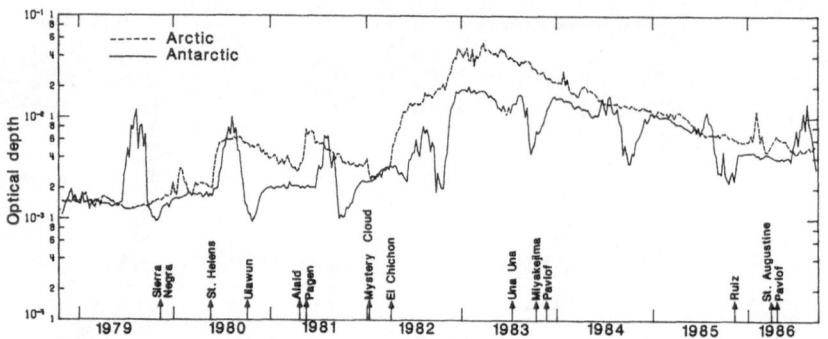

Figure 5. Eight years of aerosol optical depth measurements (weekly-averaged) at 1 μm wavelength (calculated from the tropopause plus 2 km upwards) from SAM II observations in the polar regions (McCormick and Trepte, 1987).

austral winter. These enhancements are primarily due to the formation of polar stratospheric clouds (PSC's). Note also, the minimum ($\approx 10^{-3}$) optical depth value immediately after each PSC seasonal event. It is thought that subsidence within the polar vortex and sedimentation of PSC's is very likely to be the mechanisms causing these minima. By an examination of the corresponding temperatures during the period with PSC's, it is found that PSC's are correlated with low temperatures less than 195K (weekly average), (McCormick and Trepte, 1986).

During the years from 1983 to 1986, the optical depth in the Antarctic region shows recovery from the disturbances of the El Chichon eruption. The recovery rate is somewhat less than that in the Arctic. Note again that the minimum optical depth occurred in October in each of the years from 1983 to 1985 even though enhanced volcanic loading existed. Although the occurrence of PSC's in these years before each October is not as distinct as in the years from 1979 to 1982, due to a disturbed background caused by the El Chichon eruption, the enhanced optical depth as a result of PSC's is still very noticeable. In conjunction with the recent discovery of the Antarctic ozone hole, PSC's have been suggested via heterogeneous reactions to be primary agents in the ozone removal mechanisms (Toon et al., 1986; Hamill et al., 1986.)

4.   EFFECT OF VOLCANIC ERUPTIONS

As shown in Fig. 5, several volcanic eruptions have occurred during the operating period of the SAM II, SAGE I, and SAGE II satellite instruments. Table II summarizes these events. Note that the material injected into the stratosphere by the eruption of El Chichon on April 4, 1982, is at least an order of magnitude greater than those

Table 2. Major Volcanic Eruptions 1978 - 1986

Volcano	Date	Location	Global Aerosol Mass Loading (Metric Tons)	Source
1979 Background	---	---	5.7×10^5	Kent & McCormick(1984)
Sierra Negra	11/13/79	0.8°S-91.2°W	1.6×10^5	"
Mt.St. Helens	05/18/80	46.2°N-122.2°W	5.5×10^5	"
Ulawun	10/07/80	5.0°S-151.3°E	1.8×10^5	"
Alaid	04/27/81	50.8°N-155.5°E	5.0×10^5	"
Pagan	05/15/81	18.1°N-145.8°E		
Mystery Volcano	01/82	Probably Low Latitudes	8.5×10^5	Mroz, et al. (1983)
El Chichon	04/04/82	17.3°N-93.2°W	120×10^5	McCormick (1984)
Ruiz	11/13/85	4.9°N-75.4°W	---	
St.Augustine	03/27/86	59.4°N-153.4°W	---	
Pavlov	04/18/86	55.4°N-161.9°W	---	

from the others listed, and is about 20 times that of the stratospheric background aerosol loading. The April 4, 1982 eruption of El Chichon probably produced the largest Northern Hemispheric increase in stratospheric aerosols in this century. Unfortunately, the SAGE I instrument stopped taking measurements in November 1981 and SAGE II was not launched until October 1984. Thus, only SAM II made measurements continuously, and these were only in the stratospheric Arctic and Antarctic regions (Fig. 5). Because of the magnitude of loading sent into the stratosphere by El Chichon in 1982, special attention was focused on understanding its behavior as well as its effect on the stratosphere. Three special issues of journals have been dedicated to this eruption: the November 1983 issue of Geophysical Research Letters and the April and July 1984 issues of Geofisica Internacional.

Perturbations in the stratospheric aerosols can induce changes in the radiation field in the stratosphere. Temperature increases in the stratosphere were detected (Labitzke, Naujokat, and McCormick, 1983).

Similarly, the enhanced stratospheric aerosols were capable of inter-
fering significantly with remote sensing of stratospheric gaseous
species and sea surface temperature from satellite platforms (Bandeen
and Fraser, 1982).

Since the SAM II, SAGE I, and SAGE II satellite instruments provide global scale high vertical resolution (≈1 km) aerosol extinction
measurements, it is possible to use their measurements to examine the
vertical and horizontal distributions of the injected material after a
volcanic eruption. A recent example is shown in Figures 6a and b where
optical depth before and after the eruption of the Ruiz volcano is presented. The top panel (Fig. 6a) obtained over the period October 6,
1985 to November 20, 1985 depicts a stratosphere which is still
enhanced over background, 3 1/2 years after the El Chichon eruption.
Figure 6b (from December 22, 1985 to January 31, 1986) shows a new
volcanic enhancement caused by the November 1985 eruption of Ruiz in
Columbia, South America (see Table II). This banding of fresh volcanic
material in the tropics is similar to what happened after the eruptions
of Ulawun and El Chichon.

5. DYNAMIC PROCESSES AS MANIFESTED IN AEROSOL CHANGES

The steady-state distribution of stratospheric aerosols is generally
governed by aerosol microphysics and transport processes. Under conditions not involving extreme temperature fluctuations, the measured
stratospheric aerosol extinction can be a useful tracer. It has been
recognized that the mean diabatic circulation in the lower stratosphere
is characterized mainly by a two-cell pattern with an upward branch in
the tropical regions and a downward branch at high latitudes in both
hemispheres. This mean diabatic circulation in the lower stratosphere
is largely responsible for the distribution of atmospheric conservative
tracers in that region on a climatological basis. As mentioned
earlier, the tropical lower stratosphere is evidently a source region
of stratospheric aerosols. Aerosol particles are subsequently transported poleward and downward to higher latitude regions by this mean
diabatic circulation. This view appears to be supported by the poleward and downward gradients of the aerosol data shown in Figs. 3a to
d. It should be mentioned that this interpretation of the distribution
of stratospheric aerosols requires a sink region at higher latitudes.
The possible mechanism for removing stratospheric aerosols is still not
clear (Turco et al., 1982) but the formation of PSC's (McCormick et
al., 1982) and troposphere-stratosphere exchange processes (Browell et
al., 1987) are probably important to the budgets of stratospheric aerosols. The effect of mean diabatic circulation is also evident in the
transport of volcanic material to the polar regions after they have
been injected into the stratosphere (Fig. 5).

As mentioned in the Introduction, the SAM II satellite instrument
provides continuous observations in the polar regions. Detailed daily
variations of stratospheric aerosols can be used to study the polar
vortex isolation and such events as sudden warmings. This is also true
for SAGE I and SAGE II during the period when their sampling locations

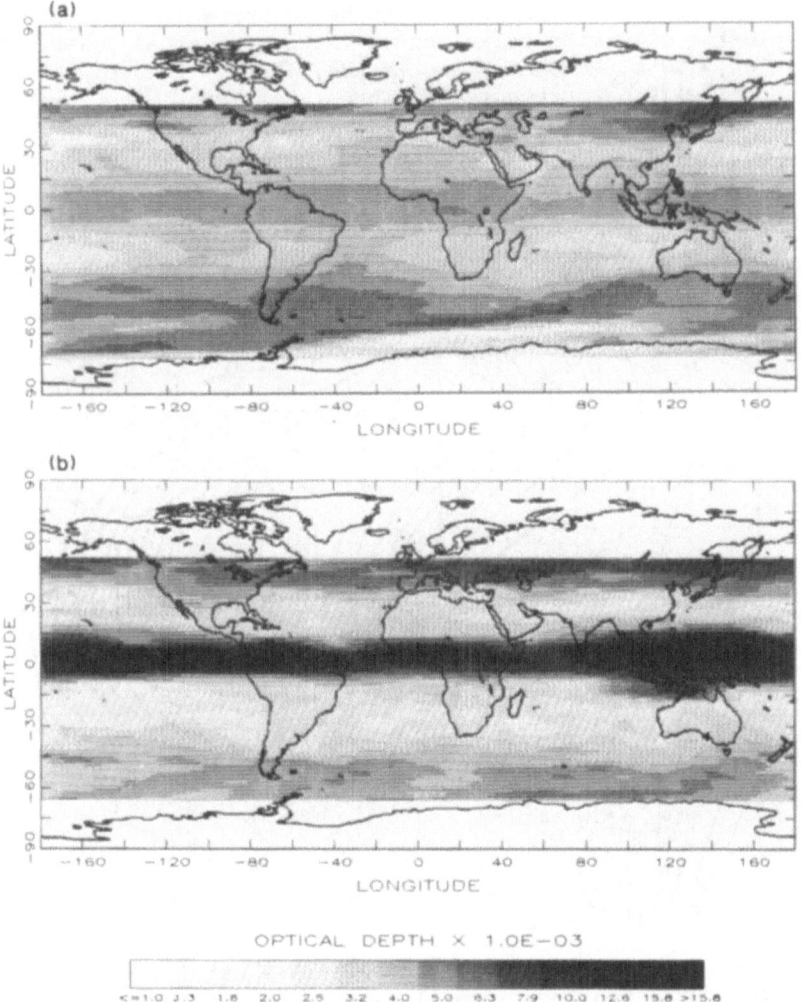

Figure 6. Global distribution of aerosol optical depth at 1 μm
wavelength (measured from tropopause plus 2 km upwards) derived from
SAGE I observations. (a) Top: From October 6, 1985 to November 20,
1985. (b) Bottom: From December 22, 1985 to January 31, 1986.

approach an extreme latitude, usually in winter. Using the SAM II
aerosol extinction and associated meteorological data during the
January-February 1979 stratospheric warming, Wang and McCormick (1985a)
showed that distinct time changes in the zonal-mean aerosol extinction
ratio occurred. They further showed that horizontal eddy transport due
to planetary waves may have played a significant role in determining
the distribution of the zonal mean aerosol extinction ratio.

The behavior of the winter stratospheric aerosols is also related
to the evolution of winter polar vortex (McCormick et al., 1984; Wang
and McCormick, 1985b; McCormick and Larsen, 1986). Figure 7 shows the
locations of SAM II measurements on February 1, 1983. The inset at the

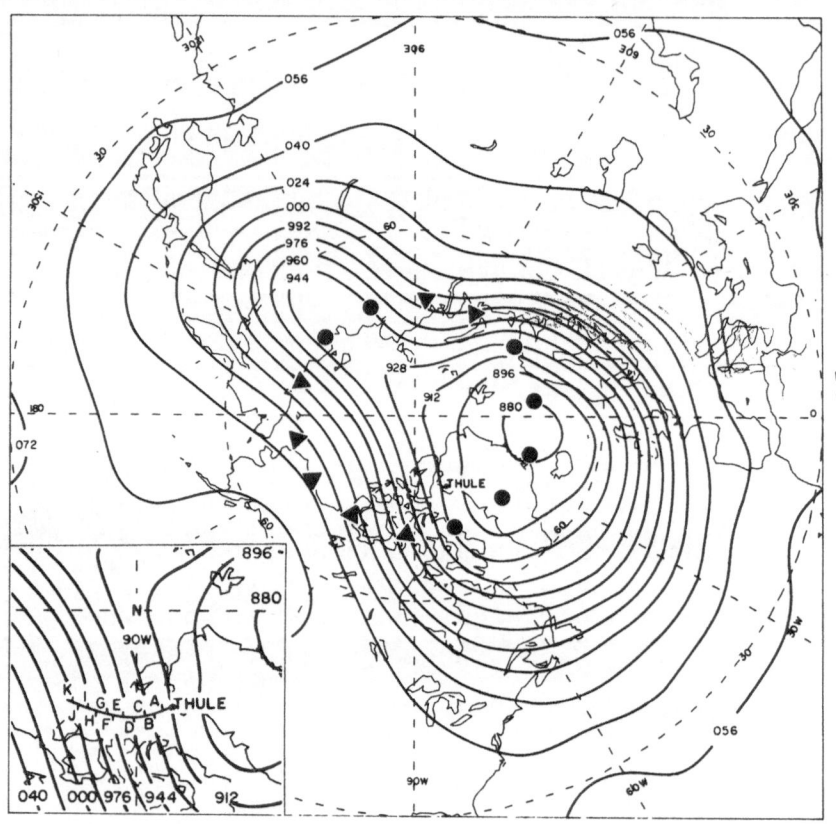

Figure 7. The 50-mb upper atmospheric analysis map for February 1,
1983, at 1200 GMT. The locations of SAM II observations are shown by
the triangles and circles. The triangles are for optical depths >10^{-3}
and the circles are for optical depths <10^{-3}. The insert shows locations of airborne lidar measurements taken on the same day (McCormick
et al., 1983).

lower left shows the track of an airborne lidar flight from Thule,
Greenland. Both SAM II and the airborne lidar data showed a substantially smaller value of aerosol extinction above ≈17 km altitude exists
inside the vortex. The airborne lidar data also indicate sharp gradients of extinction over very short distances across the vortex
(McCormick et al., 1983). Figure 8 shows data from both SAGE I and
SAM II. The variations of aerosol optical depth with longitude reveal
a similar relationship with respect to the location of the vortex even

after the vortex breaks into two parts. Note that the SAM II measurements enter one of the low pressure regions whereas SAGE I makes measurements in both. The lower column values above 50 mb clearly show
up. This relationship is also evident in the SAGE I O_3 and NO_2 data
shown in Fig. 9.

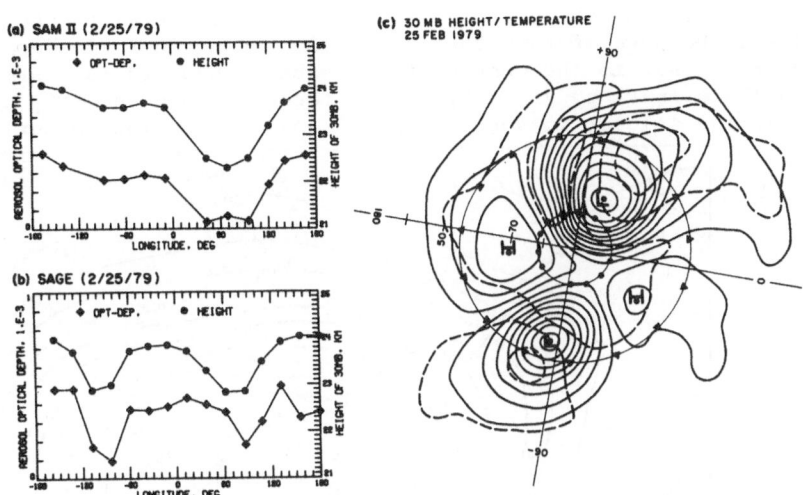

Figure 8. Longitudinal distributions of aerosol optical depth at 1 μm
wavelength above 50 mbar and the NWS geopotential height (km) of the
30-mbar pressure level at (a) SAM II and (b) SAGE measurement locations, respectively, on February 25, 1979. In (c), the solid circles
and triangles are the sampling locations of the SAM II and SAGE I
instruments, respectively (Wang and McCormick, 1985b).

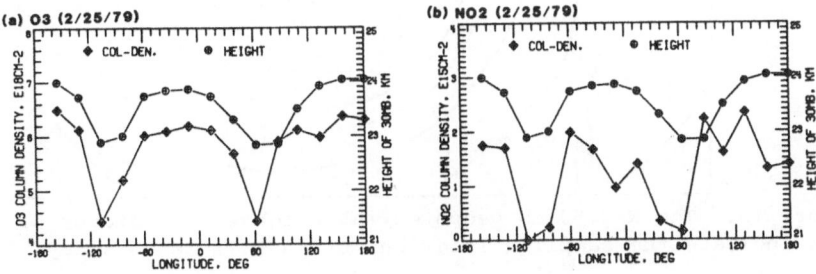

Figure 9. SAGE-derived column abundances (above 50 mbar) at the
measurement locations shown in Fig. 8c for (a) O_3 and (b) NO_2 (Wang and
McCormick, 1985b).

6. SAGE II O_3 AND NO_2 OBSERVATIONS IN THE ANTARCTIC

In this section, we present recent results of the SAGE II observations
in the Antarctic region. It is understood that the SAGE II data are

still in its preliminary stage and undergoing validation. Thus, caution must be exercised concerning the results shown here. The depletion of total ozone at Halley Bay during the Antarctic spring was first
reported by Farman et al. (1985). Figure 10 shows the SAGE II sampling
locations (dots) on October 5, 1986, along with the 50 mb pressure surface for that day. The vortex reveals a mild wave- number 3 pattern
with the vortex center shifting to a lower latitude along the Greenwich
meridian. Because of this vortex position, some SAGE II sampling locations are closer to the center of the vortex than others. Similar to
the analysis in Figs. 7 and 8, the differences in O_3 and NO_2 distributions inside (closer to vortex center) and outside the vortex, as shown

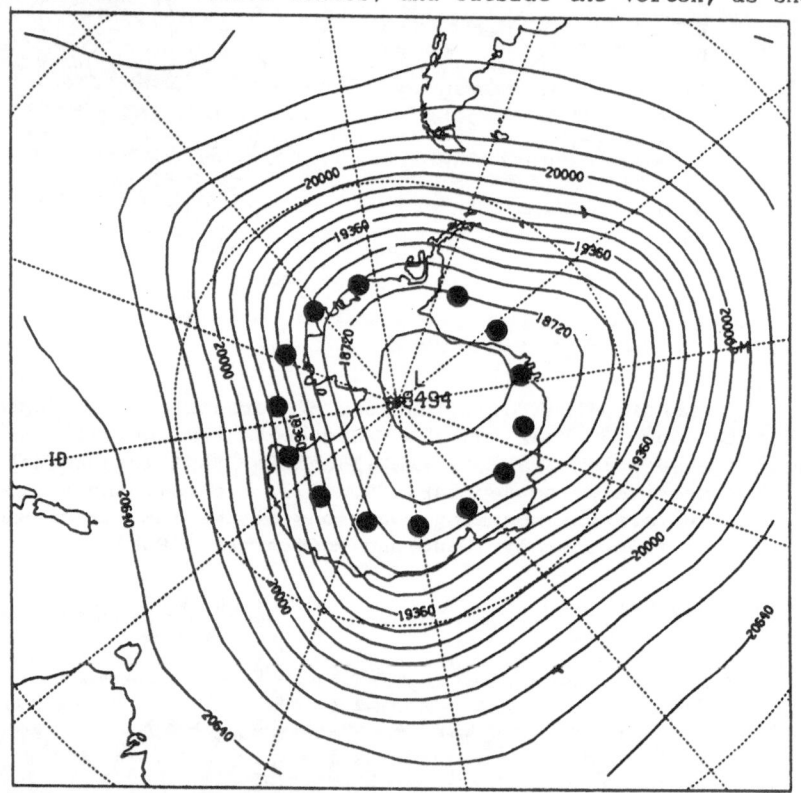

Figure 10. The NWS 50 mb geopotential heights for October 5, 1986. The
dots indicate the sampling locations of SAGE II on this day.

by the SAGE II observations are examined. Figure 11 plots two profiles
of O_3 number density with one inside the vortex (dashed line) and the
other outside (solid line). The inside profile is rather uniformly
about a factor of 2 less than that outside in the altitude range from
about 15 to 60 km. Also, in the altitude region between approximately
10 and 20 km, the profile inside the vortex shows a layer with less O_3
than that outside, with the largest difference located at an altitude
of about 20 km.

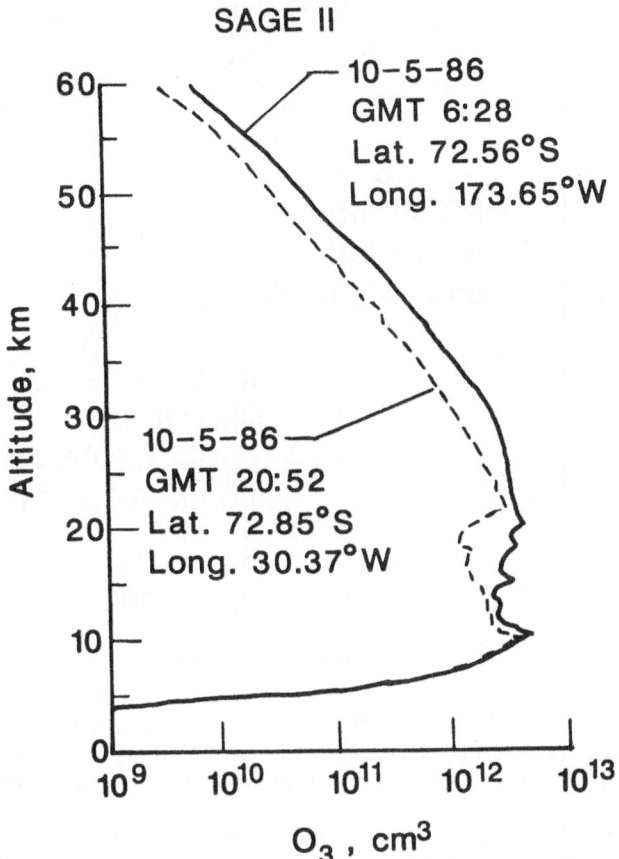

Figure 11. Vertical profiles of SAGE II-derived O_3 number density (cm^{-3}). The solid line indicates outside the vortex and the dashed line indicates inside (see Fig. 10).

The corresponding SAGE II NO_2 measurements are displayed in Fig. 12. Like the SAGE II O_3, the NO_2 profile inside the vortex shows much less NO_2 concentration than that outside.

7. CONCLUSIONS

Analyses based on stratospheric aerosol observations from the SAM II, SAGE I, and SAGE II satellite instruments reveal many interesting features. It is shown that the tropical lower stratosphere is probably a source region for the background stratospheric aerosols. Aerosols are subsequently transported poleward and downward to high latitude regions by the mean diabatic circulation. The global stratospheric

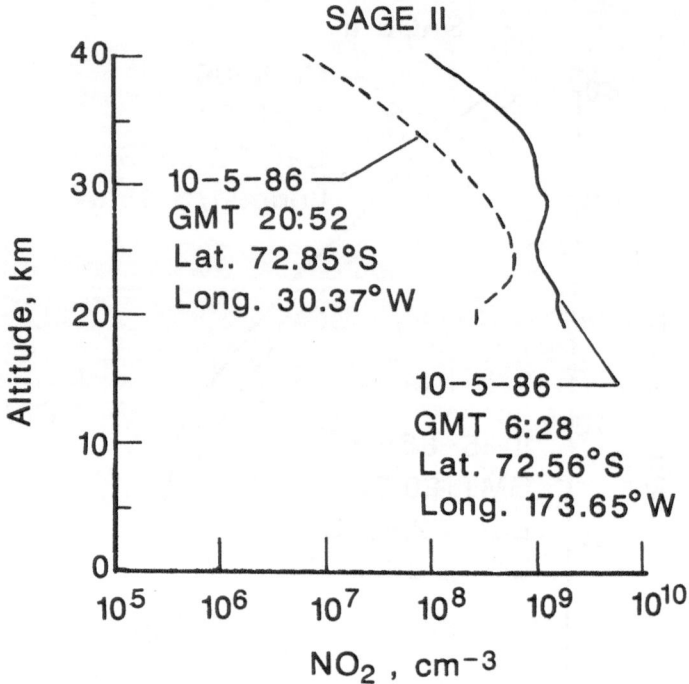

Figure 12. Vertical profiles of SAGE II-derived NO_2 number density (cm^{-3}). The solid line indicates outside the vortex and the dashed line indicates inside (see Fig. 10 for measurement locations).

background aerosol loading is about 0.5 MTonne (occurred in 1979). The long-term trend of the stratospheric aerosols reveals volcanic perturbations. An estimated 12 MTonne of volcanic material was injected into the stratosphere by the April 1982 eruption of El Chichon. In the polar region, large variability in aerosol extinction is observed. The Antarctic stratospheric aerosols especially exhibit a seasonal behavior with the wintertime recurrence of PSC's and a relative yearly minimum in October. It is shown that polar variability, in general, depends on the location of the polar vortex. The polar night jet acts as a barrier creating an isolated air mass in the polar regions above about 14 km. In addition to aerosols, the distributions of O_3 and NO_2 concentrations at high latitudes observed by the SAGE I and SAGE II satellite instruments also indicate the dependence on the location of the polar vortex. Very low values of NO_2 and O_3 were found within the polar vortex as opposed to outside the vortex on a given day.

ACKNOWLEDGMENTS

The authors gratefully acknowledge the members of the SAM II, SAGE I, and SAGE II Science Teams, and the data reduction group. We are also grateful to C. R. Trepte, J. C. Larsen, G. Kent, and S. Schaffner for helpful assistance in preparation of the manuscript. One of us (P.-H. Wang) is supported by NASA Contract NAS1-18252.

REFERENCES

Bandeen, W. R., and R. S. Fraser, 1982: 'Radiative effects of the El Chichon volcanic eruption: Preliminary results concerning remote sensing.' NASA Technical Memorandum 84959.

Browell, E. V., E. F. Danielsen, S. Ismail, G. L. Gregory, and S. M. Beck, 1987: 'Tropopause fold structure determined from airborne lidar and in situ measurements.' J. Geophys. Res., 92, 2112-2130.

Farman, J. C., B. G. Gardiner, and J. D. Shanklin, 1985: 'Large losses of total ozone in Antarctica reveal seasonal CLOχ/NOχ interaction.' Nature, 315, 207-210.

Hamill, P., O. B. Toon, and R. P. Turco, 1986: 'Characteristics of polar stratospheric clouds during the formation of the Antarctic ozone hole.' Geophys. Res. Lett., 13, 1888-1291.

Junge, C. E., C. W. Chagnon, and J. E. Manson, 1961: 'Stratospheric aerosols.' J. Meteorol., 18, 81.

Kent, G. S., and M. P. McCormick, 1984: 'SAGE and SAM II measurements of global stratospheric aerosol optical depth and mass loading.' J. Geophys. Res., 89, 5303.

Labitzke, K., B. Naujokat, and M. P. McCormick, 1983: 'Temperature effects on the stratosphere of the April 4, 1982 eruption of El Chichon, Mexico.' Geophys. Res. Lett., 10, 24-26.

McCormick, M. P., P. Hamill, T. J. Pepin, W. P. Chu, T. J. Swissler, and L. R. McMaster, 1979: Satellite studies of the stratospheric aerosol.' Bull. Amer. Meteor. Soc., 60, 1038.

McCormick, M. P., H. M. Steele, P. Hamill, W. P. Chu, and T. J. Swissler, 1982: 'Polar stratospheric cloud sightings by SAM II.' J. Atmos. Sci., 39, 1387-1397.

McCormick, M. P., C. R. Trepte, and G. S. Kent, 1983: 'Spatial changes in the stratospheric aerosol associated with the north polar vortex.' Geophys. Res. Lett., 10, 941-944.

McCormick, M. P., T. J. Swissler, W. H. Fuller, W. H. Hunt, and M. T. Osborn, 1984: 'Airborne and ground-based lidar measurements of the El Chichon stratospheric aerosol from 90°N to 56°S.' Geofis. Int., **23.2**, 187-221.

McCormick, M. P., and C. R. Trepte, 1986: 'SAM II measurements of Antarctic PSC's and aerosols.' Geophys. Res. Lett., **13**, 1276-1279.

McCormick, M. P., and J. C. Larsen, 1986: 'Antarctic springtime measurements of ozone, nitrogen dioxide, and aerosol extinction by SAM II, SAGE, and SAGE II.' Geophys. Res. Lett., **13**, 1280-1283.

McCormick, M. P., and C. R. Trepte, 1987: 'Polar stratospheric optical depth observed between 1978 and 1985.' J. Geophys. Research., 92, 4297-4306.

McMaster, L. R., 1986: 'Stratospheric Aerosol and Gas Experiment (SAGE II).' Sixth Conf. on Atmospheric Radiation, Am. Meteorol. Soc., Williamsburg, VA, May 13-16.

Mroz, E. J., A. S. Mason, and W. A. Sedlicek, 1983: 'Stratospheric sulfate from El Chichon and the mystery volcano.' Geophys. Res. Lett., **10**, 873-876.

Toon, O. B., Patrick Hamill, R. P. Turco, and J. Pinto, 1986: 'Condensation of HNO_3 in the winter polar stratosphere.' Geophys. Res. Lett., **13**, 1284-1287.

Turco, R. P., R. C. Whitten, and O. B. Toon, 1982: 'Stratospheric aerosols: Observation and theory.' Rev. Geophys. and Space Phys., **20**, 233.

Wang, P.-H., and M. P. McCormick, 1985a: 'Behavior of zonal mean aerosol extinction ratio and its relationship with zonal mean temperature during the winter 1978-1979 stratospheric warming.' J. Geophys. Res., **90**, 2360-2364.

Wang, P.-H., and M. P. McCormick, 1985b: 'Variations in stratospheric aerosol optical depth during northern warmings.' J. Geophys. Res., **90**, 10,597-10,606.

Yue, G. K., M. P. McCormick, and W. P. Chu, 1984: 'A comparative study of aerosol extinction measurements made by the SAM II and SAGE satellite experiments.' J. Geophys. Res., **89**, 5321-5327.

EVIDENCE FOR PLANETARY WAVE BREAKING FROM SATELLITE DATA:
THE RELATIVE ROLES OF DIABATIC EFFECTS AND IRREVERSIBLE MIXING

Neal Butchart
Meteorological Office
Bracknell, Berkshire
United Kingdom

ABSTRACT.   Isentropic maps of satellite observed ozone and water vapour
concentrations are shown to corroborate the evidence inferred from
isentropic maps of Ertel potential vorticity for planetary wave
breaking in the middle stratosphere. Wave breaking generally connotes
irreversible isentropic mixing of potential vorticity and constituents
and a diagnostic formalism is developed to quantify the extent of the
mixing. In addition the diabatic sources and sinks of potential
vorticity and the diabatic cross-isentropic transport of both potential
vorticity and tracers are estimated from realistic local radiative
heating and cooling rates calculated from observed temperature and
ozone data. The relative roles of irreversible mixing and diabatic
effects in determining the evolutions of the isentropic distributions
of potential vorticity and constituents are assessed.

1.   INTRODUCTION

A useful choice of coordinates for studying transport processes in the
middle atmosphere utilizes potential temperature, θ, as a measure of
vertical position. The rationale underlying this choice is presented,
for example, in Mahlman et al.'s (1984) comprehensive review of tracer
transport (see also Mahlman, 1985; Tung, 1982, 1984). Briefly, the
so-called isentropic coordinate approach has the practical merits of
Eulerian calculations yet, because the isentropic surfaces become
material surfaces for adiabatic motion, θ can be considered as a quasi-
Lagrangian coordinate. A conceptual advantage is that a number of
stratospheric phenomena which appear obscure and complicated from a
traditional viewpoint become simpler in the isentropic perspective
(again, see Mahlman et al., 1984).
     Recently Butchart and Remsberg (1986) have extended the
Lagrangian-oriented nature of the isentropic perspective with a
diagnostic technique based on contours of either constant Ertel
potential vorticity Q or species (ozone or water vapour) mixing ratio χ
on an isentropic surface. In the absence of sources and sinks of the
tracers the contours become material lines for conservative motion and
the horizontal projection of the areas they contain will remain

G. Visconti and R. Garcia (eds.), Transport Processes in the Middle Atmosphere, 121–136.

constant provided the flow is <u>also</u> non-divergent on the isentropic
surface. Systematic changes in the "areas" are possible when one or
more of the above restrictions is invalid, however, and the full
potential of this new diagnostic was only appreciated following the
identification of "planetary wave breaking" in the stratosphere by
McIntyre and Palmer (1983, 1984). This phenomenon involves the rapid,
irreversible deformation of otherwise wavy potential vorticity contours
on an isentropic surface and is accompanied by an irreversible cascade,
or mixing to unresolved scales. Because the contours become so
deformed that they are no longer adequately resolved, the horizontal
projection of the areas they contain is <u>observed</u> to change
systematically and "permanently." In particular, when the isentropic
distribution of Q (IPV distribution) is expressed as a monotonic
function of its associated areas the wave breaking or, more generally,
the irreversible isentropic mixing will be characterized by a smearing
out of the gradient of this function in the mixing region compensated
by a strengthening of the gradient at the edges of the region (see Fig.
1 of Butchart and Remsberg, 1986). Other quasi-conservative tracers
will behave similarly when they have a gradient in a mixing region.

The evolution of the so-called area diagnostic for potential
vorticity, ozone, water vapour and nitric acid at the 850 K isentropic
level in the Northern Hemisphere stratosphere was examined by Butchart
and Remsberg (1986) for the period 25 October 1978 through 2 April 1979
using LIMS (Limb Infra-red Monitor of the Stratosphere) data (Gille and
Russell, 1984). Their results show that following a weak "Canadian
warming" in early December there is a core of high Q forming a "polar
main vortex" surrounded by a region of relatively weak gradient or a
"surf zone" (cf. McIntyre and Palmer, 1983, 1984). From the time of its
formation until the end of February the main vortex appeared to be in
the process of continuous erosion while the size of the surf zone
gradually increased. Similar structure and evolution could be seen in
the ozone and water vapour data, though for ozone only the boundary
between the two distinct regions was clearly visible. Butchart and
Remsberg concluded that this picture was consistent with McIntyre and
Palmer's (1983, 1984) hypothesis that as winter progresses the main
vortex would be eroded by the action of breaking planetary-scale waves
in the surf zone and is also in agreement with the results of Schoeberl
and Smith (1986) who investigated integrated quasi-geostrophic
potential enstrophy budgets calculated from LIMS data for the same
winter. Schoeberl and Smith also noted that radiative processes could
contribute significantly to the enstrophy budget.

The contribution of diabatic heating to the evolution of the area
diagnostic was only briefly examined by Butchart and Remsberg (1986).
In a provisional calculation they found that diabatic changes to the
IPV distribution in January and February were unlikely to account for
the observed changes in the sizes of the main vortex and surf zone and
moreover they argued that it was more plausible to interpret their
results as being consistent with the view that diabatic processes would
act to restore the uniform gradient to the IPV distribution. No attempt
was made by Butchart and Remsberg to calculate the diabatic
cross-isentropic transport of the constituents. The purpose of this

paper is to provide a more detailed analysis of the role of diabatic
heating at the 850 K level in middle and high latitudes during January
and February, the most active period of the 1979 winter. While the main
aim is to assess the influence of diabatic processes on the evolution
of the area diagnostic (Section 4) selected isentropic maps of the
effective diabatic sources and sinks of Q and ozone on the 850 K
surface are also presented (Section 3). In addition Section 3 provides
the first direct comparison between the LIMS derived isentropic maps of
ozone and water vapour and the IPV maps.

2. DATA ANALYSIS AND DIAGNOSTIC FORMALISM

The observations used in this study were derived from the results of
the LIMS experiment. Further discussion of the data and references are
given in Butchart and Remsberg (1986) as is the method of calculating
the isentropic distributions of Q, ozone and water vapour. In the
extratropical middle stratosphere these tracers are quasi-conservative,
satisfying equations of the form

$$\frac{D\chi}{Dt} = G. \tag{1}$$

It can then be shown (Butchart and Remsberg, 1986) that if $A(t)$ is
the horizontal projection of the area enclosed by a contour Γ of
constant $\hat{\chi}$ on an isentropic surface

$$\frac{dA(t)}{dt} = \oint_{\hat{\Gamma}} (G - \dot{\theta}\frac{\partial\chi}{\partial\theta}) \frac{ds}{|\nabla_\theta\hat{\chi}|} + \int_{A(t)} \nabla_\theta \cdot \hat{\mathbf{v}} dA - \oint_{\hat{\Gamma}} \mathbf{v} \cdot \nabla_\theta \chi' \frac{ds}{|\nabla_\theta\hat{\chi}|} , \tag{2}$$

where ds is an element of arc length along $\hat{\Gamma}$ and dA the horizontal
projection of an area element. In (2) the symbol (ˆ) denotes observed
values while the primes denote the deviations of the observed values
from the true values (e.g., $\chi = \hat{\chi} + \chi'$). The remaining notation is
standard. The physical interpretation of the individual terms on the
right-hand side of (2) is discussed by Butchart and Remsberg (1986) and
for completeness a precis is included here. The last term represents a
flux of χ across the contour Γ at the unresolved scales and will
include the irreversible mixing processes described in the
Introduction. The second term takes account of any divergence of the
large-scale flow and will be assumed to be small. Finally, the first
term represents the combined contributions from the in-situ sources and
sinks of χ and the diabatic cross-isentropic transport of χ to the
"budget" of $A(t)$.
   One attribute of the LIMS experiment is the simultaneous
measurements of ozone concentration and temperature at relatively high
vertical resolution which can be used to calculate realistic local
diabatic heating rates consistent with the observations. A diagnostic
framework based on Equation (2) is particularly suitable for assessing
the relationship between the diabatic heating and observed
distributions of potential vorticity and constituents. Because the
approach has conceptual roots in the generalized Lagrangian mean theory

(Andrews and McIntyre, 1978) many temporary and reversible dynamical
processes which tend to mask systematic changes in the tracer
distributions are automatically eliminated from consideration (McIntyre
and Palmer, 1983, 1984; Butchart and Remsberg, 1986). In this research
diabatic heating and cooling rates have been calculated using an
extended version of the modern radiation scheme used by Austin (1986).
The scheme has a Curtis matrix calculation for the 15 μm CO_2 band, the
9.6 μm O_3 band is parameterized as in Harwood and Pyle (1975) and the
solar heating is calculated using a random band model. There are small
additional terms taken from London (1980) to account for the IR bands
in H_2O and NO_2. These heating rates are then used to calculate the
diabatic contributions to the right-hand side of Equation (2) for Q,
ozone and water vapour at the 850 K level. For the ozone and water
vapour this just involves the diabatic cross-isentropic transport
$-\dot{\theta}\partial\chi/\partial\theta$ whereas for potential vorticity it involves the net effect of
the in-situ diabatic sources and sinks of Q and the diabatic
cross-isentropic transport, that is $Q\partial\dot{\theta}/\partial\theta-\dot{\theta}\partial Q/\partial\theta$ (Haynes and McIntyre,
1987). The only other nonconservative process that is likely to be
important in the budgets of A(t) considered in this study is ozone
photochemistry. Unfortunately computational constraints have so far
prevented incorporating photochemical effects into the area diagnostic
approach, though some insight into the relative importance of
photochemistry can be found in Austin's (1987) study of air parcel
trajectories.

3.   ISENTROPIC MAPS

3.1. Ertel potential vorticity

The synoptic maps of Ertel's potential vorticity Q on the 850 K
isentropic surface, for the extratropical Northern Hemisphere during
January-February 1979 have been extensively analysed by a number of
authors. McIntyre and Palmer (1983, 1984) used data retrieved from the
Tiros-N Stratospheric Sounding Unit (SSU), Dunkerton and Delisi (1986)
used LIMS data and Grose (1984) confirmed that similar results could be
expected from the use of either dataset. The four LIMS-derived maps
presented in Fig. 1 (solid contours) show some of the significant
features emphasized by these authors. Figure 1a for 27 January shows a
core of high Q or "main vortex", with strong gradients at its edge,
displaced from the pole. Extending clockwise from the main vortex into
a broad region of much weaker gradients is a tongue of relatively high
Q values which eventually breaks off from the main vortex on 30 January
(Fig. 1b). According to McIntyre and Palmer (1983, 1984) this behaviour
would be expected in "Rossby wave breaking" situations when high Q air
would be drawn out from the circumpolar vortex and, to an extent
dependent on the detailed dynamics, irreversibly mixed into the
surrounding region to form a mixed or "surf zone." As a consequence of
any wave breaking there is likely to be a shrinking or "preconditioning" of the main vortex due to erosion and entrainment of potential
vorticity in the surf zone. A significant aspect of the preconditioning

is that it reduces the size of the region on which Rossby waves can readily propagate thereby focussing all available wave activity into a smaller region more susceptible to breakdown and the occurrence of a major sudden warming. Figure 1c for 15 February certainly shows that six days prior to the February 1979 major warming there was a very small main vortex and it has already been firmly established that the non-climatological zonal mean flow associated with this IPV

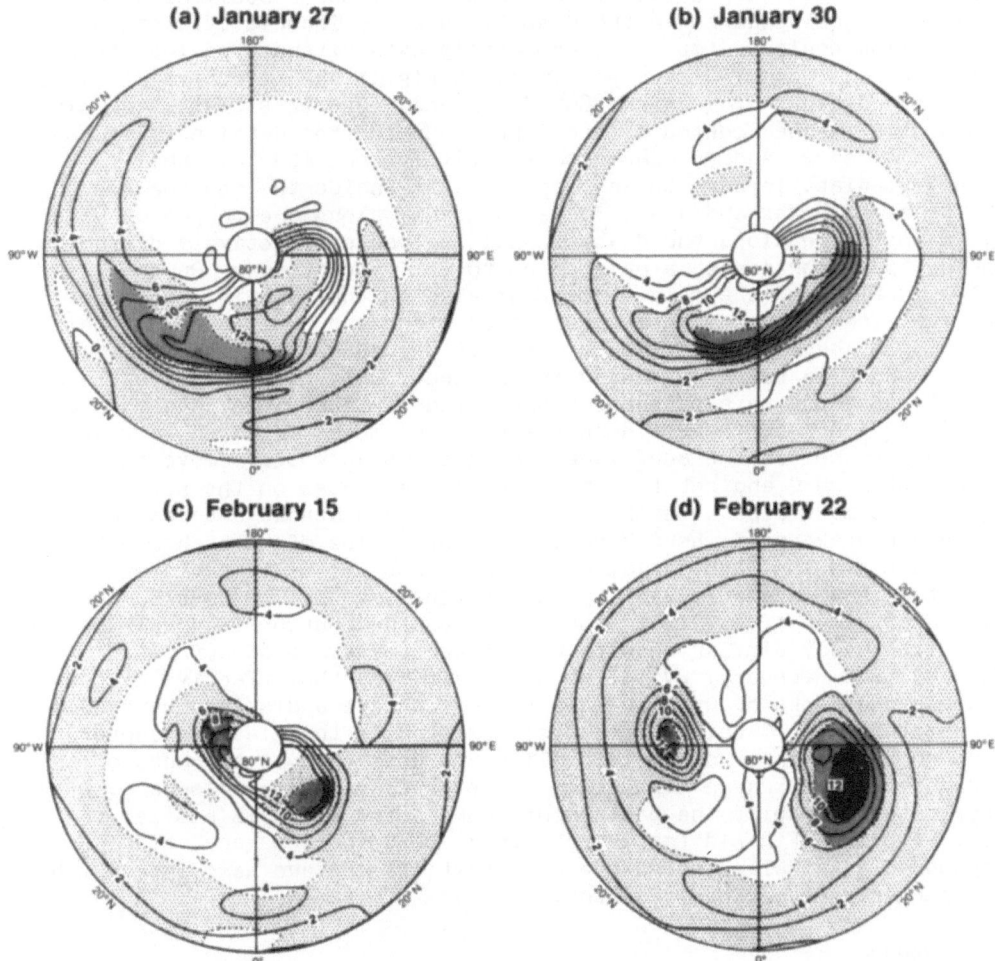

Fig. 1. The 850 K isentropic maps for (a) January 27, (b) January 30, (c) February 15 and (d) February 22, 1979. Solid contours: Ertel potential vorticity Q expressed in 'PV units' defined as g H_0 p_0^{-1} 10^{-4} K m^{-1} s^{-1} where g is the gravitational acceleration, p_0 a standard pressure (1000 mb) and H_0 a standard pressure scale height (7 km). Dashed contours: effective diabatic source of Q on the isentropic surface derived from local heating rates calculated from LIMS ozone and temperature data. Contour interval is 0.1 PV units per day. Negative values (sinks) are shaded with the heavier shading denoting the stronger sinks.

distribution was a crucial ingredient for the occurrence of that major
warming (Butchart et al., 1982; Palmer, 1981). The warming itself was
accompanied by the main vortex splitting into two with the larger of
the vortices moving into the Eastern Hemisphere (see Fig. 1d for 22
February).

     Although both Dunkerton and Delisi (1986) and McIntyre and Palmer
(1983, 1984) stressed the importance of dynamical processes in
determining the evolution of the IPV distribution during January and
February 1979, Clough et al. (1985) noted from their study of December
1981 that local changes in the diabatic heating induced by disturbances
in the flow could be important especially over extended periods of
time. For the four IPV maps presented in Fig. 1 the effective diabatic
sources and sinks of Q on the 850 K surface, $(Q\dot{\partial\theta}/\partial\theta-\dot{\theta}\partial Q/\partial\theta)$, are also
shown using dashed contours (see figure caption for details). Figures
1a and 1b for 27 and 30 January respectively indicate that the
strongest diabatic effects are located just inside the equatorward edge
of the main vortex and also, for 27 January, along the tongue of high
Q. In these locations the diabatic heating would effectively destroy Q
on the 850 K surface at a rate of up to 0.2 PV units/day. On the
poleward flank of the main vortex and also in the Aleutian region of
the surf zone where diabatic cooling is strongest Q is generally being
created on the 850 K surface by diabatic processes though at a rate
less than it is being destroyed in the previously mentioned locations.
Broadly speaking this pattern of sources and sinks would tend to reduce
the Q gradients on the southern flank of the main vortex while weakly
increasing them in the surf zone. In addition it would remove the
tongue of high Q and act to resymmetrize the vortex on the pole.
However it must also be emphasized that these diabatic sources and
sinks are weak and nowhere have an absolute value of more than 0.2 PV
units/day. As some indication of the strength of the diabatic effects
note that a source of 0.2 PV units/day would require 10 days to change
the local value of Q by the same amount as the contour interval used in
Fig. 1. Broadly similar results were found for all 8 days from 26
January to 2 February. By 15 February (Fig. 1c) the effective diabatic
sources and sinks of Q on the 850 K surface have a distinct wave number
two pattern with Q being effectively created in the Aleutian sector
through the stronger diabatic cooling and destroyed at the ends of the
elongated vortex where high Q air has been displaced equatorward. After
the vortex has split the pattern of sources and sinks is similar to
that of Q (see Fig. 1d for 22 February). Both vortices are being
reduced diabatically though the strongest effects are associated with
the larger more southerly vortex.

3.2. Ozone

Synoptic maps of LIMS derived ozone concentrations for the 850 K
isentropic surface are presented in Fig. 2 for 27 and 30 January and 15
and 22 February. Comparing these maps with the corresponding IPV maps
shown in Fig. 1 confirms that the negative correlation found by Leovy
et al. (1985) between 10 mb isobaric maps of ozone mixing ratio and
850 K IPV maps also extends to the 850 K isentropic distributions of

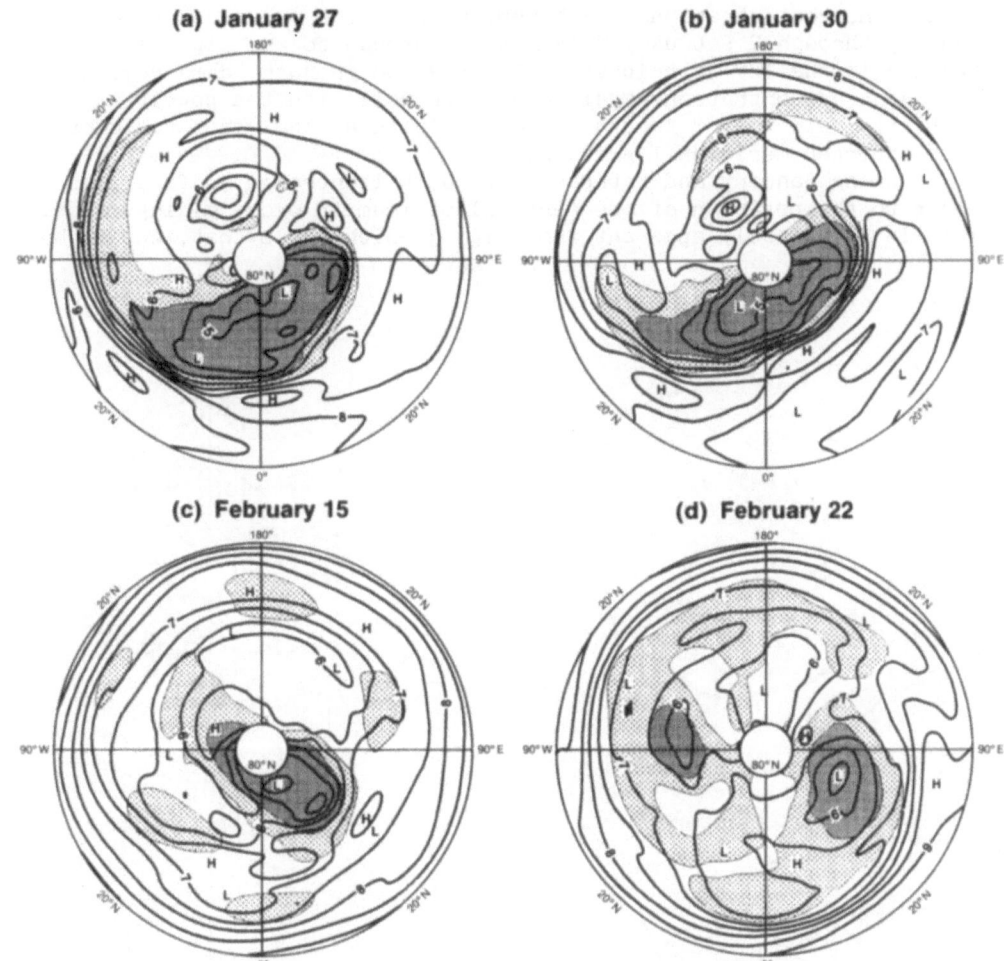

(a) January 27

(b) January 30

(c) February 15

(d) February 22

Fig. 2. Solid contours: LIMS ozone concentration (ppmv) on the 850 K isentropic surface for (a) January 27, (b) January 30, (c) February 15 and (d) February 22, 1979. Contour interval is 0.5. Regions where Q on the 850 K isentropic surface has a value greater than 4 PV units are lightly shaded and greater than 8 PV units heavily shaded (cf. Fig. 1).

both tracers. This is further illustrated by shading lightly and heavily those regions in Fig. 2 where Q at 850 K would be greater than 4 and 8 PV units respectively. On both 27 and 30 January the division of the isentropic surface into surf zone and main vortex regions is delineated in the ozone maps by a band of stronger gradients which, on 27 January, also marks the position of the equatorward edge of the high Q tongue. On the other hand, there is very little in the ozone maps to distinguish the poleward edge of this tongue. Similarly the pool of

high ozone concentrations which remains over the Aleutians from 26
January through 2 February does not correspond to any significant
feature in the IPV distribution. For the two February days regions of
low ozone concentrations again correlate well with the potential
vorticity main vortex though at lower latitudes the ozone field has a
much stronger meridional gradient.

During January and February the 850 K isentropic surface lies very
close to the position of the vertical maximum in ozone mixing ratio.
This is immediately apparent from Fig. 3 which shows that, while there

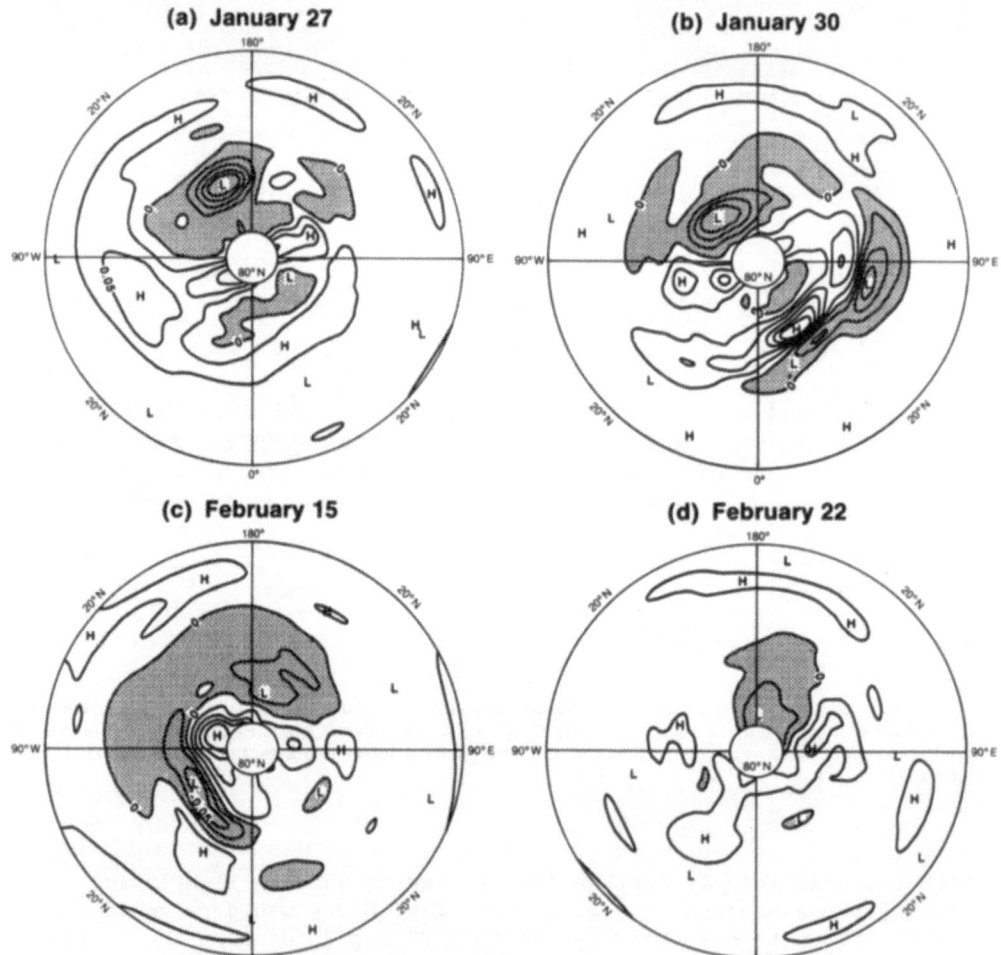

Fig. 3. Diabatic cross-isentropic transport of ozone $-\dot{\theta}\partial\chi/\partial\theta$ (ppmv/day) at the 850 K isentropic level and for
(a) January 27, (b) January 30, (c) February 15 and (d) February 22, 1979. Derived from local heating and cooling
rates calculated from LIMS ozone and temperature data. Contour interval is 0.025 with negative values shaded.

is diabatic cooling everywhere north of 20°N at 850 K, both higher and
lower values of ozone mixing ratio are being transported across the
isentropic surface. The downward transport of lower values of ozone
mixing ratio observed over the Aleutians on 27 and 30 January (see
Figs. 3a and 3b) can perhaps explain the eventual disappearance of the
pool of high ozone concentration in that region (cf. Figs. 2a and 2b)
but then leaves the question of how the pool was formed in the first
instance, that is, assuming it is real which seems likely given its
persistence (some care is required here since the Kalman filter will
also introduce some persistence). The other major feature to diabatic

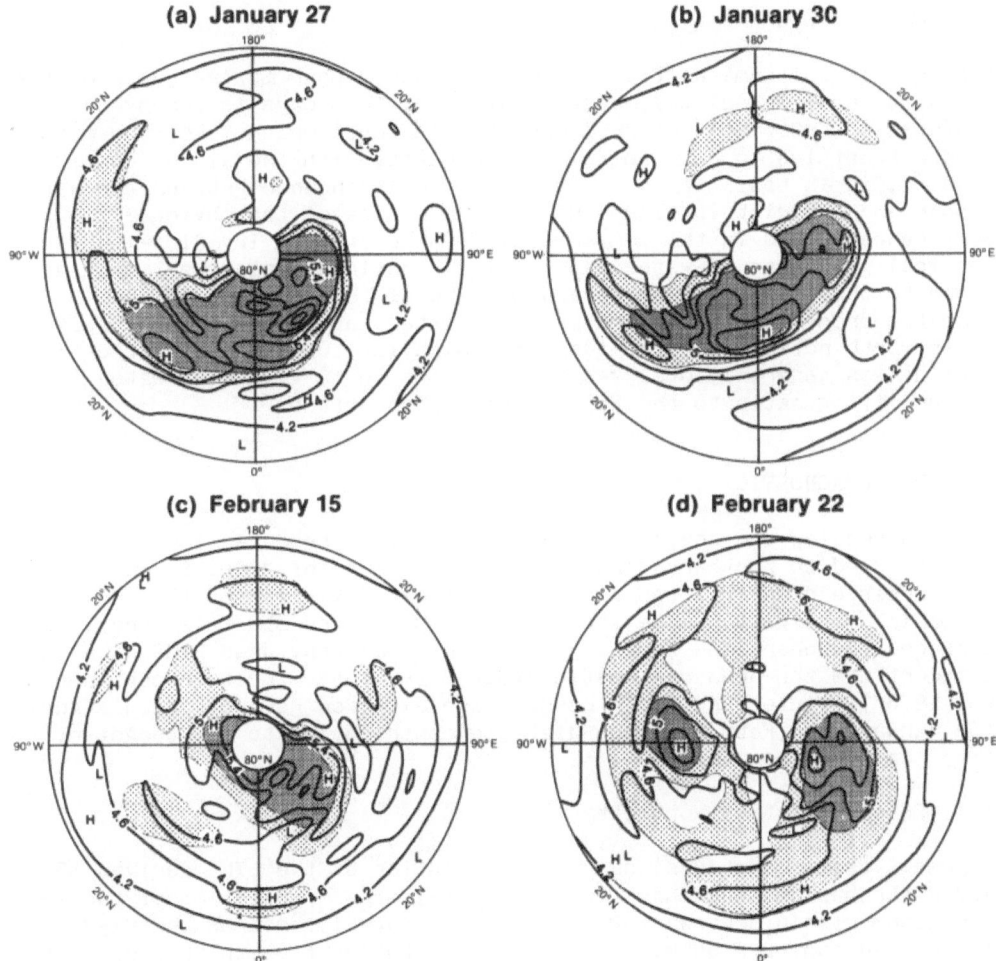

Fig. 4.   As Fig. 2 but the solid contours are for the LIMS water vapour concentration (ppmv) on the 850 K isentropic surface. Contour interval is 0.4.

transport in late January is the north-south dipole-like structure
between 0° and 120°E seen most prominently in Fig. 3b for 30 January. A
comparison with Fig. 2b indicates that this diabatic cross-isentropic
transport would reduce the band of stronger gradients in the ozone
field defining the edge of the main vortex region. Fig. 3c for 15
February shows the diabatic transport to be bringing down higher values
of ozone concentration in the section of the main vortex protruding
into the Western Hemisphere while lower values are being brought to the
surrounding region. After the major warming diabatic cross-isentropic
transport of ozone is generally weak everywhere at the 850 K level (see
Fig. 3d for 22 February).

3.3. Water vapour

Figure 4 is the same as Fig. 2 but for water vapour mixing ratio on the
850 K surface. Immediately apparent for all four days is the overall
flatness of the water vapour field at this level which makes the maps
rather sensitive to data noise. Nevertheless, with the aid of the
shading, it is possible to see a close correspondence between the water
vapour and IPV distributions and there are none of the obvious
differences found in the comparison of the ozone and IPV distributions.
Vertical gradients of water vapour mixing ratio are also very weak and
prone to the effects of noise and because of this no maps of the
diabatic cross-isentropic transport of water vapour are shown. However,
it is worth noting that in middle and high latitudes the diabatic
cooling was generally observed to transport lower values of water
vapour mixing ratio to the 850 K isentropic surface.

4.   AREA DIAGNOSTIC

In this section the so-called area diagnostic introduced in Section 2
will be used to further quantify the net effects of diabatic and
irreversible dynamical processes in systematically redistributing
potential vorticity, ozone and water vapour on the 850 K isentropic
surface. Butchart and Remsberg (1986) have already used LIMS data to
document how the area diagnostic evolved for these three tracers during
the 1978/79 winter and their results are extended here by calculating,
for January and February, the diabatic contribution to the budget of
$A(t)$ (see Equation (2)).

4.1. Ertel potential vorticity

Figure 5 (solid contours) shows the evolution of the LIMS-derived 850 K
IPV distribution as a function of the horizontal projection of the area
contained within contours of constant Q. Superimposed on the figure
are dashed contours showing the IPV distribution from the appropriate
part of an annual cycle integration of a zonally symmetric general
circulation model of the stratosphere (cf. Fig. 6 of Butchart and
Remsberg, 1986). The exclusion of planetary waves from the model
eliminates the possibility of rapid irreversible mixing so that the

gradual evolution of modelled IPV distribution is directly related to the seasonal variation in the diabatic heating. Differences between the observed and modelled IPV distributions provide a measure of the extent to which the real stratosphere has been "preconditioned." This is independent of details of the planetary waves and fluctuations in the shape and position of the vortex. However, it should not be forgotten that it is the planetary waves that are responsible for the preconditioning and, assuming that the divergence of the large-scale flow is negligible, Equation (2) suggests two principal mechanisms.

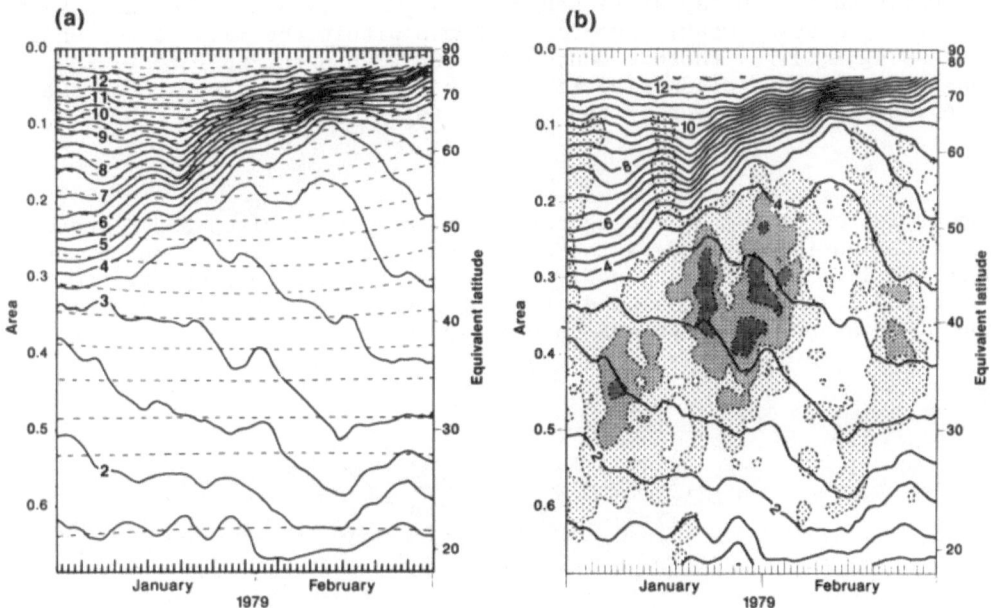

Fig. 5. Solid contours: LIMS derived 850 K isentropic distribution of potential vorticity Q for the period 1 January to 28 February, 1979 expressed as a function of the area enclosed within the contours of constant Q. In PV units (see Fig. 1). Dashed contours: (a) as solid contours but for the 850 K isentropic distribution of Q obtained from the appropriate period of an annual cycle integration of a zonally symmetric general circulation model of the stratosphere. (b) Diabatic contribution in Equation (2) applied to contours of Q on the 850 K surface and expressed as a function of the area enclosed within the contours. In units of the surface area of a hemisphere (i.e. the units shown on the left hand ordinate) per 100 days. Contour interval is 0.2. Positive values are shaded,with the heaviest denoting the largest values. The equivalent latitude marked on the right hand axis denotes the latitude circle enclosing the area indicated on the left hand axis.

Firstly, the inherent nonlinearity of the atmosphere implies that there will almost inevitably be a cascade to smaller scales and, depending on the detailed dynamics, some mixing. The broad expanding region of weak gradients (surf zone) in Fig. 5a together with the erosion of the core of high Q (main vortex) is a characteristic signature of isentropic mixing (see Introduction). Supporting this are Schoeberl and Smith's (1986) results suggesting a downscale transfer of enstrophy during this period.

Secondly, local changes in the diabatic heating induced by
disturbances in the flow could lead to significant changes in IPV
distribution which have yet to be quantified. Butchart and Remsberg
(1986) provided a preliminary estimate of the diabatic contribution to
the budget of A(t) for potential vorticity and their calculation has
been repeated here using an independent radiation scheme. The results
are presented in Fig. 5b using dashed contours. The quantity shown is
the diabatic contribution to the budget of A(t) plotted as a function
of A(t). Comparing Fig. 5b with Fig. 8 of Butchart and Remsberg (1986)
indicates that both radiative schemes give broadly similar results with
diabatic processes weakly decreasing areas within the main vortex while
increasing A(t) for those Q contours lying within the surf zone. In
general the diabatic contribution is far too small to account for the
observed rate of erosion of the main vortex.

4.2. Ozone

The evolution of the area diagnostic for the LIMS-derived 850 K ozone
distribution is depicted by the solid contours in Fig. 6. When compared
with the IPV distribution in Fig. 5b it is possible to identify, for
the ozone distribution, a band of tighter gradient following roughly
the 5.4 contour which appears to correspond to the edge of the main

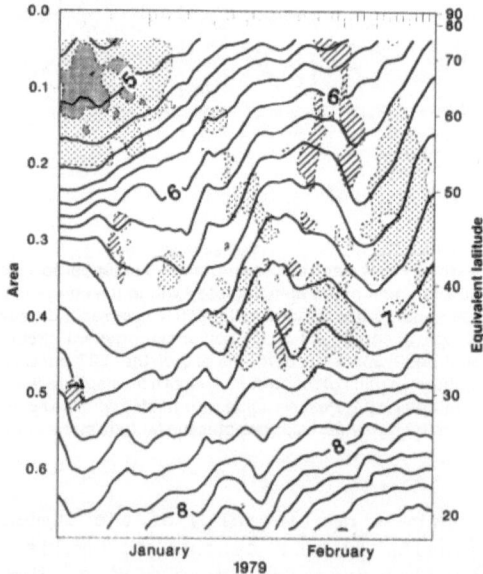

Fig. 6.   Solid contours: as Fig. 5b but for the LIMS 850 K isentropic distribution of ozone (ppmv). Dashed contours: as
Fig. 5b but for the contribution from the diabatic cross-isentropic transport of ozone in Equation (2)
applied to contours of ozone on the 850 K isentropic maps. Contour interval is now 0.5. Values less than −1.0 are lightly
stippled and less than −1.5 heavily stippled. Positive values are indicated by the striped shading.

vortex region. Otherwise the gradients of the ozone distribution are fairly uniform with an overall decrease of the areas contained by all the contours. The decrease, however, is not monotonic and is punctuated by two major increases in areas — one at the beginning of January and another in mid-February. There is also a period of rather rapid decrease of areas in late January and again in late February (see, for example, the 6.6 contour).

The dashed contours in Fig. 6 indicate the contribution of the diabatic cross-isentropic transport to the budget of $A(t)$ for ozone. Over most of the domain shown the diabatic cross-isentropic transport is calculated to decrease areas at a rate of up to 0.5 times the area of the hemisphere per 100 days. Broadly speaking this is consistent with the observed overall decrease of areas. At the apparent edge of the main vortex the calculated rate of decrease is about 0.5 and is comparable with the observed rate of erosion of the vortex. Within the region of very low ozone mixing ratios (top left corner of Fig. 6) or "ozone hole" (cf. Leovy et al., 1985) the results indicate an even stronger rate of decrease due to diabatic cross-isentropic transport and it seems likely that this had a prominent role in filling this ozone hole. Stronger rates of decrease due to diabatic cross-isentropic transport are also seen in late January and late February and could partially explain the rather rapid decreases in areas observed during those periods. In addition the two periods of increasing areas (see above) occur when the diabatic contribution becomes positive (i.e. it

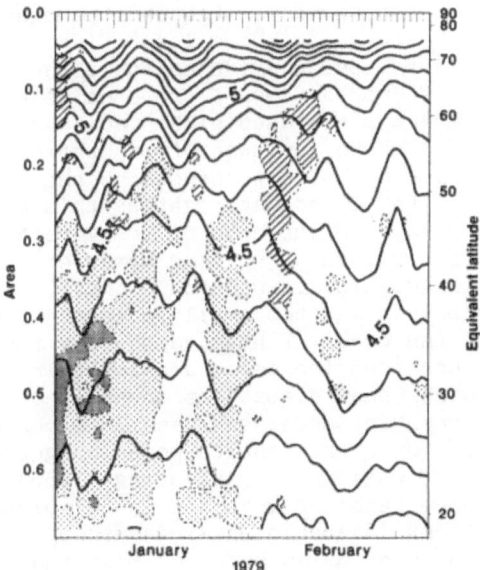

Fig. 7.   As Fig. 6 but for the LIMS 850 K isentropic distribution of water vapour (ppmv). The dashed contour interval is now 0.3 with values less than −0.3 lightly stippled and less than −0.6 heavily stippled. Positive values are again indicated by the striped shading.

would increase A(t)), again suggesting the importance of diabatic
cross-isentropic transport.

4.3. Water Vapour

Results for the LIMS-derived 850 K isentropic distribution of water
vapour are shown in Fig. 7 which is essentially the same as Fig. 6 for
ozone except the dashed contour interval is now 0.3. The solid contours
show that the water vapour distribution is similar in many respect to
that of IPV (see Fig. 5b) with a shrinking core of high values
surrounded by a region of weaker gradients. As with the IPV this
behaviour is characteristic of isentropic mixing of the water vapour.
The contribution of diabatic cross-isentropic transport of water vapour
(dashed contours) to the budget of A(t) is generally very weak apart
from the region of low water vapour concentrations in early January.
Moreover, it is generally of the wrong sign to account for the increase
of some of the areas (see, for example, the 4.5 contour) observed in
late January and February.

5. CONCLUDING REMARKS

The maps presented in Section 3 showed that for four days in January
and February 1979 there is a good correspondence between the 850 K
isentropic distributions of Q, ozone and water vapour. In actual fact
the positive correlation between Q and water vapour and negative
correlation between Q and ozone is observed throughout most of these
two months. Similar features to those identified by McIntyre and Palmer
(1983, 1984) in the IPV maps as evidence for planetary wave breaking
also appear in the isentropic maps of the species concentrations.
        Calculating the areas enclosed within contours on the isentropic
maps provides a useful means for quantifying the contributions made by
irreversible mixing and diabatic processes to the evolution of the
tracer distributions. For Q and water vapour this diagnostic displayed
the characteristic behaviour expected of irreversible mixing with a
shrinking core of high values surrounded by an expanding region of
weaker gradients. In addition it was found that, for these two tracers,
the diabatic contribution was too small to account for the observed
rates of erosion of the cores of high values and for the water vapour
would decrease, rather than increase, areas within the region of weaker
gradients. On the other hand diabatic processes could contribute
significantly to increasing areas enclosed by Q contours lying within
the surf zone. Also, for Q the results were in agreement with the
independent results of Butchart and Remsberg (1986) which suggests
that, despite the uncertainties involved in calculating the small
difference of two terms (diabatic cross-isentropic transport and
in-situ diabatic source of Q) an order of magnitude larger, the results
are meaningful. On this basis it would seem fair to say that, since
there are no other significant nonconservative processes acting on the
Q and water vapour fields at the 850 K level, irreversible mixing was
the important mechanism responsible for eroding the cores of higher

values. The 850 K isentropic distribution of ozone, on the other hand, was quite different with only a hint of a two-region structure. In this case the results of the diabatic calculation gave the impression that during January-February 1979 the 850 K ozone distribution was strongly influenced by the diabatic cross-isentropic transport of ozone. It also has to be remembered that at this level ozone photochemistry could be important and clearly there is now real need for incorporating photochemical effects into the area diagnostic approach.

ACKNOWLEDGMENTS

My thanks to John Austin for his comments and generous assistance in producing this manuscript. Helpful comments from A. O'Neill, D. E. Parker, V. D. Pope and E. E. Remsberg are also acknowledged.

REFERENCES

Andrews, D. G., and M. E. McIntyre, 1978: An exact theory of nonlinear waves on a Lagrangian-mean flow. J. Fluid Mech., 89, 609-646.
Austin, J., 1986: Comparison of stratospheric air parcel trajectories calculated from SSU and LIMS satellite data. J. Geophys. Res., 91, 7837-7851.
————, 1987: Evidence of planetary wave breaking in the stratosphere using a photochemical model along air parcel trajectories. (In this volume).
Butchart, N., and E. E. Remsberg, 1986: The area of the stratospheric polar vortex as a diagnostic for tracer transport on an isentropic surface. J. Atmos. Sci., 43, 1319-1339.
————, S. A. Clough, T. N. Palmer and P. J. Trevelyan, 1982: Simulations of an observed stratospheric warming with quasigeostrophic refractive index as a model diagnostic. Quart. J. Roy. Meteor. Soc., 108, 475-502.
Clough, S. A., N. S. Grahame and A. O'Neill, 1985: Potential vorticity in the stratosphere derived using data from satellites. Quart. J. Roy. Meteor. Soc., 111, 335-358.
Dunkerton, T. J., and D. P. Delisi, 1986: Evolution of potential vorticity in the winter stratosphere of January-February 1979. J. Geophys. Res., 91, 1199-1208.
Gille, J. C., and J. M. Russell III, 1984: The Limb Infrared Monitor of the Stratosphere: Experiment description, performance and results. J. Geophys. Res., 89, 5125-5140
Grose, W. L., 1984: Recent advances in understanding stratospheric dynamics and transport processes: Application of satellite data to their interpretation. Adv. Space Res., 4, 19-28.
Harwood, R. S., and J. A. Pyle, 1975: A two-dimensional mean circulation model for the atmosphere below 80 km. Quart. J. Roy. Meteor. Soc., 101, 723-747.

Haynes, P. H., and M. E. McIntyre, 1987: On the evolution of vorticity
    and potential vorticity in the presence of diabatic heating and
    frictional or other forces. J. Atmos. Sci., **44**, (In press).
Leovy, C. B., C-R. Sun, M. H. Hitchman, E. E. Remsberg,
    J. M. Russell III, L. L. Gordley, J. C. Gille and L. V. Lyjak,
    1985: Transport of ozone in the middle stratosphere: Evidence for
    planetary wave breaking. J. Atmos. Sci., **42**, 230-244.
London, J., 1980: Radiative energy sources and sinks in the stratos-
    phere and mesosphere. Proceedings of the NATO Advanced Study
    Institute on Atmospheric Ozone: Its Variation and Human
    Influences, Rep. FAA-EE-80-20, A. C. Aikin, Ed., DOT. FAA.
    Washington, DC, 703-721.
Mahlman, J. D., 1985: Mechanistic interpretation of stratospheric
    tracer transport. Adv. Geophys., **28A**, 301-323.
————, D. G. Andrews, D. L. Hartmann, T. Matsuno and R. G. Murgatroyd,
    1984: Transport of trace constituents in the stratosphere.
    Dynamics of the Middle Atmosphere, J. R. Holton and T. Matsuno,
    Eds., Terra Scientific, 387-416.
McIntyre, M. E., and T. N. Palmer, 1983: Breaking planetary waves in
    the stratosphere. Nature, **305**, 593-600.
————, and ————, 1984: The "surf zone" in the stratosphere. J. Atmos.
    Terr. Phys., **46**, 825-849.
Palmer, T. N., 1981: Diagnostic study of a wavenumber-2 stratospheric
    sudden warming in a transformed Eulerian-mean formalism. J.
    Atmos. Sci., **38**, 844-855.
Schoeberl, M. R., and A. K. Smith, 1986: The integrated enstrophy
    budget of the winter stratosphere diagnosed from LIMS data.
    J. Atmos. Sci., **43**, 1074-1086.
Tung, K. K., 1982: On the two-dimensional transport of stratospheric
    trace gases in isentropic coordinates. J. Atmos. Sci., **39**,
    2330-2355.
————, 1984: Modeling of tracer transport in the middle atmosphere.
    Dynamics of the Middle Atmosphere, J. R. Holton and T. Matsuno,
    Eds., Terra Scientific, 417-444.

SEASONAL VARIATION IN THE VARIANCE OF STRATOSPHERIC OZONE AND POTENTIAL TEMPERATURE OVER HOHENPEISSENBERG, F.R.G.

E.P. Röth
Institut für Physikalische und Theoretische Chemie
Universität Essen - GHS
D-4300 Essen 1
Fed. Rep. Germany

D.H. Ehhalt
Institut für Chemie 3: Atmosphärische Chemie
Kernforschungsanlage Jülich GmbH
D-5170 Jülich
Fed. Rep. Germany

Abstract

All stratospheric trace gases exhibit a temporal variance in their local concentration. For long-lived species that variance is mainly caused by transport. In those cases it is useful to express the variance as equivalent displacement height (EDH). The EDH is defined as the local mean standard deviation divided by the local mean vertical gradient of the tracer's mixing ratio. The concept of EDH will be discussed. In addition, we present the analysis of the seasonal variations of the EDH of ozone and potential temperature obtained from measurements at Mt. Hohenpeissenberg in Southern Germany (48°N) by Attmannspacher. The EDH of both tracers show the same seasonal behavior, although they differ in magnitude by a factor of five. The EDH is highest in late winter and has a minimum around August/ September.

Introduction

Measurements of stratospheric trace gas concentrations exhibit a substantial temporal variance, which varies with altitude and depends on the species investigated. The corresponding mean standard deviation is shown in figure 1 for the gases CH_4, N_2O, CF_2Cl_2, and $CFCl_3$ along with the mean mixing ratio profiles obtained from 14 balloon flights

G. Visconti and R. Garcia (eds.), Transport Processes in the Middle Atmosphere, 137–152.

over Southern France at 44°N (Volz et al.,1981;Schmidt et al.,1986). For better comparison, the mean profiles are divided by the respective concentrations at the earth's surface. To remove the influence of any secular trend, all measurements were normalized to June 1979.

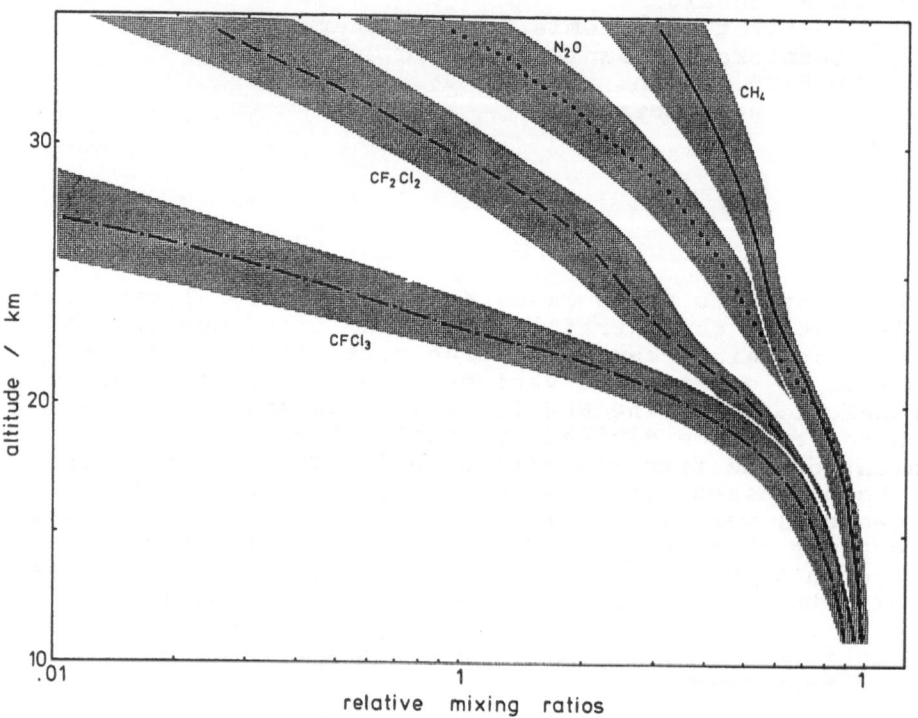

Figure 1 : Vertical profiles of the volume mixing ratio of CH₄ (———), N₂O (·····), CF₂Cl₂ (- - -), and CFCl₃ (-·-·-) averaged over 14 balloon flights in Southern France (44° N). The local mixing ratios are given relative to a tropospheric value of 1. The shaded areas indicate the mean standard deviation of the data points (Volz et al.,1981: Schmidt et al.,1986).

Obviously, the mean standard deviation (msd) increases
considerably with altitude in the stratosphere. It also
increases in the sequence CH_4, N_2O, CF_2Cl_2, $CFCl_3$. This is
clearly seen when projecting the shaded areas mapping out
the msd onto the axis of the relative mixing ratio. If we,
on the other hand, would project these areas onto the
height ordinate, the resulting msd in height would be about
equal at a given altitude for all four gases.

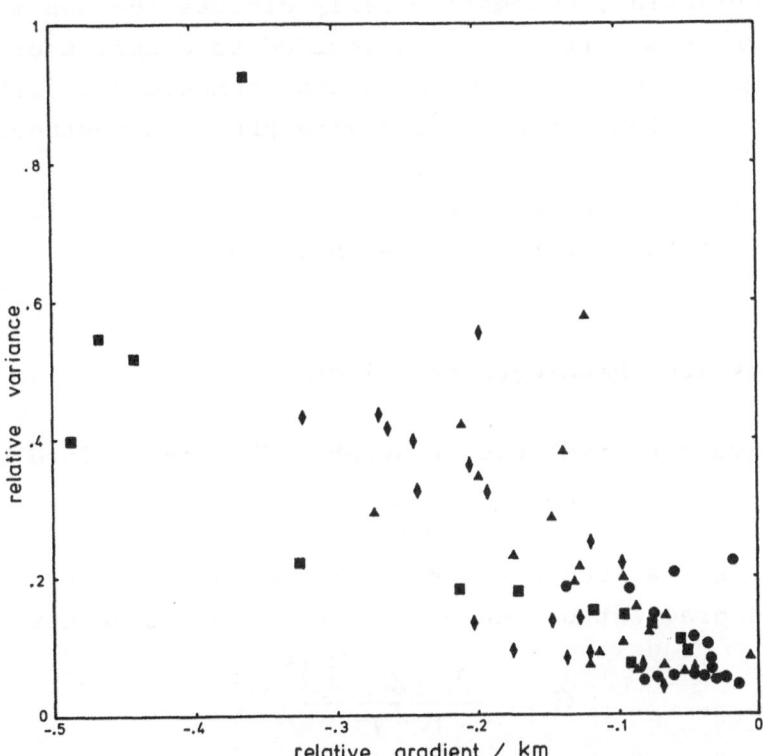

Figure 2 : Correlation between the local variance and the
local vertical gradient. To obtain a better comparison of
the data for the different species (CH_4 : ● , N_2O : ▲ ,
CF_2Cl_2 : ◆ ,and $CFCl_3$: ■) both the gradient and the
variance are divided by the local mean mixing ratio.

For this to happen, the mean standard deviation in the
mixing ratio must be proportional to the vertical gradient
of a trace gas. This is indeed the case, as is demonstrated
by the correlation of the msd with the vertical gradient at
various altitudes (figure 2). It is this observation which
led Ehhalt et al.(1983) to introduce the equivalent
displacement height, EDH. It is emphasized that, despite
the term "height", this does not mean that the temporal
variance is due to vertical motions only.
In the following, we shall briefly discuss the concept of
EDH. Then this method will be applied to a series of
measurements for ozone and potential temperature, obtained
from observations at Mt. Hohenpeissenberg in Southern
Germany (Attmannspacher, 1985). From these data,
information on the seasonal variation of the EDH is derived
for the altitudinal range between 15 and 30 km.

The Equivalent Displacement Height

The equivalent displacement height, Δ, is defined by

$$\Delta(z) = \frac{\sigma(z)}{\frac{\partial \bar{\mu}(z)}{\partial z}} \tag{1}$$

where σ is the mean standard deviation and $\partial \bar{\mu} / \partial z$ the mean
vertical gradient of the mixing ratio, μ, at an altitude z.

$$\sigma = \frac{\sum (\mu_i - \bar{\mu})^2}{n - 1} \tag{2}$$

In the data discussed below μ_i represents the individual
measurement and $\bar{\mu}$ the monthly mean. The vertical gradient
is obtained by a linear regression over an altitude window
of $\Delta z_i = \pm 2$ km.

As indicated in figure 3, equation (1) converts the mean
standard variation in the mixing ratio into a mean standard
variation in altitude, hence the name "equivalent
displacement height". As was demonstrated by Ehhalt et
al.(1983), the vertical profiles of Δ are virtually
identical for the trace gases CH_4, N_2O, CF_2Cl_2, $CFCl_3$, as

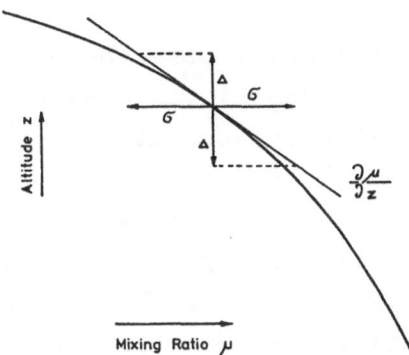

Figure 3 : Illustration of the conversion of σ to Δ.

well as for O_3. They further showed that most of the
variance Δ^2 must be caused by advection across a
meridional field of varying μ. Using the mixing length
hypothesis in a meridional frame they derived a theoretical
expression for Δ^2 :

$$\Delta^2 = \Delta y^2 \left[(\bar{\alpha} - \bar{\beta})^2 + \overline{\alpha'^2} \right]$$

(3)

which relates the locally observed temporal variance to the
transport parameters Δy, $\bar{\alpha}$, $\bar{\beta}$, and $\overline{\alpha'^2}$. Here, Δy is the
latitudinal displacement, $\beta = (\partial z / \partial y)_\mu$ is the mean
meridional gradient of the surfaces of constant mixing

ratio, and $\bar{\alpha} = \Delta z / \Delta y$ is the mean meridional slope of the
transport surface. $\overline{\alpha'^2}$ is the variance of the latter.
It was also shown that for conservative tracers $\overline{\alpha'^2}$ is
small compared to $(\tilde{\alpha} - \bar{\beta})^2$ at most altitudes, so that we can
approximate Δ by

$$\Delta^2 \approx \Delta y^2 (\bar{\alpha} - \bar{\beta})^2 \tag{4}$$

We can rewrite equation (4) to give

$$\Delta \approx \Delta y \left(\frac{\Delta z}{\Delta y} - \bar{\beta} \right) = \Delta z - \bar{\beta} \Delta y \tag{5}$$

Equation (5) relates the locally observed equivalent
displacement height to the (mean) vertical displacement Δz
and the (mean) meridional displacement Δy, simultaneously
experienced by an air mass when advected from one point to
another. Thus, if Δ is measured for two or more tracers,
whose $\bar{\beta}$ are known, the transport parameters Δz and Δy can
be deduced; i.e. the observation of Δ provides important
information on transport processes, which are otherwise
difficult to observe.
In the earlier paper, we derived a vertical profile of Δ
from CH_4, N_2O, CF_2Cl_2, $CFCl_3$, and O_3 data. Since most of
the measurements were made during June and September, this
profile is characteristic for the summer months, during
which the stratosphere is relatively quiet. Yet , the
observed variance Δ was substantial. This led Hess and
Holton (1985) to argue that the variance observed in June
was " frozen in" variance, generated by the final stratospheric warming during early spring. The authors supported
their argument with a model calculation of Δ.
There are presently not enough measurements of long-lived
gases to experimentally test these calculations. There are,
however, sufficient measurements of O_3 and potential
temperature over Mt.Hohenpeissenberg, Southern Germany, to
provide a seasonal variation of Δ at altitudes from 10 to

30 km. In the following, we will present these data, derive seasonal profiles of the equivalent displacement height and compare them to Hess' and Holton's calculated values.

Evaluation of the Hohenpeissenberg-Data

The meteorological observatory of the German Weather Service at Hohenpeißenberg, Bavaria (48°N latitude) is monitoring stratospheric O_3 since November 1966 (cf. Attmannspacher, 1985). The vertical profiles of the O_3 concentration are obtained by balloon-borne electrochemical ozone sondes. Temperature and pressure are also measured by balloon-borne instruments. All three quantities are monitored continuously during ascent. Since the response time of the O_3 instrument is about 60 s and the ascent rate approximately 3 m/s, the vertical resolution of the O_3 measurement is about 200 m. Similar numbers are obtained for pressure and temperature.

The early flight program until October 1977 operated at a rate of one balloon flight per week. Ever since, the O_3 sondes were launched twice or three times a week. Thus, altogether more than 1600 O_3 profiles are available – over 130 per month.

The mean monthly O_3 profiles are shown in figure 4a, the mean monthly profiles of potential temperature, Θ, calculated from pressure and temperature, are given in figure 4b.

$$\Theta = T \left(\frac{1000}{P} \right)^{.286}$$

$$(6)$$

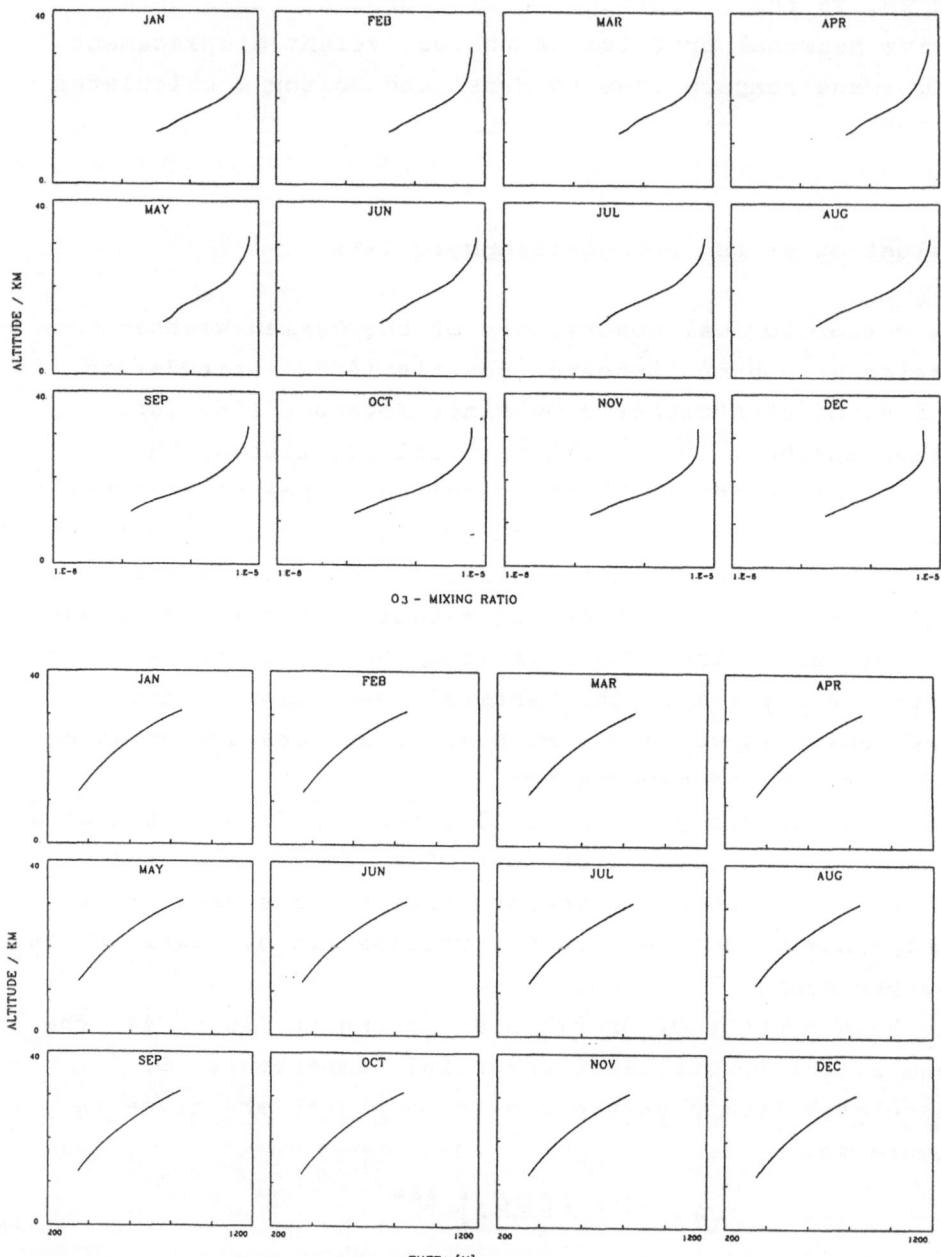

Figure 4 : Monthly mean vertical profiles of the mixing
ratios of ozone and of the potential temperature derived
from a 18 years' series of measurements over Mt.
Hohenpeissenberg (48° N).

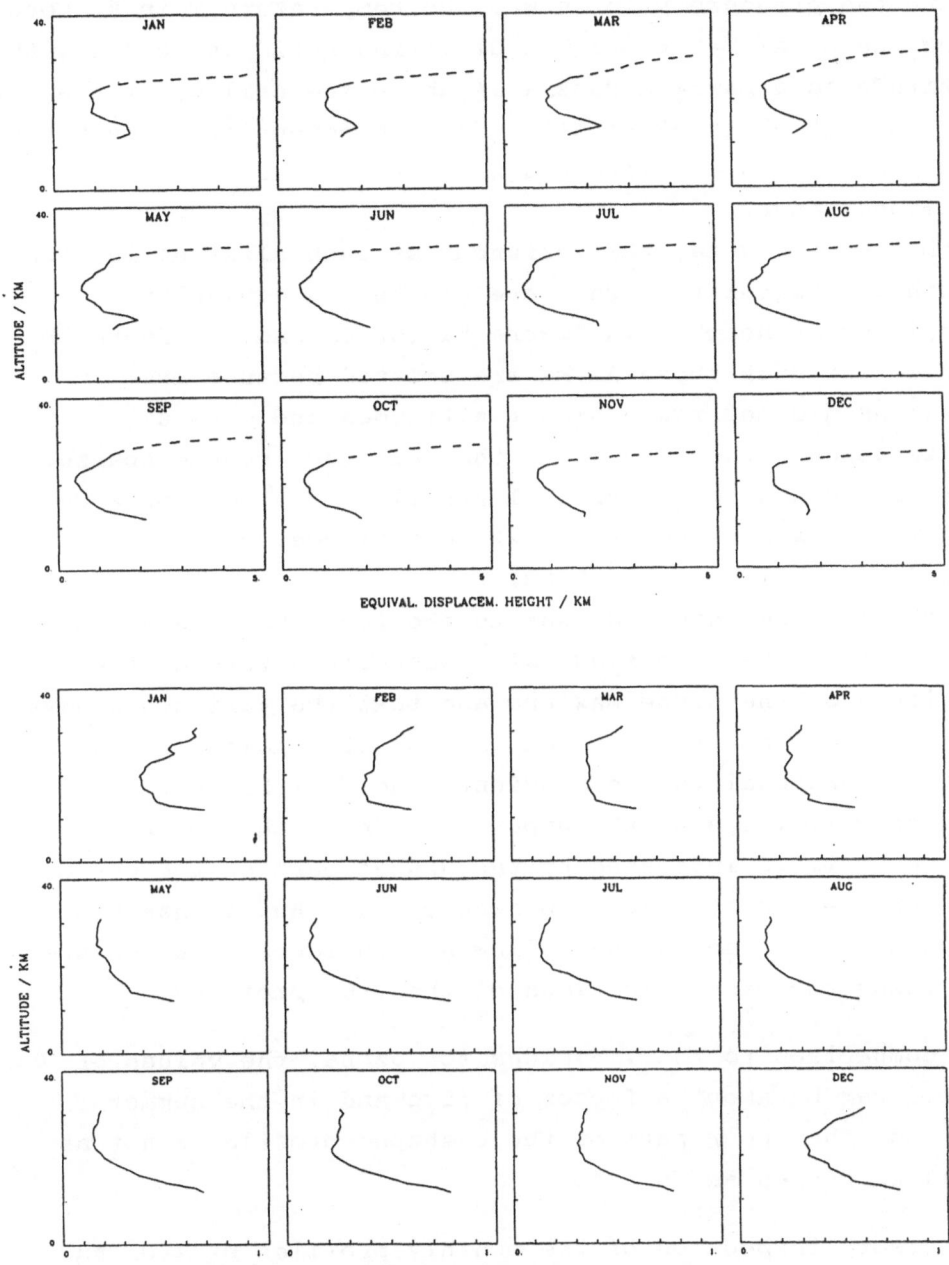

Figure 5 : Monthly mean vertical profiles of the equivalent displacement heights of ozone and potential temperature derived from a 18 years' series of measurements over Mt.

where the pressure p is in mb, the temperature T in K. Both
show the usual behavior: The O_3 mixing ratio increases with
altitude to a maximum generally above the ceiling of the
balloons at about 32 km altitude; the potential temperature
increases at an accelerating rate over the observed
altitude range.

From the same data, the variances at each altitude in each
month are calculated. They are plotted as equivalent
displacement heights in figure 5a for O_3 and in figure 5b
for Θ. The data above 32 km are omitted because most
balloons did not reach higher altitudes and - as a
consequence - the data above that altitude become sparse.
For all seasons, the vertical profiles of ΔO_3 show a C-
shape with a minimum around 22 km altitude, as already
noted by Ehhalt et al. (1983).

We should take note that due to the fact that the vertical
gradient in the O_3 mixing ratio approaches zero at the
altitude of the ozone maximum and that the variance always
contains a contribution from instrumental noise, which is
not proportional to the gradient, the ΔO_3 becomes
excessively large at the upper portion of the observed
range. This is indicated by the dashed part of the curves.
A proper error analysis can remedy that. But it has not
been done yet and in the following considerations, we are
omitting the dashed portions of the ΔO_3 profiles.

In comparison to the Δ-values for ozone, the values of Δ_Θ
are lower by about a factor of five and in the summer
period, the upper part of the C-shaped profile is not as
well expressed as for ΔO_3.

A cursory inspection of the monthly profiles of ΔO_3 and
Δ_Θ already indicates a seasonal variation, which seems to
be similar for both cases. To examine this further, the
seasonal variation of the Δ-values at 16, 20, 24, and 28
km altitude are plotted in figure 6.

Quite clearly there is a seasonal variation for both, ΔO_3 and Δ_θ at all altitudes with low Δ-values in summer and high values in winter. Moreover, for both ΔO_3 and Δ_θ, the amplitude of the seasonal variation increases with altitude. There is very little phase shift between the two. It appears, however, that ΔO_3 peaks in February, whereas Δ_θ peaks in January at all altitudes.

These observed seasonal variations in Δ can be compared to the time dependence in the Δ, predicted by Hess and Holton (1985) for Θ and for a conservative tracer ($N_2 O$) after the final spring warming. Their results for 15 km and 25 km altude given by the dotted curves in figure 6.

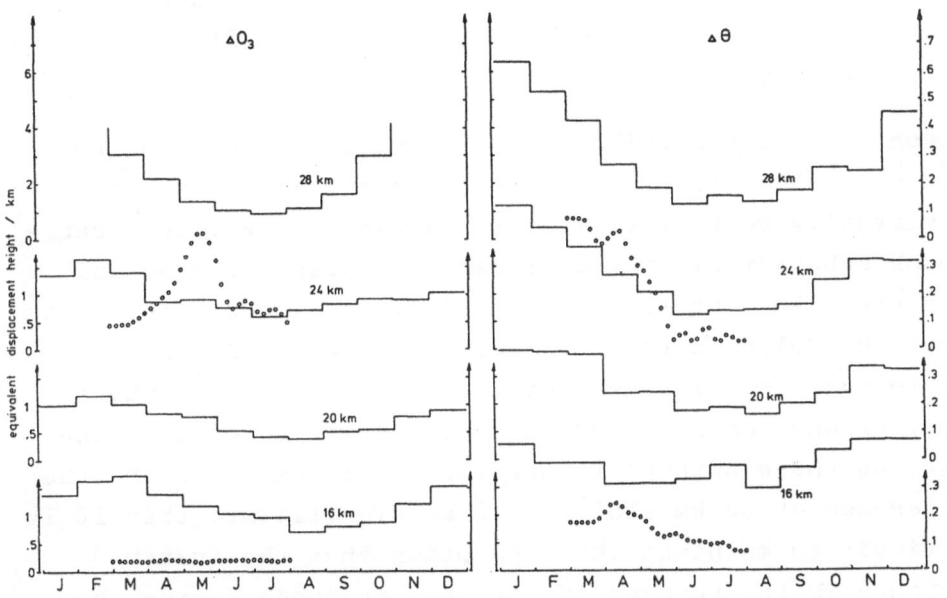

Figure 6 : The seasonal variation of the equivalent displacement heights for ozone and potential temperature at four different altitudes. The dotted curves represent the calculations by Hess and Holton (1985) at 15 and 25 km altitude (their day 0 corresponds to March 1st).

We first compare Δ_θ. Here the agreement between prediction
and measurement, both in amplitude and trend, is quite
reasonable. The experimental values at 24 km altitude and
the predicted values at 25 km trace each other quite
closely. The observed data at 16 km and the predicted
values at 15 km altitude do not agree quite as well.
However, these differences are readily explained. At that
altitude, wave energy is supplied all the year around from
the troposphere, which is not accounted for in the model
calculations, but is present in the observed data. In
addition, the model allows for only one, the final,
stratospheric warming, whereas the real polar stratosphere
experiences several warmings during a winter. Thus the
experimental data exceed significantly those modelled at
all times except during April, i.e. shortly after the onset
of the warming.

In contrast, the modelled Δ for an inert tracer and the
observed values for ΔO_3 differ dramatically, although O_3
is virtually conservative over the observed altitude range.
Though substantial, the difference between observed and
predicted ΔO_3, at 15 km altitude is not incompatible with
Hess' and Holton's thesis, again because the lowest
stratosphere remains subject to more or less continuous
input of wave energy from the troposphere throughout the
year, as these authors themselves point out. However, the
difference at 24 km altitude is so significant, that it is
difficult to maintain the assumption that the summer
variance is the remnant of variance produced during the
final warming. Hess and Holton predict a relatively sharp
maximum in Δ about 15 days after the zero wind line
crossed 25 km altitude. In contrast, the observed values of
ΔO_3 show a steady decline during that time (May) from
earlier, higher values. In fact, as pointed out above, the

observed ΔO_3 and Δ_θ show a parallel trend, which is not
predicted. From the additional observation that the decay
of the higher variance in winter to the lower values in
summer proceeds with a time constant of about 80 days,
which is much longer than the time constant for the
Newtonian cooling of about 20 days, which should dominate
the decay of Δ_θ, we would argue that variance is generated
throughout the year albeit at lower rates during summer,
when tropospheric wave activity is blocked from penetrating
into the stratosphere. In addition, our seasonal variations
of Δ indicate that the maximum generation of variance
occurs in midwinter and not in spring. In short, we argue
that the seasonal variation in Δ reflects seasonal pattern
of generation of variance, which varies relatively smoothly
with time and is not caused by a singular event like the
final warming.

Although the seasonal variation in figure 6 is an average
over 18 years, it is quite close to that of each individual
year. Therefore a smoothing due to varying time of
occurance of the final warming in the individual years
cannot account for the difference between the prediction by
Hess and Holton and the seasonal trend derived from the
whole period of measurements.

Hess and Holton (1985) predict another parameter, namely
the actual displacement, Δz, from their model. In the
following, we attempt to derive this parameter from the
observed ΔO_3 and Δ_θ using eq.(5) given above. We note,
however, that for Δ_θ this approach is not quite justified
because the potential temperature is not an inert tracer
and secondly, in the case of Θ, $\overline{\alpha'^2}$ in eq.(3) is not
negligible. This is due to the fact that advection is
nearly isentropic, i.e. $\overline{\alpha} \approx \overline{\beta}$ so that $(\overline{\alpha}-\overline{\beta})^2$ in eq.(3)
nearly vanishes.

However, to obtain a first guess, we proceed with the
simplified approach of eq.(5) keeping in mind that the thus
derived Δz might physically not be quite identical with
the one predicted by Hess and Holton (1985).

The required slopes $\bar{\beta}$ of the mixing surfaces of O_3 and of
the equipotential temperature surfaces were taken from the
literature, the latitudinal dependence of Θ from Cole and
Kantor (1978) and the O_3 mixing ratios from the LIMS-SBUV
data (McPeters et al., 1984). Cole and Kantor presented
tables of monthly mean data at 15° latitude intervals with
altitude steps of 2 km. McPeters et al. tabulated monthly

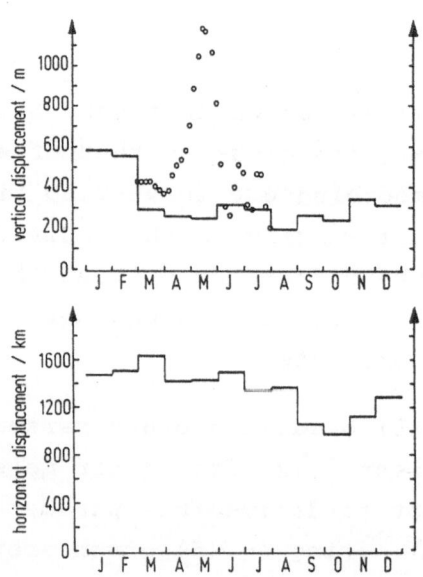

Figure 7 : The seasonal variations of the vertical and the
horizontal displacements at 25 km altitude. The solid lines
represent the absolute values of the displacements, the
values of Hess and Holton (1985) are indicated by the
dotted line.

averaged profiles of ozone mixing ratios depending on pressure. Their data are given for latitudinal bands of $10°$.

In figure 7 the empirically derived seasonal variations of Δy and Δz are plotted for 25 km altitude. For comparison, the 5 months' time sequence of Δz predicted by Hess and Holton is shown by the dotted line. Apart from the May maximum in the predicted Δz, which is again not seen in the observational data, the agreement is reasonable. The Δz values are of the order of 500 m in winter and about half that value in summer.

Also shown in figure 7 are the observed meridional displacements Δy. They are of the order of 1000 km, in good agreement with the values derived by Ehhalt et al. (1983) for that altitude.

Conclusion

We have shown Δ to vary seasonally with a maximum in winter and a minimum in summer, as would be expected from stratospheric wave activity. The detailed trend is not compatible, however, with the notion that the final spring warming is responsible for the variance remaining in summer. Other process must be at work. The preliminary attempt to derive vertical and meridional displacements leads to reasonable results, but much more work needs to be done to remove uncertainties in Δo_3, Δy, and Δz. In particular, seasonal data for other, conservative tracers would be useful to improve the basis for the interpretation of Δ.

Acknowledgment

We want to express our thanks for helpful discussions to
Dr. J. Gille and Dr. R. Garcia from NCAR, Boulder and to
Dr. U. Schmidt from the KFA, Jülich. The programming was
carried out by Mrs. H. London, which is gratefully
acknowledged. The evaluation of the Mt. Hohenpeissenberg
data was sponsored by the German Ministry for Science and
Technology.

References

Attmannspacher, W., 1985: Sonderbeobachtungen des
Meteorologischen Observatoriums Hohenpeißenberg. Ergebnisse
der aerologischen und bodennahen Ozonmessungen 1966-1985.

Cole, A.E., and A.J. Kantor, 1978: Air Force reference
atmospheres, Air Force Geophysics Laboratory Hanscom, AFGL-
TR-78-0051.

Ehhalt, E.H., E.P. Röth and U. Schmidt, 1983: On the
temporal variance of stratospheric trace gas
concentrations, J. Atmos. Chem. 1, 27-51.

Hess, P.G., and J.R. Holton, 1985: The origin of temporal
variance in long-lived trace constituents in the summer
stratosphere, J. Atmos. Sci. 42, 1455-1464.

McPeters, R.D., D.F. Heath, and P.K. Bhartia, 1984: Average
ozone profiles for 1979 from the NIMBUS 7 SBUV instrument,
J. Geophys. Res. 89, 5199-5214.

Schmidt, U., C. Jebsen, F.J. Johnen, A. Khedim, E. Klein,
D. Knapska, G. Kulessa, J. Rudolph, G. Schumacher, and E.
Schunck, 1986: Stratospheric observations of long-lived
trace gases at midlatitudes 1982-1985, Berichte der
Kernforschungsanlage Jülich, JÜL-Spez-375.

Volz, A., U. Schmidt, J. Rudolph, D.H. Ehhalt, F.J. Johnen,
and A. Khedim, 1981: Vertical profiles of trace gases at
midlatitudes, Berichte der Kernforschungsanlage Jülich,
JÜL-1742.

SATELLITE OBSERVATIONS OF THE ANTARCTIC OZONE DEPLETION

Mark R. Schoeberl
Code 616
NASA/Goddard Space Flight Center
Greenbelt, MD 20771

ABSTRACT. The morphology of satellite total ozone and temperature
observations in the polar and subpolar Antarctic regions are reviewed.
An analysis of the total ozone and 70 mb temperature data in the
depletion region for for October and September shows similar secular
trends. These trends result from changes which primarily occur in
late September in both the minimum total ozone and temperature. The
strong correlation between ozone and temperature suggests an important
dynamical component to this phenomenon. A simple computation shows
that a vertical velocity field of 0.06 cm/sec can account for both the
secular ozone and temperature trends in the minimum region.

1. INTRODUCTION

The rapid secular decline in total ozone in September and October in
the Antarctic first noted in the Halley Bay observations by Farman et
al. (1985) has ignited both scientific and public concern. Farman et
al. suggested that the total ozone decrease might be directly related
to anthropogenically produced chlorocarbon compound increases in the
stratosphere. However, it was quickly realized that most of the
depletion in the ozone layer appeared to be taking place in the lower
stratosphere below 30 km (e.g., McCormick and Larsen, 1986).
Unfortunately, stratospheric assessment models predict that
significant chlorine catalytic destruction of ozone will take place
above 30 km. Therefore ozone-chlorine catalysis, in order to account
for the ozone decrease, would have to be quite different from the
standard formulation perhaps involving heterogeneous reactions
(Solomon et al., 1986) and bromine (McElroy et al., 1986).

Aside from the chlorine hypothesis, two other mechanisms for the
total ozone change have been suggested. Callis and Natarajan (1986)
proposed that an increase in odd nitrogen compounds associated with
the last solar cycle could have produced the depletion, while Tung et
al. (1986), and Mahlman and Fels (1986) have proposed that upwelling
of air in the spring polar region possibly associated with a change in
the aerosol distribution or an increase in polar stratospheric clouds

G. Visconti and R. Garcia (eds.), Transport Processes in the Middle Atmosphere, 153–165.

could bring about the decrease in total ozone.

The purpose of this paper is to outline some of the more significant observational constraints that the satellite data places on these
hypothesis. Most of the observations reported here are from the Total
Ozone Mapping Spectrometer (TOMS) which has been taking data since
November 1978 from the polar orbiting Nimbus 7. The instrument and
data quality are discussed by Fleig et al. (1986). Temperature data
discussed here is post processed NMC data (Geller et al. 1983; Newman
and Schoeberl, 1986).

2. TOTAL OZONE DISTRIBUTION

The change in the total ozone distribution described by Farman et al.
(1985) is confined to the polar and sub polar south latitudes during
the spring equinox period. The minimum total ozone amounts in the
south polar region are roughly the same for the years 1979-1985 at the
end of August. For the years 1980-1985, the minimum declines from the
beginning of August until the beginning of October. Total ozone
values then remain low until late October or early November when they
return to "normal" with the break up of the polar night jet or polar
vortex. In the Northern hemisphere, the polar vortex breaks up near
equinox, which is a month or more sooner than in the Southern
hemisphere. During the polar vortex breakup, the strong circumpolar
belt of stratospheric westerlies is replaced by periods of cross polar
flow and weak easterlies. Mid-latitude minor constituents are mixed
into the polar region. Therefore in the Northern hemisphere the
spring sun rises on fairly well mixed constituent distribution, while
in the Southern hemisphere, the sun rises on constituents which have
been relatively isolated from midlatitudes throughout the polar
night. However, it is doubtful that these constituents have remained
in darkness over the entire winter period since the polar vortex
meanders considerably. In the rest of this paper we will concentrate
on the September-October period.

Figure 1 shows the monthly mean October distribution of total
ozone for 1979 and 1985 and illustrates the striking decline which
occurred between those years. The lowest values of total ozone are
found over the eastern Antarctic region usually nearer to the Syowa
(69°S, 40°E) and Halley Bay (76°S, 27°W) stations than to McMurdo
(78°S, 167°E). However, the pattern of total ozone is highly variable
throughout the month and total ozone amounts over single stations may
vary by 100% in a few days. Surrounding the minimum in total ozone is
a C-shaped (wave one) pattern of very high total ozone, with the
maximum located in the southern Pacific. Since there are virtually no
ground stations in the region, the structure of the maximum can only
be determined by satellite observations.

The seasonal development of the high and low total ozone regions
in the southern hemisphere is shown in Figure 2 where April and
October zonal means are plotted. The fall distributions for both
hemispheres are remarkably similar. The spring distribution in the
Southern Hemisphere clearly shows the midlatitude maximum and the

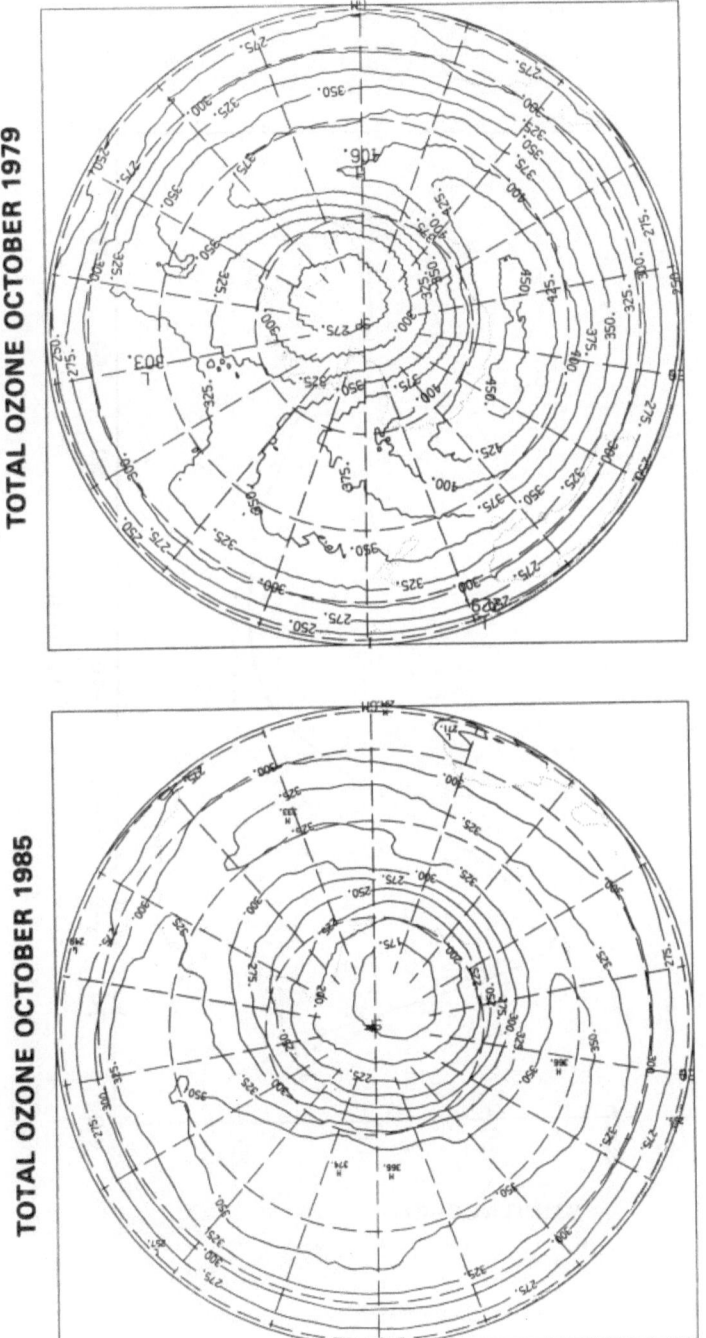

FIG. 1. The monthly mean October distribution of total ozone in Dobson Units for 1979 and 1985. Latitude circles are 10°, 30°, 50° and 70°S.

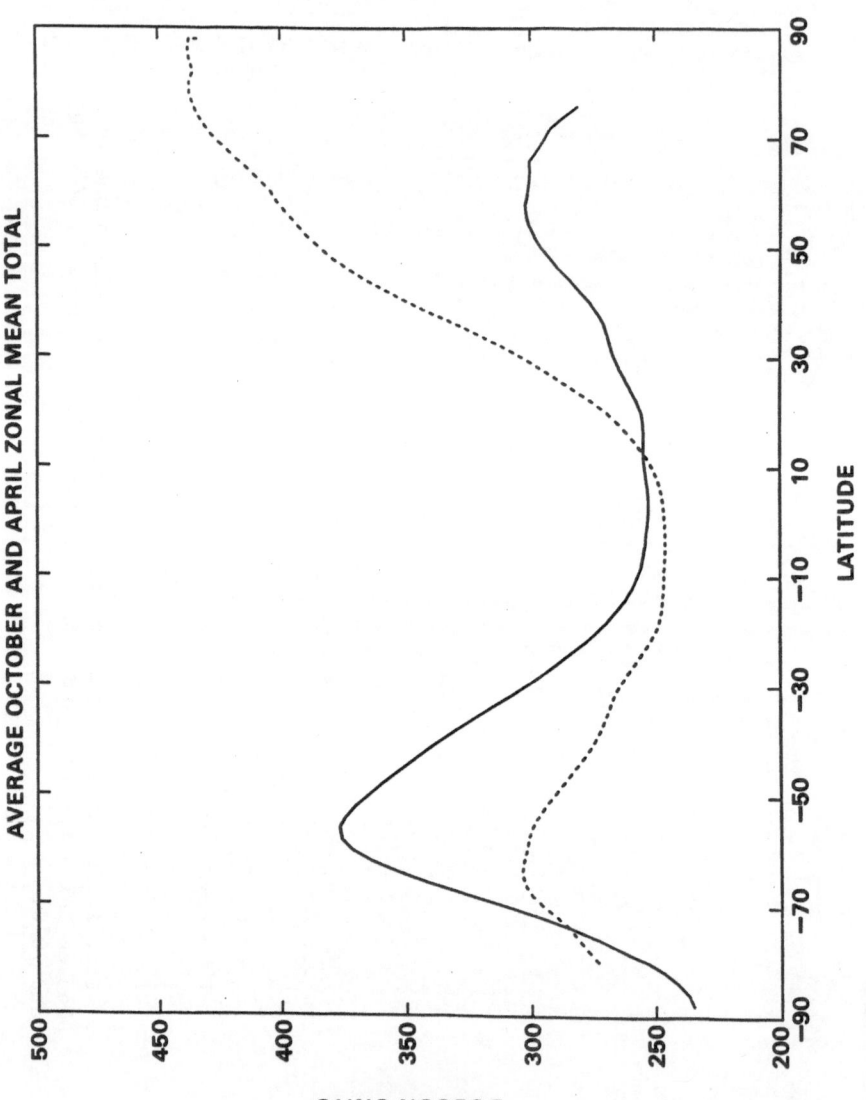

FIG. 2. The April (dotted line) and October (solid line) zonal mean, monthly mean total ozone distribution obtained by averaging 1979–1982 values.

South Polar minimum, no such minimum is seen in the Northern
Hemisphere. The total ozone distribution is also remarkably well
correlated with the 70 mb temperature distribution (Newman and
Schoeberl, 1986). The evolution of the total ozone from the fall to
spring shows clearly that the hemispheric ozone distribution is not
conserved. Since cross hemispheric flow in the stratosphere is
extremely weak, this nonconservation must be a result of changes in
the photochemical production and storage of ozone.

The differences between the two hemispheres can be mostly
explained by the the differences between the transport circulation as
illustrated in Figure 3. The transport circulation shown is computed
from the observed temperatures using a modern radiative transfer
code. In the Northern Hemisphere spring, the strongest downward
motion in the lower stratosphere is near the polar region, while in
the Southern Hemisphere spring the strongest downward motion is at
midlatitudes. The actual total ozone distribution (Figure 2) is
fairly consistent with the computed transport circulation. The
development of a strong midlatitude descending cell in late winter and
spring in the Southern Hemisphere will bring ozone from the production
region in the upper stratosphere to the lower stratosphere where the
lifetime is long, and will produce the hemisphere increase in total
ozone over the winter-spring period as observed (Rood, 1985).

Figure 4 shows the October monthly mean, map and zonal mean
changes in total ozone from 1979 to 1986. The map values are the
minimum and maximum values seen on a southern hemisphere map south of
30°S. This figure clearly shows the year to year general decline in
total ozone in both the maximum and minimum. The recently obtained
1986 values are higher than in 1985, but are close to the range of
fluctuation seen in earlier years. Since the ozone hole is quite near
the pole, the zonal mean and map minimum values are similar, but the
presence of the significant wave one disturbance at subpolar latitudes
(see Figure 1) produces a large difference between the zonal mean and
map maximum values. The larger maximum map value decline relative to
the zonal mean maximum indicates that the amplitude of the stationary
wave in the stratosphere has declined as well. This decline is also
seen in the temperature data, both for the zonal mean and the map
maximum and minimum values (Newman and Schoeberl, 1986).

Year to year differences in the total ozone amounts at subpolar
latitudes can be largely explained by the changes in the transport
circulation pattern. That is, because the transport circulation is
determined by the level and distribution of wave activity in the
stratosphere (see Atmospheric Ozone: 1985, Chapter 4; 1986), a
decline in wave activity would produce a weaker descending cell at
midlatitudes, relatively colder temperatures and less total ozone.
All of this is qualitatively consistent with observations.

At polar latitudes, total ozone decreases in September, even
though the temperature is rising all through this period (Newman,
1986; Farrara and Mechoso, 1986). Figure 5a shows the day-to-day
change in the map minimum total ozone from September through October
smoothed by a 3 day running mean. Each year starts with roughly the
same total ozone value. The secular changes in polar ozone amounts in

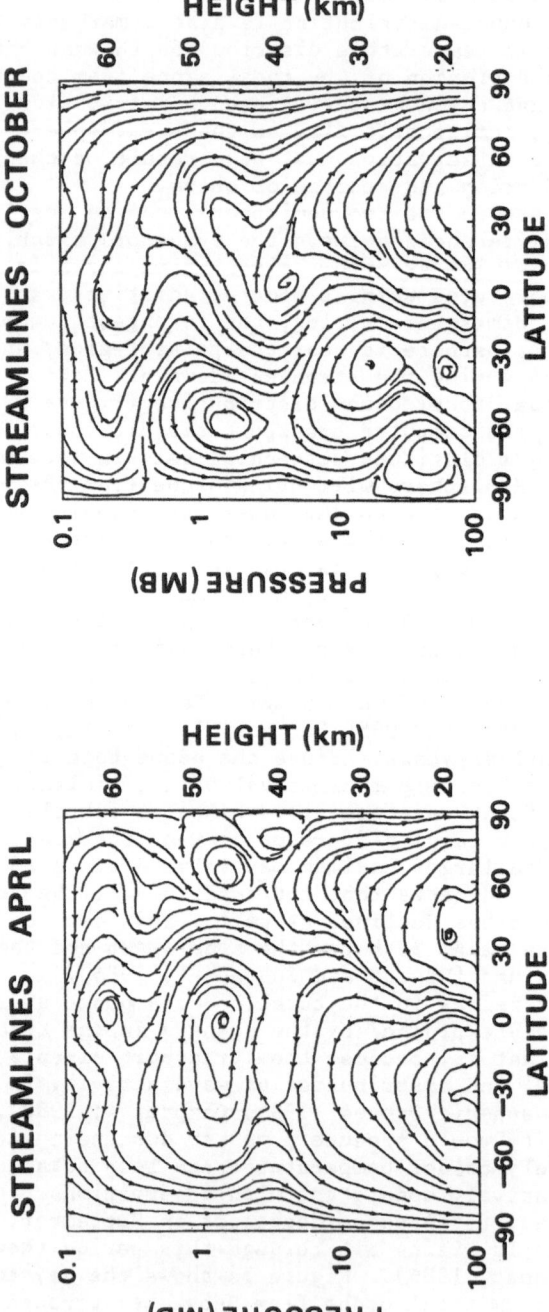

FIG. 3. The diabatic circulation in the Southern hemisphere from Rosenfield et al. (1987) for October and April.

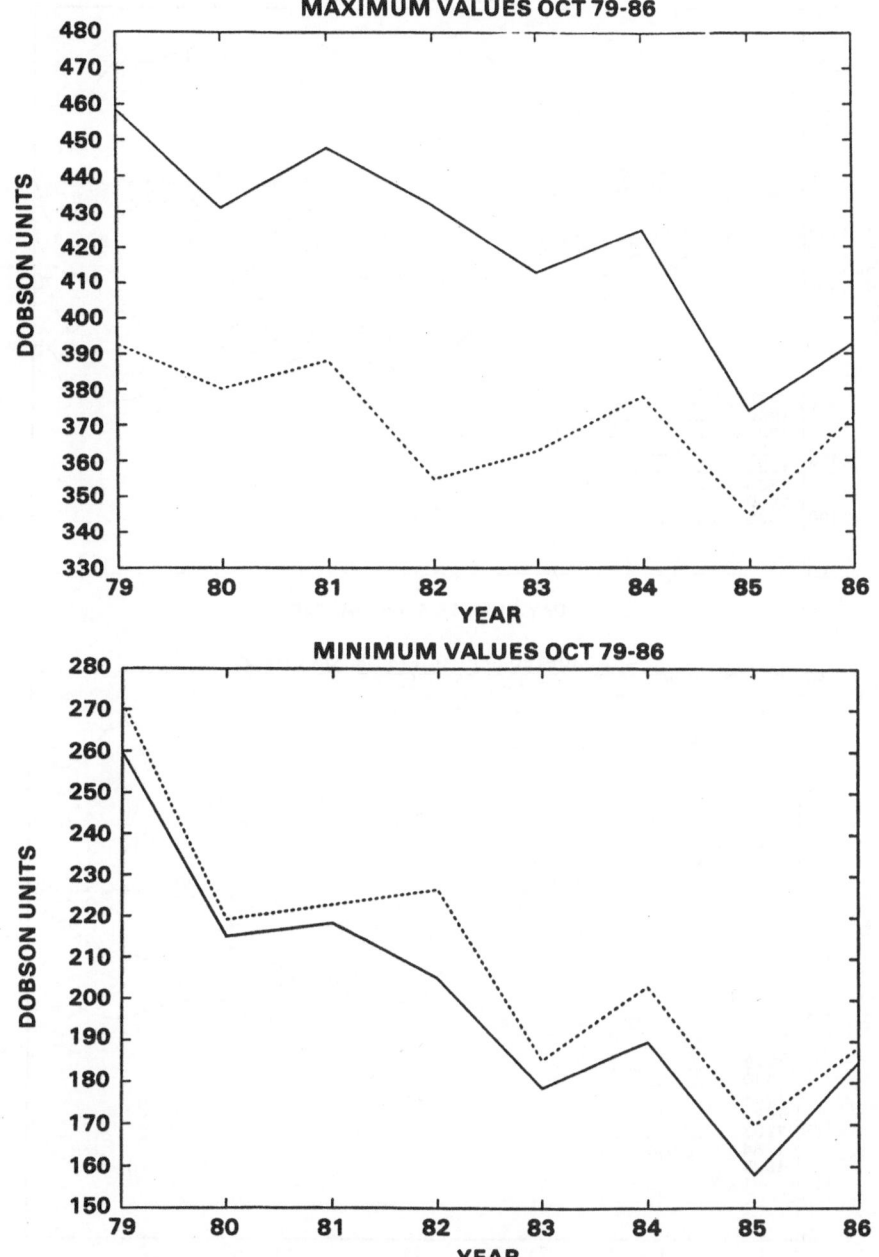

FIG. 4. The October monthly mean, map and zonal mean changes in
total ozone 1979–1986. Minimum/maximum values shown are the
minimum/maximum zonal mean and map total ozone amounts south of 30°S.

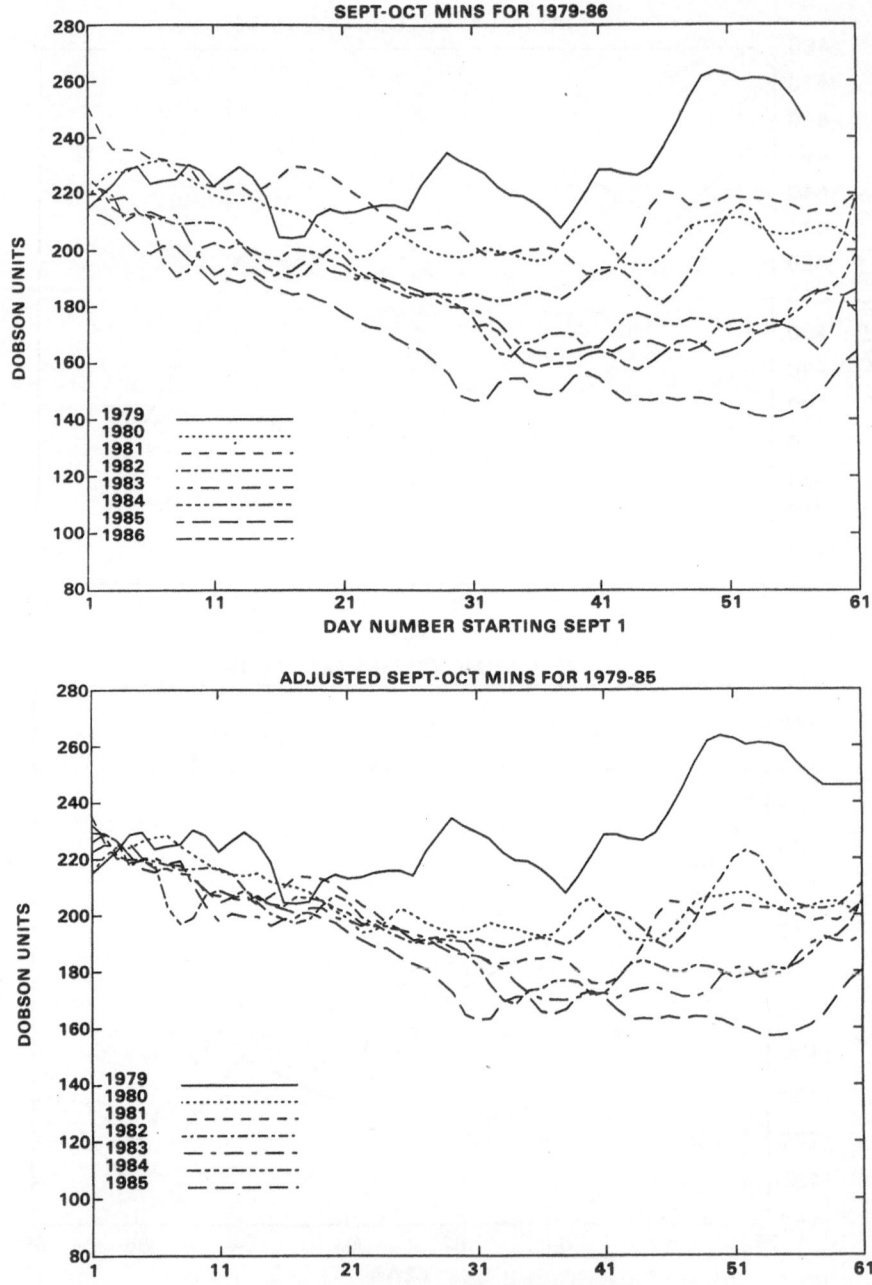

FIG. 5. Part a shows the total ozone changes at the map minimum in September and October at 70 mb. Part b shows the same results with curves adjusted to the 1979 September 1-6 average.

October are therefore a result of the changes which occur in late
September. In 1979 almost no decrease occurs through September while
in subsequent years a noticable decrease is seen. The decrease ends
roughly in the first week of October. The secular trends are a result
of the changes in the late September decrease, and this conclusion is
supported by surface station observations as well (Komhyr et al.,
1986). Also shown in Figure 5a are 1986 observations for October, as
of this writing, the September, 1986 data has not been completely
processed.
    Figure 5b shows the trends in map minimum total ozone values
after initial values are adjusted to correct for early September
differences. The first six days of September are averaged and total
ozone amounts for 1980-1985 are adjusted to 1979 average values for
those days. Except for 1979, all of the years now show similar
decreases of about 1.35 DU per day up to the last week of September (a
slightly greater decrease for 1985). The curves then separate near
the first of October; 1980, 1982 level off with the other years
declining further. Thus the secular differences in the minimum total
ozone for the years 1980-1985 appear to arise from 1) initial differences before September, and 2) changes in the date at which the
"uniform decline" in September ends. The exception is 1979 which
shows no real change over this period. Newman (1986) has pointed out
the 1979 was a very dynamically active year compared to subsequent
years.
    During the September decline in total ozone, the temperature in
the depletion region continuously rises. This temperature increase
does not appear to be due to descending motion, but to radiative
forcing (Rosenfield and Schoeberl, 1986). In order to produce a
decrease in total ozone dynamically, rising motion must be present.
Therefore, cooling due to upward motion in the polar region is also
consistent with a relatively colder polar stratosphere since 1979.
The following simple model, similar to that used by Tung (1986),
illustrates this mechanism. The mixing ratio of ozone, μ, is governed
by

$$D\mu/Dt + wd\mu_o/dz = S \qquad (1)$$

where $D\mu/Dt$ follows the horizontal motion on a pressure surface and
the other terms have their usual definitions. Total ozone,
Ω, is $\int \mu\rho dz$. The potential temperature, Θ, is determined by a
similar equation,

$$D\Theta/Dt + wd\Theta_o/dz = Q = D\Theta_r/Dt \qquad (2)$$

where Q is the net heating rate, written as $D\Theta_r/Dt$. $D\Theta_r/Dt$ is the
diabatic rate of change in potential temperature. If we assume that
the motion in the lower stratosphere is nearly uniform (at least in
the region between the tropopause and about 10 mb, the zone where the
depletion occurs) then total ozone and ozone will have similar
behavior. Further if we assume that the chemical source, S, is small
compared to the advection terms on the l.h.s. of (1), then eliminating

w between (1) and (2) gives

$$D\mu/Dt = \mu_o(h_\mu d\theta_o/dz)^{-1} D(\theta-\theta_r)/Dt \tag{3}$$

where μ_o is assumed to have the form $\mu_a \exp(z/h_\mu)$.

Equation (3), while very approximate, gives a simple relation between changes in ozone and lower stratospheric temperature. Furthermore, it shows that if θ increases slower than θ_r, total ozone will decrease and temperature increase. Eq. (3) also predicts that the temperature increase during September in the ozone minimum region in recent years should be slower because of the appearance of vertical motion. Figure 6 shows the temperature changes in September and October at 70 mb at the temperature minimum which is roughly colocated with the total ozone minimum. A slow increase in the temperatures occurs throughout the month, but the fastest increase is in 1979, and the slowest in 1985, consistent with Eq. (3). In addition, the largest differences in temperature between years appears after September similar to the adjusted total ozone (Figure 5b).

Using the ideas above, we can check the consistency of a dynamically produced ozone depletion mechanism. The increase in temperature in 1979 is about 0.4 K/day from Figure 6 while for the years 1980-1985 it is roughly 0.21 K/day. Since total ozone stays roughly constant in the spring of 1979, Eq. (3) suggests that the diabatic rate of change is 0.4 K/day. The slower increase in temperature after 1979 would be due to added adiabatic cooling of 0.19 K/day associated with ascending motion. This translates into roughly 0.044 cm/sec vertical velocity. In order to produce a total ozone decrease of 1.35 DU/day with a vertical scale height of lower stratospheric ozone of 8 km, the motion field must be roughly 0.059 cm/sec. Thus, it is apparent even with this very simplified computation that dynamical processes can account for a large part of the total ozone decrease.

3.   DISCUSSION

The distribution of total ozone in the Southern Hemisphere and the associated temperature distribution are consistent with dynamical hypothesis that the secular changes at the pole are due to increased upward motions, and that secular changes at midlatitudes are due to decreased downward motions. Both ozone and temperature data show that the stationary wave amplitude in the stratosphere have declined since 1979. Still unanswered is what has produced these changes, and what is the role of chemistry?

Any change in the stratospheric vertical motion field must be a result of an imbalance in the thermodynamic or momentum equations. Tung et al. (1986) suggested that increases in aerosols, and a subsequent rise in aerosol heating could produce an increased polar spring upward motion. Mahlman and Fels (1986) also discussed this mechanism with regard to polar stratospheric clouds. However, aerosol measure-

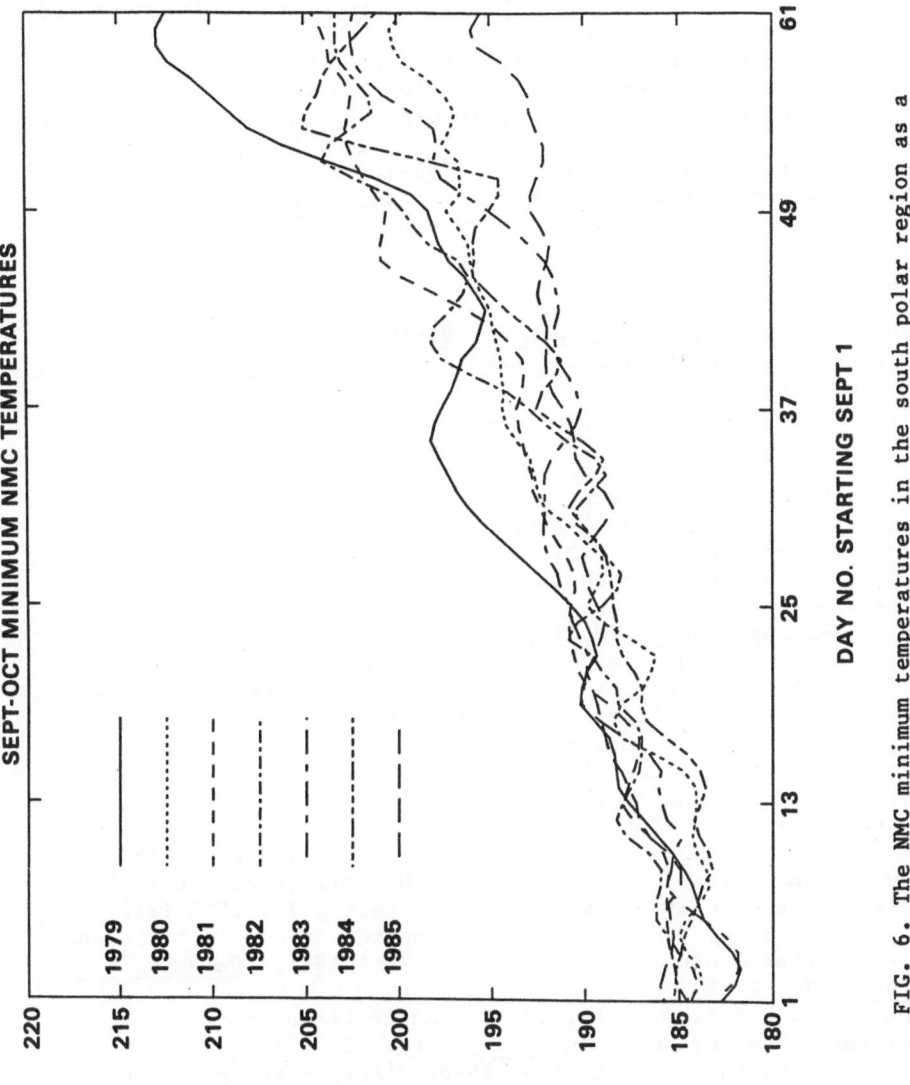

FIG. 6. The NMC minimum temperatures in the south polar region as a function of time, filtered with a three day running mean. Missing or bad data has been filled by linear interpolation. Most of the missing days are in 1979; 9/7-9/12, 9/17-9/19, 9/26-9/30, 10/2-10/5, 10/22.

ments by SAGE (McCormick and Trepte, 1986) and by balloon (Hofmann et al., 1986) are not consistent with an aerosol powered secular trend in temperature. The alternative is that changes in the eddy forcing of the stratosphere has shifted the momentum balance, and that the vertical motion field has been altered through the continuity equation.

Another real possibility is that midlatitude and polar declines in total ozone are the result of different mechanisms, the later having some chemical component while the former being mostly dynamical. However, it is difficult to imagine such a process retaining the good polar ozone temperature correlation discussed here and by Schoeberl et al., (1986) as well as the integral conservation discussed by Stolarski and Schoeberl (1986). A better assessment of the chemistry awaits NASA's U-2 mission planned for September 1987.

REFERENCES

Atmospheric Ozone 1985, Assessment of our Understanding of the Processes Controlling its Present Distribution and Change, 1-3, WMO Report 16, 1986.

Callis, L. C., and M. Natarajan, 1986: 'The antarctic ozone minimum: Relationship to odd chlorine, the final warming and the 11-year solar cycle,' J. Geophys. Res., 91, 10,771-10,796.

Farman, J. C., B. G. Gardiner, and J. D. Shanklin, 1985: 'Large losses of total ozone in Antarctica reveal seasonal ClO_x/NO_x interaction,' Nature, 315, 207-210.

Farrara, J. D. and C. R. Mechoso, 1986: 'An observational study of the final warming in the southern hemisphere stratosphere,' Geophys. Res. Lett., 13, 1232.

Fleig, A. J., P. K. Bhartia, C. G. Wellemeyer, and D. S. Silberstein, 1986: 'Sevens years of total ozone from the TOMS instrument--A Report on Data Quality,' Geophys. Res. Lett., 13, 1355-1358.

Geller, M. A., M. F. Wu, and M. E. Gelman, 1983: 'Troposphere-Stratosphere (surface-55km) monthly winter general circulation statistics for the northern hemisphere - four year averages,' J. Atmos. Sci., 40, 1334-1352.

Hofmann, D. J., J. M. Rosen, J. A. Harder, and S. R. Rolf, 1986: 'Ozone and aerosol measurements in the springtime antarctic stratosphere in 1985,' Geophys. Res. Lett., 13, 1252-1255.

Komhyr, W. D., R. D. Gross, and R. K. Leonard, 1986: 'Total ozone decrease at South Pole, Antarctica, 1964-1985,' Geophys. Res. Lett., 13, 1248.

Mahlman, J. D. and S. B. Fels, 1986: 'Antarctic ozone decreases: A dynamical cause?' Geophys. Res. Lett., 13, 1316-1319.

McCormick, M. P. and J. C. Larson, 1986: 'Antarctic springtime measurements of ozone, nitrogen dioxide and aerosol extinction by SAM II, SAGE, and SAGE II,' Geophys. Res. Lett., 13, 1280-1283.

McCormick, M. P. and C. R. Trepte, 1986: 'SAM II measurements of Antarctic PSC's and aerosols,' Geophys. Res. Lett., 13, 1276-1279.

McElroy, M. B., R. J. Salawitch, S. C. Wofsy, and J. A. Logan, 1986:

'Antarctic ozone: Reductions due to synergistic interactions in chlorine and bromine,' Nature, 321, 759-762.

Newman, P. A., 1986: 'The final warming and polar vortex disappearance during the Southern Hemisphere spring,' Geophys. Res. Lett., 13, 1228-1231.

Newman, P. A. and M. R. Schoeberl, 1986: 'October Antarctic temperature and total ozone trends from 1979-1985,' Geophys. Res. Lett., 13, 1206-1209.

Rood, R. B., 1983: 'Transport and the seasonal variation of ozone,' PAGEOPH, 121, 1049-1064.

Rosenfield, J. E., M. R. Schoeberl, and M. A. Geller, 1987: 'A computation of the stratospheric diabatic circulation using an accurate radiative transfer model,' J. Atmos. Sci., 44, 859-876.

Schoeberl, M. R., A. J. Krueger, and P. A. Newman, 'The morphology of Antarctic total ozone as seen by TOMS,' Geophys. Res. Lett., 13, 1217-1220, 1986.

Solomon, S., R. R. Garcia, F. S. Rowland, and D. Wuebbles, 1986: 'On the depletion of Antarctic ozone,' Nature, 321, 755-758.

Stolarski, R. S. and M. R. Schoeberl, 1986: 'Further interpretation of satellite measurements of Antarctic total ozone,' Geophys. Res. Lett., 13, 1210-1212.

Tung, K. K., M. K. W. Ko, J. M. Rodriguez, and N. D. Sze, 1986: 'Are Antarctic ozone variations a manifestation of dynamics or chemistry,' Nature, 322, 8811-8813.

3. TRANSPORT IN THE MIDDLE ATMOSPHERE: THEORY AND MODELING

TRANSPORT MECHANISMS IN THE MIDDLE ATMOSPHERE : AN INTRODUCTORY SURVEY

David G. Andrews
Meteorological Office Unit
Hooke Institute
Clarendon Laboratory
Parks Road
Oxford   OX1 3PU
United Kingdom

ABSTRACT. Basic ideas concerning the transport of dynamical and chemical tracers in the middle atmosphere are presented. Tracer transport models in one, two and three dimensions are discussed and some of the conceptual and practical advantages and disadvantages of each type are indicated.

1. INTRODUCTION

In this section we present some basic definitions pertaining to atmospheric transport and give some examples of transported quantities. The most fundamental concept is that of a <u>materially conserved tracer</u> (often called a <u>conservative tracer</u>), the amount of which remains constant in time following each moving mass element of air (or 'air parcel'). If χ is the 'amount' of tracer per unit mass of air then the material conservation property has the mathematical expression

$$\frac{D\chi}{Dt} \equiv \frac{\partial \chi}{\partial t} + \underset{\sim}{u} \cdot \nabla \chi = 0, \qquad (1.1)$$

where u is the fluid velocity and t is time. In practice, no physical tracer~is precisely conserved and (1.1) must be replaced by

$$\frac{D\chi}{Dt} = S, \qquad (1.2)$$

where S represents sources and sinks of tracer. However, for many atmospheric tracers S is weak in the sense that it only leads to small relative changes in χ following an air parcel, over relevant advection timescales: χ is then called a <u>quasi-conservative</u> or <u>long-lived</u> tracer.
    A second way of classifying tracers is to distinguish between <u>dynamically-active</u> tracers, whose dynamical, thermodynamic or chemical properties allow them to influence the flow field that transports them, and <u>dynamically-passive</u> tracers, which have no such effect. In principle

169

G. Visconti and R. Garcia (eds.), Transport Processes in the Middle Atmosphere, 169–181.
© 1987 by D. Reidel Publishing Company.

the latter could be <u>chemically</u> active.

The most well-known atmospheric tracer is the potential temperature θ, which is related to temperature T and pressure p by the relation

$$\theta \equiv T(p/p_s)^{-\kappa} , \tag{1.3}$$

where p_s is a reference pressure (e.g. 1000 mb) and $\kappa = R/c_p \simeq 2/7$, R being the gas constant for dry air and c_p the specific heat at constant pressure. (The entropy per unit mass is then $c_p \ln\theta$ + constant.) This satisfies the equation

$$\frac{D\theta}{Dt} = \frac{J}{c_p}\left(\frac{p}{p_s}\right)^{-\kappa} \equiv Q, \tag{1.4}$$

where J is the net heating rate per unit mass. θ is quasi-conservative for sufficiently weak heating rates — for example if J represents net radiative heating in a region where the radiative relaxation time is long compared to the advection time. Since θ is a basic thermodynamic variable, it is clearly a dynamically-active tracer.

Somewhat less familiar is Ertel's potential vorticity P, defined by

$$P \equiv \rho^{-1}(\underset{\sim}{\omega} + 2\underset{\sim}{\Omega}) \cdot \nabla\theta , \tag{1.5}$$

where ρ is atmospheric density, $\underset{\sim}{\omega} \equiv \nabla \times \underset{\sim}{u}$ is the relative vorticity and $\underset{\sim}{\Omega}$ is the Earth's rotation vector. If a body force $\underset{\sim}{X}$ per unit mass acts on the atmosphere then

$$\frac{DP}{Dt} = \rho^{-1}(\underset{\sim}{\omega} + 2\underset{\sim}{\Omega}) \cdot \nabla Q + \rho^{-1}\nabla\theta \cdot \nabla \times \underset{\sim}{X} , \tag{1.6}$$

where Q is defined in (1.4). P is quasi-conservative if the heating term Q and the body force X are sufficiently weak. P is also a dynamically-active tracer in general: this follows from the 'invertibility princ-iple', which states that the field of P at any instant determines the velocity and temperature fields at that instant, given suitable boundary conditions and the assumption that the atmosphere is in a 'balanced' state. [For discussion of this principle and many other aspects of the use of Ertel's potential vorticity, see the review by Hoskins, McIntyre and Robertson (1985).]

A further broad class of atmospheric tracers comprises the minor chemical species, with χ given by the mass or volume mixing ratio. Many of these species are long-lived: examples include ozone in the lower stratosphere and methane and nitrous oxide throughout most of the strat-osphere. Much of the current interest in tracer transport is stimulated by the need for better understanding of the ways in which these chem-icals are carried from one part of the atmosphere to another, and esp-ecially how they move in the latitudinal and vertical directions from source regions to sinks and reservoirs. Other papers in this volume are devoted to specific examples of these processes.

2. SOME BASIC TRANSPORT MECHANISMS

To illustrate some of the ways in which passive, conservative or quasi-conservative tracers can be transported by atmospheric motions, we now consider some very idealised flows. We first restrict attention to two-dimensional flows u = (u(x,y), v(x,y)) in the horizontal plane, where x is distance eastwards and y is distance northwards from some origin. In the simplest case in which u = constant and v = 0, the flow is a uniform eastward advection and a blob of passive, conservative tracer released into the fluid will be carried bodily eastward, with no distortion from its initial shape. If however u varies with y (but v = 0 still), the blob will become stretched out in longitude by the shear, some parts being transported more rapidly than others. The longitudinal extent of the blob grows linearly with time and sharp y-gradients of tracer mixing ratio develop; however, no latitudinal transport can take place.
    Suppose now that the tracer is weakly diffusive, rather than precisely conservative, with the source term S in (1.2) given by

$$S = \lambda \nabla^2 \chi \ ,$$

where λ is a small, constant diffusion coefficient. For a blob released into the zonal flow u = (u(y), 0), the combination of the y-shear and the latitudinal diffusion tends to lead to a rapid effective zonal diffusion (Taylor, 1953); hence the tracer tends to homogenise in the x-direction. Transport in the y-direction is still weak, however.
    A further useful idealisation is to allow for a slow latitudinal flow v = constant, in addition to the zonal flow u(y) and diffusion. If v is small enough, the longitudinal homogenisation occurs before v has had much effect. Eventually, however, v does lead to significant meridional transport of the zonally-symmetrized tracer field.
    Flows that are not zonally-symmetric are much more effective at bringing about latitudinal transport. Consider for example the steady flow depicted in Fig.1, which includes a meandering 'jet stream'. If tracer is injected at a low-latitude source region and removed in a high-latitude sink, a systematic poleward transport is performed by this wavy flow in conjunction with non-conservative processes (Pyle and Rogers, 1980).
    Of course, the real middle atmosphere -- especially the northern winter stratosphere -- is much more complex still, with rapid meridional transport by transient, large-amplitude planetary-wave motions. This is vividly illustrated by maps of quasi-conservative tracers derived from satellite measurements: see for example McIntyre and Palmer (1983, 1984) and Clough, Grahame and O'Neill (1985) for maps of Ertel's potential vorticity on isentropic surfaces, Jones (1984) for methane maps and Leovy et al.(1985) for ozone maps. (In particular, the latter paper indicates how ozone is transported polewards, from photochemical source regions in low latitudes to high northern latitudes, in narrow tongues associated with the transient major and minor warming events occurring in the northern winter.) Much current work is being devoted to attempts to quantify the information provided by such maps, and interpret them in

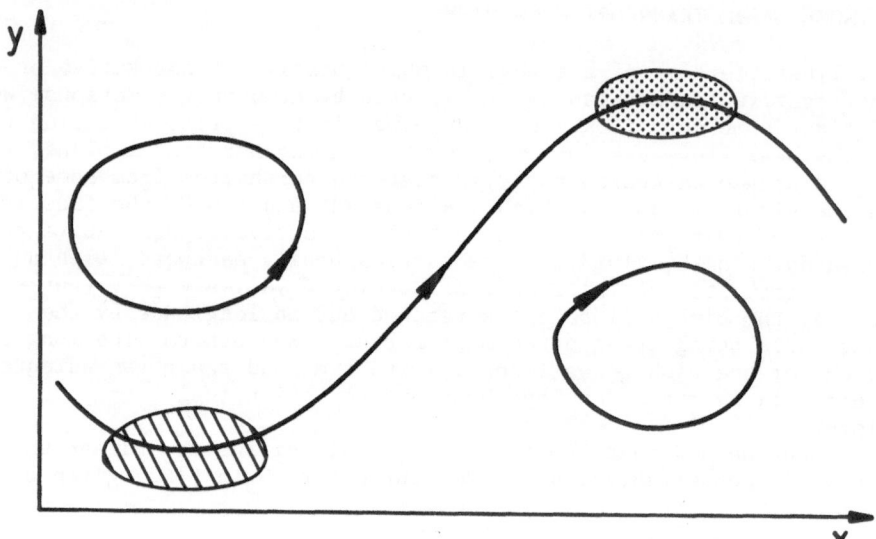

Figure 1. Illustrating how a steady meandering jet stream can transport tracer meridionally from a low-latitude source (shown hatched) to a high-latitude sink (shown stippled).

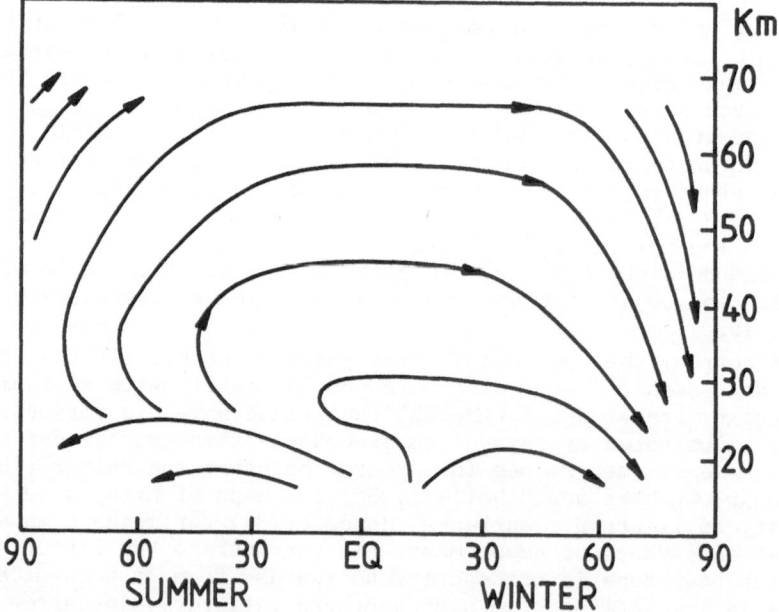

Figure 2. Schematic representation of Murgatroyd and Singleton's diabatic circulation at the solstices. (After Dunkerton, 1978.)

dynamically-meaningful ways (e.g. Butchart and Remsberg, 1986).
    In addition to the quasi-horizontal, large-scale transport depicted
by these maps, significant vertical transport of tracers may be brought
about by other types of wave, notably breaking gravity waves in the
upper mesosphere: see, e.g., Garcia and Solomon (1985).

3. MODELS OF TRANSPORT

3.1 One-dimensional models

One-dimensional models have been much used by atmospheric chemists to
predict the vertical structure of constituents. To demonstrate this
approach we introduce the log-pressure vertical coordinate

$$z \equiv -H\ln(p/p_s),$$ (3.1)

where H is a constant, representative scale height (typically about 7
km in the middle atmosphere). Then the mass-continuity equation in
Cartesian coordinates is

$$\frac{\partial u}{\partial x} + \frac{\partial v}{\partial y} + \rho_0^{-1}\frac{\partial(\rho_0 w)}{\partial z} = 0,$$ (3.2)

where $w = Dz/Dt$ and $\rho_0(z) = p/gH \propto e^{-z/H}$, g being the acceleration of
gravity. Using (3.2) the tracer continuity equation (1.2) can be written
in 'flux form'

$$\frac{\partial \chi}{\partial t} + \frac{\partial}{\partial x}(\chi u) + \frac{\partial}{\partial y}(\chi v) + \rho_0^{-1}\frac{\partial}{\partial z}(\rho_0 \chi w) = S .$$ (3.3)

The horizontal global integral of any quantity ψ is

$$<\psi> \equiv \iint_{globe} \psi(x,y,z,t) \; dx \; dy \; ;$$ (3.4)

$<\psi>$ is a function of z and t. On applying (3.4) to (3.3) and noting
that the global integral of the horizontal divergence vanishes, we
obtain

$$\frac{\partial <\chi>}{\partial t} = -\rho_0^{-1}\frac{\partial}{\partial z}(\rho_0<\chi w>) + <S> .$$ (3.5)

The term $<\chi w>$ represents the globally-integrated vertical flux of
tracer. To obtain a prognostic equation for $<\chi>$ it is usually assumed
that this flux can be represented by diffusion down the (vertical)
gradient of $<\chi>$; thus

$$<\chi w> = - K(z)\frac{\partial <\chi>}{\partial z}$$ (3.6)

where $K(z)$ is an empirical diffusion coefficient. On substituting (3.6) into (3.5) we obtain a diffusion equation for $\langle \chi \rangle$, with sources and sinks:

$$\frac{\partial \langle \chi \rangle}{\partial t} = \rho_0^{-1} \frac{\partial}{\partial z} \{ \rho_0 K \frac{\partial \langle \chi \rangle}{\partial z} \} + \langle S \rangle \ . \tag{3.7}$$

This approach can be generalised to allow for many constituents χ_n ($n = 1, 2, \ldots N$) where the relevant sources $\langle S_n \rangle$ may depend on all the $\langle \chi_m \rangle$; in general a different K_n should be used for each constituent.
    The advantage of the one–dimensional approach is that the representation of transport is so simple that the available computer power can be devoted mainly to consideration of a large number of chemical species and many reactions. Models of this kind are therefore valuable for testing and comparing photochemical theories. However, since such models deal with global (or sometimes hemispheric) integrals, it is difficult to justify their comparison with local measurements of constituents in the real atmosphere. It is also difficult to give a convincing justification for the 'diffusive flux' parameterisation (3.6), although recent papers by Holton (1986) and Mahlman, Levy and Moxim (1986) have provided some rationale for such an approach.

3.2 Two–dimensional models

A less restrictive approach is to employ the zonal (Eulerian) average $\overline{\psi}(y,z,t)$. Although this is again a nonlocal quantity, the zonal-homogenisation process mentioned above may frequently be relevant to the middle atmosphere, in the absence of strong planetary–wave motions. Moreover, satellite measurements of composition are often averaged zonally to improve signal–to–noise levels. A study of the transport of $\overline{\chi}$ in the meridional plane is therefore of some interest.
    We denote deviations from the zonal mean (loosely referred to as the 'eddy' or 'wave' contributions) by primes; (3.2) and (3.3) then yield the 'Eulerian mean' equation

$$\frac{\partial \overline{\chi}}{\partial t} + \overline{v} \frac{\partial \overline{\chi}}{\partial y} + \overline{w} \frac{\partial \overline{\chi}}{\partial z} = - [\frac{\partial}{\partial y} (\overline{v'\chi'}) + \rho_0^{-1} \frac{\partial}{\partial z} (\rho_0 \overline{w'\chi'})] + \overline{S} \tag{3.8}$$

after zonal averaging. This equation shows that the rate of change of $\overline{\chi}$ is determined not only by the zonal-mean sources and sinks and the advection of $\overline{\chi}$ by the Eulerian-mean wind ($\overline{v}$, $\overline{w}$), but also by an 'eddy flux divergence' (in square brackets). In practice, it is often found that the 'mean advection' and 'eddy flux' terms almost cancel: the reason is that the Eulerian mean circulation ($\overline{v}$, $\overline{w}$) is not independent of the presence of the eddies -- indeed it is usually largely determined by them. One should therefore not attach much physical significance to the labels 'mean transport' and 'eddy transport' that are often given to these expressions. In fact it turns out that there are other, quite different, ways of arranging eq.(3.8) into 'mean advective transport' and 'eddy flux divergence' terms: these will be discussed below. Before doing this, however, we recall some historical details.

Some early ideas about meridional transport of tracers in the lower stratosphere were introduced by Brewer (1949) and Dobson (1956), who suggested that observed distributions of water vapour and ozone might qualitatively be accounted for by advection by a hypothetical zonally-symmetric circulation. This flow consists of a rising branch passing through the tropical tropopause, and poleward and downward flow in the extratropics; it has come to be known as a 'Brewer-Dobson circulation'.

A few years later, Murgatroyd and Singleton (1961) performed a diagnostic calculation of a 'possible meridional circulation' in the middle atmosphere, here denoted by $(\overline{v}_D, \overline{w}_D)$, from the spherical-geometry equivalents of the zonal-mean thermodynamic equation

$$\frac{\partial \overline{T}}{\partial t} + \overline{v}_D \frac{\partial \overline{T}}{\partial y} + \overline{w}_D \left(\frac{\partial \overline{T}}{\partial z} + \frac{\kappa \overline{T}}{H} \right) = \frac{\overline{J}}{c_p} \qquad (3.9)$$

and mass-continuity equation

$$\frac{\partial \overline{v}_D}{\partial y} + \rho_0^{-1} \frac{\partial}{\partial z} (\rho_0 \overline{w}_D) = 0, \qquad (3.10)$$

together with boundary conditions $\overline{v}_D = 0$ at the poles. They obtained the requisite $\overline{T}(y,z,t)$ from observations and the net heating rate $\overline{J}$ from a radiative calculation using $\overline{T}$ and observed ozone fields. However, they did not include any eddy heat flux divergence terms in the thermodynamic equation (3.9). A sketch of the resulting circulation at the solstices is given in Fig.2: in the lower stratosphere it resembles the Brewer-Dobson circulation, while higher up it consists of rising motion in the summer hemisphere, cross-equatorial flow in the mesosphere and descent in the winter hemisphere. A circulation diagnosed by this method is now often called a 'diabatic circulation'. (The $\partial \overline{T}/\partial t$ term is usually small, and is sometimes omitted from the calculation.)

Murgatroyd and Singleton noted that their results, being based on zonal-mean observations, necessarily omitted the 'eddy heat flux convergence' term

$$C \equiv - \frac{\partial}{\partial y} (\overline{v'T'}) - \rho_0^{-1} \frac{\partial}{\partial z} (\rho_0 \overline{w'T'}) - \frac{\kappa \overline{w'T'}}{H} \qquad (3.11)$$

that would appear on the right of (3.9) if $(\overline{v}_D, \overline{w}_D)$ were to be replaced by $(\overline{v}, \overline{w})$. Thus their diabatic circulation differs from the Eulerian-mean circulation in regions where C is significant, such as the northern winter stratosphere: in this region Eulerian-mean ascent, rather than descent, is found.

We now return to the alternative formulations of the zonal-mean tracer equation (3.8) mentioned above. One of these is the so-called 'transformed Eulerian-mean' approach, in which the 'residual circulation' $(\overline{v}^*, \overline{w}^*)$, defined by

$$\overline{v}^* \equiv \overline{v} - \rho_0^{-1} \frac{\partial}{\partial z} (\rho_0 \Psi), \quad \overline{w}^* \equiv \overline{w} + \frac{\partial \Psi}{\partial y} , \qquad (3.12)$$

is introduced, where

$$\Psi \equiv \overline{v'T'} / \left(\frac{\partial \overline{T}}{\partial z} + \frac{\kappa \overline{T}}{H} \right) \tag{3.13}$$

(Andrews and McIntyre, 1976, 1978a). Then in place of (3.8) we have

$$\frac{\partial \overline{\chi}}{\partial t} + \overline{v}^* \frac{\partial \overline{\chi}}{\partial y} + \overline{w}^* \frac{\partial \overline{\chi}}{\partial z} = E + \overline{S} \tag{3.14}$$

where $\rho_0 E$ is the divergence of a mean quadratic function of eddy quantities. The relevance of the residual circulation in the present context is that it satisfies not only a mass continuity equation identical in form to (3.10) but also a thermodynamic equation

$$\frac{\partial \overline{T}}{\partial t} + \overline{v}^* \frac{\partial \overline{T}}{\partial y} + \overline{w}^* \left(\frac{\partial \overline{T}}{\partial z} + \frac{\kappa \overline{T}}{H} \right) = G + \frac{\overline{J}}{c_p} \tag{3.15}$$

with a mean quadratic eddy term G. In contrast to the quantity C defined in (3.11), G is formally negligible for quasigeostrophic motions. In such a case, and indeed under a wider variety of circumstances as well, $(\overline{v}_D, \overline{w}_D) \simeq (\overline{v}^*, \overline{w}^*)$; thus Murgatroyd and Singleton's diabatic circulation is a good approximation to the residual circulation in many cases of atmospheric interest. It is also quite closely related to the 'generalised Lagrangian-mean' circulation (Andrews and McIntyre, 1978b) defined, roughly speaking, as an average following sets of moving fluid parcels rather than along lines of constant latitude and log-pressure altitude, as with the Eulerian mean. More details of the relationships between these circulations are given in Section 6.5.2 of WMO (1985).

We now briefly mention the specification of two-dimensional (2D) models for transport and chemical studies. (A comprehensive account is to be found in Chapter 12 of WMO, 1985.) These tend to use either (a) the Eulerian-mean formulation or (b) the residual, diabatic or similar formulation. In each case some form of eddy flux convergence term generally appears in the tracer continuity equation. (An exception is the generalised Lagrangian-mean formulation, in which, however, a mass-continuity equation of the form (3.10) does not hold.) This eddy flux must be parameterised in terms of the zonal-mean state with which the model is working; this is usually done by means of a two-dimensional 'diffusion tensor' $\underset{\sim}{K}$, relating the vector eddy flux to the mean gradient $(\partial \overline{\chi}/\partial y, \partial \overline{\chi}/\partial z)$.

A parameterisation of this type is difficult to justify rigorously except in one formulation of type (b), based on the 'effective transport velocity' introduced by Plumb (1979). This is defined by

$$\overline{v}^T \equiv \overline{v} - \rho_0^{-1} \frac{\partial}{\partial z}(\rho_0 \Xi) , \quad \overline{w}^T \equiv \overline{w} + \frac{\partial \Xi}{\partial y} , \tag{3.16}$$

where

$$\Xi \equiv \tfrac{1}{2}(\overline{\eta' w'} - \overline{\zeta' v'}) \tag{3.17}$$

and (η', ζ') are the northward and vertical eddy parcel displacements. The diffusion tensor is given by

$$\underset{\sim}{K} \equiv \frac{1}{2}\frac{\partial}{\partial t}\left(\begin{array}{cc} \overline{\eta'^2} & \overline{\eta'\zeta'} \\ \overline{\eta'\zeta'} & \overline{\zeta'^2} \end{array}\right) \tag{3.18}$$

in the absence of eddy sources and sinks of tracer (i.e. $S' = 0$). Given these definitions, the tracer continuity equation becomes

$$\frac{\partial\overline{\chi}}{\partial t} + \overline{v}^T\frac{\partial\overline{\chi}}{\partial y} + \overline{w}^T\frac{\partial\overline{\chi}}{\partial z} = \rho_0^{-1}\nabla\cdot(\rho_0\underset{\sim}{K}\cdot\nabla\overline{\chi}) + \overline{S} , \tag{3.19}$$

with errors that are cubic in eddy amplitude, and the mass–continuity equation is analogous to (3.10). In this formulation the rate of change of $\overline{\chi}$ thus depends on advection by the effective transport velocity, diffusion by $\underset{\sim}{K}$, and the mean sources and sinks. Note that the diffusion is generally anisotropic and may even take the form of 'anti–diffusion' if, for example, all parcel displacements are decreasing with time. If S' is nonzero but weak a similar formulation applies, with additional terms appearing in $\underset{\sim}{K}$.

Owing to the Lagrangian nature of Ξ and $\underset{\sim}{K}$, it is in general difficult to estimate them from atmospheric observations. However, Plumb and Mahlman (1987) derived $(\overline{v}^T, \overline{w}^T)$ and $\underset{\sim}{K}$, month–by–month, from a three-dimensional general circulation model of the troposphere and lower stratosphere. They then used these quantities in a 2D model incorporating (3.19), which was found to reproduce quite well the zonal–mean evolution of the 3D model's tracer fields. (The errors incurred by neglecting the terms cubic in amplitude were thus mostly fairly small in this case.) Further discussion of this approach is given in Section 6.5 of WMO (1985), where it is noted, among other things, that $(\overline{v}^T, \overline{w}^T)$ and $\underset{\sim}{K}$ must be specified in advance for the 2D model, and cannot self-consistently be allowed to respond to changes in the 2D model fields: a full 3D model would be needed to calculate the required changes in the eddy statistics.

A related approach that has also been used in 2D transport models is the isentropic formulation, in which θ is used as a vertical coordinate. Under quasi–adiabatic conditions the 'vertical' velocity and displacement are small. This method has been employed by Tung (1982, 1986) and Ko et al. (1985).

Advantages of 2D models over 1D models include the facts that they can in principle represent meridional transport in a fairly self-consistent manner and can be validated against zonal–mean observations at different latitudes. Furthermore, their dynamical formulation is still sufficiently simple to allow a large proportion of the computing resources to be devoted to detailed chemistry. On the other hand they do suffer from the essentially 'non–interactive' nature of the eddy parameterisations mentioned above, which can be fully circumvented only by use of 3D models.

3.3 Three-dimensional models

We conclude this section with a few brief remarks on 3D models of
middle atmosphere transport. In principle, these may treat the coupling
between fully three-dimensional dynamics, radiation and photochemistry.
A major advantage of 3D models is that they explicitly describe the
dynamics of the large-scale planetary waves and the transport associated
with them. Of course, such models still need to parameterise the effects
of disturbances of smaller scale than the grid size, including gravity
waves. 3D models can also represent comparatively localised events and
handle the transport of local injections of tracer. This may make for
more meaningful validation of the models against observational data.
Unlike the 2D or 1D models, no artificial separation is made into zonal
or global means and departures therefrom.

Owing to the large computer resources required for dealing with
the dynamics and radiation, few 3D models have yet included any but the
simplest chemical schemes. An approach that is often adopted is to
perform the tracer studies 'off-line': the model's velocity fields are
stored, and are used in a separate calculation to advect the tracer
around. This allows dynamically-passive tracers to be treated, with or
without active chemistry. However, dynamically-active tracers, which
influence the motion or temperature fields, cannot be studied in this
way, but must be incorporated in a fully interactive 'on-line' calculation. Examples of the off-line method include those of Mahlman and his
co-workers [see, e.g., Mahlman and Moxim (1978), Mahlman, Levy and
Moxim (1980, 1986), Mahlman (1985) and Plumb and Mahlman (1987)], and
those of Grose and his colleagues [Grose, Turner and Nealy (1985); also
this volume]. An example of the on-line approach, including a simple
but accurate parameterisation of ozone photochemistry, is that of
Cariolle and Déqué (1986).

Finally, it should not be forgotten that large 3D models generate
vast quantities of data, which must be suitably organised and compressed
if meaningful comparisons with observations are to be made, and physical
interpretation carried out. There is a clear need for further discrimin-
ating diagnostics to aid this process.

4. CONCLUSION

This brief tutorial review has attempted to give a flavour of the fundamentals of atmospheric transport phenomena, and some idea of the types
of model that are currently being used to study problems of chemical
transport in the stratosphere and mesosphere. Other papers in this
volume will consider specific applications in more detail.

REFERENCES

Andrews, D.G. and McIntyre, M.E., 1976: Planetary waves in horizontal
and vertical shear: the generalized Eliassen-Palm relation and the mean
zonal acceleration. J. Atmos. Sci. 33, 2031-2048.

Andrews, D.G. and McIntyre, M.E., 1978a: Generalized Eliassen-Palm and Charney-Drazin theorems for waves on axisymmetric mean flows in compressible atmospheres. J. Atmos. Sci. 35, 175-185.

Andrews, D.G. and McIntyre, M.E., 1978b: An exact theory for nonlinear waves on a Lagrangian-mean flow. J. Fluid Mech. 89, 609-646.

Brewer, A.W., 1949: Evidence for a world circulation provided by the measurements of helium and water vapour distribution in the stratosphere. Quart. J. Roy. Meteor. Soc. 75, 351-363.

Butchart, N. and Remsberg, E.E., 1986: The area of the stratospheric polar vortex as a diagnostic for tracer transport on an isentropic surface. J. Atmos. Sci. 43, 1319-1339.

Cariolle, D. and Déqué, M., 1986: Southern hemisphere medium-scale waves and total ozone disturbances in a spectral general circulation model. J. Geophys. Res. 91, 10,825-10,846.

Clough, S.A., Grahame, N.S. and O'Neill, A., 1985: Potential vorticity in the stratosphere derived using data from satellites. Quart. J. Roy. Meteor. Soc. 111, 335-358.

Dobson, G.M.B., 1956: Origin and distribution of the polyatomic molecules in the atmosphere. Proc. Roy. Soc. Lond. A 236, 187-193.

Dunkerton, T.J., 1978: On the mean meridional mass motions of the stratosphere and mesosphere. J. Atmos. Sci. 35, 2325-2333.

Garcia, R.R. and Solomon, S., 1985: The effect of breaking gravity waves on the dynamics and chemical composition of the mesosphere and lower thermosphere. J. Geophys. Res. 90, 3850-3868.

Grose, W.L., Turner, R.E. and Nealy, J.E., 1985: Transport processes in the stratosphere: model simulations and comparisons with satellite observations. Middle Atmosphere Program: Handbook for MAP, vol. 18, pp. 381-385.

Holton, J.R., 1986: A dynamically based transport parameterization for one-dimensional photochemical models of the stratosphere. J. Geophys. Res. 91, 2681-2686.

Hoskins, B.J., McIntyre, M.E. and Robertson, A.W., 1985: On the use and significance of isentropic potential vorticity maps. Quart. J. Roy. Meteor. Soc. 111, 877-946.

Jones, R.L., 1984: Satellite measurements of atmospheric composition: three years' observations of CH_4 and N_2O. Adv. Space Res. 4, No.4, 121-130.

Ko, M.K.W., Tung, K.K., Weinstein, D.K. and Sze, N.D., 1985: A zonal-mean model of stratospheric tracer transport in isentropic coordinates: numerical simulations for nitrous oxide and nitric acid. J. Geophys. Res. 90, 2313-2329.

Leovy, C.B., Sun, C.-R., Hitchman, M.H., Remsberg, E.E., Russell, J.M., Gordley, L.L., Gille, J.C. and Lyjak, L.V., 1985: Transport of ozone in the middle atmosphere: evidence for planetary wave breaking. J. Atmos. Sci. 42, 230-244.

McIntyre, M.E. and Palmer, T.N., 1983: Breaking planetary waves in the stratosphere. Nature 305, 593-600.

McIntyre, M.E. and Palmer, T.N., 1984: The 'surf zone' in the strato-sphere. J. Atmos. Terrest. Phys. 46, 825-849.

Mahlman, J.D., 1985: Mechanistic interpretation of stratospheric tracer transport. Adv. Geophys. 28A, 301-323.

Mahlman, J.D., Levy, H. and Moxim, W.J., 1980: Three-dimensional tracer structure and behavior as simulated in two ozone precursor experiments. J. Atmos. Sci. 37, 655-685.

Mahlman, J.D., Levy, H. and Moxim, W.J., 1986: Three-dimensional simu-lations of stratospheric N_2O: predictions for other trace constituents. J. Geophys. Res. 91, 2687-2707.   Also corrigenda, 91, 9921.

Mahlman, J.D. and Moxim, W.J., 1978: Tracer simulation using a global general circulation model: results from a midlatitude instantaneous source experiment. J. Atmos. Sci. 35, 1340-1374.

Murgatroyd, R.J. and Singleton, F., 1961: Possible meridional circula-tions in the stratosphere and mesosphere. Quart. J. Roy. Meteor. Soc. 87, 125-135.

Plumb, R.A., 1979: Eddy fluxes of conserved quantities by small-amplit-ude waves. J. Atmos. Sci. 36, 1699-1704.

Plumb, R.A. and Mahlman, J.D., 1987: The zonally-averaged transport characteristics of the GFDL general circulation/transport model. J. Atmos. Sci., 44, 298-327.

Pyle, J.A. and Rogers, C.F., 1980: Stratospheric transport by stationary planetary waves -- the importance of chemical processes. Quart. J. Roy. Meteor. Soc. 106, 421-446.

Taylor, G.I., 1953: Dispersion of soluble matter in solvent flowing slowly through a tube. Proc. Roy. Soc. Lond. A 219, 186-203.

Tung, K.K., 1982: On the two-dimensional transport of stratospheric trace gases in isentropic coordinates. J. Atmos. Sci. 39, 2330-2355.

Tung, K.K., 1986: Nongeostrophic theory of zonally averaged circulation. Part I: Formulation. J. Atmos. Sci., to appear.

WMO, 1985: World Meteorological Organization: Global Ozone Research and Monitoring Project. Report No. 16. Atmospheric Ozone, 1985.

A COUPLED MODEL OF ZONALLY AVERAGED DYNAMICS, RADIATION AND CHEMISTRY

K. K. Tung
Department of Mathematics and Computer Science
Clarkson University
Potsdam, NY 13676 U.S.A.

ABSTRACT. A brief review is given of the three generations of 2-D models used in chemical transport studies, with emphasis on the "third generation", coupled models in both isobaric and isentropic coordinates.

1. INTRODUCTION

Excellent reviews on the development of zonally averaged models (or the so-called 2-D models) of chemical transport for the middle atmosphere can be found in WMO (1986), chapters 6 and 12. The present lecture concentrates on recent developments and on a specific "third generation" 2-D model in isentropic coordinates that we have recently put into operation.

1.1 Some General Preliminary Remarks

Let χ_i be the mass mixing ratio of species i and S_i be its source. The three-dimensional transport equation is

$$\frac{d}{dt} \chi_i = S_i (\chi_j) , \qquad \begin{array}{l} i = 1,2, \ldots., N \\ \\ j = 1,2, \ldots., N \end{array} \qquad (1)$$

for N interacting chemical species, where

$$\frac{d}{dt} \equiv \frac{\partial}{\partial t} + u \frac{\partial}{\partial x} + v \frac{\partial}{\partial y} + w \frac{\partial}{\partial z}$$

Let an overhead bar denote average over longitude λ (dx = $a\cos\phi d\lambda$, ϕ = latitude, y = $a\sin\phi$), and prime denote the deviation from the zonal average, i.e.

G. Visconti and R. Garcia (eds.), Transport Processes in the Middle Atmosphere, 183–198.

$$\overline{(\quad)} \equiv \frac{1}{2\pi} \int_0^{2\pi} (\quad) \, d\lambda$$

$$(\quad)' \equiv (\quad) - \overline{(\quad)} \, .$$

By taking the zonal average of Eq. (1), one obtains the 2-D equation of transport, which is, in log-pressure coordinates for example:

$$\frac{\partial}{\partial t} \, \overline{\chi}_i + \overline{v}\cos\phi \, \frac{\partial}{\partial y} \, \overline{\chi}_i + \overline{w} \, \frac{\partial}{\partial z} \, \overline{\chi}_i + \frac{\partial}{\partial y} \, \overline{v'\chi_i' \cos\phi} +$$

$$+ \frac{1}{\rho_o} \, \frac{\partial}{\partial z} \, \rho_o \, \overline{w'\chi_i'} = S_i(\chi_j) \tag{2}$$

1.2.   Treatment of Eddy Fluxes:

The presence of eddy flux terms $(\overline{v'\chi_i'}$ and $\overline{w'\chi_i'})$ in Eq. (1) poses a closure problem for the 2-D equations. The adopted approach using linear wave theory has been described, in Plumb (1979), Matsuno (1980), Holton (1980, 1981), Tung (1982, 1984) and WMO (1986). By defining the eddy displacement fields:

$$(\frac{\partial}{\partial t} + \overline{u} \, \frac{\partial}{\partial x})\eta' = v', \quad (\frac{\partial}{\partial t} + \overline{u} \, \frac{\partial}{\partial x})\zeta' = w'$$

and $$(\frac{\partial}{\partial t} + \overline{u} \, \frac{\partial}{\partial x})\sigma_i' = S_i' \, ,$$

and using the linear perturbation form of Eq. (1):

$$(\frac{\partial}{\partial t} + \overline{u} \, \frac{\partial}{\partial x})\chi_i' + v'\cos\phi\frac{\partial}{\partial y} \, \overline{\chi}_i + w' \, \frac{\partial}{\partial z} \, \overline{\chi}_i = S_i' \, ,$$

one "solves" for χ_i' as

$$\chi_i' = - \, \eta'\cos\phi \, \frac{\partial}{\partial y} \, \overline{\chi}_i - \zeta' \, \frac{\partial}{\partial z} \, \overline{\chi}_i + \sigma_i' \tag{3}$$

From it one forms:

$$
\begin{bmatrix}
\overline{v'\chi_i'}\cos\phi \\
\\
\\
\overline{w'\chi_i'}
\end{bmatrix}
= - \; \mathbb{K}\cdot\nabla\;\overline{\chi}_i +
\begin{bmatrix}
\overline{v'\sigma_i'}\cos\phi \\
\\
\\
\overline{w'\sigma_i'}
\end{bmatrix} , \tag{4}
$$

where $\nabla \equiv \vec{J}\,\dfrac{\partial}{\partial y} + \vec{k}\,\dfrac{\partial}{\partial z}$ here.

The last term in (4), involving perturbations in the chemical source term (σ_i'), is called chemical eddy fluxes. Its parameterization has been described in Tung (1982) for general chemical reactions and in Plumb (1979), Matsuno (1980), and Tung (1984) for a simple relaxation form for S_i, and will not be discussed here. The "K-tensor",

$$
\mathbb{K} \equiv
\begin{bmatrix}
K_{yy}\cos^2\phi & K_{yz}\cos\phi \\
\\
K_{zy}\cos\phi & K_{zz}
\end{bmatrix}
\equiv
\begin{bmatrix}
\overline{v'\eta'}\cos^2\phi & \overline{v'\zeta'}\cos\phi \\
\\
\overline{w'\eta'}\cos\phi & \overline{w'\zeta'}
\end{bmatrix} \tag{5}
$$

is independent of the concentration of the minor species, χ_i, being transported, although some authors have, erroneously, called it the "chemical eddies".

The K-tensor term, when substituted into Eq. (2), gives rise to a diffusive transport of $\overline{\chi}_i$ if the tensor is symmetric and an advective transport if it is antisymmetric.

2.   EVOLUTION OF 2-D MODELS.

The first generation, classical Eulerian models generally followed the parameterization of Reed and German (1965), in which they took instead of (3) , the definition:

$$
v' = \frac{1}{\Delta t}\,\eta' \quad \text{and} \quad w' = \frac{1}{\Delta t}\,\zeta' \quad ,
$$

with the consequence that

$$
K_{yz} = \frac{1}{\Delta t}\,\overline{\eta'\zeta'} = K_{zy} \quad .
$$

Hence the K-tensor used in these models were taken to be symmetric and so diffusive in nature. Advective transport was provided solely by the Eulerian mean advection $(\overline{v}, \overline{w})_p$ in pressure coordinates.
The importance of the advective transport provided by the antisymmetric part was recognized and incorporated in the second

generation of 2-D models (see references quoted in section 1.2).
This advective transport is included in the definition of the
" transport mean circulation", $(\bar{v}_T, \bar{w}_T)_p$, used in place of $(\bar{v}, \bar{w})_p$,
in Eq. (2), resulting in (see WMO (1986)):

$$\frac{\partial}{\partial t} \bar{\chi}_i + \bar{v}_T \cos\phi \frac{\partial}{\partial y} \bar{\chi}_i + \bar{w}_T \frac{\partial}{\partial z} \bar{\chi}_i \qquad (6)$$

$$= \frac{1}{\rho_o} \nabla \cdot \rho_o \, \mathbb{K}_s \cdot \nabla \bar{\chi}_i + \bar{S}_i + \text{(chemical eddies)}$$

Here $\mathbb{K}_s$ is the symmetric part of $\mathbb{K}$ and represent diffusive
transport by transient eddies only.

In implementation, the transport velocities are usually
approximated by the "residual mean circulation", $(\bar{v}^*, \bar{w}^*)_p$, and
$\mathbb{K}_s$ by simple Fickian diffusion coefficients with values generally
smaller than those used in the first generation models.   To
deduce $(\bar{v}^*, \bar{w}^*)_p$, most models further adopted Dunkerton's approximation (Dunkerton,   1978):

$$\bar{w}^* = \bar{Q}/\Gamma,$$

of the thermodynamics equation, with $\bar{Q}$ being the net diabatic heating
rate and Γ the static stability parameter.   The circulation thus
deduced is called the "diabatic circulation".

The somewhat confusing situation of having to deal with three
different mean circulations —— Eulerian, residual and diabatic
mean circulations —— does not arise when the zonal averaging is
performed along isentropic surfaces.   The Eulerian mean $(\bar{v}, \bar{w})$ is
the same as the diabatic mean circulation and there is no need to
artificially define a residual mean circulation.   Furthermore in
isentropic coordinates the diffusion tensor, $\mathbb{K}_s$, in Eq. (6) is
dominated in the lower stratosphere by one component, $K_{yy} = \partial/\partial t$
$\eta'2$, which represents mixing by transient waves along isentropes
(Mahlman et al., 1984; Tung, 1984; Ko et al., 1985), i.e.

$$\mathbb{K}_s \cong \begin{bmatrix} K_{yy} \cos^2\phi & 0 \\ & \\ 0 & 0 \end{bmatrix}$$

in isentropic coordinates in the absence of significant cross
isentropic mixing by breaking gravity waves.

Thus the second generation models have narrowed the transport
parameters to two fundamental quantities: the net diabatic heating
rate $\bar{Q}$, which produces the advective transport, and the isentropic
mixing coefficient K_{yy}, which smooths out the mean gradients of
tracers generated by the diabatic circulation.   In most operating
2-D models of this kind, $\bar{Q}$ is based on the radiative calculations
of Dopplick (1979) below 25km and some scaled version of Murgatroyd

and Singleton (1961) above (These include Miller et al., (1981),
Guthrie et al., (1984), Ko et al., (1985), Stordal et al., (1985),
while the model of Garcia and Solomon (1983) used $\bar{Q}$ calculated using
the Newtonian cooling parameterization and residual circulations
calculated from energy and momentum equation using Rayleigh
friction parameterization of the eddy momentum sources.) For K_{yy},
most adopted small values in the range of 2 to 4 x 10^9 cm^2/s based
on gross estimates of Kida (1983) and Tung (1984) and most adopted
negligible values for K_{yz} and K_{zz} in the lower stratosphere although
there were exceptions.

   Although it was recognized that in reality K_{yy} should be a
function of space as well as time, and that it should ultimately be
related to the other transport quantity $\bar{Q}$ (see comments in Ko et al.,
(1985)), a constant fixed value for K_{yy} was adopted in the absence of
a known algorithm for relating the two. This has to await the
development of the third generation 2-D models.

   In the meantime there were some efforts to deduce K_{yy} from the
outputs of three dimensional models. Notable is the work of Pitari
and Visconti (1985) , who deduced directly from definition K_{yy} and
other diffusion coefficients (including even the chemical eddy terms)
from a 3-D quasi- geostrophic model. Although the magnitudes of K_{yy}'s
were sensitive to initialization and time averaging procedures adopted,
they are generally of comparable magnitudes as those used in the 2-D
models, at least in the lower stratosphere. The strong latitudinal
dependence of K_{yy} was emphasized by Plumb and Mahlman (1987), whose
K_{yy}, deduced by running tracer experiments using wind fields from
a general circulation model, showed an order of magnitude larger
values in a "surf zone" centered around the tropical zero-wind
line than outside. These and other 3-D experiments are the subjects
of review by Plumb (1987) in this volume. Other efforts using
atmospheric data have followed and some of these will be discussed
later. It appears that, with the limited amount of results now
available, the "surf zone" behavior is found only in the GFDL GCM
used by Plumb and Mahlman (1987) and it is probably premature to
conclude that it is representative of the behavior in the real
atmosphere.

   Given the large variability of the observed atmospheric mean
circulation from month to month and from year to year (Geller et al.,
1984) it is probably meaningless to define a "typical" profile for
K_{yy}. Instead a better defined approach will be to determine the
K_{yy} distribution that is consistent with the other transport
parameter, e.q. $\bar{Q}$, that is being used. This, then, forms the
goal of the third generation models

3. THIRD GENERATION OF 2-D MODELS.

The ideas for achieving self-consistency between $\bar{Q}$ and K_{yy} existed
much earlier (see Dickinson (1969) and Edmon et al., (1980)). They
were utilized in the context of 2-D modeling first by Plumb and
Mahlman (1987) and Newman et al., (1986) under the quasi-geostrophic

approximation. Tung (1986) recently gave a more general nongeostrophic formulation in isentropic coordinates.

The additional constraint (between $\bar{Q}$ and K_{yy}) is provided by the zonal momentum equation, which has not been used so far in most second generation models of transport utilizing the diabatic circulation. This equation can be written, in pressure coordinates, as

$$\left(\frac{\partial}{\partial t} + \bar{v}^*\cos\phi\frac{\partial}{\partial y} + \bar{w}^*\frac{\partial}{\partial z}\right)\bar{u}\cos\phi - f\bar{v}^*\cos\phi = \frac{1}{\rho_o}\nabla\cdot F \quad , \tag{7}$$

where $\nabla\cdot F$ is the Eliassen–Palm flux divergence, and represents the net eddy forcing of the zonal mean flow. Under the assumptions of quasi-geostrophy and adiabaticity for the eddies, the E–P flux divergence can be shown to be related to the meridional eddy flux of quasi- geostrophic potential vorticity q:

$$\nabla\cdot F = \rho_o \overline{v'q'}\cos\phi \tag{8}$$

(see Edmon et al., (1980, and references therein).

If one further assumes that q is conserved, i.e.

$$\frac{d}{dt}q = 0 \ .$$

Then by replacing χ_i with q and setting $S_i = 0$, the same treatment of eddy fluxes as discussed in section 1.2 can be applied to the "tracer" q, resulting in

$$\overline{v'q'\cos\phi} = - K_{yy}\cos^2\phi \frac{\partial}{\partial y}\bar{q} - K_{yz}\cos\phi \frac{\partial}{\partial z}\bar{q} \quad , \tag{9}$$

where the coefficients K_{yy} and K_{yz} are the same as those defined in (5) for tracer transports.

In pressure coordinates, on which (9) is based, the last two terms in Eq. (9) are usually of the same magnitude even if mixing can be assumed to act along isentropic surfaces (see Tung(1984)). To obtain a relationship between K_{yy} and K_{yz}, the single assumption that mixing acts along isentropic surfaces does not appear to be sufficient for a formulation in pressure coordinates. Conservation of potential temperature θ during eddy displacement (i.e. d/dt θ = 0) leads to, from (3) with χ_i replaced by θ,

$$\theta' = - \eta'\cos\phi \frac{\partial}{\partial y}\bar{\theta} - \zeta'\frac{\partial}{\partial z}\bar{\theta} \quad .$$

Therefore $K_{yz}\cos\phi/K_{yy}\cos^2\phi \equiv \overline{v'\zeta}/\overline{v'\eta'}\cos\phi$

$$= - \bar{\theta}_y/\bar{\theta}_z - \overline{v'\theta'}/(\overline{v'\eta'}\cos\phi\ \bar{\theta}_z) \tag{10}$$

The last term   represents the effect of meridional heat flux by
adiabatic waves and is usually not negligible. It has nevertheless
been dropped in some previous work with the assumption that eddy
displacements are along the surfaces of <u>zonally averaged</u> isentropes.
    This and the assumption that q is conserved on isobaric surfaces
are just part of inaccuracies that contribute to problems arising in
practice from using the above mentioned formulations. (The other
inaccuracy comes from using geostrophically derived winds (Robinson
(1986)). A resulting symptom is that the calculated K_{yy} is found to
have large regions with negative values even in the winter stratosphere
where one expects large-scale mixing to occur.
    In implementing the ideas mentioned above into an operating
2-D model of global extent, we have found it convenient to use the
formulation of Tung (1986) in isentropic coordinates. It appears to
require less assumptions, and leads to K_{yy}'s with fewer problems. The
formulation is, in addition, nongeostrophic, which is more consistent with the global formulation of the transport equation.

4. NONGEOSTROPHIC FORMULATION IN ISENTROPIC COORDINATES.

In this formulation, Ertel's potential vorticity,

$$\Pi \equiv \rho_\theta^{-1} \; [(\frac{\partial}{\partial x} v - \frac{\partial}{\partial y} u\cos\phi) + 2\Omega\sin\phi] \tag{11}$$

$$(\rho_\theta \equiv \rho \frac{\partial z}{\partial \theta} = -\frac{1}{g} \frac{\partial p}{\partial \theta})$$

is used instead of the quasi-geostrophic potential vorticity q.
Tung (1986) and Andrews (personal communication, 1986) have shown
that, in an isentropic coordinate formulation, the E-P flux
divergence is related to the meridional flux of Ertel's potential
vorticity in the following Taylor formula:

$$\nabla \cdot F = \overline{v'' \cos\phi(\rho_\theta \Pi'')} \tag{12}$$

$$(h'' \equiv h - \overline{\rho_\theta h}/\overline{\rho_\theta}) \quad .$$

This relationship is obtained with no restriction to geostrophy
or small amplitudes. It is an identity for adiabatic waves, and
has better than 10% accuracy for waves whose radiative damping
time scales are 5 days or longer.
    The nongeostrophic form of the vorticity equation and mass
continuity equation can be combined to give a dynamical equation for
the evolution of Π as

$$\frac{d}{dt}\Pi = \Pi \frac{\partial}{\partial \theta} \overset{\circ}{\theta} \quad ,$$

Treating Π as a chemical tracer (replacing χ_i by Π and S_i by $\Pi \ \partial/\partial\theta \ \overset{\circ}{\theta}$), the procedure outlined in section 1.2 can be used to obtain

$$\overline{v''\cos\phi(\rho_\theta\Pi'')} = - \ \bar\rho_\theta \ K_{yy}\cos^2\phi \ \frac{\partial}{\partial y} \ \bar\Pi \ - \ \bar\rho_\theta \ K_{y\theta}\cos\phi \ \frac{\partial}{\partial\theta} \ \bar\Pi$$

$$+ \ \bar\rho_\theta \ \bar V_E\cos\phi \ \bar\Pi \tag{13}$$

For eddy displacements that are predominantly along isentropes, the last two terms in Eq. (13) vanish. Thus

$$\frac{1}{\bar\rho_\theta} \nabla \cdot F \cong - \ \bar\rho_\theta \ K_{yy}\cos^2\phi \ \frac{\partial}{\partial y} \ \bar\Pi \tag{14}$$

Eq. (14) is the required parameterization for the Eliassen-Palm flux divergence. The assumption of quasi-adiabatic eddy displacements that is used in deriving it is expected to fail above approximately 40km where significant breaking-gravity-wave mixing may occur. In principle this problem can be solved with a parameterization for gravity wave mixing and with the left—hand side of Eq. (14) replaced by the difference between $1/\rho_\theta \ \nabla\cdot F$ and the parameterized momentum diffusion by gravity waves. However, at present such a parameterization is not accurate enough to be incorporated in our procedure. We choose instead to use (14) and restrict our attention to below 40km.

Although our formulae are nongeostrophic, one type of equatorial waves, the Kelvin waves, is not treated well at all. Because Kelvin waves produce no (almost no) meridional displacements, they do not enter into any term in (13) or (14). The effect of Kelvin waves should enter as an extra term on the right-hand side of Eq. (12). Without incorporating Kelvin waves in the present parameterization, the semi-annual oscillation cannot be properly treated. Again this phenomenon is important above 40km.

Restricting our attention to below 40km, and incorporating the zonal momentum equation with the eddy forcing term parameterized as in (14), we now have a procedure for coupling $\bar Q$ and K_{yy}. One algorithm is shown schematically in Figure 1. Instead of having to input two fields: $\bar Q$ and K_{yy}, as in the second generation 2-D models, only the quantity, $\bar T$, is needed as an input. From $\bar T$, the zonal momentum is known using the thermal wind (or the balanced wind) relationship. The amount of eddy forcing, $1/\rho_\theta \ \nabla\cdot F$, that is needed to maintain such a zonal momentum can then be inferred. From it, one can deduce K_{yy} using the parameterization (14). Such a pair of consistently deduced $\bar Q$ and K_{yy} enters into the chemical tracer equation as advective and diffusive transports, respectively. An alternative algorithm is that depicted in Figure 2. Again only one input function is required. This is, however, taken to be, K_{yy}, representing the level of 3-D wave activity. Knowing K_{yy} and using (14) the E-P flux divergence can be deduced given an initial zonal mean wind field. The evolution of zonal mean wind can be predicted

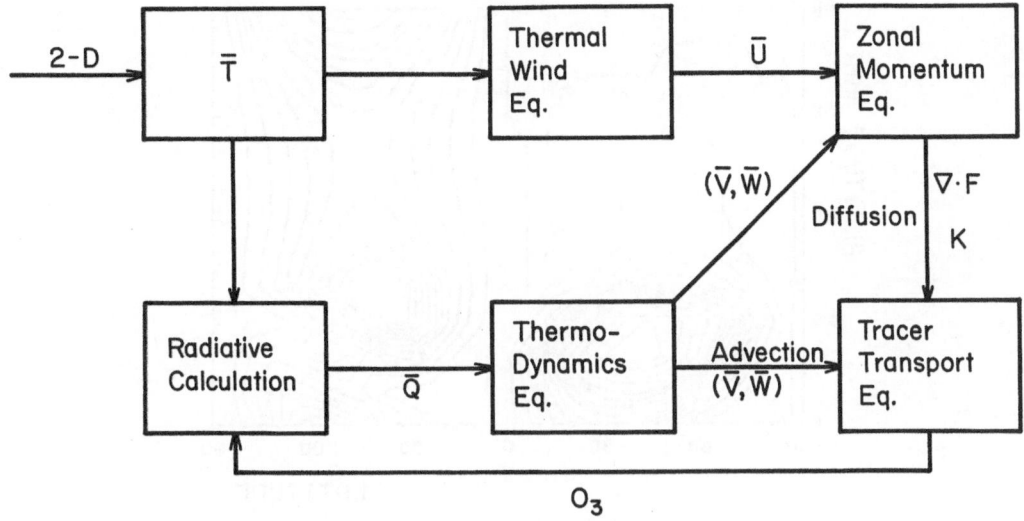

Figure 1.   Schematic diagram showing an algorithm for determining the
coupled set of transport parameters with prescribed $\bar{T}$ as
input.

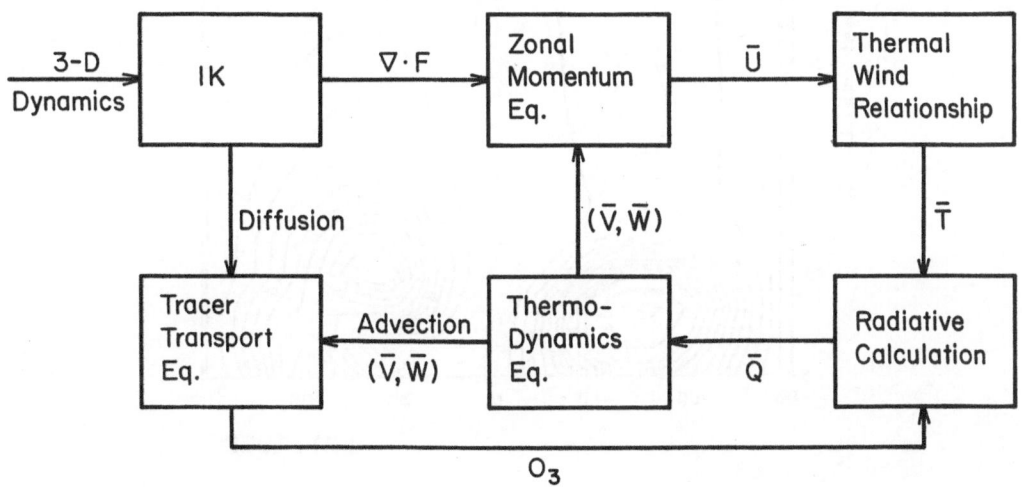

Figure 2.   Schematic diagram showing an algorithm for determining
the coupled set of transport parameters with prescribed $\mathbb{K}$ as
input.

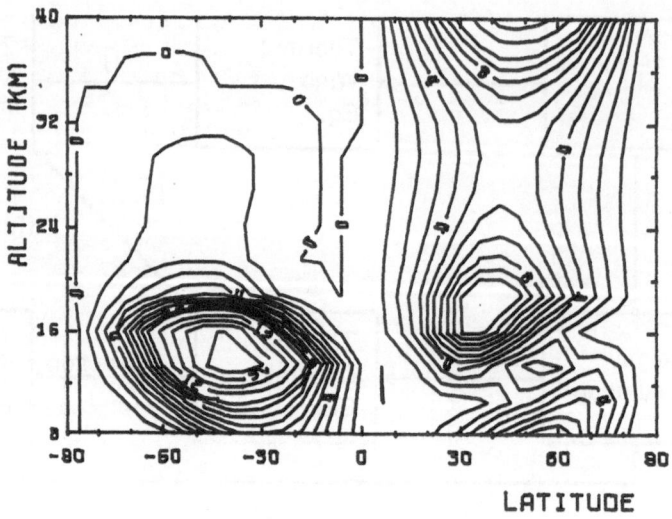

$K_{YY} \times COS(\phi) \times \times 2$ IN JAN. $(10^5 M^2/S)$

Figure 3. $K_{yy} \cos^2\phi$ in January, in units of 10^5 m²/s.

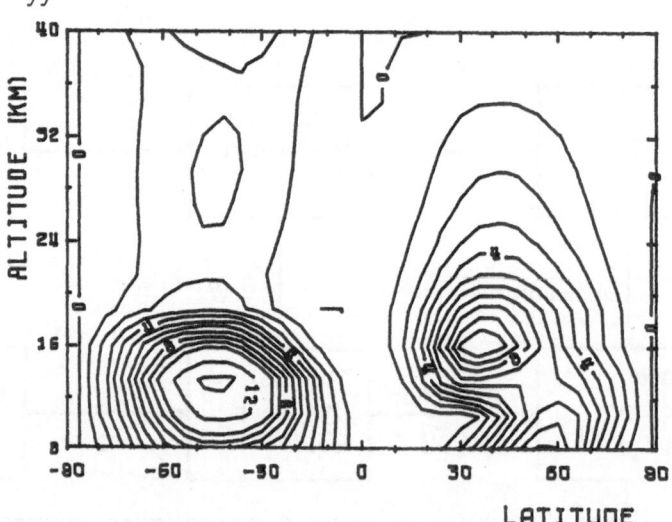

$K_{YY} \times COS(\phi) \times \times 2$ IN APRIL $(10^5 M^2/S)$

Figure 4. $K_{yy} \cos^2\phi$ in April, in units of 10^5 m²/s.

from the prognostic zonal momentum equation. Zonal mean temperature $\bar{T}$ can be deduced from the zonal momentum via thermal wind relationship. Knowing $\bar{T}$, $\bar{Q}$ can be calculated consistent with the input K_{yy}.

Theoretically, the second algorithm would appear to be more appealing as it is the irreversible eddy events which are responsible for driving the zonal wind and the departure from radiative equilibrium. However, the determination of K_{yy} from 3-D data is far from straight forward. Nevertheless, for some assessment scenarios one can combine the two algorithms. First the initial observed $\bar{T}$ is used to deduce K_{yy} using the first algorithm. Then one assumes, as part of the scenario, that the 3-D wave activity is unchanged from the initial value, and uses the second algorithm to predict the temperature $\bar{T}$ and heating $\bar{Q}$ incorporating all the chemical-temperature feedbacks.

6. SOME SAMPLE K_{yy} FIELDS.

The first algorithm was adopted by Yang and Tung (1987). With $\bar{T}$ prescribed from climatology for January, April, July, and October, $\bar{Q}$ and K_{yy} are consistently deduced for these four months. It should be first noted that since $\bar{T}$ is known to be quite variable in the real atmosphere, the deduced K_{yy} can easily vary by a factor of two or more. Secondly, the accuracy of K_{yy} is dependent on our ability to accurately calculate $\bar{Q}$ from the input $\bar{T}$. In a coupled model, a theoretical advancement often carries with it the price that in practice our imperfect knowledge of one physical process can significantly impact the other processes coupled with it. A better model will require the advancement in our knowledge in all fronts.

With these caveats in mind, we now show the deduced K_{yy} field for the four months in Figures 3 to 6, taken from Yang and Tung (1987). In general, positive mixing is found in westerly regions, as expected. In the summer upper stratosphere where easterlies prevail, large-scale stationary waves are known to be prevented from propagating into this region. The K_{yy} field there should be negligible and the deceleration of the easterly jet required above 40km should be provided by the breaking gravity waves not considered here.

In the lower stratosphere, where isentropic mixing is important in helping to determine the slope of isopleths for most chemical tracers, the deduced values for $K_{yy}\cos^2\phi$ fall into the range of 1 to 4 x 10^9 cm^2/s, with the smaller value appearing in the summer hemisphere and in equinoxes. [Note that the quantity $K_{yy}\cos^2\phi$ was called Dyy in Tung (1982, 1984) and it is this combination that shows up in the tracer transport equation with y = $a\sin\phi$ used as the horizontal coordinates.]

These values are comparable to those in use in current (second generation) 2-D models.

No "surf zone" in K_{yy} is apparent in these and other cases we have examined, nor is it a feature of the K_{yy} field that Newman et al., (1986) deduced from observed wave fields. This does not however, exclude the possibility that if the temperature field, $\bar{T}$,

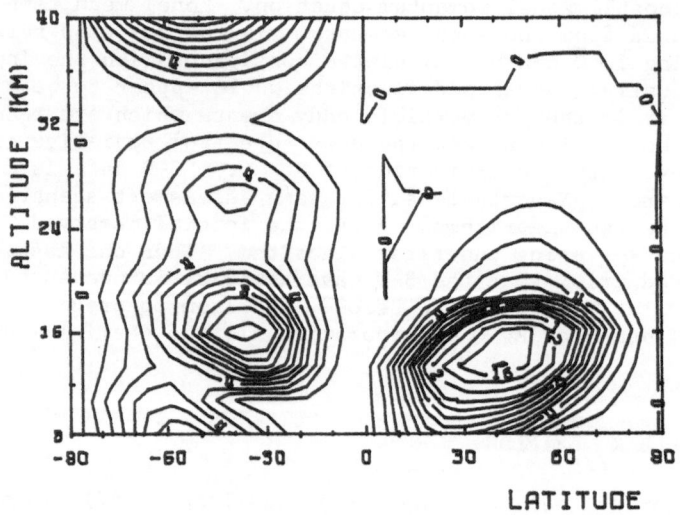

$K_{YY} \times COS(\phi) \times \times 2$ IN JULY $(10^5 M^2/S)$

Figure 5. $K_{yy} \cos^2\phi$ in July, in units of 10^5 m²/s.

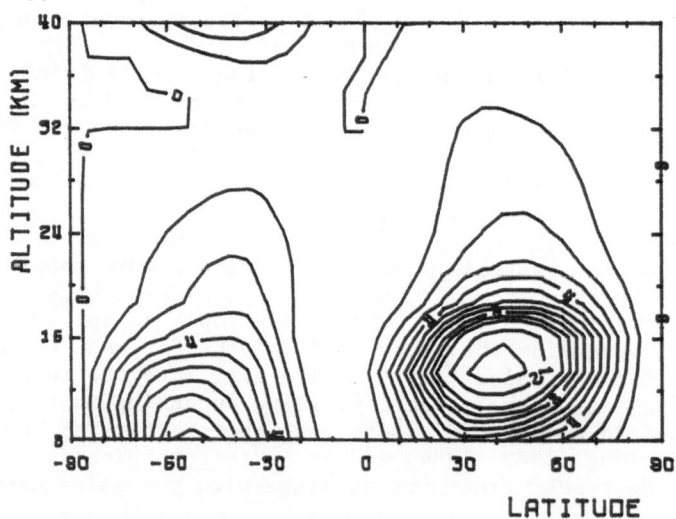

$K_{YY} \times COS(\phi) \times \times 2$ IN OCT. $(10^5 M^2/S)$

Figure 6. $K_{yy} \cos^2\phi$ in October, in units of 10^5 m²/s.

consistent with that of the GFLD GCM is used, the surf zone may
become an important freature.

7. CONCLUDING REMARKS.

The third generation of 2-D models, in which the important transport
parameters are consistently determined from physical input, is best
suited for diagnostic studies. Using the first algorithm, one can
input the observed $\bar{T}$, from which the transport parameters are determined.   These are used to predict the chemical species χ_i in the
2-D model, which can then be compared with the observed χ_i in the
atmosphere. Our knowledge about the chemical and dynamical processes
in the atmosphere is critically tested and hopefully improved in the
process.
          It has often been said that 1- or 2-D models are not suitable
for assessment studies because the dynamics in these models, unlike
those in the comprehensive GCM's, are fixed and not allowed to
change in response to changes in atmospheric composition. The
third generation of 2-D models removes part of the objection by
imposing a consistency constraint among the chemistry, radiation
and dynamics components. It ensures that changes in radiation and
dynamics are reflected in the chemical composition and vice versa.
It however, does not predict such changes, especially if they arise
from dynamical factors not modelled e.g. changes in sea-surface
temperature which can affect wave activity and hence the mixing of
chemical species. It is therefore necessary that in defining a
scenario, one specifies not only the anticipated changes in the
emission of chemical species such as the CFC's and CH_4, but also
states explicitly that the assessment is based on the assumption
that the large-scale wave activity is to either remain as in the
present day atmosphere, or to change in a prescribed manner.
Prediction of this dynamical component requires a coupled model of
the atmosphere and the oceans in three dimensions. Even a comprehensive 3-D GCM with prescribed sea-surface temperatures may not be
adequate.   Careful and creative design of assessment scenarios is
an important part of any prognostic modelling, whether the model used
is 2-D or 3-D.

8. ACKNOWLEDGMENT

The research in 2-D modeling is supported by NASA, under Grant
NAGW-910.

9. REFERENCES

Dickinson, R.E., 1969: Theory of planetary wave-zonal flow inter-
    action. *J. Atmos. Sci.*, 26, 73-81

Dopplick, T.G., 1979: Radiative heating of the global atmosphere:
    Corrigendum. *J. Atmos. Sci.*, 36, 1812-1817

Dunkerton, T., 1978: On the mean meridional mass motions of the
    stratosphere and mesosphere. *J. Atmos. Sci.*, 35, 2325-2333

Edmon, H.J., B.J. Hoskins and M.E. McIntyre, 1980: Eliassen-Palm
    cross-sections for the troposphere. *J. Atmos. Sci.*, 37,
    2600-2616; Corrigendum, 38, 1115

Garcia, R.R., and S. Solomon, 1983: A numerical model of the
    zonally averaged dynamical and chemical structure of the
    middle atmosphere. *J. Geophys. Res.*, 88, 1379-1400

Geller, M.A., M.-F. Wu and M.E. Gelman, 1984: Troposphere-
    stratosphere (surface- 55km) monthly winter general circulation
    statistics for the Northern Hemisphere - Interannual variation.
    J. Atmos. Sci., 41, 1726-1744

Guthrie, P.D., G.H. Jackman, J.R. Herman, and C.J. McQuillan, 1984:
    A Diabatic circulation experiment in a two-dimensional
    photochemical model. *J. Geophys. Res.*, 89, 9589-9602

Holton, J.R., 1980: Wave propagation and transport in the middle
    atmosphere. *Phil. Trans. Roy. Soc. London,* A296, 11909-11994

Holton, J.R. , 1981: An advective model for two-dimensional transport
    of stratospheric trace species. *J. Geophys. Res.*, 86,
    11989-11994

Kida, H., 1983: General circulation of air parcels and transport
    characteristics derived from a hemispheric GCM, II. Very
    long-term motions of air parcels in the troposphere and
    stratosphere. *J. Met. Soc. Japan,* 61, 510-523

Ko, M.K.W., K.K. Tung, D.K. Weisenstein and N.D. Sze, 1985: A
    zonal mean model of stratospheric tracer transport in
    isentropic coordinates: Numerical simulations for nitrous
    oxide and nitric acid. *J. Geophy. Res.*, 90, 2313-2329

Mahlman, J.D., D.G. Andrews, D.L. Hartmann, T. Matsuno and
    R.G. Murgatroyd, 1984: Transport of trace constituents in
    the stratosphere, in *Dynamics of the Middle Atmosphere,*
    387-416.   J.R. Holton and T. Matsuno, editors.   Terra
    Scientific Publishing Company, Tokyo, Japan

Matsuno, T., 1980: Lagrangian motion of air parcels in the stratosphere in the presence of planetary waves. *Pue and Appl. Geophys.*, 118, 189-216

Miller, C., D.L. Filkin, A.J. Owens, J.M. Steed, and J.P. Jesson, 1981: A two-dimensional model of stratospheric chemistry and transport, *J. Geophys. Res.*, 86, 12039-12065

Murgatroyd, R.J., and F. Singleton, 1961: Possible meridional circulation in the stratosphere and mesosphere. *Quart. J. Roy. Met. Soc.*, 87, 125-135

Newman, P.A., M.R. Schoeberl, and R.A. Plumb, 1986: A computation of the horizontal mixing coefficients calculated from NMC data. *J. Geophys. Res.*, 91, 7919-7924

Pitari, G., and G. Visconti, 1985: Two-dimensional tracer transport: Derivation of residual mean circulation and eddy transport tensor from a 3-D model data set, *J. Geophys. Res*, 90, 8019-8032

Plumb, R.A., 1979: Eddy fluxes of conserved quantities by small-amplitude waves. *J. Atmos. Sci.*, 36, 1699-1704

Plumb, R.A., and J.D. Mahlman, 1987: The zonally-averaged transport characteristics of the GFDL general circulation/tracer model, *J. Atmos. Sci.*, 44, 298-327

Reed, R.J., and K.E. German, 1965: A contribution to the problem of stratospheric diffusion by large-scale mixing. *Mon. Wea. Rev.*, 93, 313-321

Robinson, W.A., 1986: The application of the quasi-geostrophic Eliassen-Palm flux to the analysis of stratospheric data. *J. Atmos. Sci.*, 43, 1017-1023

Stordal, F., I.S.A. Isaksen, and K. Hortveth, 1985: A diabatic circulation two-dimensional model with photochemistry: Simulations of ozone and ground released tracers, *J. Geophy. Res.*, 90, 5757-5776

Tung. K.K., 1982: On the two-dimensional transport of stratospheric trace gases in isentropic coordinates, *J. Atmos. Sci.*, 39, 2330-2355

Tung, K.K., 1984: Modeling of tracer transport in the middle atmosphere, in *Dynamics of the Middle Atmosphere*, J.R. Holton and T. Matsuno, editors, 417-4444

Tung, K.K., 1986: Nongeostrophic theory of zonally averaged
     circulation. Part I: Formulation, *J. Atmos. Sci.*,
     43, 2600-2618

WMO, 1986: *Atmospheric Ozone,* 1985, assessment of our understanding
     of the processes controlling its present distribution and
     change.   WMO, Global Ozone Research and Monitoring Project –
     Report No. 16

Yang, H., and K.K. Tung, 1987: Nongeostrophic theory of zonally
     averaged circulation.   Part II: Eliassen–Palm flux divergence
     and isentropic mixing coefficient, in preparation

THE INFLUENCE OF THE SEMI-ANNUAL AND QUASI-BIENNIAL OSCILLATIONS ON EQUATORIAL TRACER DISTRIBUTIONS

L. J. Gray
Rutherford Appleton Laboratory
Chilton, Didcot
Oxon., OX11 0QX
United Kingdom

J. A. Pyle
Physical Chemistry Department
Cambridge University
Lensfield Road
Cambridge, CB2 1EP
United Kingdom

ABSTRACT. The influence of equatorial dynamics on the distribution of chemical tracers is investigated in a two-dimensional model study. The semi-annual oscillation is shown to be responsible for the double peak structure observed in measurements of methane, nitrous oxide and other trace gases. Preliminary results are presented from a model which included a parametrization of the quasi-biennial oscillation (QBO) in addition to the SAO. The evolution of the modelled zonal wind at the equator is compared with observations and the associated meridional circulations at various stages of the oscillation are discussed.

1. INTRODUCTION

Measurements of constituent fields by the NIMBUS 7 satellite have revealed some unexpected structures (Jones and Pyle 1984, Jones 1984). For example, at certain times of the year the SAMS methane (CH_4) and nitrous oxide (N_2O) fields exhibit a 'double peak' structure: maximum mixing ratios are present on a constant pressure surface in subtropical latitudes, with a local minimum at the equator. The feature is prominent for several months around the northern hemisphere spring equinox and again six months later, but in a much weaker form. At other times of the year there is just one mixing ratio maximum situated at low latitudes in the summer hemisphere. Associated structures are also evident at equinox in the LIMS measurements of ozone (Russell 1984), water vapour (Remsberg et al. 1984, 1987(this volume)) and night-time nitrogen dioxide (Gray and Pyle 1986). In a two-dimensional modelling study, Jones and Pyle (1984)

October. The study confirmed the ideas of Gray and Pyle and demonstrated
that advection by the mean meridional motions associated with the SAO is
important in determining the distribution of chemical tracers in
equatorial latitudes.

A subsequent diagnostic study also supported the proposed mechanism
for the double peak (Solomon et al. 1986). Mean meridional circulations
were deduced from the net radiative heating rates based on LIMS measurements of temperature and radiatively active trace species. These showed
a circulation in the same sense as indicated in figure 1 at the
appropriate times of the year. The circulations, in conjunction with
photochemical loss rates for CH_4 and N_2O, (which were also obtained
from the satellite data), were used to reproduce the temporal behaviour
of the tracer fields, including the formation of double peak structures.

In an extension to their preliminary study, Gray and Pyle (1987)
included a parametrization of the SAO in their model which was more
physically based, and also added a parametrization of the momentum
deposition associated with gravity waves breaking in the mesosphere.
Both of these led to substantially improved accuracy in modelling the
wind and tracer fields. The main features of the two schemes are
described in the next section and some results from the new version of
the model are presented. For a more detailed discussion of the
parametrization schemes and the parameter values employed, see Gray and
Pyle (1987). Finally, in section 3 some preliminary results are shown
from a model that included not only the SAO but also the quasi-biennial
oscillation (QBO).

2. THE SEMI-ANNUAL OSCILLATION

2.1. Westerly Phase

The two dimensional Eulerian model developed by Harwood and Pyle (1975,
1977,1980) was employed in the study. The model extends from pole to
pole and from the ground to approximately 100 km. It calculates the
zonal mean values of temperature, wind components and chemical
constituent mixing ratios with a resolution of $\pi/19$ in latitude, half a
scale height in the vertical (approximately 3.5 km) and a six hour
timestep. A second order partial differential equation is solved for the
meridional streamfunction given the forcing by radiative and other
diabatic heating and eddy heat and momentum fluxes. The dynamical and
radiative formulation is that described by Haigh and Pyle (1982). A full
photochemical scheme is included, with up-to-date kinetics data (Gray and
Pyle 1987).

The westerly phase of the SAO was modelled using a WKB approximation
for the mean flow acceleration associated with the thermal damping of a
vertically propagating Kelvin wave. The scheme is essentially that
employed by Dunkerton (1979) in a one dimensional model study and by
Takahashi (1984) and Dunkerton (1985) in two dimensional studies. An
extra term is included in the model momentum equation:

$$\frac{d\bar{u}}{dt} = A \exp \left[\frac{z-z_o}{H}\right] R(z) \exp(-P(z))$$

$$(1)$$

reproduced the gross features of the N_2O and CH_4 distributions but were unable to reproduce the double peak. Gray and Pyle (1986) proposed that the double peak may be associated with the observed semi-annual oscillation (SAO) of zonal winds and temperature in equatorial latitudes (Reed 1966, Hopkins 1975), a feature that was not present in the model of Jones and Pyle. The oscillation consists of alternating easterlies and westerlies and has a six month period with peak westerlies present around each equinox. The damping of vertically propagating Kelvin waves is believed to give rise to the westerly phase, although gravity waves may also be important (Hamilton 1987, this volume).

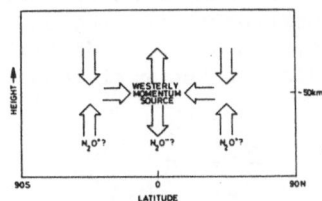

Figure 1. Latitude-height cross-section showing the expected induced meridional circulation during the westerly phase of the semi-annual oscillation.

There were two main factors that suggested a connection between the double peak structure and the SAO. Firstly, during the westerly acceleration phase of the oscillation, which occurs just before each equinox, an induced meridional circulation results in the downward motion of air at the equator with upward motion at subtropical latitudes (see figure 1). As both N_2O and CH_4 mixing ratios decrease with height, this is precisely the circulation that would give rise to a local minimum at the equator and hence a double peak structure at the times of the year when they are observed. Secondly, the easterly phase of the oscillation is thought to be due, at least in part, to equatorially propagating planetary waves from the winter hemisphere (Hopkins 1975). It is well known that the level of planetary wave activity is greater in the northern hemisphere winter than in the southern hemisphere winter. Hence, there exists the possibility of an asymmetry in the easterly phase of the SAO, which in turn would affect the amplitude of the westerly acceleration phases of the SAO. This might explain the observed asymmetry in the amplitude of the double peaks, which are much stronger from March to May than six months later.

In a preliminary study, Gray and Pyle (1986) used the same model as Jones and Pyle to test the ideas outlined above. They forced a semi-annual reversal of the equatorial winds in the model by specifying a westerly momentum forcing which was switched on and off at the appropriate times of the year. Suitable height and latitude distributions were chosen for the forcing. No attempt was made in that study to model the momentum deposition associated with waves propagating in equatorial regions. A reasonable representation of the SAO in the modelled equatorial zonal winds and temperatures was achieved and the N_2O and CH_4 fields exhibited a double peak structure in April and again in

$$\text{where} \quad R(z) = \frac{\alpha(z)N}{k(\bar{u}-c)^2} \quad \text{and} \quad P(z) = \int_{z_o}^{z} R(z')dz'$$

Parameter values appropriate to a Kelvin wave were employed. A is the vertical momentum flux at z_o (=16 km), specified as 7.0×10^{-3} $m^2 s^{-2}$, $\alpha(z)$ the thermal damping rate was chosen to be the 'slow' damping rate of Dunkerton (1979) which peaks at approximately 2×10^{-6} s^{-1} at 50 km. N is the Brunt-Vaisala frequency and H is the scale height. A phase speed $c=30$ ms^{-1} and zonal wavenumber $k=1$ were chosen to be characteristic of the dominant Kelvin wave that is damped in the height region of the SAO. All other symbols have their usual meaning and an overbar represents a zonal mean quantity. A gaussian distribution about the equator was applied to the forcing in the meridional plane, with an e-folding width

$$Y_L = \left| \frac{2v}{\beta k} \right|^{1/2} \quad \text{where} \quad \frac{v}{k} = c \tag{2}$$

(following Holton (1975), and also employed by Takahashi (1984)). This westerly forcing associated with thermally damped Kelvin waves was applied at all times of the year.

2.2. Easterly Phase

As a result of the presence of the westerly forcing at all times of the year, it was found that the modelled zonal winds at the equator were always westerly; nonlinear advection of easterlies from the summer hemisphere was not sufficient to produce an easterly phase. Therefore, an additional easterly momentum was imposed in the model around each solstice. The modelled zonal winds in equatorial latitudes were relaxed towards $-50 ms^{-1}$ on a timescale of 12.5 days over the height range 20-45 km. The forcing was varied in time like sin $(4\pi d/365)$ where d the day number was chosen so that the maximum easterlies occurred at each solstice. Only the negative half of the cycle was used. During the positive half of the cycle the forcing was set to zero. A latitudinal gaussian distribution of the forcing about the equator, identical to that employed in the westerly phase (see equation 2), was also applied.

2.3. Gravity Waves

A parametrization of the momentum deposition associated with gravity waves breaking in the mesosphere was included in the model, and replaced the Rayleigh friction that was previously employed. The method of Lindzen (1981) and Holton (1982) was followed, where an extra forcing term F_x is included in the model momentum equation such that:

$$\text{where} \quad F_x = \frac{N^2 D}{(\bar{u}-c)} \quad z \geqslant z_b \tag{3}$$

$$D = \frac{K(\bar{u}-c)^4}{2N^3} \left[\frac{1}{H} \frac{-3\frac{d\bar{u}}{dz}}{(\bar{u}-c)} \right]$$

and z_b, the breaking level of the wave is calculated using the equation

$$z_b = z_o + 3H \ln \left(\frac{\bar{u}-c}{u^*}\right)$$

A spectrum of waves with phase speeds $0, \pm10, \pm20, \pm30, \pm40$ was specified. The amplitude of the wave at z_o (=16 km) was determined through the parameter u^*, defined by

$$(u^*)^3 = \frac{B^2 N^2}{K^2(\bar{u}(z_o)-c)} = \frac{2N}{K}\overline{u'w'}\Big|_o$$

where B is the amplitude of the vertical velocity perturbation at z_o. The value of u^* for each wave was assigned using the formulation of Garcia and Solomon (1985)

$$\overline{u'w'}\Big|_o = 10^{-2}\left(\exp\left(\frac{-c_i}{30}\right)\right)^2 \quad m^2 s^{-2}$$

for the ith component of the spectrum, so that

$$u_i^* = \left[\frac{2N}{K} \times 10^{-2} \times \left(\exp\left(\frac{-c_i}{30}\right)\right)^2\right]^{1/3} \tag{4}$$

These values of u_i^* were such that the waves achieved breaking levels in the mesosphere. Unlike in the study of Garcia and Solomon (1985), equation (4) has no latitude dependence, so that the forcing is present in equatorial latitudes. (There is no need in our model to introduce an arbitrary latitude dependence so, in the absence of adequate measurements, we chose to use a constant forcing at the lower boundary). Acceleration of the zonal flow close to the equator can produce inertial instability. If this is detected in the model, the mean wind field is adjusted back to a neutral inertial stability flow whilst preserving the mean angular momentum. To avoid discontinuities at the breaking level z_b, the forcing derived by equation (3) was specified to decrease as $\exp((z-z_b)/H)$ below z_b. A single horizontal wavenumber was specified for the waves, corresponding to a wavelength of 3000 km. Following the practice of earlier studies (Holton 1982, 1983, Holton and Zhu 1984, Garcia and Solomon 1985) a so-called efficiency factor was used so that the wavelength of 3000 km may be considered to represent wavelengths of around 300 km that are present for only 10% of the time. Finally, a Rayleigh friction which peaked at approximately 0.3 day^{-1} at the top level of the model was applied above 90 km, to represent the forcing associated with the breaking of the solar diurnal tide in the tropics (Lindzen 1981) and ion drag (Matsuno 1982, Garcia and Solomon 1985).

2.4. Results

The introduction of the parametrization schemes outlined in the previous sections have led to substantial improvements to the modelling of the SAO compared with our earlier studies. Figure 2 shows the zonal wind time series at the equator from a model run that included both the gravity wave and SAO schemes. A strong semi-annual signal is present at all

levels from 20-60 km. The oscillation has its largest amplitude at about
40 km with winds varying between approximately ±30 ms^{-1}. The dependence
of both formulations on the background wind flow, via the term $(\bar{u}-c)$, has
led to the gradual descent of the westerly phase through the atmosphere
with time, which agrees well with observations (Hitchmann and Leovy
1986). The main advantage of the gravity wave scheme is that the flow at
mesospheric levels is forced towards the phase speed of the waves instead
of towards zero, as in a Rayleigh friction scheme. This has permitted
the development of an SAO in the mesosphere as well as in the
stratosphere. The mesospheric SAO is out of phase with the stratospheric
SAO, again in good agreement with observations (Hirota 1978).

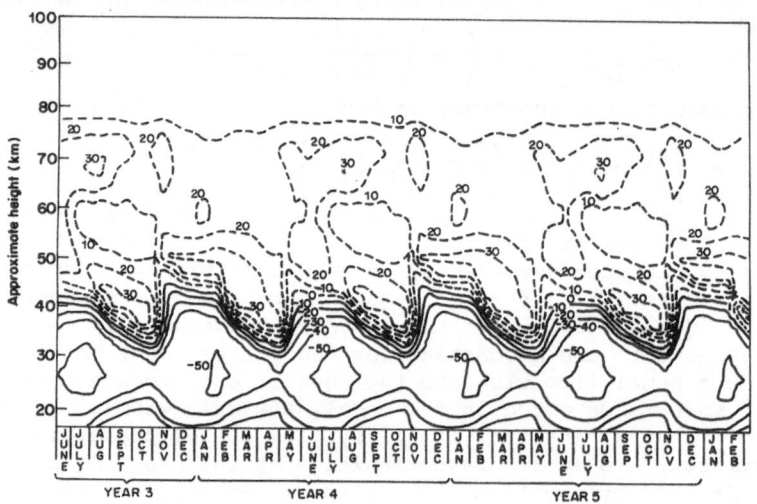

Figure 2. Time-height cross-section of zonal wind at the equator from
a model run which included both gravity wave parametrization and the
semi-annual oscillation. Contour interval 10 ms^{-1}, dashed contours
denote westerlies.

Figure 3 shows the latitude-height cross-sections of pressure-
weighted velocity vectors in March from two model runs: run A
(figure 3(i)) included the gravity wave parametrization only, while run B
(figure 3(ii)) included both the gravity wave and SAO parametrizations.
Model run A has two Hadley-like circulations cells in equatorial
latitudes with rising motion at the equator and descent in mid-latitudes.
In run B, on the other hand, the circulation in the lower stratosphere
has been disturbed by the presence of the SAO. Descending motion is now
present over the equator at approximately 30 km, which is in agreement
with the circulation shown in figure 1. In accordance with the theory,
one would expect this circulation to gradually descend through the
atmosphere with time, and this was indeed the situation in model
run B.

The circulation at high latitudes in run B (figure 3(ii)) is quite
different to that in run A. The reverse cell at high latitudes in the

(i)

(ii)

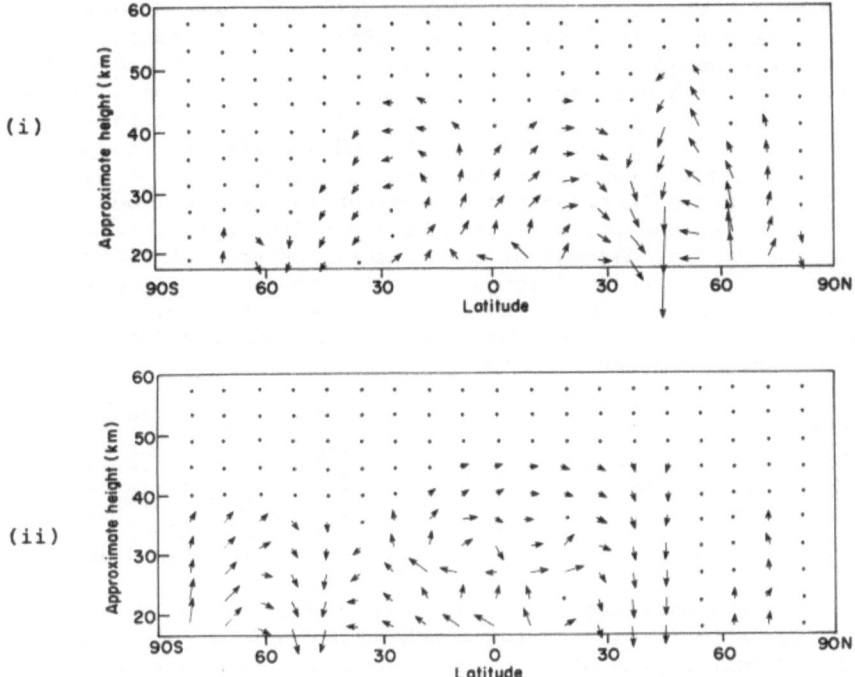

Figure 3. Latitude-height cross-section of the pressure-weighted meridional velocity vectors in March from (i) a model run with the gravity wave parametrization only, and (ii) a model run with both the gravity wave and the semi-annual oscillation parametrizations.

northern hemisphere is substantially reduced while the corresponding cell in the southern hemisphere has grown in amplitude. The circulation near the poles in run B is practically identical with that of run A approximately two weeks further into the run (i.e.in early April); the changes in circulation at high latitude associated with the June solstice have been allowed to develop rather more quickly in run B because of the change in the equatorial circulation. The changes at high latitudes were due to a change in temperature in equatorial latitudes which affected the ozone concentration and hence the heating rates. This shows an interesting mechanism by which the equatorial dynamics may influence the circulation on a global scale.

The inclusion of both the gravity wave and SAO formulations had a considerable effect on the distribution of chemical tracers in equatorial regions of the model. Figure 4 shows a latitude-height cross-section of N_2O in March from three model runs. In figure 4(i) the model included neither of the two parametrization schemes. This is essentially the model of Jones and Pyle (1984). Note that there is a single local maximum in mixing ratio over the equator. Figure 4(ii) shows the

corresponding plot from a model run that included the gravity wave
parametrization. The isolines of mixing ratio have been deformed at the
equator above about 50 km, and a local minimum is now present over the
equator at these heights. Finally, figure 4(iii) shows the distribution
from a model run with both gravity wave and SAO formulations. A strong
double peak structure is now present, extending throughout the depth of
the atmosphere down to 35 km. Note that both gravity wave and SAO
schemes were necessary to obtain the depth of structure seen in figure
4(iii). The double peak structure is present in the model during the
months March to May and again six months later from September to
November. The gradual descent of the westerly phase of the SAO in this
model led to a double peak of longer duration than in the preliminary
study of Gray and Pyle (1986), and this compares favourably with the
observations. A comparison of the modelled CH_4 with the SAMS
observations shows similar results.

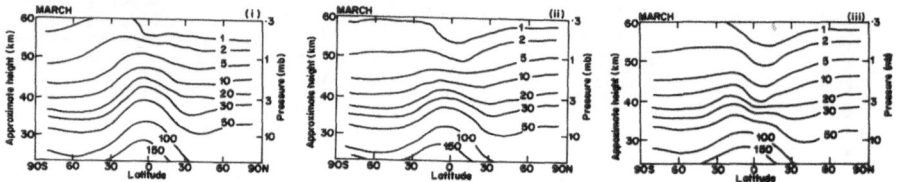

Figure 4. Monthly averaged latitude-height cross-section of N_2O
volume mixing ratio (ppbv) for the month of March from (i) a model
run which had neither the gravity wave parametrization nor the semi-
annual oscillation, (ii) a model run with the gravity wave
parametrization but not the semi-annual oscillation, and (iii) a model
run with both the gravity wave and the semi-annual oscillation
parametrizations.

     In the preliminary study of Gray and Pyle (1986) a connection
was proposed between the asymmetry in the amplitude of the double
peak from March to May compared with that present from September to
November and the level of planetary wave activity in the northern and
southern hemispheres. This idea was tested in a model run in which a
marked asymmetry was imposed on the strength of the easterly phase of
the SAO. Around the December solstice the equatorial zonal wind was
forced towards -50 ms^{-1}, as in all previous model runs (see section
2.2), but during the months around the June solstice it was forced
towards only -20 ms^{-1}, to reflect the weaker level of planetary wave
activity in the southern hemisphere. This resulted in a marked reduction
in the strength of the induced meridional circulation associated with the
westerly acceleration phase of the SAO in the second half of the year and
hence a marked reduction in the amplitude of the double peak from
September to November compared with that present from March to May (Gray
and Pyle 1987). This experiment therefore supports the suggestion that
the larger amplitude of the double peak observed by SAMS from March to
May is associated with the higher level of planetary wave activity during
the northern hemisphere winter.

3. THE QUASI-BIENNIAL OSCILLATION

In an extension to the SAO study outlined above, a parametrization of
some of the possible waves that give rise to the quasi-biennial
oscillation (QBO) was included in the model. The QBO has its maximum
amplitude at about 25 km and consists of easterlies and westerlies
oscillating between +25 ms^{-1} with a variable period between 22 and 34
months. For an excellent review of the characteristics of the QBO and a
discussion of some of the possible mechanisms that give rise to it, see
Plumb (1984). The QBO is of particular interest as it is the dominant
equatorial feature in the stratosphere below about 30 km, which is the
region of maximum ozone concentration. A QBO signal in column ozone is
observed in the atmosphere, and has been well documented (Angell and
Korshover 1973,1978, Tolson 1981, Hilsenrath and Schlesinger 1981,
Oltmans and London 1982, Hasebe 1983). Additionally, because the
momentum deposition associated with vertically propagating waves is
dependent on the vertically integrated background flow of the atmosphere,
it is important to model the winds in the lower layers of the
stratosphere as accurately as possible. The introduction of the QBO
is thus expected to influence the nature of the SAO in the model.
     One of the proposed mechanisms for the QBO is the thermal damping
of vertically propagating equatorial waves (Lindzen and Holton 1968).
In this theory, Kelvin waves are proposed to give rise to the westerly
phase, in much the same manner as in the westerly phase of the SAO, and
Rossby-gravity waves are proposed to account for the easterly phase.
Other possible mechanisms have been suggested, for example, the role of
equatorially propagating planetary waves in the easterly phase has been
examined by Dunkerton (1983) and contributions to both westerly and
easterly phases from gravity waves are also likely (Hamilton 1987, this
volume). The purpose of the present study is primarily to examine the
influence of the QBO on the distribution of chemical tracers and not the
detailed dynamical mechanisms that cause the QBO. For simplicity, the
parametrization scheme has been limited to Kelvin and Rossby-gravity wave
momentum deposition only.
     Both of the parametrization schemes for the thermal damping of
Kelvin and Rossby-gravity waves were based on equation (1), using the
parameter values listed in table 1. The values are those appropriate to
the two waves as summarised by Wallace (1973). Note that the sign of the
phase speed determines the sign of the momentum deposition. The smaller
Rossby-gravity wave forcing is consistent with observations and has been
used in previous studies (Dunkerton 1985). The damping of the waves has
been restricted to thermal damping, as in the SAO parametrization; an
identical damping rate profile $\alpha(z)$, was used above 30 km, and below this
level a constant value of 0.35×10^{-6} s^{-1} was specified. Although no
mechanical damping was explicitly employed in the parametrization scheme,
the model includes a vertical diffusion operating on the wind fields with
a constant value of $K_{zz} = 1.0$ m^{2}s^{-1} at all latitudes and heights.
     The time-series of zonal winds at the equator from a model run that
included the QBO as well as the SAO is shown in figure 5. The six year
integration displays three periods of the oscillation with no perceptible
sign of a drift with time. The period of the oscillation is

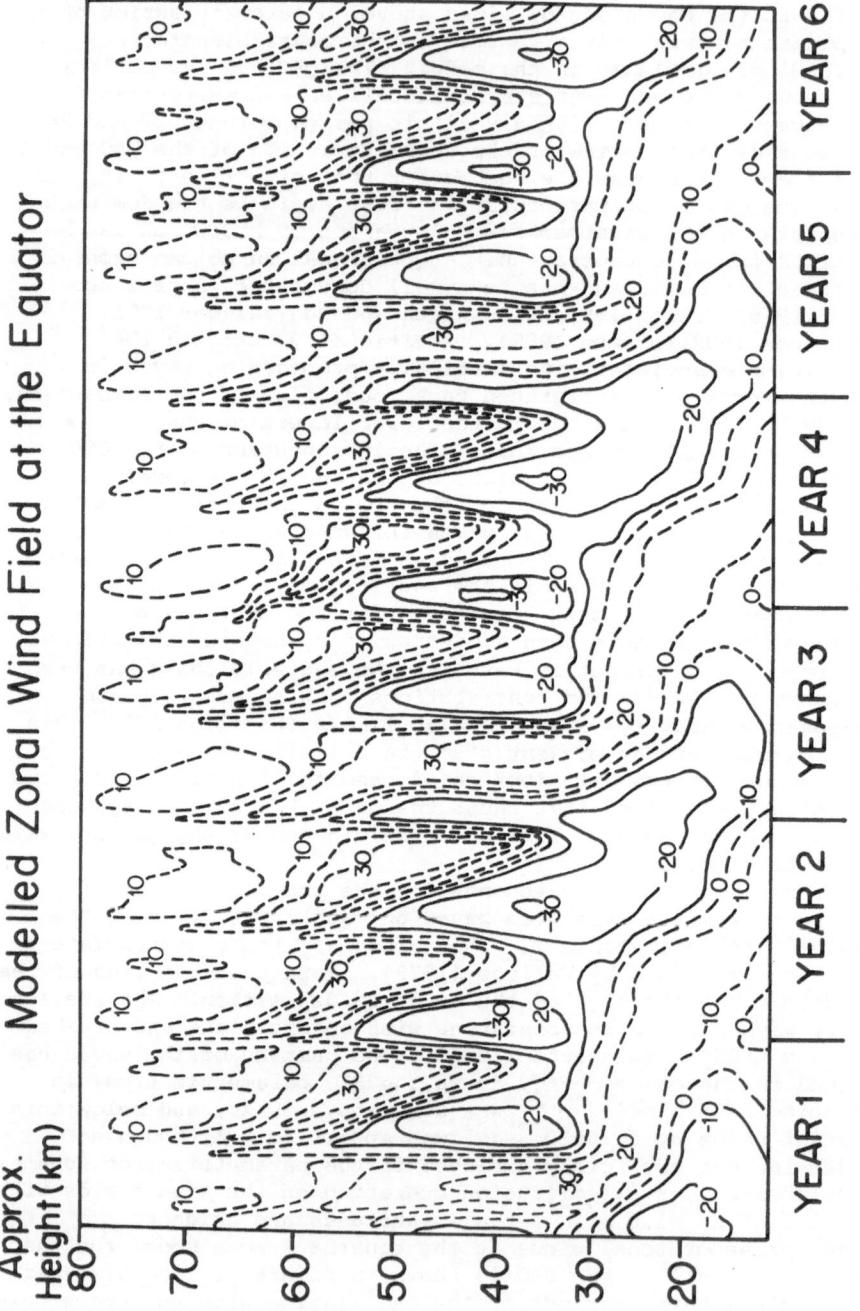

Figure 5. Time-height section of zonal wind at the equator from a model run which included both the semi-annual oscillation and the quasi-biennial oscillation. Dashed contours denote westerlies.

approximately 24 months, in reasonable agreement with observations. It
should be noted that the period was not imposed on the model (as in the
case of the SAO) but was purely a result of the model parameters and
assumptions. It was found possible to vary the period by changing the
relative forcing of the two waves while keeping other parameters
constant. Similarly, a change in the phase speeds or damping rate would
be expected to alter the period of the oscillation. There was a marked
tendency for the period to vary in multiples of six months in the height
region 25-30 km, in synchronization with the phases of the SAO, and this
is discussed in more detail below. At lower levels, however, this was
the not case. A large alteration in the ratio of the forcing of the two
waves resulted in the disappearance of the oscillation, with perpetual
westerlies overlying easterlies, or vice versa.

TABLE 1.

	Kelvin wave	Rossby-gravity wave
Phase speed c	$+25 \text{ ms}^{-1}$	-25 ms^{-1}
Zonal wavenumber k	1	4
A, amplitude of vertical momentum flux at z_0	$12.5 \times 10^{-3} \text{ m}^2\text{s}^{-1}$	$8.0 \times 10^{-3} \text{ m}^2\text{s}^{-1}$
e-folding width of gaussian latitudinal distribution	$\left\|\dfrac{2v}{\beta k}\right\|^{1/2}$	$\left\|\dfrac{2v}{\beta(\frac{\beta}{v}-k)}\right\|^{1/2}$

The amplitude of the oscillation was a maximum at approximately
25-30 km, with winds varying from 20ms^{-1} to -23ms^{-1} at 25 km. The
westerly acceleration in the region of 25 km is larger than the easterly
acceleration, in good agreement with observations (Wallace 1973) although
this asymmetry reverses at lower levels. In general, the easterly wind
maximum is larger than the corresponding westerly maximum at a given
level. The maximum easterly and westerly accelerations at the upper
limits of the oscillation (i.e. around 25-30 km) are significantly
correlated with the phases of the SAO. For example, the maximum westerly
acceleration always occurs as an extension of the SAO extending down
through the atmosphere. This behaviour compares well with observations,
for example the zonal wind time series measured at Ascension Island
(Hirota 1978). Similarly, the maximum easterly acceleration at 25-30 km
occurs at about the same time as an easterly phase of the SAO at upper
levels. The formulation of the momentum deposition associated with the
thermal damping of Kelvin and Rossby-gravity waves (equation (1)) is such
that the mean flow acceleration at any level depends on the parameter
values at that level and at lower levels only. It has therefore been
argued (Plumb 1977) that the SAO, occurring at levels higher than the
QBO, cannot influence the nature of the QBO. However, this model study

has indicated a possible influence of the SAO on the QBO via its control
of the term ($\bar{u}$-c) in the region of 30 km. The bias towards a period of
either 24 or 30 months in the model may be a clue to the large variation
in the period of the observed QBO period, from 22 to 34 months. Such a
large variability is difficult to understand simply in terms of the year
to year variability in the forcing of the waves that influence the QBO.
It is more easily understood in terms of a variability around two
preferred periods, which are themselves six months apart.

The latitudinal structure of the QBO in the model is illustrated in
figure 6, which shows the induced vertical velocity due to the QBO alone,
for two different phases of the QBO. Figures 6(i) and 6(ii) are from
March of years three and four of the model run, respectively. In year
three the westerly phase of the QBO is just beginning to dominate in the
region 25-30 km (see figure 5). Hence large westerly forcing exists in
this height region, with easterly forcing still present at lower levels.
Figure 6(i) indicates a circulation with downward motion over the equator
and upward motion at mid-latitudes in the region of 30 km, associated
with the westerly forcing (c.f. figure 1). The circulation in the
opposite sense in the height region 15-25 km is associated with the
easterly forcing. In year four, with the easterly phase about to
dominate at 30 km, the situation is reversed and so too are the
directions of the meridional circulations (figure 6(ii)).

The presence of these induced circulations is expected to influence
the distribution of ozone in the model, both directly, by advection (the
lifetime of ozone is comparable to the advective timescale in the lower
stratosphere) and also indirectly (in the higher regions of the QBO) via
changes in temperature. The model displayed a strong QBO in column
ozone, not only in equatorial regions but also at high latitudes. The
details of this QBO in ozone are still being investigated and will be
reported in a future paper.

4. SUMMARY

The modelling of the observed SAO in zonal winds at the equator has been
improved by the introduction, in a two-dimensional model, of a
parametrization of (a) the westerly momentum deposition associated with
thermally damped Kelvin waves, (b) the easterly momentum deposition
associated with equatorially propagating planetary waves and (c) the
momentum deposition associated with gravity waves breaking in the
mesosphere. An SAO is now present in the modelled mesosphere and is out
of phase with the stratospheric SAO. The dependence of the SAO forcing
on the background wind field has substantially improved the modelling of
the westerly phase of the SAO, which gradually descends through the
atmosphere with time, in good agreement with observations. This has led
to an improvement in the modelling of the double peak structures, which
now extend throughout a greater depth of the modelled atmosphere and are
present for several months around each equinox. The importance of both
the SAO and gravity waves in the depth of the double peak structure has
been illustrated. The mean meridional circulation in the model was
altered by the inclusion of the SAO. Changes occurred not only in

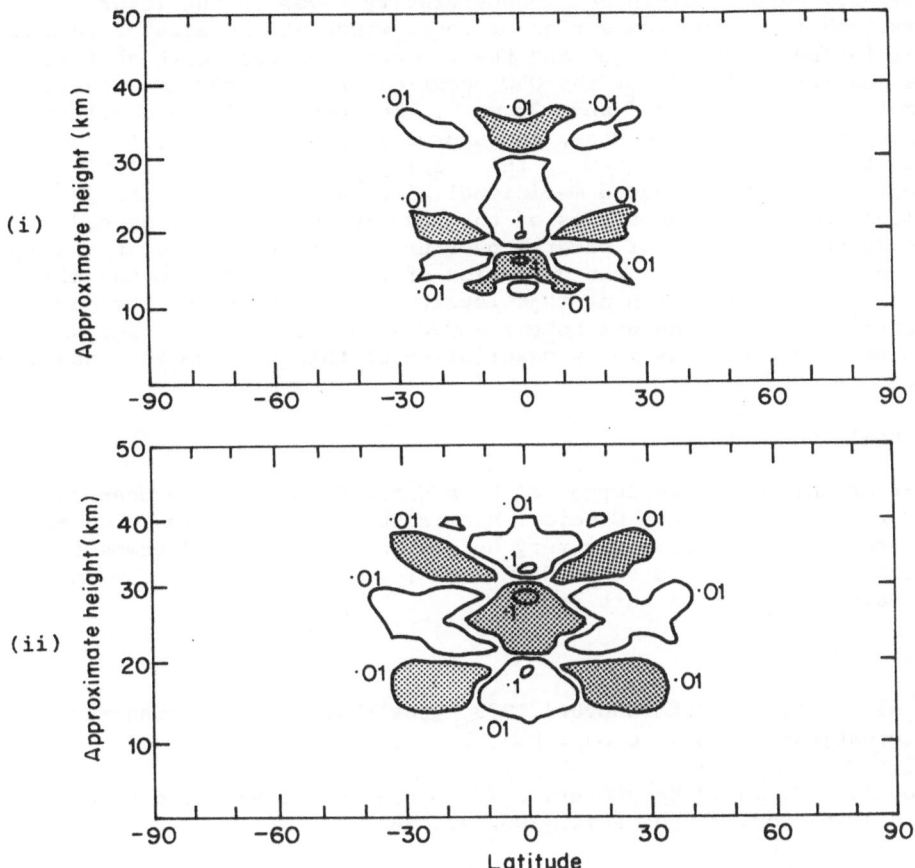

Figure 6. Latitude-height cross-section of the induced vertical velocity (mm s^{-1}) due to the presence of the quasi-biennial oscillation at the March equinox from (i) year 3 and (ii) year 4 of the model run that included both the semi-annual and the quasi-biennial oscillations. Shaded areas represent downward motion greater than 0.01 mm s^{-1}.

equatorial regions, as expected, but also at higher latitudes as a result of changes to the ozone distribution affecting the calculated heating rates. A model experiment was described in which the asymmetry in planetary wave activity between the northern and southern hemisphere winters was investigated. This supported the theory that it plays an important part in the production of the asymmetry between the double peak amplitude observed from March to May compared with that observed from September to November.

The quasi-biennial oscillation (QBO) has also been reproduced in the model by the inclusion of a parametrization of momentum deposition due to thermally damped Kelvin and Rossby-gravity waves in the lower stratosphere. The time-series of zonal winds at the equator in a model that included both the QBO and the SAO compared well with observations. The maximum amplitude of the QBO occurred around 25-30 km with winds varying between 20 ms^{-1} and -23 ms^{-1}. The period of the oscillation was approximately 24 months, and showed a marked tendency to synchronize with the phases of the SAO in the overlap of the two oscillations at about 30 km. The induced meridional circulations due to the QBO were calculated at various stages of the modelled QBO. These agreed well with theory and consisted of several overturning cells centred at the equator, one on top of another, the direction of which depended on the sign of the zonal wind acceleration at that level. The influence of the QBO on the distribution of ozone was noted; a QBO signal in column ozone has been produced by the model and a description of this will be the subject of a future paper.

Acknowledgements

Part of this work was supported by a grant from the Fluorocarbon Programme Panel of the Chemical Manufacturers Association. We are also pleased to acknowledge the very helpful discussions and comments of Michael McIntyre, Bob Harwood and the reviewers of previous papers that we have published on this topic.

References

Angell, J.K. and J.Korshover, 1973. Quasi-biennial and long-term fluctuations in total ozone. Mon. Weath. Rev., 101, 426-443.

Angell, J.K. and J.Korshover, 1978. Global ozone variations: an update into 1976. Mon. Weath. Rev., 106, 725-737.

Dunkerton, T.J., 1979. On the role of the Kelvin wave in the westerly phase of the semi-annual oscillation. J.Atmos.Sci., 36, 32-41.

Dunkerton, T.J., 1983. Laterally propagating planetary waves in the easterly phase of the quasi biennial oscillation. Atmosphere-Ocean, 21, 55-68.

Dunkerton, T.J., 1985. A two-dimensional model of the quasi-biennial oscillation. J.Atmos.Sci., 42,1151-1160.

Garcia, R.R. and S. Solomon, 1985. The effect of breaking gravity waves on the dynamics and chemical composition of the mesosphere and lower thermosphere. J. Geophys. Res., 90,3850-3868.

Gray, L.J. and J.A. Pyle, 1986. The semi annual oscillation and equatorial tracer distributions. Quart. J. Roy. Met. Soc., 112, 387-407.

Gray, L.J. and J.A. Pyle, 1987. Two dimensional model studies of equatorial dynamics and tracer distributions. Quart. J. Roy. Met. Soc., (to appear).

Haigh, J. and J.A. Pyle, 1982. Ozone perturbations in a two-dimensional circulation model. Quart. J. Roy. Met. Soc., 108, 551-574.

Harwood, R.S. and J.A. Pyle, 1975. A two-dimensional circulation model for the atmosphere below 80 km. Quart. J.R. Met. Soc., 101, 723-748.

Harwood, R.S. and J.A. Pyle, 1977. Studies of the ozone budget using a zonal mean circulation model and linearized photochemistry. Quart. J.R. Met. Soc., 103, 319-343.

Harwood, R.S. and J.A. Pyle, 1980. The dynamical behaviour of a two-dimensional model of the stratosphere. Quart. J.R. Met. Soc., 106, 395-420.

Hasebe, F., 1983. Interannual variations of global total ozone revealed from NIMBUS 4 BUV and ground based observations. J. Geophys. Res., 88, 6819-6834.

Hilsenrath, E. and B.M. Schlesinger, 1981. Total ozone seasonal and interannual variations derived from the 7 year NIMBUS 4 BUV data set. J. Geophys. Res., 86, 12087-12096.

Hirota, I., 1978. Equatorial waves in the upper stratosphere and mesosphere in relation to the semi-annual oscillation of the zonal wind. J. Atmos. Sci., 35, 714-722.

Hitchmann, M.H. and C.B. Leovy, 1986. Evolution of the zonal mean state in the equatorial middle atmosphere during October 1978 - May 1979. J. Atmos. Sci., 43, 3159-3176.

Holton, J.R., 1975. An introduction to Dynamic Meteorology. Academic Press.

Holton, J.R., 1982. The role of gravity wave induced grag and diffusion in the momentum budget of the mesosphere. J. Atmos. Sci., 39, 791-799.

Holton, J.R., 1983. The influence of gravity wave breaking on the general circulation of the middle atmosphere. J. Atmos. Sci., 40, 2497-2507.

Holton, J.R. and X. Zhu, 1984. A further study of gravity wave induced drag and diffusion in the mesosphere. J. Atmos. Sci., 41, 2653-2662.

Hopkins, R.H., 1975. Evidence of polar-tropical coupling in upper stratospheric zonal wind anomolies. J. Atmos. Sci., 32, 712-719.

Jones, R.L., 1984. Satellite measurements of atmospheric composition: three years' observations of CH_4 and N_2O. Adv. Space Res., 4, 121-130.

Jones, R.L. and J.A. Pyle, 1984. Observations of CH_4 and N_2O by the NIMBUS 7 SAMS: A comparison with in situ data and two dimensional numerical model calculations. J. Geophys. Res., 89, 5263-5279.

Lindzen, R.S. and J.R. Holton, 1968. A theory of the quasi biennial oscillation. J. Atmos. Sci., 25, 1095-1107.

Lindzen, R.S., 1981. Turbulence and stress owing to gravity wave and tidal breakdown. J. Geophys. Res., 86, 9707-9714.

Matsuno, T., 1982. A quasi one-dimensional model of the middle atmosphere circulation interacting with internal gravity waves. J. Meteor. Soc. Japan, 60, 215-226.

Oltmans, S.J. and J. London, 1982. The quasi biennial oscillation in atmospheric ozone. J. Geophys. Res., 87, 8981-8989.

Plumb, R.A., 1977. The interaction of two internal waves with the mean flow: implications for the theory of the quasi-biennial oscillation. J. Atmos. Sci., 34, 1847-1858.

Plumb, R.A., 1984. The quasi biennial oscillation, in Dynamics of the Middle Atmosphere. Ed. by J.R. Holton and T. Matsuno. Terra Scientific Publishing Company, 217-251.

Reed, R.J., 1966. Zonal wind behavior in the equatorial stratosphere and lower mesosphere. J. Geophys. Res., 71, 4223-4233.

Remsberg, E.E., J.M. Russell III, L.L. Gordley, J.C. Gille and P.L. Bailey, 1984. Implications of stratospheric water vapour distribution as determined from the NIMBUS 7 LIMS experiment. J. Atmos. Sci., 41, 2934-2945.

Russell, J.M. III, 1984. The global distribution and variability of stratospheric constituents measured by LIMS. Adv. Space Res., 4, 107-116.

Solomon, S., J.T. Kiehl, R.R. Garcia and W. Grose, 1986. Tracer transport by the diabatic circulation deduced from satellite observations. J. Atmos. Sci., 43, 1603-1617.

Takahashi, M., 1984. A two dimensional numerical model of the semi-annual oscillation, in Dynamics of the Middle Atmosphere. Ed. by J.R. Holton and T. Matsuno. Terra Scientific Publishing Company, 253-269.

Tolson, R.H., 1981. Spatial and temporal variations of monthly mean total columnar ozone derived from 7 years of BUV data. J. Geophys. Res., 86, 7312-7330.

Wallace, J.M., 1973. General circulation of the tropical lower stratosphere. Rev. Geophys. Space Phys., 11, 191-222.

THE EFFECT OF BREAKING GRAVITY WAVES ON THE DISTRIBUTION OF TRACE SPECIES IN THE MIDDLE ATMOSPHERE

Guy Brasseur and Matthew Hitchman
National Center for Atmospheric Research
P.O. Box 3000
Boulder, CO 80307
USA

ABSTRACT. Gravity wave temperature amplitudes will increase upward into the middle atmosphere due to decreasing ambient density and will grow rapidly approaching a critical level. Waves will break when their amplitudes are so large that they become convectively unstable. Above the breaking level turbulence will mix tracers in the vertical and the mean flow will accelerate toward the phase speed of the wave as the wave is absorbed. A parameterization of these effects is employed in a two dimensional model of the middle atmosphere. Results from this coupled dynamical-radiative-photochemical model are used to review the effects of breaking gravity waves on the mean circulation and on the meridional distribution of trace species.

1. INTRODUCTION

Radiative equilibrium temperature calculations do not correctly reproduce the observed cold summer mesopause (160 K) or the relatively warmer winter mesopause (220 K). By using observed temperatures, Murgatroyd and Singleton (1961) showed that radiative heating rates are $\sim$ 10 K/day in the summer mesosphere and $\sim$ -10 K/day in the winter mesosphere. They derived the meridional circulation required to balance this: ascent at $\sim$ 1 cm/s and adiabatic cooling in the summer mesosphere, descent and warming in the winter mesosphere, and a summer to winter flow of $\sim$ 1-5 m/s. Haurwitz (1961) recognized that this meridional circulation is frictionally induced and determined comparable velocities by balancing the Coriolis torque with a diffusion of the zonal wind. Leovy (1964) successfully modeled the observed temperature and zonal wind distributions by decelerating the zonal flow with a Rayleigh friction having a time scale of $\sim$ 10 days. In recent years more detailed radiative calculations (e.g., Wehrbein and Leovy, 1982) have yielded significantly larger heating and cooling rates, which necessitate the use of much larger Rayleigh friction coefficients in numerical models, corresponding to mesospheric deceleration rates of $\sim$ 50-100 m/s-day^{-1} (Schoeberl and Strobel, 1978; Holton and Wehrbein, 1980; Apruzese et al., 1982; Wehrbein and Leovy, 1982; Garcia and Solomon, 1983). By describing the circulation of the middle atmosphere in terms of "vertical motions required to balance heating

G. Visconti and R. Garcia (eds.), Transport Processes in the Middle Atmosphere, 215–227.
© *1987 by D. Reidel Publishing Company.*

rates" and "drag required to decelerate the zonal flow" one may be left with the impression that the primary cause of the circulation is differential radiative heating. It is now generally agreed, however, that gravity waves are responsible for the reversed winter-summer mesospheric temperature difference (Holton, 1983), greatly accelerate the meridional circulation, and are a major factor in controlling the shapes of zonal jets.

Gravity waves can affect the mean flow in a permanent sense only if they are absorbed. Since time scales for the propagation of gravity wave energy are much smaller than radiative decay time scales, it is believed that wave breaking is the usual precursor to absorption. The concept of gravity wave breaking (Pitteway and Hines, 1963; Lindzen 1967, 1968; Hodges, 1969) is based on conservation of wave action during ascent through shear and decreasing ambient density: eventually wave amplitudes will increase to the point that the local temperature lapse rate becomes superadiabatic. Convective overturning will mix air parcels in the vertical and thereby reduce wave amplitudes. Since wave amplitudes are self-limited above the breaking level, there will be a flux convergence of momentum and the mean flow will be driven toward the phase speed of the wave, which may be far from zero.

Lindzen (1981) introduced a parameterization of breaking gravity waves which allows calculation of vertical profiles of the vertical eddy diffusivity and torque on the zonal flow, given a zonal wind profile and wave phase speeds. Holton (1982) used this parameterization in a numerical model, assumed phase speeds of -20, 0 and 20 m/s in the zonal direction only and obtained a very realistic circulation. Independently, Matsuno (1982) applied a formulation for gravity wave amplitude attenuation due to background turbulence to investigate the effect of steady waves with five discrete phase speeds and four angles of orientation on the mean circulation. Dunkerton (1982) showed that similar results are obtained for transient, stochastic waves. The Lindzen/Holton formulation was employed by Garcia and Solomon (1985) in their two-dimensional model to assess the effect of gravity waves on the meridional distribution of chemical constituents.

The purpose of this paper is to review recent work dealing with the effect of gravity waves on the distribution of trace species in the middle atmosphere in the framework of a two-dimensional model containing interactive dynamics, chemistry and radiation (Brasseur et al., 1987). First we describe the wave breaking parameterization (section 2) and the two-dimensional model (section 3), then present model results in the context of previous studies (section 4). Section 5 contains concluding remarks.

2.   THE LINDZEN PARAMETERIZATION

Lindzen (1981) considered a gravity wave propagating below its critical level, z_c, where the mean flow, $\bar{u}$, equals the phase speed of the wave, c. Following Holton (1982) we consider only oscillations in the altitude-longitude plane. If the perturbation vertical velocity is assumed to have the form

$$w'(x, y, z, t) = \tilde{w}(z) \ e^{z/2H} \ e^{ik(x-ct)}, \tag{1}$$

the vertical structure equation becomes

$$\frac{d^2\tilde{w}}{dz^2} + \lambda^2 \tilde{w} = 0, \tag{2}$$

where

$$\lambda^2 = \frac{N^2}{(\overline{u} - c)^2} \tag{3}$$

and the symbols have their usual meteorological meaning (*e.g.*, N is the buoyancy frequency). Assuming that the zonal wind varies slowly relative to wave phase, (2) has the approximate WKBJ solution

$$\tilde{w}(z) = \tilde{w}_o \left(\frac{\overline{u} - c}{\overline{u}_o - c} \right)^{1/2} exp \left[i \int_{z_o}^z \lambda dz' \right] \tag{4}$$

where the subscript zero refers to the excitation level. Once generated, a gravity wave will keep the same phase speed relative to the ground. A wave with non-zero phase speed can arise from transient flow over topography, convective activity or 'geostrophic adjustment'. If the wave energy reaches a level where $\overline{u} \to c$, phase lines come closer together in the vertical ($\lambda^2 \to \infty$ from (3)) and oscillations tend to become more horizontal ($\tilde{w} \to 0$ from (4)). Gravity waves stay below their critical level.

The atmosphere is convectively unstable when

$$\frac{\partial T'}{\partial z} + \Gamma \leq 0, \tag{5}$$

where $\Gamma = g/c_p + d\overline{T}/dz$. Alternatively it can be shown that neutral stability is reached when

$$u' = |c - \overline{u}| \tag{6}$$

(Fritts, 1984). Since T' is related to w' through the thermodynamic energy equation

$$-ik(c - \overline{u})T' + \Gamma w' = 0, \tag{7}$$

the altitude at which the wave breaks, z_b, can be derived. If the temperature amplitude is large enough some portion of the wave will overturn. Substituting (1), (3) and (4) into (7), differentiating, assuming WKBJ, and removing oscillatory dependences by taking the absolute value, (5) gives the altitude above which the atmosphere is locally convectively unstable:

$$z_b = 3 \ H \ ln \left[\frac{|\overline{u} - c|}{\tilde{u}} \right], \tag{8}$$

where

$$\tilde{u} = \left(\frac{\tilde{w}_o N}{k|\overline{u}_o - c|^{1/2}} \right)^{2/3}. \tag{9}$$

Waves with larger initial amplitudes, longer zonal scales, and those which approach critical levels will tend to break at lower altitudes than other waves. If the formation qualities are parameterized by a specified value of $\tilde{u}$, the breaking height may be calculated for a given phase speed. The wave will break only if z_b is less than or equal to an altitude corresponding to a value of $\overline{u}$ used in (8).

For $z_b \leq z \leq z_c$ wave amplitudes are self-limited by vertical mixing. In this region (7) should include dissipation:

$$-ik(c - \overline{u})T' + \Gamma w' = K_{zz}\frac{\partial^2 T'}{\partial z^2}, \tag{10}$$

from which it can be shown that the vertical eddy diffusivity due to breaking gravity waves is

$$K_{zz} = A\frac{(\overline{u} - c)^4}{N^2}\left[1 - 3H\frac{\partial\overline{u}/\partial z}{\overline{u} - c}\right] \tag{11}$$

where $A = \gamma k/2HN$ and γ is the portion of a latitude circle occupied by waves. Since the eddy continuity equation is

$$w' = -\frac{k}{\lambda}u', \tag{12}$$

from (6),

$$\rho\,\overline{u'w'} = -\frac{1}{2}\,\gamma\,\rho\,\frac{k}{\lambda}\,(c - \overline{u})^2 \tag{13}$$

for $z \geq z_b$, so that, using (3),

$$F_g = -\frac{1}{\rho}\frac{\partial}{\partial z}\rho\,\overline{u'w'} = \frac{-k\,\gamma\,(\overline{u} - c)^2}{2N}\left[\frac{(\overline{u} - c)}{H} - 3\frac{\partial\overline{u}}{\partial z}\right] \tag{14}$$

is the body force per unit mass on the zonal flow (Fritts, 1984). By comparing (11) and (14) it is seen that

$$F_g = \frac{N^2}{(c - \overline{u})}K_{zz}. \tag{15}$$

A is a second parameter which, when specified, allows determination of the profiles of K_{zz} and F_g. Note that K_{zz} and F_g get smaller approaching a critical level. If a spectrum of gravity waves ascends in westerly shear, westerly waves will not be found above their critical levels, leaving easterly waves to break at higher levels with much higher amplitudes. This will cause easterly acceleration and close off the westerly jet. From (11) it can be seen that K_{zz} should decrease above the jet cores. The expected lower mesospheric maximum in K_{zz} is consistent with the upward decrease near the mesopause derived from atomic oxygen profiles (Brasseur and Offermann, 1986).

3.   TWO-DIMENSIONAL MODEL

Our model extends from pole to pole and from the surface to 85 km altitude. Chemical species and entropy are advected by a residual (transformed Eulerian) mean meridional circulation formulated in log-pressure coordinates. The residual mean meridional streamfunction is forced by spatial gradients in wave driving and in radiative heating or cooling (see Garcia and Solomon, 1983). Zonal winds are derived from the thermal wind law. In the results reported here, insolation absorption by ozone is calculated with the parameterization of Schoeberl and Strobel (1978), while infrared transfer is approximated by Newtonian relaxation.

The model can be run with a variety of dynamical forcings. Most simply, momentum drag is parameterized by a Rayleigh friction and eddy diffusivities vary with altitude only. A more complex representation of the effects of breaking gravity waves is based on the Holton/Lindzen parameterization. Five phase speeds are chosen and parameters are slightly tuned (see Table I) to yield residual circulations comparable to those inferred from LIMS observations (Hitchman and Leovy, 1986). Profiles of K_{zz} and F_g are smoothed in the vertical to approximate various effects such as nonlinear interaction among waves and radiative damping. A parameterization of Rossby wave driving is also introduced to provide a self-consistent determination of their torque on the zonal flow and meridional tracer dispersion (Hitchman and Brasseur, 1987). The influence of Rossby waves is primarily in the lower stratosphere, but there are feedbacks through the filtering effects of zonal winds on gravity wave transmission.

Approximately 40 species are included. These belong to the oxygen, carbon, hydrogen, nitrogen and chlorine families. The reaction rate constants are taken from the JPL compilation (De More et al., 1985). The diurnal average of the photodissociation coefficients is approximated by a 4 point integral between sunrise and sunset (see Cunnold et al., 1975).

The family grouping technique used to solve the tracer continuity equations avoids the numerical problems associated with the stiffness of the system. Centered space differences are used, while the time integration is done with an implicit 'alternating direction' method (Carnahan et al., 1969).

4.   MODEL RESULTS

Differential radiative heating during the approach to solstice will cause a weak upward drift in the summer hemisphere, sinking in the winter hemisphere, with Coriolis torques on the summer to winter drift giving rise to summer easterlies and winter westerlies. We expect that westerly gravity waves will be filtered out in the lower winter hemisphere, leaving easterly gravity waves to impart a westward torque on the flow at upper levels. Figure 1 shows the body force per unit mass on the zonal flow due to gravity waves near the austral winter solstice after one year of model integration. Easterly accelerations as high as 80 m/s-day are seen above the westerly jet core (not shown). Near the stratopause the drag is of the order of 20 m/s-day, in good agreement with values derived from the LIMS experiment (Smith and Lyjak, 1985; Hitchman and Leovy, 1986).

These accelerations may be balanced by advection of high angular momentum from the equator to southern latitudes and of low angular momentum from the north pole to northern midlatitudes. Figures 2 and 3 show the meridional circulation induced by gravity wave absorption. The strong meridional winds transport long-lived

Table I. Parameters used to determine the breaking
altitudes, vertical eddy diffusivity and momentum drag
(eqns. 8, 11, and 14).

Zonal phase speed (c) [m/s]	Amplitude coefficient (A) $[10^{-9}m^{-2}s^{-1}]$	Breaking level coefficient ($\tilde{u}$) [m/s]
-40	0.25	3
-20	0.5	3
0	1	3
+20	0.5	3
+40	0.25	3

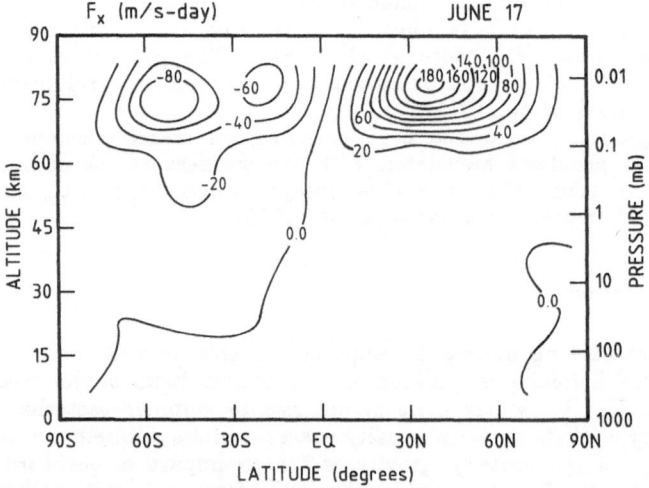

Figure 1. Latitude-altitude section of the body force per unit mass $(m\text{-}s^{-1}\text{-}day^{-1})$ on the zonal flow due to gravity waves near the boreal summer solstice after one year of model integration.

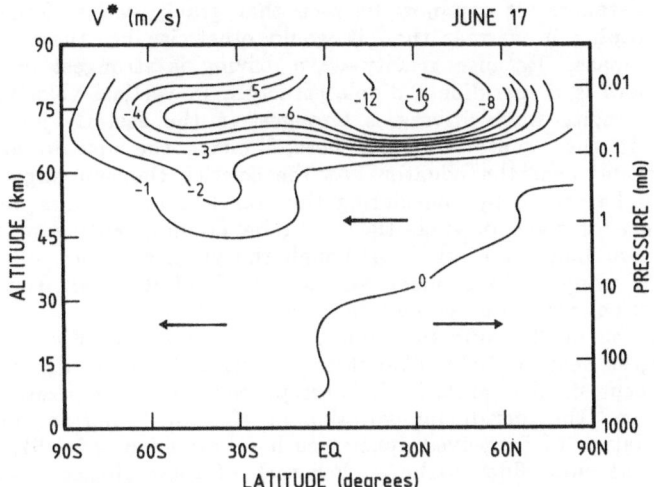

Figure 2. As in Fig. 1, except of the meridional component of the residual circulation (m/s).

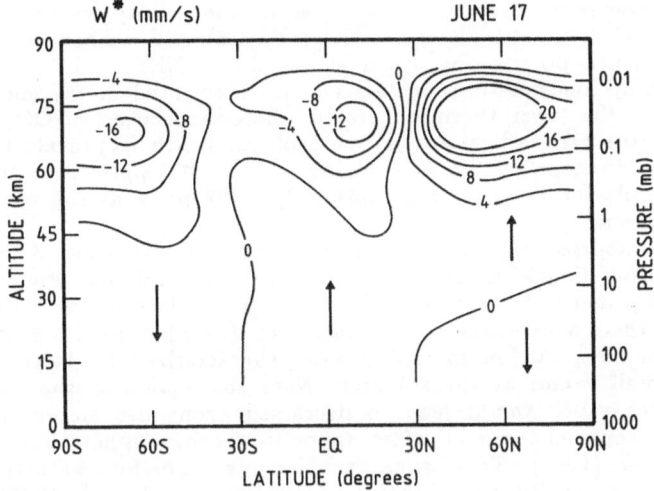

Figure 3. As in Fig. 2, except of the vertical component (mm/s).

tracers over long distances. It can now be seen that gravity waves force descent over the winter pole, keeping it warmer than it would otherwise be, the region radiating its excess heat to space. Because gravity wave driving is strongest in the summer and winter jets, there is a meridional dipole structure in vertical velocity around each jet. At this time gravity wave driving is strongest in the easterly jet and there is downward motion in the equatorial mesosphere, limiting the upward extent of the 'stratospheric fountain' over the equator. At the solstice the sun angle is changing very slowly. It can be shown by considering the transformed Eulerian mean equations that, in the absence of wave driving, there will be no net heating at the solstice, consequently no meridional circulation. Although the gross structure of the circulation can be obtained by using a global mean vertical profile of Rayleigh friction, important latitudinal and vertical variations cannot be accounted for.

The distribution of K_{zz} due to gravity waves is shown in Fig. 4. As expected, the values are largest near the jets. The strongest vertical mixing due to gravity wave breaking should occur in the midlatitude mesosphere and upper stratosphere. Both vertical advection and the meridional variation in K_{zz} are important in determining the meridional variation of long-lived species such as methane (Fig. 5), nitrous oxide, water vapor, and the chlorofluorocarbons. For each of these species we see a 'double peak' near the stratopause in our model results which is due strictly to gravity wave driving. Such a double peak has been observed in water vapor (Gordley et al., 1985) and in methane and nitrous oxide (Jones and Pyle, 1984). This has been interpreted in terms of advection associated with the semiannual oscillation (Gray and Pyle, 1986; Solomon et al., 1986; Gille et al., 1987). We are currently working on a parameterization of Kelvin waves with the purpose of discerning the relative roles of advection and diffusion in creating this pattern.

Molecules which form in the thermosphere and are stable in the polar night are readily advected downward over the winter pole by the gravity wave driven circulation. Note the large values of odd nitrogen (NO_y; Fig. 6) and carbon monoxide (CO; Fig. 7) over the south pole in the upper stratosphere. Nitric oxides are produced in the thermosphere by ionospheric processes and predissociated in the sunlit mesosphere. CO is produced in the lower thermosphere by photodissociation of CO_2 and destroyed by reaction with the OH radical in the mesosphere, which is prevalent in the sunlit portion. Near the stratopause the CO mixing ratio is 160 ppbv over the winter pole, compared to 30 ppbv at the equator, while NO_y is 27 ppbv at the winter pole and 8 ppbv in other regions.

As equinox approaches, the zonal jets weaken, gravity wave driving weakens, hence so do the meridional circulation and vertical eddy mixing. Figure 8 shows the distribution of K_{zz} near the vernal equinox, when the latitudinal variation is also significantly less than near solstice. Thomas et al. (1984) reported a strong seasonal variation in ozone near 80 km in midlatitudes characterized by large values at the equinoxes and small values at the solstices. Near the equinoxes the reduced upward transport of water vapor would lead to decreased ozone destruction by hydrogen radicals. Such seasonal changes in water vapor in the mesosphere have been observed by Bevilacqua et al. (1985). This ozone variation may therefore be attributable to the seasonal variation in gravity wave breaking (Garcia and Solomon, 1985).

Atmospheric tides will produce a diurnally-varying background flow which should strongly influence the time and altitude of gravity wave breaking. Bjarnason et al. (1987) have shown that K_{zz} can vary by an order of magnitude during a day, and that the vertical diffusion of species such as atomic oxygen and hydroxyl radicals are significantly modulated by this effect. Such dynamical effects are believed to be important in the diurnal behavior of the airglow emission (Meinel bands).

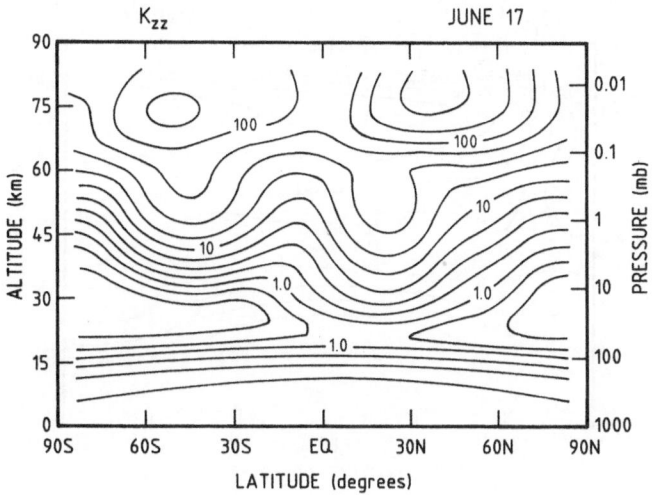

Figure 4. As in Fig. 1, except of vertical eddy diffusivity (m²/s). The values are derived from the Lindzen parameterization (see text).

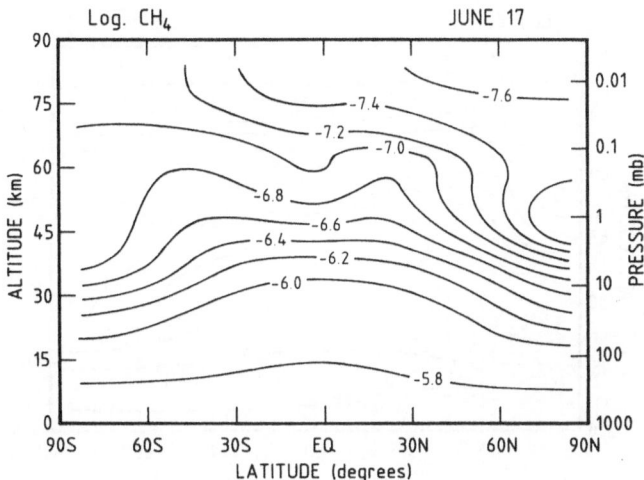

Figure 5. As in Fig. 1, except of the logarithm of methane mixing ratio (-6.0 corresponds to 1 ppmv).

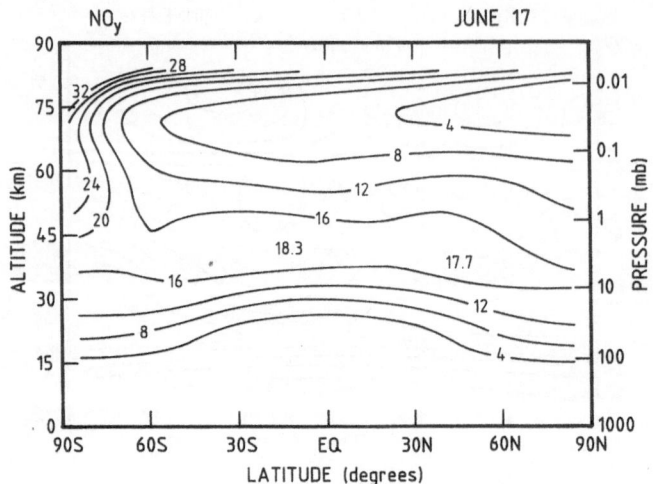

Figure 6. As in Fig. 1, except of total odd nitrogen ($NO_y = N + NO + NO_2 + NO_3 + 2 \times N_2O_5 + HNO_3 + HO_2NO_2 + ClONO_2$) in ppbv.

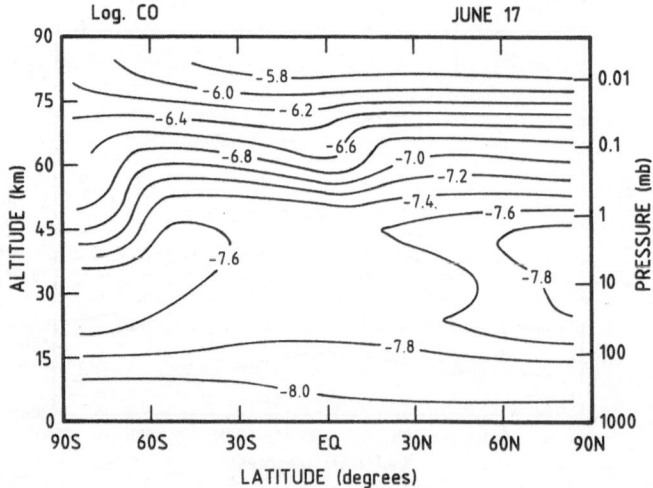

Figure 7. As in Fig. 5, except of carbon monoxide.

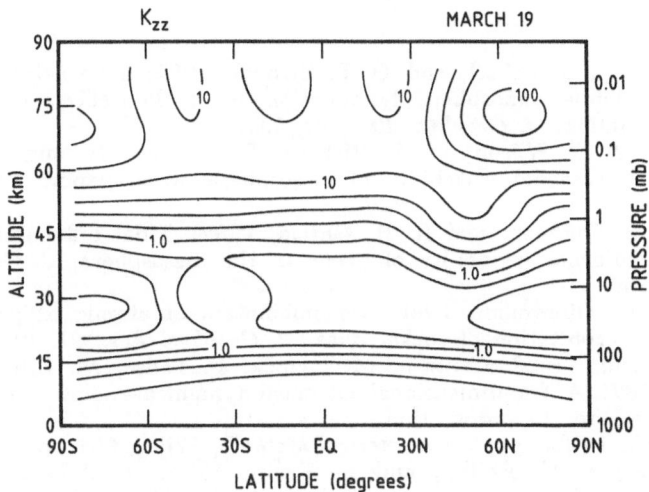

Figure 8. As in Fig. 4, except near the vernal equinox.

5. CONCLUDING REMARKS

The distribution of trace species in the middle atmosphere appears to be very sensitive to the distribution of breaking gravity waves, which in turn depends on the distribution of zonal winds and sources of gravity wave action. Since sources are rather variable in space and time, strong variations in species such as water vapor and nitric oxides should be observed in the mesosphere. In the current model source characteristics do not vary with latitude. It is possible that future observations of trace species could help fill the large gap in our understanding of the temporal and spatial distribution of gravity wave sources.

The Lindzen parameterization provides a workable means of estimating the effects of breaking gravity waves which vary self-consistently with the distribution of zonal wind. The parameterization requires specification of phase speeds, source altitudes, initial amplitudes, zonal scales, and occurrence frequencies. The enormous influence of gravity waves on a wide variety of phenomena warrants considerable observational effort toward establishing these values. To explain observed variations in trace species it is important to include the physics of gravity wave breaking in two-dimensional models. In turn, the models themselves may be helpful in establishing gravity wave parameters through tuning to obtain agreement with new observations.

REFERENCES

Apruzese, J. P., M. R. Schoeberl, and D. F. Strobel, **1982**: Parameterization of IR cooling in a middle atmosphere dynamics model, I. The effects on the zonally averaged circulation. *J. Geophys. Res., 87*, 8951.

Bevilacqua, R. M., W. J. Wilson, W. B. Ricketts, P. R. Schwartz, and R. J. Howard, **1985**: Possible seasonal variability of mesospheric water vapor. *Geophys. Res. Lett., 12*, 397.

Bjarnason, G. G., S. Solomon, and R. R. Garcia, **1987**: Tidal influences on vertical diffusion and diurnal variability of ozone in the mesosphere. *J. Geophys. Res., 92*, 5609-5620.

Brasseur, G., and D. Offermann, **1986**: Recombination of atomic oxygen near the mesopause: Interpretation of rocket data. *J. Geophys. Res., 91*, 10818.

Brasseur, G., C. Brühl, P. J. Crutzen, M. Dymek, E. Falise, M. Hitchman, and M. Pirre, **1987**: A two-dimensional chemical-dynamical-radiative model of the middle atmosphere. In preparation.

Carnahan, B., *et al.*, **1969**: *Applied Numerical Methods*, Wiley, 604 pp.

Cunnold, D., F. Alyea, N. Phillip, and R. Prinn, **1975**: A three-dimensional dynamical-chemical model of atmospheric ozone. *J. Atmos. Sci., 32*, 170.

DeMore, W. B., *et al.*, **1985**: Chemical kinetics and photochemical data for use in stratospheric modeling, evaluation number 7. *JPL Publication 85-37*, 226 pp., Jet Propulsion Lab., Pasadena, CA.

Dunkerton, T. J., **1982**: Stochastic parameterization of gravity wave stresses. *J. Atmos. Sci., 39*, 1711.

Fritts, D. C., **1984**: Gravity wave saturation in the middle atmosphere: A review of theory and observations. *Revs. Geophys. Space Phys., 22*, 275.

Garcia, R. R., and S. Solomon, **1983**: A numerical model of the zonally averaged dynamical and chemical structure of the middle atmosphere. *J. Geophys. Res., 88*, 1379.

Garcia, R. R., and S. Solomon, **1985**: The effect of breaking gravity waves on the dynamics and chemical composition of the mesosphere and lower thermosphere. *J. Geophys. Res., 90*, 3850.

Gille, J. C., L. V. Lyjak, and A. K. Smith, **1987**: The global residual mean circulation in the middle atmosphere for the northern winter period. *J. Atmos. Sci., 44*, 1437-1452.

Gordley, L. L., J. M. Russell, and E. E. Remsberg, **1985**: Global lower mesospheric water vapor revealed by LIMS observations. In *Atmospheric Ozone, Proc. Quadrennial Ozone Symposium*, Reidel.

Gray, L. G., and J. A. Pyle, **1986**: The semiannual oscillation and equatorial tracer distributions. *Quart. J. Roy. Meteor. Soc., 112*, 387.

Haurwitz, B., **1961**: Frictional effects and the meridional circulation in the mesosphere. *J. Geophys. Res., 66*, 2381.

Hitchman, M. H., and C. B. Leovy, **1986**: Evolution of the zonal mean state in the equatorial middle atmosphere during October 1978-May 1979. *J. Atmos. Sci., 43*, 3159-3176.

Hitchman, M. H., and G. Brasseur, **1987**: Rossby wave action as an interactive tracer in a 2-D model: parameterization of wave driving and eddy diffusivity. To be submitted to *J. Geophys. Res.*

Hodges, R. R., **1969**: Eddy diffusion coefficients due to instabilities in internal gravity waves. *J. Geophys. Res., 74*, 4087.

Holton, J. R., **1982**: The role of gravity wave induced drag and diffusion in the momentum budget of the middle atmosphere. *J. Atmos. Sci., 39*, 791.

Holton, J. R., **1983**: The influence of gravity wave breaking on the general circulation of the middle atmosphere. *J. Atmos. Sci., 40*, 2497.

Holton, J. R., and W. M. Wehrbein, **1980**: A numerical model of the zonal mean circulation of the middle atmosphere. *Pageoph, 118*, 284.

Jones, R. L., and J. A. Pyle, **1984**: Observations of CH_4 and N_2O by the Nimbus 7 SAMS: A comparison with in-situ data and two-dimensional numerical model calculations. *J. Geophys. Res., 89*, 5263.

Leovy, C., **1964**: Simple models of thermally driven mesospheric circulation. *J. Atmos. Sci., 21*, 327.

Lindzen, R. S., **1967**: Thermally driven diurnal tide in the atmosphere. *Quart. J. R. Meteorol. Soc., 93*, 18.

Lindzen, R. S., **1968**: The application of classical atmospheric tidal theory. *Proc. R. Soc. London Ser. A, 303*, 299.

Lindzen, R. S., **1981**: Turbulence and stress owing to gravity wave and tidal breakdown. *J. Geophys. Res., 86*, 9707.

Matsuno, T., **1982**: A quasi one-dimensional model of the middle atmosphere circulation interacting with internal gravity waves. *J. Meteor. Soc. Japan, 60*, 215.

Murgatroyd, R. J., and F. Singleton, **1961**: Possible meridional circulations in the stratosphere and mesosphere. *Quart. J. R. Meteor. Soc., 87*, 125.

Pitteway, M.L.V., and C. O. Hines, **1963**: The viscous damping of atmospheric gravity waves. *Can. J. Phys., 43*, 2222.

Schoeberl, M. R., and D. F. Strobel, **1978**: The zonally averaged circulation of the middle atmosphere. *J. Atmos. Sci., 35*, 577.

Smith, A. K., and L. V. Lyjak, **1985**: An observational estimate of gravity wave drag from the momentum balance in the middle atmosphere. *J. Geophys. Res., 90*, 2233.

Solomon, S., J. T. Kiehl, R. R. Garcia, and W. Grose, **1986**: Tracer transport by the diabatic circulation deduced from satellite observations. *J. Atmos. Sci., 43*, 1603.

Thomas, R. J., C. A. Barth, and S. Solomon, **1984**: Seasonal variations of ozone in the upper mesosphere and gravity waves. *Geophys. Res. Lett., 11*, 673.

Wehrbein, W. M., and C. B. Leovy, **1982**: An accurate radiative heating and cooling algorithm for use in a dynamical model of the middle atmosphere. *J. Atmos. Sci., 39*, 1532.

MODELING THE TRANSPORT OF CHEMICALLY ACTIVE CONSTITUENTS IN THE STRATOSPHERE

W. L. Grose, J. E. Nealy, R. E. Turner, W. T. Blackshear
Atmospheric Sciences Division
NASA Langley Research Center
Hampton, Virginia 23665-5225

ABSTRACT. A three-dimensional, spectral, primitive equation, atmospheric model incorporating comprehensive chemistry has been used to study dynamics and transport processes and to simulate the distribution of ozone and other trace constituents in the stratosphere. Preliminary results from a simulation of the seasonally varying evolution of several important constituents are presented. Comparisons of simulated species distributions with data obtained from satellite experiments demonstrate good agreement in many instances. Of particular interest is the occurrence of incursions or tongues of ozone-rich air parcels from lower latitudes into the polar cap region associated with the displaced polar vortex during a mid-winter stratospheric warming. During the period of enhanced dynamical activity, the model successfully simulates many aspects of observed ozone behavior as well as features described as wave-breaking and irreversible mixing observed in isentropic distributions of potential vorticity inferred from satellite temperature data. Examination of the evolving constituent distributions suggests that episodic transport of ozone into the polar region during wave-breaking events culminates with the production of a high-latitude, spring maximum in total column ozone.

1. INTRODUCTION

    Active interest in the transport of trace constituents in the stratosphere began to evolve some 60 years ago, principally after recognition that the distribution of ozone was apparently strongly influenced as a result of poleward and downward transport away from the region of photochemical production. Mahlman et al. (1984) have provided a concise historical perspective on research related to understanding stratospheric transport processes. The advent of atmospheric nuclear weapons testing approximately 30 years ago stimulated widespread concern over the transport and lifetime of radioactive material in the atmosphere. During the last decade and a half, the possibility of ozone depletion in the stratosphere as a result of aircraft emissions, fertilizers, halocarbons, and various other chemicals further demonstrated the need to understand the complex processes which determine the spatial and temporal distribution of stratospheric trace constituents. The scientific

229

G. Visconti and R. Garcia (eds.), Transport Processes in the Middle Atmosphere, 229–250.

community and the general populace have become increasingly aware that
technological advancement and the associated effects of an ever-
increasing population may alter the composition of the atmosphere with
the potential for adverse effects on environmental quality and global
climate.

The rationale for three-dimensional model studies of
stratospheric dynamics and transport processes has been well
established (e.g. WMO, 1986, and refs.). Notably during Northern
Hemisphere winter, the stratospheric circulation is highly
three-dimensional. Air parcels undergo significant meridional and
vertical excursions in a matter of days (e.g. Leovy et al., 1985)
during a succession of large-amplitude, transient disturbances.
Despite their known deficiencies, three-dimensional atmospheric models
do provide a means of simulating such behavior for detailed study.
However, at the present time, long-term simulations with a general
circulation model (GCM) which incorporates a comprehensive and fully
interactive formulation of radiation, chemistry, and dynamics are not
feasible and are probably unwarranted considering known deficiencies
in current understanding of both dynamical and chemical processes.
For the present, this situation has dictated our approach to
stratospheric modeling using more simplified three-dimensional models.

In the sections to follow, we will briefly describe the structure
of our model. Some preliminary results from a simulation experiment
conducted with the model are presented and compared with observational
satellite data. Primary emphasis is placed upon a discussion of the
transport of ozone during a mid-winter stratospheric warming.

2. DESCRIPTION OF THE MODEL

A spectral, primitive equation, general circulation model (see
Hoskins and Simmons, 1975, Grose et al., 1984, and Blackshear et al.,
1987, for description of the model) was used for the studies described
herein. The sigma coordinate system of Phillips (1957) is adopted in
the vertical dimension with 12 levels extending from the surface to
approximately 60 km. Spectral representation is used in the
horizontal dimension with the transform method discussed by Orszag
(1970) and Eliasen et al. (1970) employed in the evaluation of nonlinear terms. The model is global in extent, with triangular
truncation (zonal wavenumber 16) of the spherical harmonic
expansions. The method of Lacis and Hansen (1974) is used for the
calculation of heating by absorption of solar radiation. Longwave
radiation is treated with a Newtonian approximation. The Earth's
orography is incorporated by a smoothed spectral representation of the
one-degree grid values given by Gates and Nelson (1975). Time
integration is accomplished with the semi-implicit technique (Robert
et al., 1972) using a 30-minute time step. Periodically, a 1-2-1
centered time filter is used on two consecutive time steps to control
odd-even time splitting.

Transport model simulations are performed with an "off-line"
technique, similar to that described by Mahlman and Moxim (1978) and

Mahlman et al. (1980), in which time-dependent wind and temperature fields are generated separately with the GCM and then used as input to the set of mass continuity equations for the tracers. Evaluation of the net chemical source (sink) term required in the continuity equations is described in the following section. The formulation of the transport model is essentially identical to that of the GCM. A scale-selective, biharmonic diffusion is used to parameterize horizontal, sub-grid-scale processes. Vertical sub-grid diffusion is not included, following Mahlman and Moxim (1978). However, special consideration is given to the vertical transport in the model when negative mixing ratios occur. Analysis of earlier experiments with an inert tracer indicated that vertical transport in the presence of strong gradients was the primary mechanism for producing negative mixing ratios. The generation of such negative mixing ratios, resulting from the use of second-order central differencing approximations to the vertical transport term, was alleviated by switching to upstream differencing only when negative values occur. The switch to upstream differencing is accomplished (in a mass-conserving way) by addition of a flux divergence term equal to the difference between an upstream difference and a central difference approximation to the vertical advection term. The effectiveness of upstream differencing for suppressing negative mixing ratios is a consequence of the positive definite advection property of this method.

The complexity of the chemistry of the stratosphere with numerous relevant species and reactions makes explicit transport of each individual species an unwieldy option. In addition, the extremely fast chemistry characteristic of some species would dictate integration time steps so small as to be impractical for conducting long-term simulations with the transport model. To circumvent these difficulties, a number of related species which undergo relatively rapid chemical transformations are grouped into a "family" such that the family has a characteristic chemical lifetime comparable to that for dynamics and much greater than for the individual species of the family. This concept has been frequently used in both one- and two-dimensional models of the stratosphere (e.g. Brasseur and Solomon, 1984, Kurzeja, 1975, and Blake and Lindzen, 1973).

In the present version of the model, the mixing ratios of the families O_x $\{O_3 + O(^3P) + O(^1D)\}$, NO_y $\{NO + NO_2 + NO_3 + N + HNO_2 + 2N_2O_5 + HNO_4\}$, $C\ell_x$ $\{C\ell + C\ell O + C\ell ONO_2 + C\ell O_2 + HOC\ell + HC\ell\}$ and the individual species HNO_3 are determined by explicit integration of their respective mass continuity equations. Partitioning families after each time step into individual species for evaluation of the net source (sink) term is accomplished through various equilibrium relationships. In order to assess the impact of using these relationships, detailed production and loss analyses have been performed for various zenith angles and latitudes and intercomparisons made with more complete mechanisms of a one-dimensional model. Largest differences appear in those individual species which have chemical lifetimes on the order of a day or so, and deviate significantly from equilibrium due to diurnal effects. Invoking the equilibrium assumption for most of these species has little impact on the net family source term (e.g. H_2O_2, HO_2, $ClNO_3$). Special treatment has been given to N_2O_5, however, since its contribution to the NO_y family can be significant at certain altitudes (middle and low stratosphere). The concentration of this species is obtained essentially by equating the daytime loss of N_2O_5 to its nighttime production which leads to a value which is bracketed by the daytime minimum and nighttime maximum values. This approximation has been found to be entirely satisfactory with regard to the evaluation of the NO_y net source term.

In a rigorous evaluation of photolysis rates required in the chemical code, it is necessary to perform integrations with respect to wavelength of the products of photodissociation cross-section, quantum efficiency, and local solar flux. Evaluation of these integrals numerically consumes a significant fraction of the computational time in a typical chemical calculation. As an additional approximation for the chemical code used in the model, all required photolysis rates are parameterized as a function of O_2 and O_3 absorber amounts. A satisfactory degree of approximation has been insured by comparison with values determined in the rigorous calculation. Ordinary rate coefficients are taken from DeMore et al. (1985).

Certain chemical species treated in the model include long-lived species which are essential to the ozone problem in that they represent sources of NO_y, Cl_x, and odd hydrogen species. These chemicals (H_2O, H_2, CH_4, N_2O, CH_3Cl, CF_2Cl_2, $CFCl_3$, CH_3CCl_3, and CO) are specified as a function of altitude, latitude, and season (as appropriate) based on observations or the two-dimensional model results of Garcia and Solomon (1983).

Polar night chemistry is a difficult problem because the chemical time constants become long for all chemicals present in significant amounts. Thus, partitioning of the families cannot be accomplished by equilibrium relations. In the present version of the model, O_x, NO_y, HNO_3, and Cl_x are transported in the polar night, but no chemistry is allowed. This approach results in negligible errors for the four transported entities, but it renders individual family member concentrations indeterminate.

3. DISCUSSION OF RESULTS

The transport model simulation discussed in the following sections consisted of a 120-day integration commencing on model day January 1. Zonally symmetric distributions were assumed for the initialization of the various chemical constituents using values either inferred from December monthly mean data from the Limb Infrared Monitor of the Stratosphere (LIMS) experiment (Gille and Russell, 1984) or from two-dimensional model simulations (Garcia and Solomon, 1983). Wind and temperature fields used in the transport model simulation were obtained from a multi-year integration of the circulation model.

Subsequent discussion will largely concentrate on the interactions between photochemical and transport processes which occurred during a mid-winter stratospheric warming. The dynamical evolution of the warming event is discussed in some detail in Blackshear et al. (1987). To provide some perspective on the rapidly varying conditions during this period, the time history of the zonal mean temperature, $\overline{T}$, at 70N, 2 mb is shown in Fig. 1. During early January, there was a strong warming pulse of approximately 30 K. The principal effects of this event occurred at the 10 mb level and above with reversal of the meridional gradient of $\overline{T}$ poleward of 60N and a marked reduction in the strength of the westerly jet. A period of recovery followed with a relaxation to conditions similar to the prewarming state of December. In the latter half of January, a general warming trend developed in high latitudes culminating in a second warming event with maximum polar temperatures occurring on February 9. After recovery from this second warming pulse, the stratospheric circulation for the remainder of the winter was characterized by a well-developed Aleutian anticyclone and a nearly symmetric polar vortex somewhat offset from the pole.

A strong correlation between zonal mean temperature and the amplitude of zonal harmonic height wave 1 is apparent in Fig. 1. Prior to both warming pulses, the wave 1 amplitude exceeded 1500 meters, a condition consistent with the observational analyses of Labitzke (1981). The corresponding synoptic situation is characterized by an intensified anticyclone with elongation of the polar vortex and displacement off the pole allowing strong cross-polar flow (e.g. Clough et al., 1985).

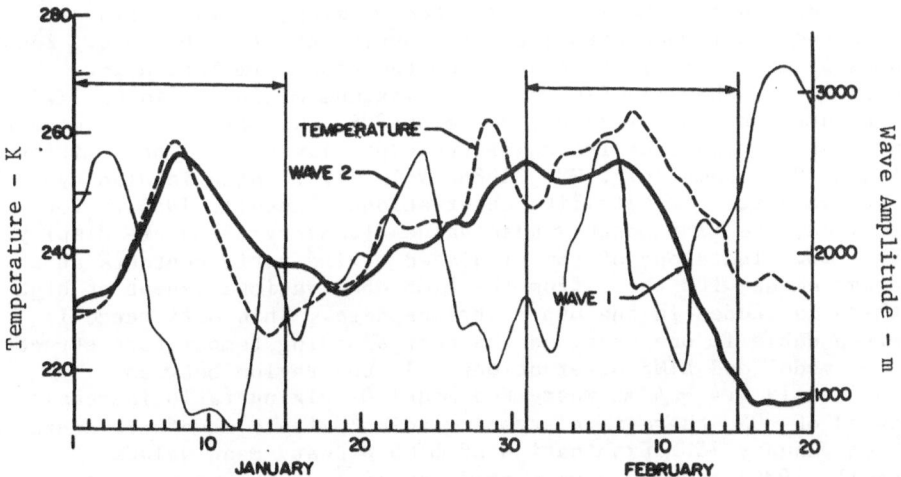

Figure 1.   Time histories of the amplitude of geopotential height wavenumbers one and two and the zonal mean temperature at 70°N latitude and 2.0 mb pressure. (Wavenumber two amplitude scaled by a factor of 5.)

Shown in Fig. 2 are latitude-height cross-sections of the zonal mean winds for January 1 and February 9. The stratospheric westerlies shown in Fig. 2a weakened such that wind speeds were 30 m/s or less throughout the stratosphere at the peak of the January warming event. During the post-warming period in mid-January, the westerly jet recovered, but the axis had been displaced 5° poleward of its position on January 1. The second warming event culminated on February 9 with a strong reversal of the meridional gradient of $\overline{T}$ in the region poleward of 60N at the 10 mb level and above. Figure 2b displays the concomitant reversal of winds with zonal easterlies at high latitudes. The post-warming state exhibited a westerly jet of narrower width with the axis displaced 20° poleward of the position at the beginning of January.

Preliminary comparisons of simulated constituent distributions with observations are generally encouraging. However, if one considers interannual variability (e.g. Geller et al., 1984), the paucity of data for some constituents, and the limited duration of the model simulation, the results must be interpreted in the appropriate context. The comparisons presented herein have been made with LIMS data because this data set provides simultaneity of measurement of temperature, ozone, nitrogen dioxide, nitric acid, and water vapor on a near-global scale. The primary drawback of the data set is the rather limited duration of 7 months. In addition, it is generally conceded that the winter stratosphere circulation for the LIMS period during 1978-1979 was somewhat atypical. Additional comparisons have been made with various other data, but the limited scope of the present paper precludes their inclusion.

Before consideration of the marked zonally asymmetric behavior which occurs during the model simulated warming, it is instructive to examine the zonal mean structure of constituent distributions. Zonal mean ozone mixing ratio contours from the model simulation for January 15 are shown in Fig. 3a. The maximum mixing ratio of 11.1 ppmv occurred at approximately 32 km at 20S. Two-dimensional models incorporating comprehensive chemistry typically exhibit peak mixing ratios of 9-12 ppmv (e.g. WMO, 1986, Vol. III), which is also typical of the range seen in satellite observations (Russell, 1986). For comparison, the LIMS monthly mean values (January 1979) are displayed in Fig. 3b. The slope of the simulated mixing ratio contours show good agreement with those from the LIMS observations except at high northern latitudes in the upper stratosphere. This difference is understandable if one examines the corresponding temperature structure for the model and LIMS observations. In the region between approximately 0.4 - 4 mb where the model O_3 mixing ratio increases poleward of 60N, there is a strong gradient with a monotonic decrease in $\overline{T}$ on January 15. Examination of LIMS January mean values of $\overline{T}$ (Russell, 1986) reveals a weak gradient of $\overline{T}$ as the ozone mixing ratio decreases poleward of 60N in the upper stratosphere. The model simulated polar temperatures are 30-40K colder than the corresponding LIMS January mean values. Comparison of vertical profiles of O_3 mixing ratio at 30N from the simulation with representative observed

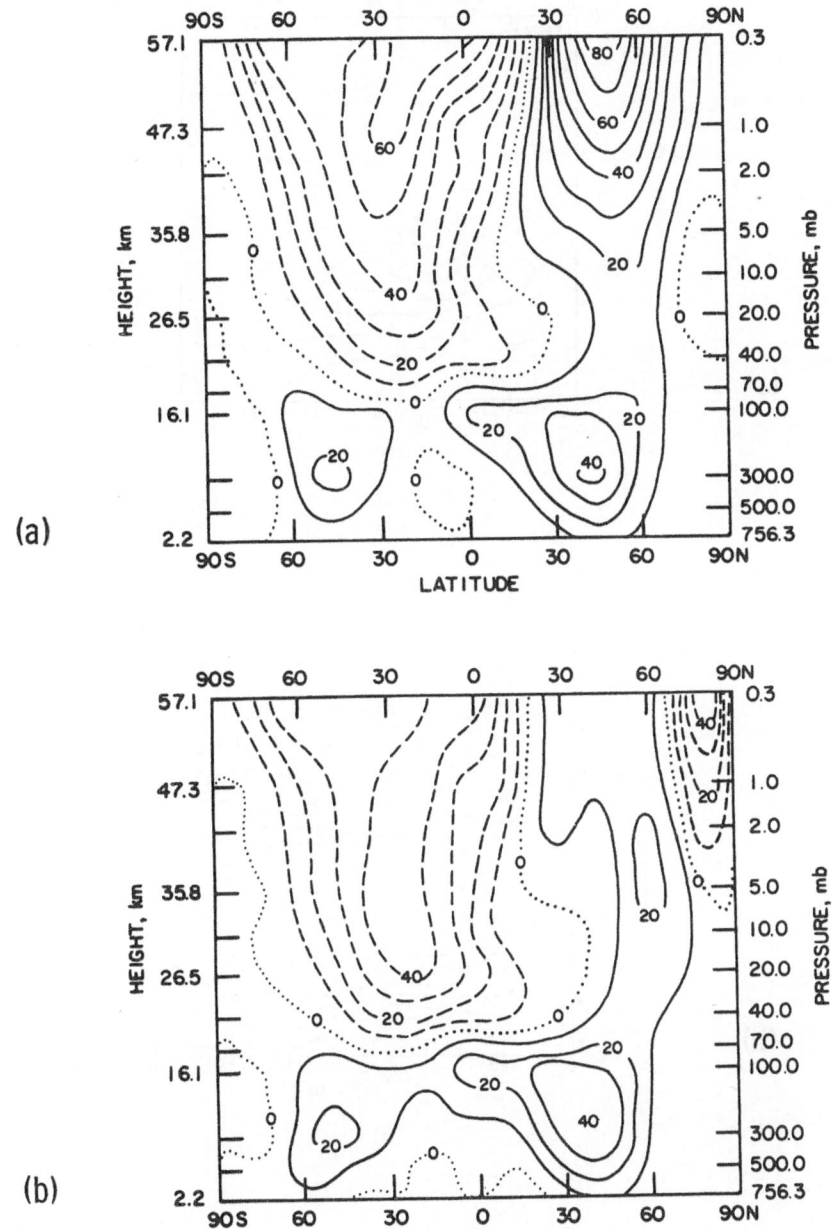

Figure 2.   Zonal mean wind (m/s) cross-sections for LaRC model
simulation for (a) January 1 and (b) February 9.

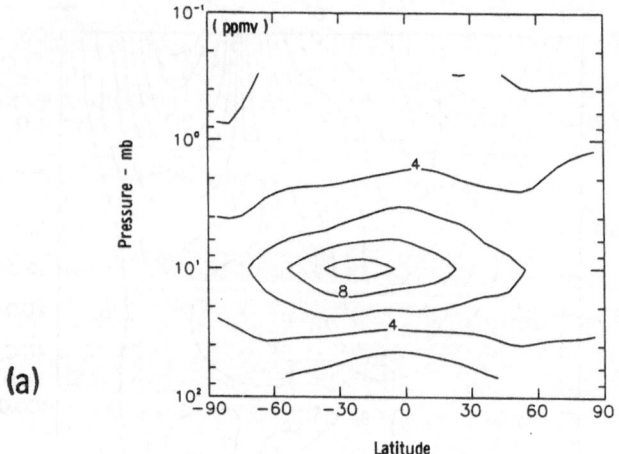

(a)

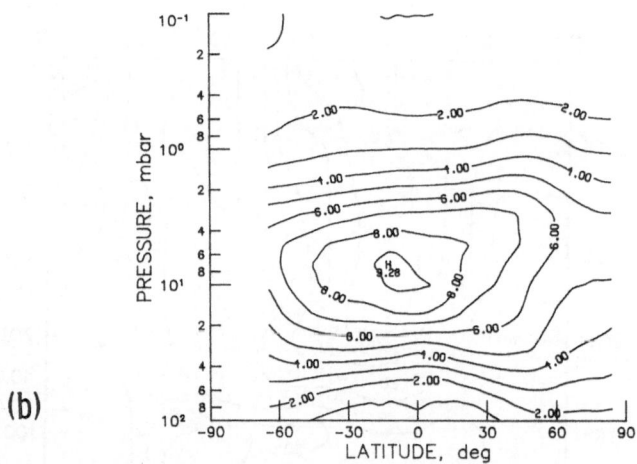

(b)

Figure 3. Zonal mean ozone (ppmv) cross-sections for: (a) LaRC model
simulation, January 15 and (b) LIMS monthly mean for January 1979.

profiles at 30N have been made. In general, the vertical gradient at
30N between 2 - 10 mb is weaker than observed. Above 2 mb, the
vertical gradient agrees reasonably well with observed values.

Zonal mean mixing ratio contours for nighttime NO_2 are shown in
Fig. 4a from the model (January 15) and in Fig. 4b from the LIMS data
(January monthly mean). The two distributions exhibit good general
agreement with respect to peak mixing ratio, the vaulted structure in
the equatorial lower stratosphere, and the high-latitude decrease
toward the winter pole in the upper stratosphere. The maximum in the
LIMS distribution has broader latitudinal extent and is located
slightly lower than that from the simulation.

The diurnal variation of NO_2 is depicted in Fig. 5 on the 10 mb
surface. The results are from the model for 1200 GMT on February 1.
The perspective view of the western hemisphere depicts the variation
of NO_2 in the region of the day-night terminator.

Some representative results from the model simulation for O_3
mixing ratio on the 10 mb surface are presented in Fig. 6 to
illustrate the marked zonally asymmetric motions which accompanied the
early February warming period. The distribution on February 1 is
characterized by an ozone minimum of 5-6 ppmv centered near the
Greenwich meridian and superimposed on the elongated polar vortex. A
secondary minimum exists in the region of the Aleutian anticyclone
(see Blackshear et al., 1987, for a discussion of the synoptic
development of the warming). This secondary minimum is largely
comprised of ozone which was eroded from the periphery of the poleward
flank of the main vortex and mixed into the so-called "surf zone" (see
McIntyre and Palmer, 1983 and 1984) during the warming in January as
the Aleutian anticyclone intensified and a strong cross-polar jet
developed. Ozone mixing ratios in the Tropics are on the order of 10
ppmv with weak zonal gradients.

By February 5, there are prominent incursions (noted by the bold
arrows) of relatively high ozone mixing ratio being drawn around the
polar vortex from lower latitude toward higher latitudes. Similar
incursions or "tongues" are evident in the LIMS ozone data for the
period during the January - February 1979 stratospheric warmings and
were noted by Leovy et al. (1985). These tongues are seen to extend
completely across the polar cap on February 7 and 9. A cutoff minimum
(8 ppmv) is seen in the eastern Atlantic sector between 35-45N on
February 7. This feature is associated with the development of a
second anticyclone in this sector during the February warming period.

Diagnosis of the ozone tendency equation on the 850K isentropic
surface (Grose et al., 1987) for the period of the simulation reveals
that the tongues (denoted in the figures by the bold arrows) are the
result of quasi- horizontal advection. Essentially no contribution
from vertical advection is seen. A corresponding analysis on pressure
surfaces is less useful, revealing a near balance between the vertical
and horizontal contributions with the net ozone tendency being a
smaller difference between those terms.

The diagnostic analysis also indicates that the transport
processes exhibit strong nonlinear mixing to smaller scales on the
poleward flanks of the vortex where the erosion occurs during the

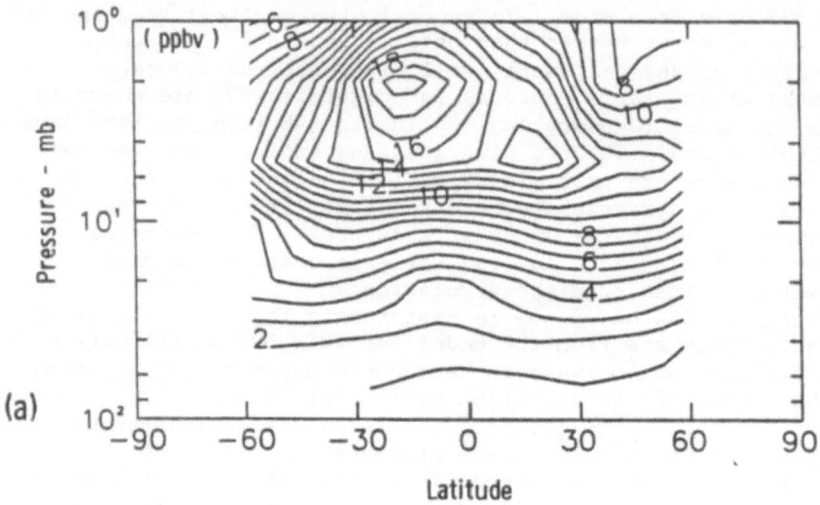

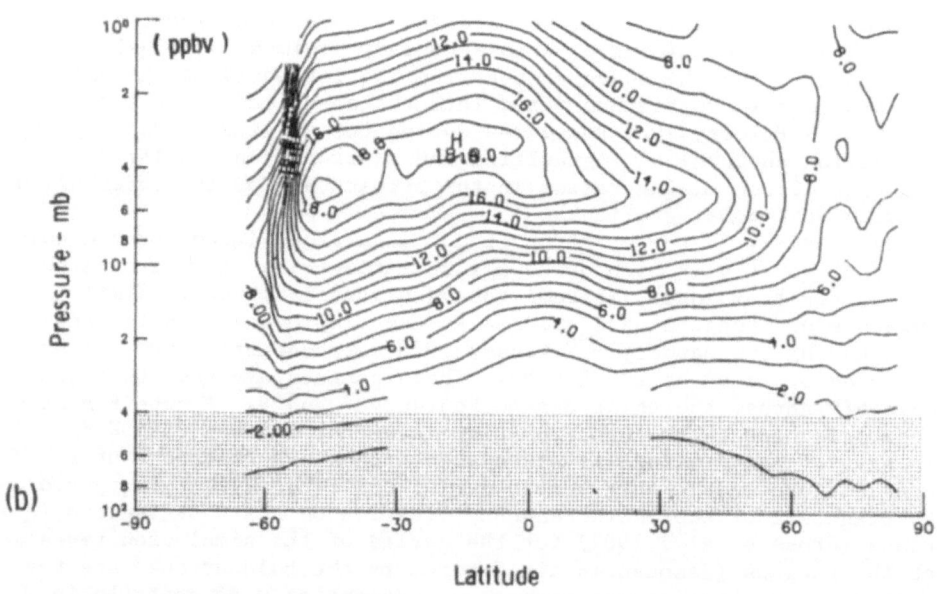

Figure 4. Zonal mean nitrogen dioxide (ppbv) nighttime cross-sections
for: (a) LaRC model simulation, January 15 and (b) LIMS monthly mean
for January]979.

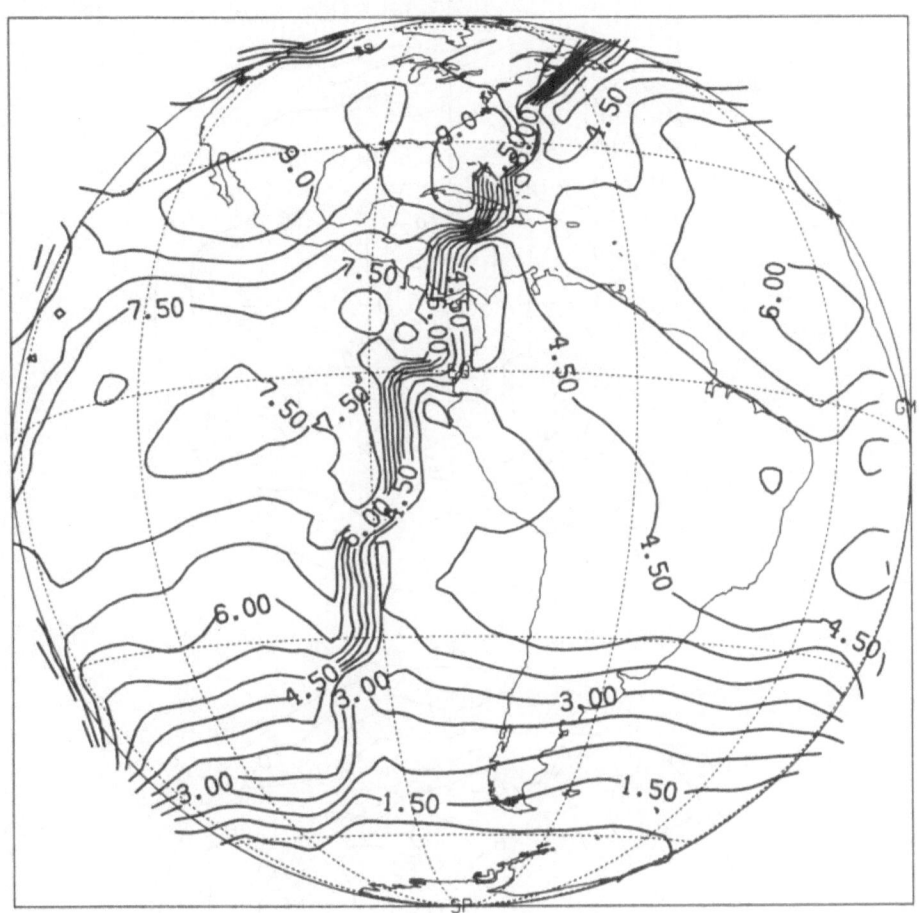

Figure 5. Nitrogen dioxide mixing ratio (ppbv) on the 10 mb pressure
level, 1200 GMT, February 1, for the LaRC model simulation.
Perspective of the Western Hemisphere illustrating the diurnal
variation in the region of the day-night terminator.

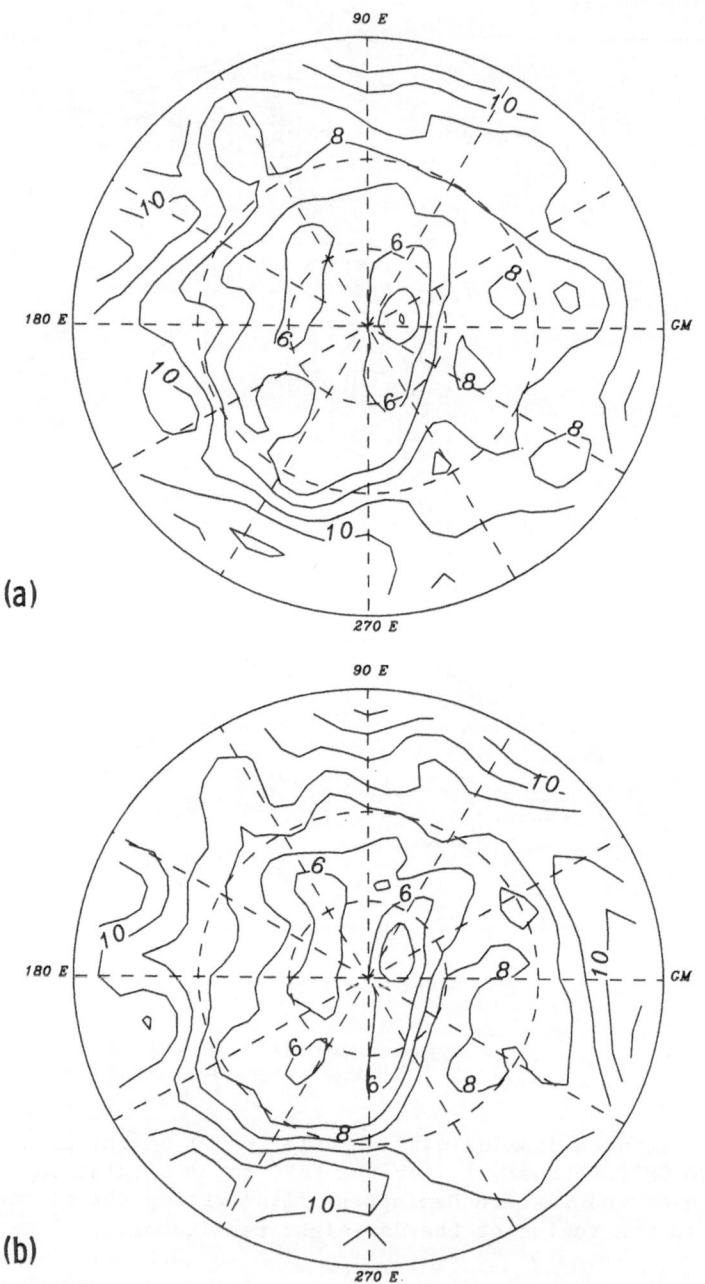

Figure 6.   Ozone mixing ratio (ppmv) on the 10 mb pressure level of the Northern Hemisphere for the LaRC model simulation: (a) February 1, (b) February 3, (c) February 5, (d) February 7, and (e) February 9.

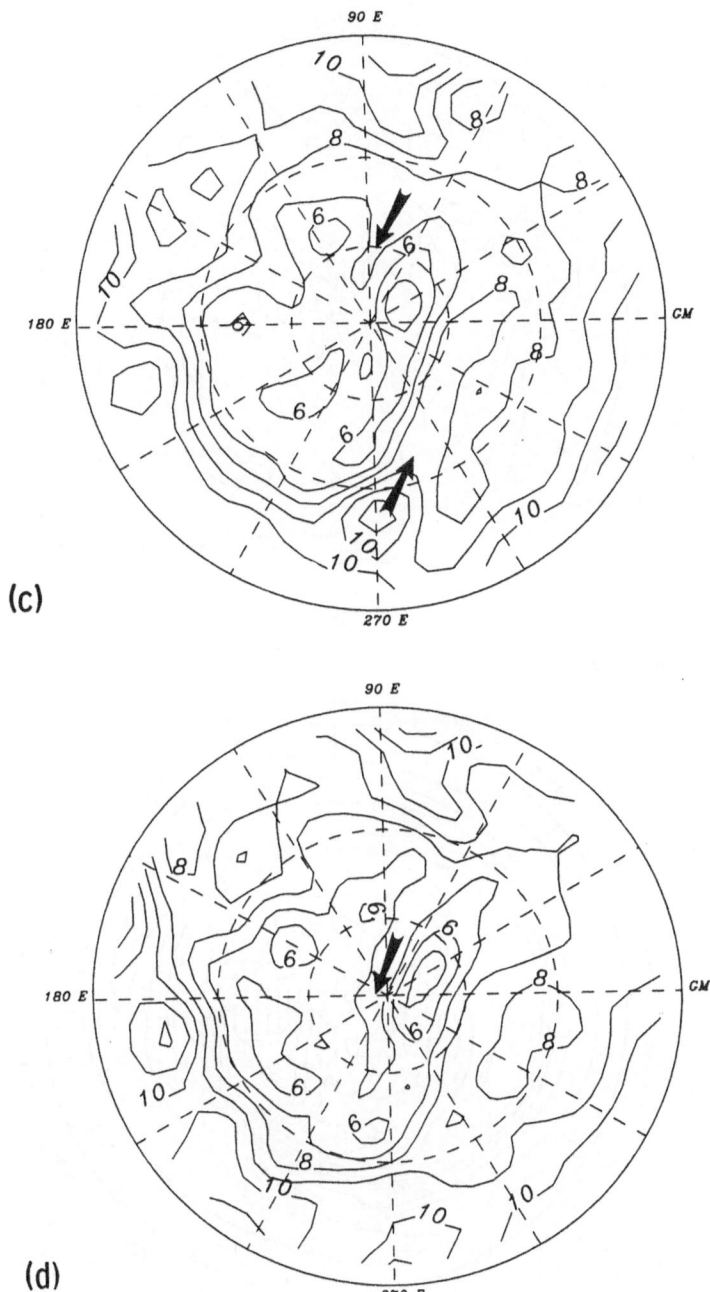

Figure 6. continued

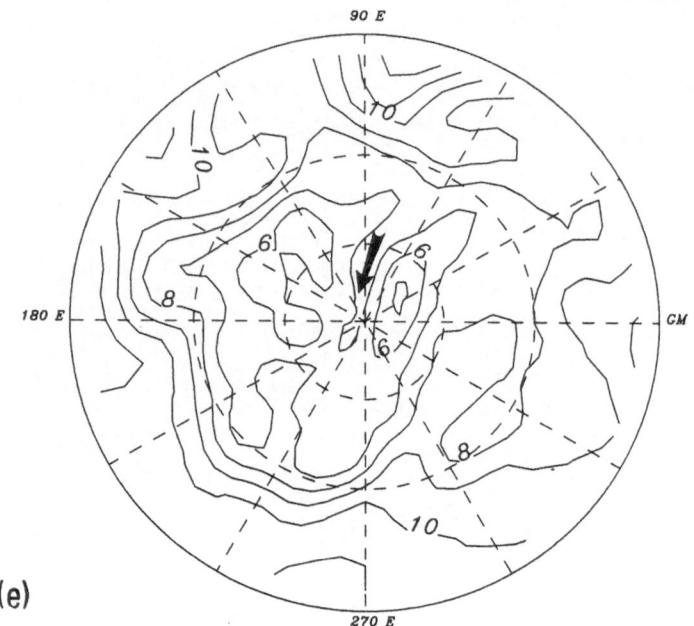

(e)

Figure 6.   continued

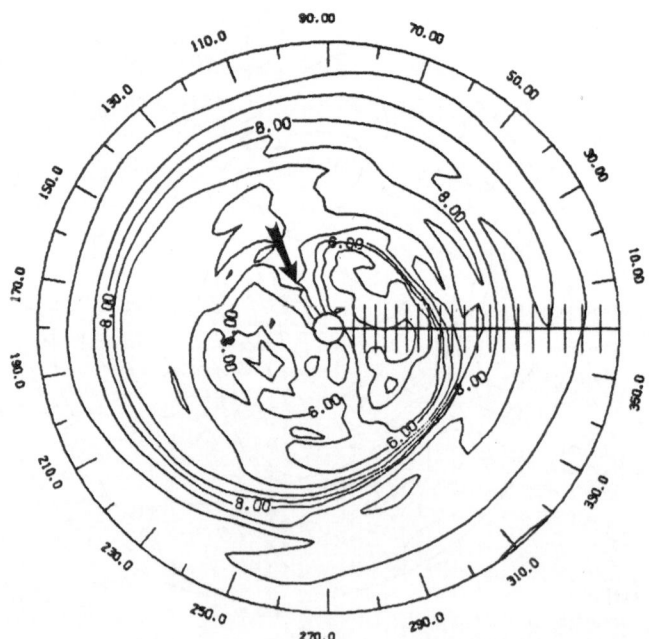

Figure 7.   Ozone mixing ratio (ppmv) on the 10 mb pressure level of
the Northern Hemisphere from the LIMS experiment, January 26, 1979.

warming events. The results are generally supportive of the concepts
of "wave-breaking" and irreversible mixing advanced by McIntyre and
Palmer (1983, 1984) in their analysis of isentropic distributions of
potential vorticity inferred from satellite data.

The ozone mixing ratio on the 10 mb surface of the Northern
Hemisphere for January 26, 1979, obtained from the LIMS experiment is
shown in Fig. 7 for comparison with the model results presented in
Fig. 6. This particular day during the minor warming of late January
was chosen because the observed circulation was quite similar to that
of the model simulated warming. A comprehensive discussion on the
transport of ozone based upon LIMS data interpretation is contained in
the study of Leovy et al. (1985). Minimum ozone mixing ratios of 5
ppmv are observed within the polar vortex. There is a prominent
tongue (denoted by the bold arrow) of relatively high ozone mixing
ratio extending over the polar cap. Low mixing ratios are also
observed in the surf zone region. Mixing ratios in the Tropics are
10-15 percent less than found in the model results. The close
resemblance of the observed and simulated distributions are readily
evident.

For contrast with the chemical tracer, ozone, distributions of
Ertel's potential vorticity, Q, on the 850K surface are presented in
Fig. 8. Note that the potential vorticity is obtained from the
long-term simulation conducted with the circulation model and is
independent of the transport simulation. Hence, the potential
vorticity is determined by the simulated velocity and temperature
fields. Alternatively, as argued by Hoskins et al. (1985), potential
vorticity being the more fundamental quantity, the velocity and
temperature fields are determined from the potential vorticity
distribution under suitable balance conditions and boundary conditions
(the invertibility principle).

At the 850K level of the stratosphere ($\approx$30 km), both ozone and Q
are quasi-conservative tracers for periods of a week or so.
Comparison of Figs. 6a and 6d with Figs. 8a and 8b, respectively,
demonstrates a strong negative correlation between O_3 and Q which
has been previously noted (e.g. Leovy et al., 1985, and Grose, 1984).
The close correlation evident by comparison of these figures provides
supportive evidence that the ozone has adjusted to the imposed
circulation, despite the limited duration of the simulation. The Q
distributions shown in Fig. 8 display the main vortex/surf zone
structure seen in the distributions of Q inferred from satellite
data by McIntyre and Palmer (1983, 1984), Clough et al. (1985), Grose
(1984), and Dunkerton and Delisi (1986), among others.

Examination of the evolving potential vorticity distributions
during the simulated warming provides a vivid illustration of the
phenomenon which McIntyre and Palmer (1983, 1984) have characterized
with the term wave breaking. The distributions illustrate tongues of
relatively high Q values being drawn out of the main vortex around
the anticyclone with pieces or "blobs" (see McIntyre and Palmer, 1983
and 1984) ultimately breaking off from the tongues. It is precisely
this irreversible buckling of the Q contours which provides the

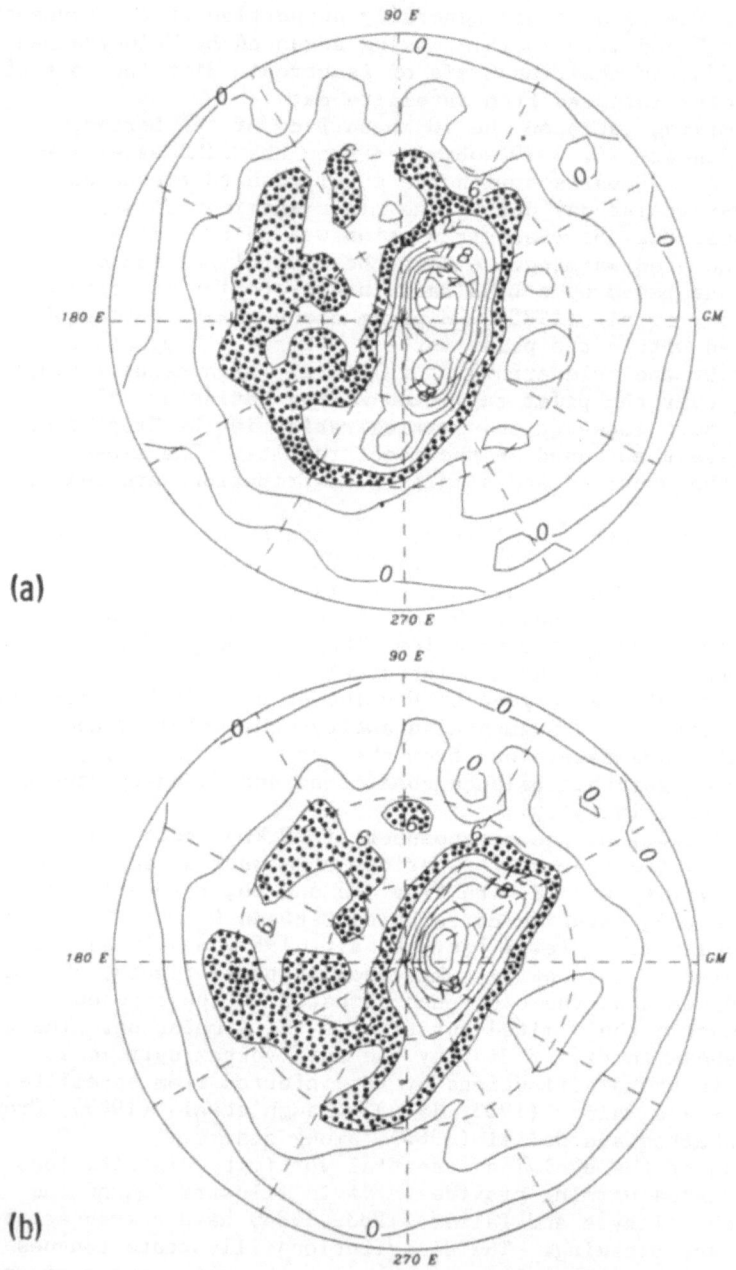

Figure 8. Distribution of Ertel's potential vorticity $(10^{-4} K\ m^{-1}s^{-1})$ on the 850K isentropic surface of the Northern Hemisphere for the LaRC model simulation: (a) February 1 and (b) February 7.

criterion for when waves are said to be breaking. Analysis of the   Q
distributions suggests that the wave-breaking process is quasi-
horizontal (on isentropic surfaces) and strongly non-linear,
consistent with the results from the ozone analysis.

As a final illustration of the model performance during the
simulation period, the zonal mean total column ozone above 100 mb is
shown in Fig. 9 as a function of latitude and time for comparison with
the corresponding LIMS data. The simulated total ozone distribution
displays a high latitude spring maximum that is in good agreement with
that obtained from the LIMS data. Both distributions exhibit an
equatorial minimum with the model values of 220 Dobson units
approximately 15 percent lower than LIMS values. The LIMS data does
not extend beyond 64S and cannot be compared with the model results.
The model distribution shows an early fall minimum over the southern
polar cap. The observations presented by Dutsch (1969) show the
occurrence of a somewhat later fall minimum near 75S. The model
structure reproduces the mid-latitude maximum observed in the Southern
Hemisphere.

Analysis of both model and LIMS column ozone reveal that the
buildup of the high latitude spring maximum in the Northern Hemisphere
occurred episodically as the large-amplitude disturbances associated
with the warmings produced rapid, irreversible meridional transport of
ozone into high latitudes. Mahlman et al. (1980) discuss the buildup
of the spring maximum in total ozone in their simulation experiments.
Their results show a spring maximum near 60N. Mahlman et al. (1980)
suggest a possible reason for this discrepancy being the lack of a
stratospheric warming in their model simulation.

4. CONCLUDING REMARKS

Some representative results from a three-dimensional model
simulation of the transport of chemically active constituents have
been presented. The model incorporates a comprehensive representation
of stratospheric chemical processes by grouping related species into
families which are transported as entities. A number of comparisons
of the simulated species distributions with data (both measured and
inferred from measurements) from satellite and other experiments have
been conducted. The results are largely encouraging and suggest that
off-line transport of families is a useful and reasonable alternative
to a fully coupled three-dimensional model. However, model/data
comparisons are at best difficult because of interannual variability
and the paucity of simultaneous, global data of the relevant
constituents. Much further assessment of the model is required.

The present model demonstrates modest success in simulating
transport processes during winter stratospheric sudden warmings.
Diagnostic analysis of the transport equations provide an
interpretation of the transport processes that is consistent with
current theory based upon satellite data and more simplified models.

Continuation of this research will focus on more extensive
comparisons with data, longer simulation experiments, and the addition

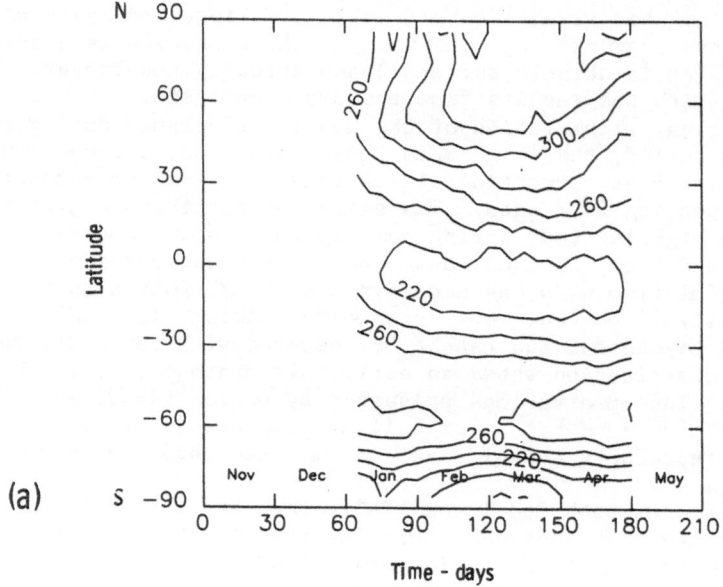

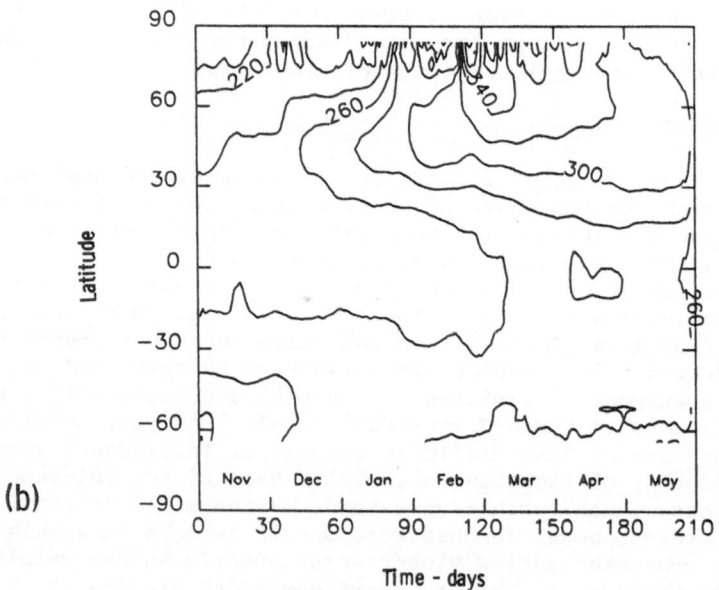

Figure 9. Variation of zonal mean total column ozone (Dobson units)
above 100 mb for: (a) LaRC model simulation and (b) LIMS experiment.

of more families to permit study of polar night chemistry in the stratosphere.

5. ACKNOWLEDGMENTS

The authors gratefully express appreciation to Drs. S. Solomon and R. Garcia for supplying two-dimensional model results and many invaluable discussions on our research. We further thank Drs. M. E. McIntyre, A. O'Neill, E. E. Remsberg, J. M. Russell III, and M. Geller for valuable comments and discussions. Thanks are due to G. Lingenfelser for figure preparation and S. D. Johnson for typing the manuscript.

6. REFERENCES

Blackshear, W. T., W. L. Grose, and R. E. Turner, 1987: 'Simulated sudden stratospheric warming: Synoptic evolution.' Quart. J. Roy. Meteor. Soc., 113, in press.

Blake, D., and R. Lindzen, 1973: 'Effect of photochemical models on calculated equilibria and cooling rates in the stratosphere.' Mon. Wea. Rev., 101, 783-802.

Brasseur, G., and S. Solomon, 1984: Aeronomy of the Middle Atmosphere. D. Reidel and Co., Dordrecht, Holland.

Clough, S. A., N. S. Grahame, and A. O'Neill, 1985: 'Potential vorticity in the stratosphere derived using data from satellites.' Quart. J. Roy. Met. Soc., 111, 335-358.

DeMore, W. B., J. J. Margitan, M. J. Molina, R. T. Watson, D. M. Golden, R. F. Hampson, M. J. Kurylo, C. J. Howard, and A. R. Ravishankara, 1985: 'Chemical kinetics and photochemical data for use in stratospheric modeling, Evaluation Number 7." JPL publication 85-37, 226 pp., Jet Propulsion Lab., Pasadena, CA.

Dunkerton, T. J., and D. P. Delisi, 1986: 'Evolution of potential vorticity in the winter stratosphere of January-February 1979.' J. Geophys. Res., 91, 1199-1208.

Dutsch, H. U., 1969: Atmospheric Ozone and Ultraviolet Radiation. World Survey of Climatology, 4, Climate of the Free Atmosphere, Elsevier, 383-432.

Eliasen, E., B. Machenaver, and E. Rasmussen, 1970: 'On a numerical method for integration of the hydrodynamical equations with a spectral representation of the horizontal fields.' Univ. of Copenhagen, Institute of Theoretical Meteorology, Report No. 2.

Garcia, R., and S. Solomon, 1983: 'A numerical model of the zonally averaged dynamical and chemical structure of the middle atmosphere.' J. Geophys. Res., 88, 1379-1400.

Gates, W. L., and A. B. Nelson, 1975: 'A new (revised) tabulation of the Scripps topography on a one degree global grid. Part I. Terrain heights.' The Rand Corp., R-1276-1-ARPH, Santa Monica, CA.

Geller, M. A., M. F. Wu, and M. E. Gelman, 1984: 'Troposphere-stratosphere (surface - 55 km) monthly winter general circulation statistics for the northern hemisphere--interannual variations.' J. Atmos. Sci., 41, 1726-1744.

Gille, J. C. and J. M. Russell III, 1984: 'The Limb Infrared Monitor of the Stratosphere experiment description, performance, and results.' J. Geophys. Res., **89**, 5125–5140.

Grose, W. L., 1984: 'Recent advances in understanding stratospheric dynamics and transport processes: Application of satellite data to their interpretation.' Adv. Space Res., **4**, 19–28.

Grose, W. L., W. T. Blackshear, and R. E. Turner, 1984: 'The response of a non-linear, time-dependent, baroclinic model of the atmosphere to tropical thermal forcing.' Quart. J. Roy. Meteor. Soc., **110**, 981–1002.

Grose, W. L., J. E. Nealy, and R. E. Turner, 1987: 'Modeling stratospheric transport processes.' American Meteorological Society Sixth Conference on the Dynamics and Chemistry of the Middle Atmosphere, Baltimore, Maryland.

Hoskins, B. J., and A. J. Simmons, 1975: 'A multi-layer spectral model and the semi-implicit method.' Quart. J. Roy. Meteor. Soc., **101**, 637– 655.

Hoskins, B. J., M. E. McIntyre, and A. W. Robertson, 1985: 'On the use and significance of isentropic potential vorticity maps.' Quart. J. Roy. Met. Soc., **111**, 877–946.

Kurzeja, R. J., 1975: 'The diurnal variation of minor constituents in the stratosphere and its effect on the ozone concentration.' J. Atmos. Sci., **32**, 899–909.

Labitzke, K., 1981: 'The amplification of height wave 1 in January 1979: A characteristic precondition for the major warming in February.' Mon. Wea. Rev., **109**, 989–989.

Lacis, A. A., and J. Hansen, 1974: 'A parameterization for the absorption of solar radiation in the Earth's atmosphere.' J. Atmos. Sci., **31**, 118–133.

Leovy, C. B., C.-R. Sun, M. H. Hitchman, E. E. Remsberg, J. M. Russell III, L. L. Gordley, J. C. Gille, and L. V. Lyjak, 1985: 'Transport of ozone in the middle stratosphere: Evidence for planetary wave breaking.' J. Atmos. Sci., **42**, 230–244.

Mahlman, J. D., and W. J. Moxim, 1978. 'Tracer simulation using a global general circulation model: Results from a mid-latitude instantaneous source experiment.' J. Atmos. Sci., **35**, 1340–1374.

Mahlman, J. D., H. Levy II, and W. J. Moxim, 1980: 'Three-dimensional tracer structure and behavior as simulated in two ozone precursor experiments.' J. Atmos. Sci., **37**, 655–685.

Mahlman, J. D., D. G. Andrews, H. U. Dutsch, D. L. Hartmann, T. Matsuno, and R. J. Murgatroyd, 1984: 'Transport of trace constituents in the stratosphere.' Dynamics of the Middle Atmosphere, D. Reidel and Co., Dordrecht, Holland, 387-416.

McIntyre, M. E. and T. N. Palmer, 1983: 'Breaking waves in the stratosphere.' Nature, 305, 593-600.

McIntyre, M. E. and T. N. Palmer, 1984: 'The 'surf zone' in the stratosphere.' J. Atmos. Terr. Phys., 46, 825-850.

Orszag, S. A., 1970: 'Transform method for calculation of vector coupled sums: Application to the spectral form of the vorticity equation.' J. Atmos. Sci., 27, 890-895.

Phillips, N. A., 1957: 'A coordinate system having some special advantages for numerical forecasting.' J. Met., 14, 184-185.

Robert, A. J., J. Henderson, and C. Turnbull, 1972: 'An implicit time integration scheme for baroclinic modes of the atmosphere.' Mon. Wea. Rev., 100, 329-335.

Russell, J. M. III, 1986: "Middle atmosphere composition revealed by satellite data.' MAP Handbook, 22, University of Illinois, Urbana, IL 61801.

World Meteorological Organization, 1986: 'Atmospheric Ozone 1985: Global Ozone Research and Monitoring Project.' Report No. 16.

COMPREHENSIVE MODELING OF THE MIDDLE ATMOSPHERE:
THE INFLUENCE OF HORIZONTAL RESOLUTION

J.D. Mahlman and L.J. Umscheid
Geophysical Fluid Dynamics Laboratory/NOAA
Princeton University, P.O. Box 308
Princeton, New Jersey 08542, U.S.A.

ABSTRACT. Results are presented from the GFDL "SKYHI" general
circulation model that illuminate the effects of increasing horizontal
grid resolution. The model experiments reveal a very pronounced
improvement of simulation skill at the highest resolution tested (1°
latitude grid size). The zonal wind climatology is particularly
affected with dramatic improvements in the stratospheric winter
westerlies and the positions of the tropospheric subtropical jet
streams. These improvements are apparently related to increases in
tropospheric wave forcing, specifically the Eliassen-Palm flux. The 1°
resolution model has successfully simulated a major mid-stratospheric
sudden warming event. Some preliminary analysis and insights on the
causes and character of the sudden warming event are presented.

1. INTRODUCTION

Over the past decade large advances have been made in understanding the
dynamics of the middle atmosphere. Previously, this region was viewed
as a place where linear theory of waves on an independently determined
zonal mean structure is appropriate. This view has given away
gradually to a realization that many middle atmosphere phenomena may be
intrinsically nonlinear in the sense of large amplitudes and strong
irreversibility. These include sudden warmings, planetary wave
breaking, seasonal transitions, gravity wave large-scale interaction,
and critical layer processes. The above realizations have led to an
increasing role for comprehensive general circulation models (GCM's)
that include the middle atmosphere. This requires realistic model
climatologies, an elusive goal because some phenomena have not yet been
well simulated. Most notably, in our GCM, the problems included
excessively strong polar night westerlies and a missing quasi-biennial
oscillation (Mahlman and Umscheid, 1984; hereafter MU84).
      It has not yet been established whether these model deficiencies
are due to insufficient computational resolution or to distortions in
physical processes. We present some early results on the effects of
horizontal resolution increases. We will emphasize the influence of

G. Visconti and R. Garcia (eds.), Transport Processes in the Middle Atmosphere, 251–266.
© *1987 by D. Reidel Publishing Company.*

horizontal resolution on the zonal mean structure and the dynamics of
the wintertime high latitudes. Also, a preliminary analysis of a
simulated major sudden stratospheric warming will be highlighted.

2. MODEL EXPERIMENTS

The model used is the GFDL "SKYHI" GCM described in Fels et al. (1980),
Andrews et al. (1983), and references therein. The model has 40
vertical levels from the earth's surface to the mesopause. This
version of the model prescribes sea surface temperature, ozone, clouds,
and "permanent" land ice according to their observed climatic values.
      The model in this form has been run through several annual cycles
at horizontal resolutions of 9° latitude by 10° longitude, 5° by 6°,
and 3° by 3.6°, respectively. A version of the model at 1° by 1.2°
resolution has been run from Oct. through Feb. for a single year. At
this highest resolution, the model becomes expensive to run. However,
the "1°" model allows exploration of a number of scientific questions
central to simulation and understanding the atmosphere. Specifically,
are the GCM deficiencies at low and moderate resolution rectified or
aggravated by resolution increases?

3. MODEL CLIMATOLOGY

One of the most vexing problems in modeling the stratosphere has been
the intense westerlies at the periphery of the winter polar dark zone
and the concomitant excessively cold polar temperatures.
Traditionally, the only "cures" to this problem have been to introduce
an ad hoc drag on the zonal flow or to reduce the efficiency with which
model radiation damps temperatures back toward the so-called radiative
temperature.
      Shown in Fig. 1 are the Dec. mean temperature profiles at 62°N for
the 9°, 5°, 3°, and 1° resolution models. At the far left is the
"radiative" temperature. The boxes represent observed temperatures
from Hamilton (1982). The figure shows a dramatic improvement of the
polar cold bias as horizontal resolution is increased. In fact the 1°
version actually looks to be a bit on the warm side; this is not the
case. In the model polar vortex the minimum temperatures appear to be
nearly 10°C too cold in the 100 mb to 10 mb region.
      The monthly mean zonal wind for the 1° model is shown in Fig. 2a.
This figure can be compared directly with Fig. 2 in MU84 calculated for
5° resolution. In the troposphere the 1° simulation is significantly
improved in most respects. The magnitudes and positions of the
subtropical jets are now close to observed. In the tropical upper
troposphere the zonal winds are near zero in contrast to the nearly
10 m s^{-1} westerlies previously. A deficiency that has appeared in the
1° model is somewhat intense surface winds. This may be due to
insufficient surface drag or to some boundary layer modeling problems.
      In the stratosphere very large zonal wind differences are
observed. While MU84 obtained maximum zonal winds of 180 m s^{-1} near

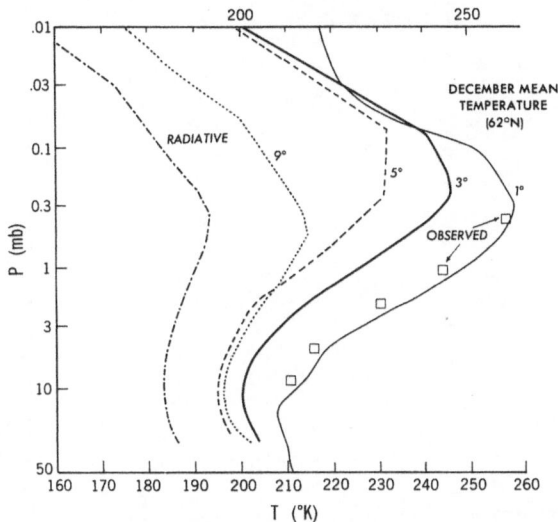

Figure 1   December temperature profiles at 62°N for 9°, 5°, 3°, and 1° resolutions.   "Radiative" profile is from Fels (1985); observed squares from Hamilton (1982).

0.1 mb, Fig. 2a shows maximum westerlies of 50 m s^{-1} at 2 mb.   At 10 mb the maximum wind has decreased from 75 to 45 m s^{-1}.   Also, the 100 mb minimum denoting the closing off of the subtropical jet has decreased by approximately 5 m s^{-1}.   In the Southern Hemisphere, the summertime easterly jet has begun to close off for the first time (without parameterized zonal drag); the maximum changes from −100 m s^{-1} previously to about −60 m s^{-1} currently.   This is still stronger than observed, with not enough decay of the easterlies into the upper mesosphere.   In the Southern Hemisphere lower stratosphere, for the first time the easterlies are quite realistic, including the transition from tropospheric westerlies near 50 mb.

We now begin to examine the basic causes of these improved zonal winds.   A natural way to compare the two versions of the model is through use of Eliassen–Palm diagnostics.   In the transformed zonal momentum equation, the Eliassen–Palm (E–P) flux divergence (EPFD) appears as a body force per unit mass exerted on the zonal flow (Andrews and McIntyre, 1978).   Moreover the E–P flux vector F provides a measure of the magnitude and direction of wave activity propagation (Edmon et al., 1980).   Details of our method of calculating the primitive equation E–P diagnostics are given in Andrews et al. (1983).

Shown in Fig. 3a is a meridional height cross section of E–P flux direction with the length of the arrows proportional to Fp^{-1}.   The values of EPFD are given in units of 10^{-5} m s^{-2}.   The magnitude of EPFD in Fig. 3a for the higher latitude Northern Hemisphere can be compared with the same quantity in Fig. 4 of MU84.   Here the magnitudes of EPFD in the mesosphere and upper stratosphere have approximately doubled.   The lower and middle stratosphere now have negative values larger than -4×10^{-5} m s^{-1} in higher latitudes.   Note that there is no "spot" of

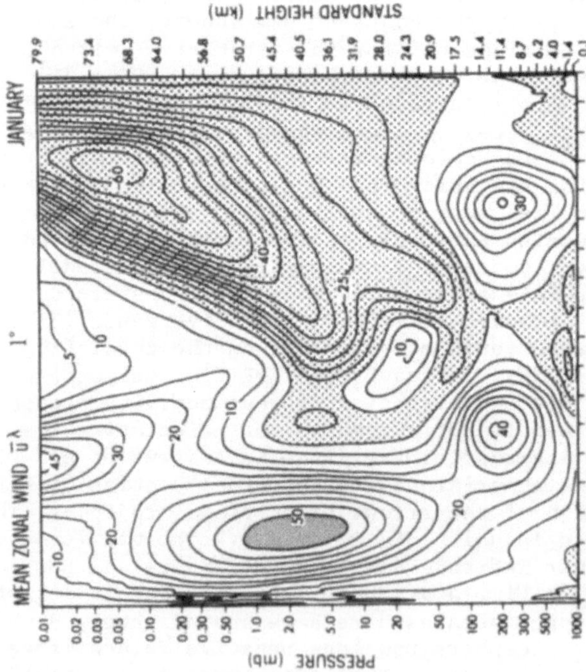

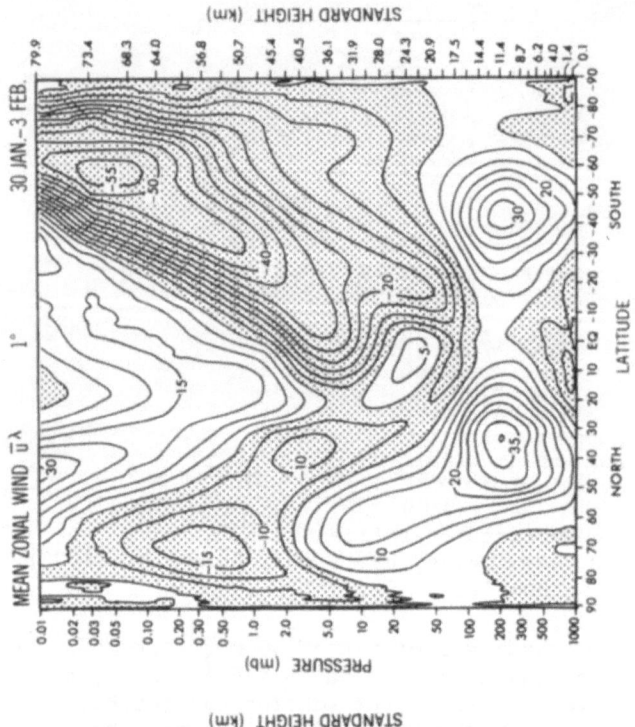

Figure 2a   Zonally averaged zonal wind in
m s⁻¹ for the 1° model for Jan. Samples
were taken twice per model day. Dark
shading indicates westerlies greater than
50 m s⁻¹; light shading indicates easterlies.

Figure 2b   Same as 2a except for the 5-day
period 30 Jan. - 3 Feb.

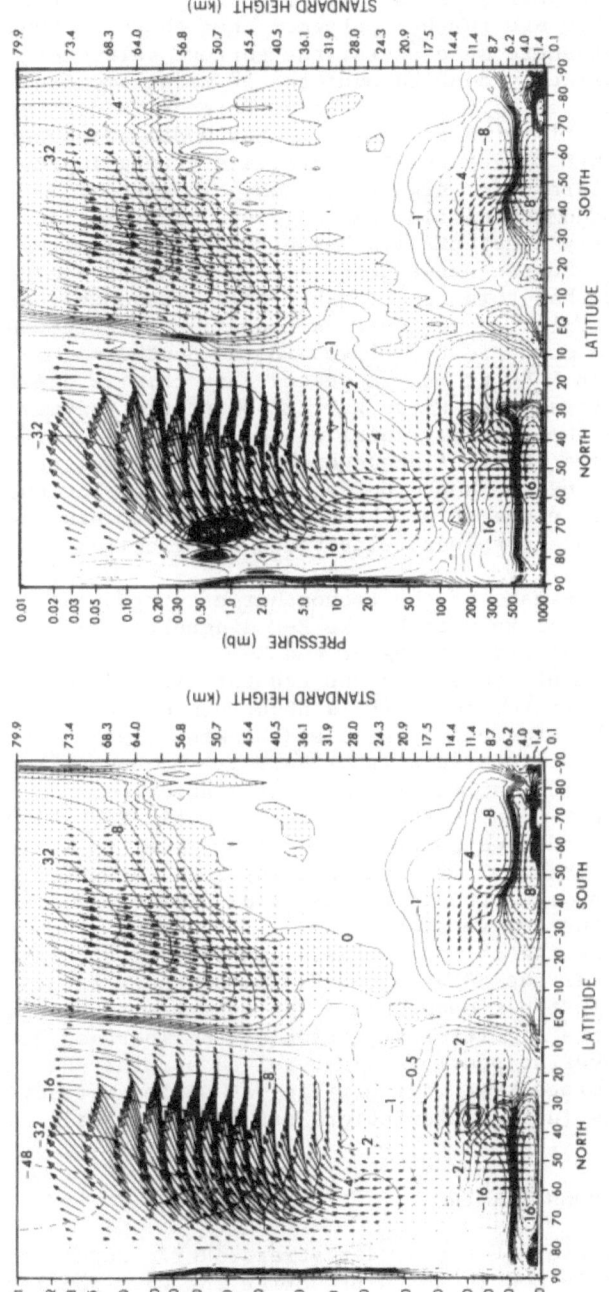

Figure 3a Meridional cross section for 1°
model for Jan. of Eliassen-Palm flux
directions (arrows). Length of arrows is
proportional to Fp^{-1} and contours of
Eliassen-Palm flux divergence (EPFD)
normalized as zonal force per unit mass
(10^{-5} m s^{-2}). Regions of positive EPFD are
shaded.

Figure 3b Same as 3a except for the 5-day
period 15-19 Jan.

positive EPFD there in contrast to the earlier study and to numerous quasi-geostrophic observational studies. Importantly, the largest magnitudes of negative EPFD are further poleward than those of MU84.

In the Southern Hemisphere mesosphere, the current values of positive EPFD are approximately doubled over the 5° model and the mesosphere maximum is shifted poleward. This difference is consistent with the 40-50 m s^{-1} easterly wind decreases in the model mesosphere.

The scaling of the E-P flux F by p^{-1} provides a strong impression that regions above 10 mb are under significant wave influence. McIntyre (1982, personal communication) has suggested that regions where Fp^{-1} is approximately constant with height may be indicative of substantial wave breaking in the sense described by McIntyre and Palmer (1983). Note in Fig. 3a that this relationship holds reasonably well in both easterly and westerly jet regions. A conspicuous effect in Fig. 3a is the pronounced downward E-P fluxes in the Southern Hemisphere easterlies. In MU84 it was suggested that this is due to a predominance of upward propagating waves having dominantly eastward phase speeds relative to the easterly zonal winds. This was verified in the analysis of Miyahara et al. (1986) which showed a wide specrum of gravity waves propagating southward out of the major low-latitude convection centers. After propagating through the lower stratosphere, only waves with eastward phase speeds relative to the growing easterlies remain.

The significant improvement of the tropospheric subtropical jet in Fig. 2a also can be explained in part by the E-P diagnostics of Fig. 3a. Note the upward-equatorward bending of the E-P flux directions out of the mid-latitude baroclinic zone into the equatorward side of the subtropical jet. In that region the EPFD is about -2x10^{-5} m s^{-2}. This diagnostic agrees with the argument by Held and Hoskins (1985) that the subtropical jet speed, shear, and location result from a competition between this EPFD and the low latitude Hadley circulation.

It is also of interest to note that the tropospheric EPFD "dipole" is somewhat stronger and centered poleward of the 5° model in MU84. This allows a preferred "leakage" of wave activity into the polar cap in the current simulation, even though the E-P flux directions are essentially the same in the two versions.

We thus conclude that the 1° model middle atmospheric simulation improved significantly over the 5° version of MU84 because of the increased wave activity in the troposphere. The analysis of Miyahara et al. (1986) showed that a significant portion of the increase comes from internal gravity waves at previously unresolved spatial scales. However, they also determined that wave activity at scales already resolved by the lower resolution calculations are larger as well.

4. A SIMULATED MAJOR SUDDEN WARMING

In earlier studies simulation of the sudden stratospheric warming have been modestly successful in models which prescribe the wave forcing at the tropopause and use weak thermal damping (e.g., Matsuno, 1971). Because of this it has been a concern that satisfactory sudden warming simulations have not been achieved in GCM's with a self-determined troposphere, an annual insolation cycle, a complete stratosphere, and realistic radiative transfer. We show here a preliminary look at our first such successful simulation.

Fig. 2b gives zonal wind (as in Fig. 2a) in the 1° model for the period 30 Jan. - 3 Feb. This is after a major sudden warming episode beginning in mid Jan. Note the obliteration of the westerly jet and the change to easterlies from the middle stratosphere to the middle mesosphere. Even in the lower stratosphere, the zonal winds have decreased markedly relative to those shown in Fig. 2a.

The easterlies of Fig. 2b have a clear maximum in the lower mesosphere with a corresponding decrease in magnitude into the middle mesosphere. This is associated with a significant cooling of the polar middle mesosphere temperature for the 30 Jan. - 3 Feb. period relative to the Jan. mean (up to 30°C). This may be compared with the mid-stratosphere polar warming temperature by 30-35°C. These model results are similar to observations of major sudden stratospheric warmings accompanied by significant mesospheric polar cooling.

We now address some aspects of this major warming event. Fig. 3b shows an E-P cross section as in Fig. 3a, but for 15-19 Jan. Superficially, Fig. 3b looks very much like Fig. 3a. A closer inspection, however, reveals some very significant differences. In the Northern Hemisphere upper troposphere, the EPFD region of -16×10^{-5} m s^{-2} of Fig. 3a is replaced by a region that now extends noticably toward the north pole in Fig. 3b. The most significant difference though is the appearance of an EPFD region of -16×10^{-5} m s^{-2} near 20 mb. This is 4 times the monthly average in Fig. 3a and is a central causal mechanism for the sudden warming that followed. Fig. 3b also shows the Fp^{-1} vertical arrows to be considerably stronger in the lower stratosphere during this disturbed period.

It is interesting to note that the Fig. 3b E-P flux directions are not much different than those of Fig. 3a. Thus, we tentatively conclude that the warming episode was initiated by a higher amplitude "burst" of E-P wave activity in the high latitude troposphere. The resulting enhanced westward body force produced a pronounced deceleration of the polar night jet and associated polar warming. The enhanced E-P absorption in the middle stratosphere appears to have reduced the penetration of E-P activity to higher altitudes. The associated reduction of the negative EPFD apparently causes the mesosphere to relax toward the extremely cold polar night radiative temperatures.

We next look at the sudden warming from the viewpoint of instantaneous synoptic maps. Figs. 4a-4f show 10 mb height fields at 5-day intervals. On 15 Jan. (Fig. 4a) the Aleutian high is well developed, but the polar vortex is centered near the pole with an intense circulation. By 20 Jan. (Fig. 4b), the Aleutian high has moved poleward while the polar low has elongated and developed the characteristic "comma tail" typical of a breaking wave signature (McIntyre and Palmer, 1983). Accompanying this change is a well developed cross-polar flow. A second high has developed near Spain.

On 25 Jan. (Fig. 4c) the Aleutian high is further developed and the cross polar flow is very pronounced. The secondary high has

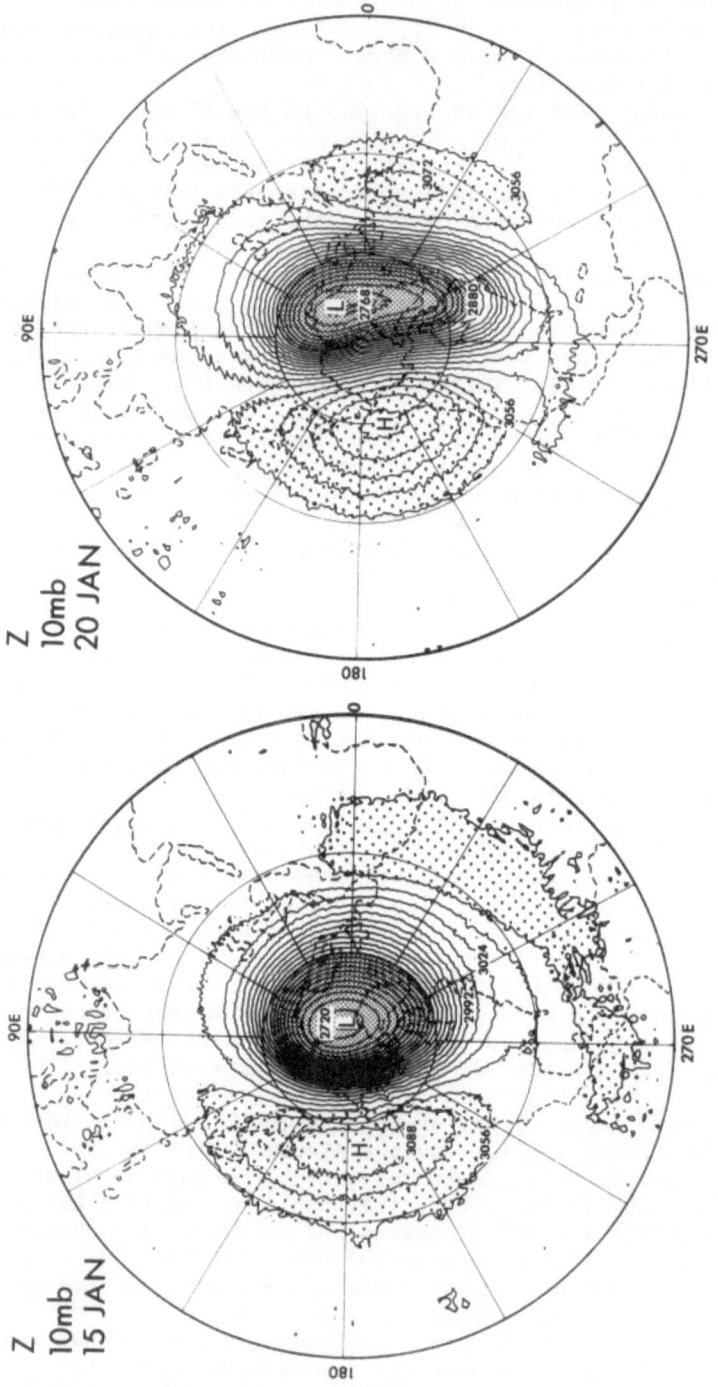

Figure 4b Same as 4a except for 20 Jan.

Figure 4a 15 Jan. 10 mb height. Contour
interval is 16 decameters. Light shading
indicates values greater than 3,056 decameters. Dark shading indicates values less
than 2,880 decameters.

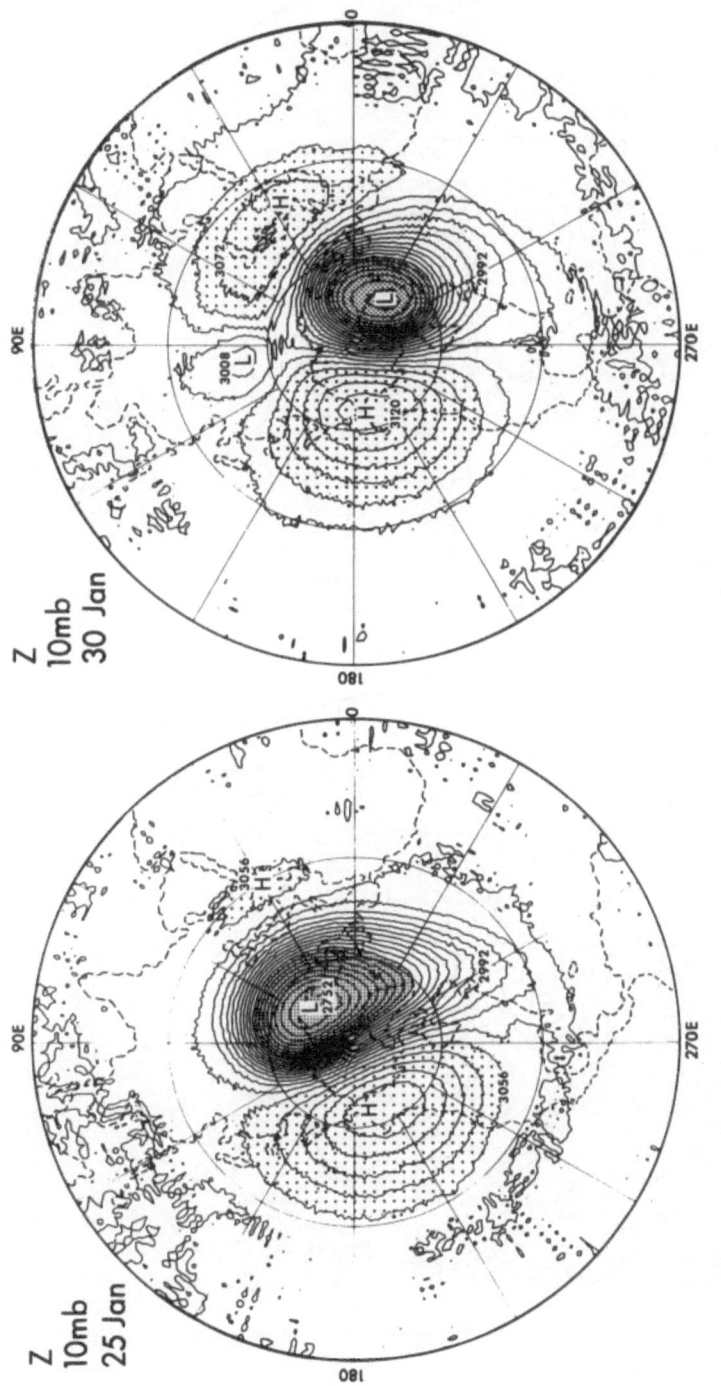

Figure 4c Same as 4a except for 25 Jan.

Figure 4d Same as 4a except for 30 Jan.

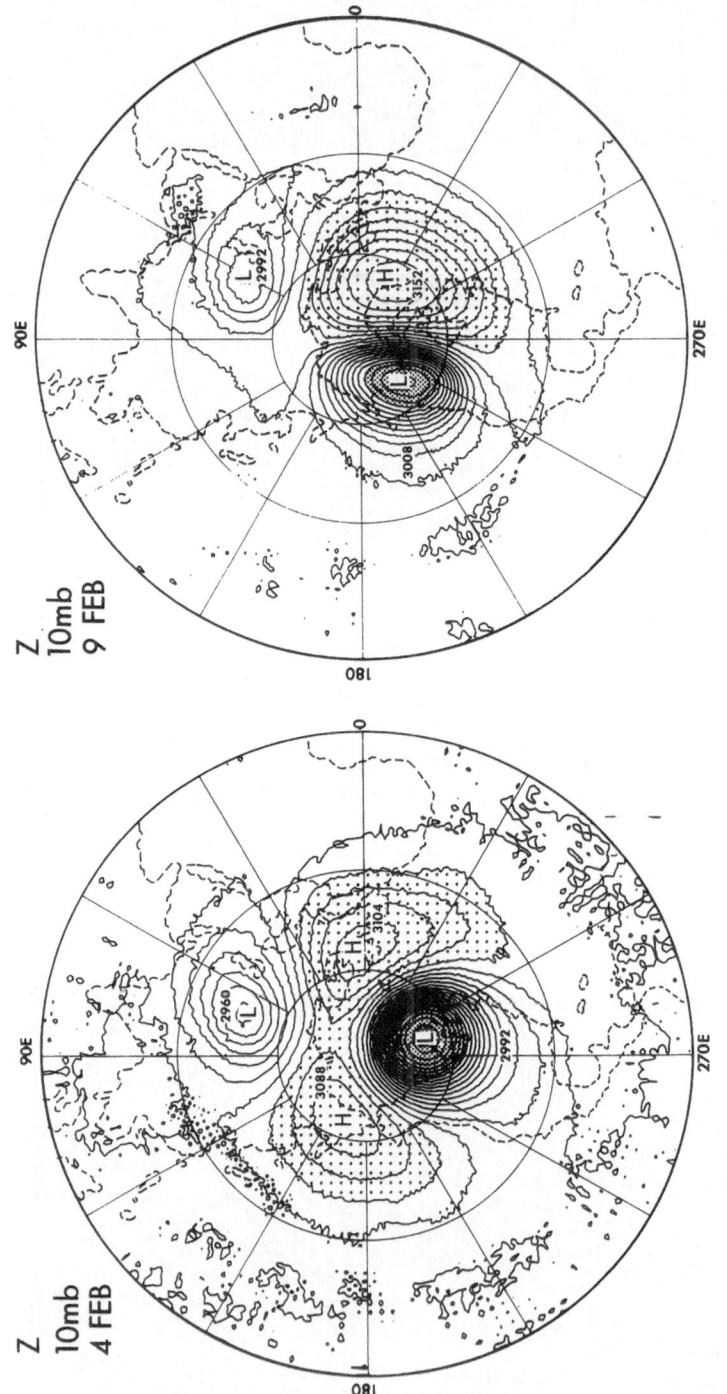

Z
10mb
9 FEB

Z
10mb
4 FEB

Figure 4f   Same as 4a except for 9 Feb.

Figure 4e   Same as 4a except for 4 Feb.

drifted eastward into the Middle East region. Note also that the
trailing "tail" is now more pronounced. This date marks the peak in
the rate of polar warming. By 30 Jan. (Fig. 4d) the secondary high has
grown and a secondary low has appeared over Siberia. On 4 Feb.
(Fig. 4e) the polar region is dominated by split highs of about equal
strength. The original polar low is now centered near Hudson Bay. The
vortex is considerably smaller than on 15 Jan. This marks the period
of highest polar temperatures and strongest easterlies.

By 9 Feb. (Fig. 4f) the original polar low has migrated westward
to Alaska. The formerly secondary high is now dominant and the
Aleutian high has disappeared. Interestingly, on 14 Feb. (not shown)
the pattern is similar to that of 9 Feb., but the secondary low becomes
the stronger of the two. Overall, however, both lows continue to
weaken.

The detailed influence of tropospheric planetary-scale disturbances on sudden warmings remains unclear even though the association with
"blocking" anticyclones has long been noted. Fig. 5 shows a 500 mb
geopotential height map for 15 January, the start of the sudden warming
sequence. This figure shows the mature stage of a very high amplitude
ridge-trough pattern over North America. This pattern (with its
enhanced poleward eddy heat flux) appears to have triggered the large

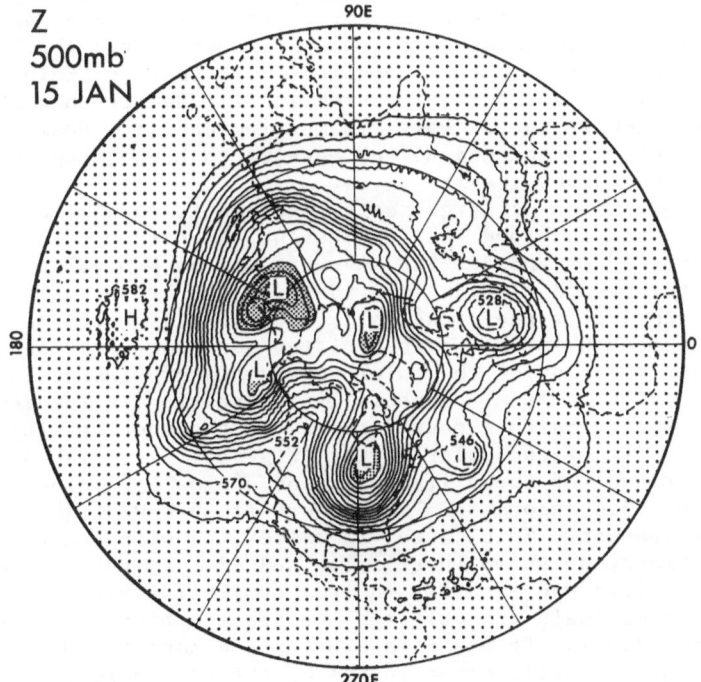

Figure 5 15 Jan. 500 mb height. Contour interval is 6 decameters.
Light shading indicates values greater than 570 decameters. Dark
shading indicates values less than 486 decameters.

burst of E-P activity shown in Fig. 3b.

We now examine some details of the peak of the warming. Figs. 6a and 6b show 10 mb temperatures at 25 and 30 January. The temperatures are smoothed by a horizontally symmetric 1-4-6-4-1 operator to reduce the confusion produced by the rich spectrum of gravity waves resolved by the model. On 25 January, Fig. 6a shows a "hot spot" of warm temperatures being advected poleward in Central Siberia, while the 30 January map in Fig. 6b shows the high temperatures have spread out and moved over the polar cap. It is this process that destroys the zonal mean westerlies and eventually weakens the polar vortex itself.

Fig. 7 shows a 10 mb smoothed map of total wind speed isolines for 25 January. The heavy solid line denotes the jet axis (that closely approximates a flow trajectory). Note the very strong implied particle accelerations of nearly 100 m s^{-1} along the jet axis from the subtropical Pacific to Northern Europe. Thus, during this high amplitude state, the polar vortex flow is far from gradient wind balance.

This point is made dramatically evident in Figs. 8a and 8b which show contours of nitrous oxide (N_2O) for 10 mb on 25 Jan. N_2O is carried as a passive variable in SKYHI to allow analysis of the model's transport simulations. Fig. 8a is an unsmoothed map, while Fig. 8b is the same field, but smoothed with the filter. A comparison of Fig. 8a with Fig. 7 shows that the most intense N_2O gradients are aligned along the jet axis and surround the polar vortex as shown in Fig. 4c. The large region of low N_2O values and weak gradients in the north Pacific region may be identified as the so-called "surf zone" of McIntyre and Palmer (1983). The small scale features in Fig. 8a are characterized by long strands suggestive of wave filamenting and/or breaking, while the more isotropic "background noise" appears to be related to the model's rich spectrum of gravity waves.

The interpretation of large-scale features is more evident in Fig. 8b, especially when combined with the height, temperature and wind fields of Figs. 4c, 6a and 7. The movement of low N_2O values out of the polar vortex toward Florida is evident, while higher values have penetrated southward in the cross polar flow towards North America. The low values of N_2O in the North Pacific "surf zone" may be a combination of air torn out of the polar vortex and of sinking in the poleward flow off East Asia. Especially interesting is the region of strongly reversing gradients near Europe. In the accelerating poleward flow around the rim of the polar vortex (Fig. 4c) very high N_2O values are seen. Just east of these lies a region of low N_2O values near Spain. East of this second N_2O "low" there is an N_2O "high" near the Eastern Mediterranean. This "high" is associated with the transient developing anticyclone shown in Figs. 4c and 4d. As this anticyclone developed, it pulled N_2O rich air in from tropical latitudes. The secondary low near Spain probably resulted from a blob of N_2O poor air being torn (shredded?) off the polar vortex and advected around the developing secondary anticyclone in the Mediterranean region.

Thus, to the east of the accelerating jet in Fig. 7, there are three adjacent regions where the N_2O contrasts are roughly comparable to the original equator to pole N_2O contrast. This can only occur as a result of very high amplitude disturbances producing large deformation

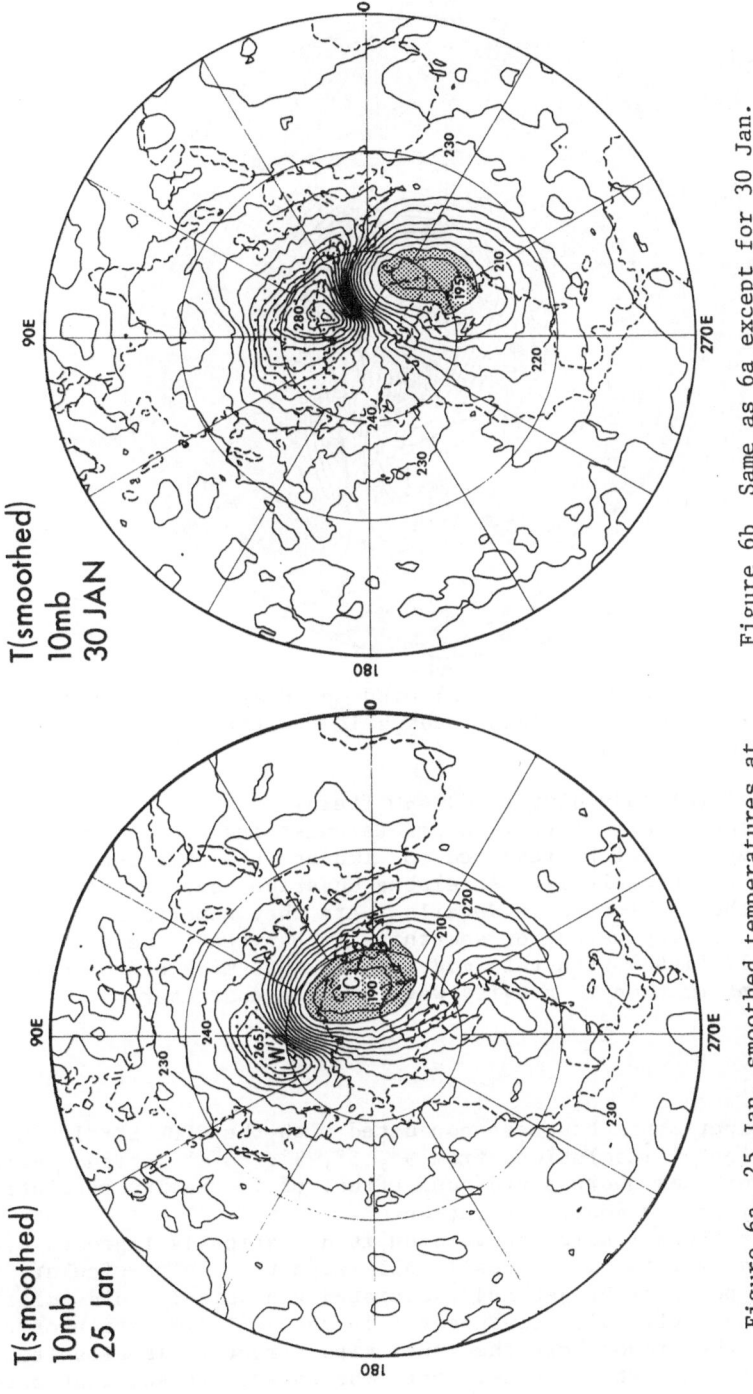

Figure 6a 25 Jan. smoothed temperatures at 10 mb. Contour interval is 5°K. Light shading indicates values greater than 250°K. Dark shading indicates values less than 200°K.

Figure 6b Same as 6a except for 30 Jan.

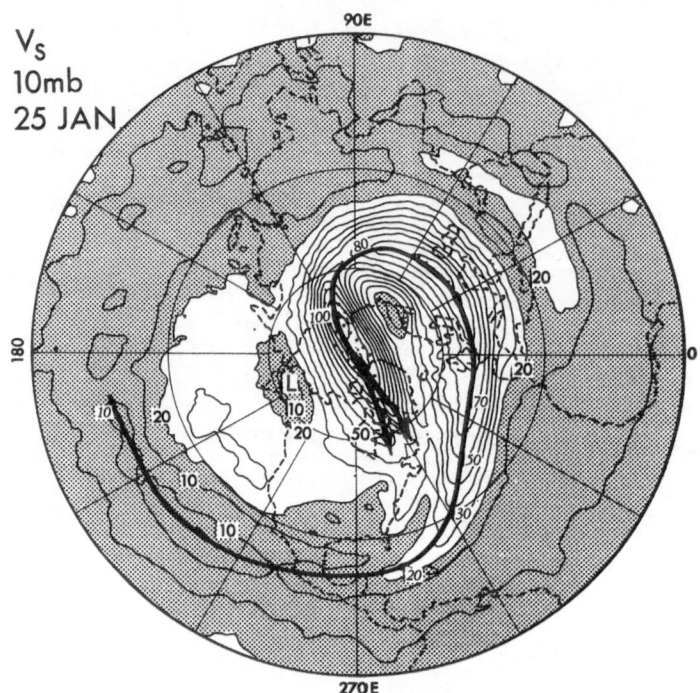

Figure 7 25 Jan. smoothed total wind speed at 10 mb. Contour
interval is 10 m s^{-1}. Light shading is greater than 100 m s^{-1}.
Dark shading is less than 20 m s^{-1}.

of the original (already disturbed) N_2O field.

The extreme detail of Fig. 8a leaves questions as to how much
spatial resolution will be required to simulate the stratospheric circulation more quantitatively. A related question concerns the
character of the smaller scale details. Are they numerical noise, physically sound gravity waves, or something in between? The work by
Miyahara et al. (1986) with this model shows that much of the higher
wave number information is gravity-wave like in character.

5. SUMMARY

A series of experiments has been conducted with the GFDL SKYHI GCM at a
range of horizontal resolutions from 9°, 5°, 3°, to 1° latitude grid
size. This work has shown a profound effect of increasing resolution
on the quality of the model simulation.

The 1° latitude resolution version is dramatically improved over
the 5° latitude version in virtually all respects. In particular, the
overly strong polar night jet and associated excessively cold polar
temperatures have virtually disappeared in the Jan. simulation of the
1° model. In the troposphere the zonal wind structure is close to the
observed except near the surface, where the speeds are somewhat strong.

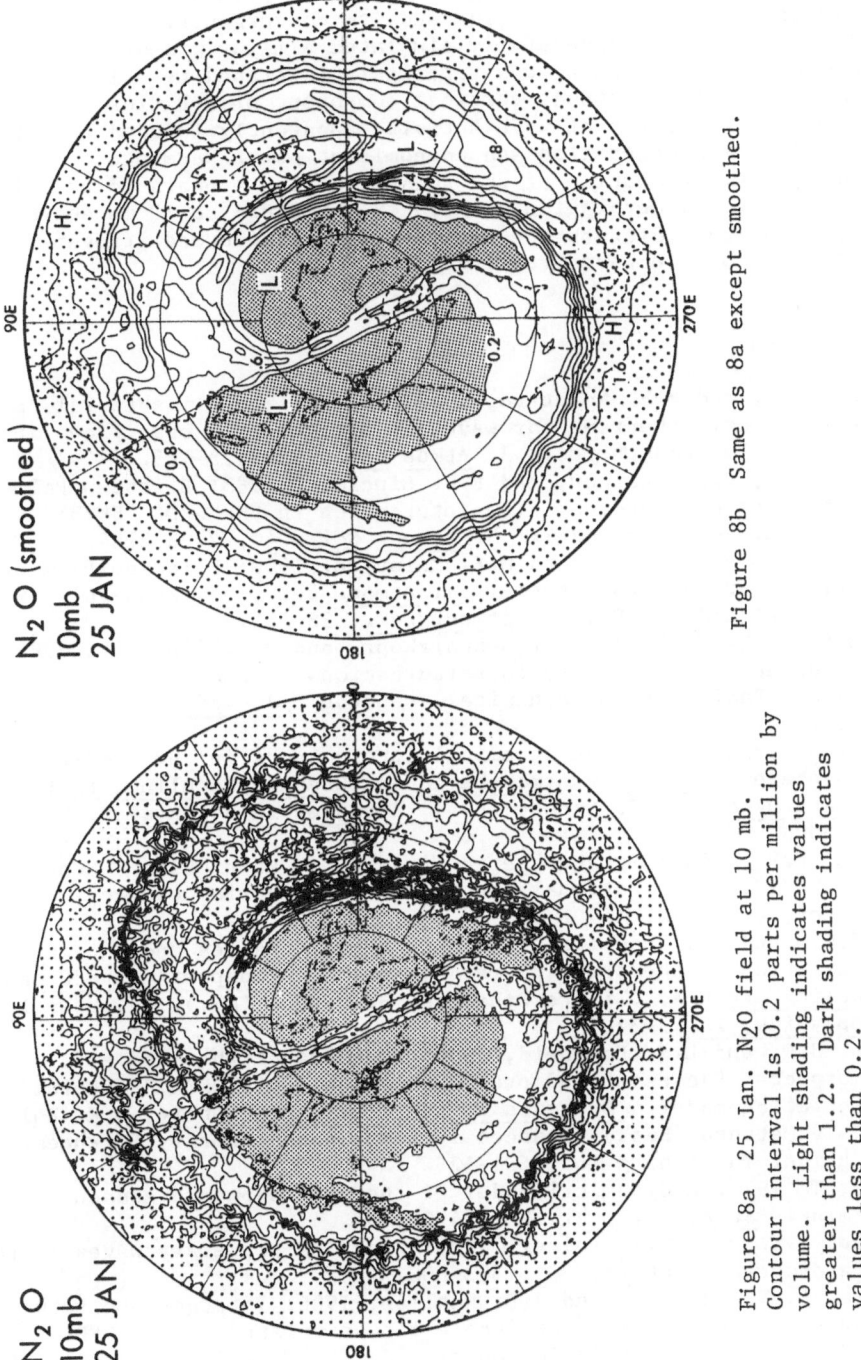

N₂O (smoothed)
10mb
25 JAN

N₂O
10mb
25 JAN

Figure 8b Same as 8a except smoothed.

Figure 8a 25 Jan. N₂O field at 10 mb.
Contour interval is 0.2 parts per million by
volume. Light shading indicates values
greater than 1.2. Dark shading indicates
values less than 0.2.

The improved zonal wind climatology is apparently related to an
increased magnitude of the E-P flux and its convergence.

The 1° model has successfully simulated a major mid-stratospheric
sudden warming. The warming was apparently initiated by a large
increase in the high latitude vertical component of E-P flux out of the
troposphere. This burst of wave activity apparently induced a very
large zonal flow deceleration and concomitant polar warming.

The synoptic evolution of the sudden warming appears to be quite
realistic. The flow evolution and tracer structure at 10 mb shows a
very high amplitude disturbance activity with pronounced irreversibility and probable wave breaking.

6. REFERENCES

Andrews, D.G., and M.E. McIntyre, 1978: Generalized Eliassen-Palm and
    Charney-Drazin theorems for waves in axisymmetric mean flows in
    compressible atmospheres. J. Atmos Sci., 35, 175-185.
Andrews, D.G., J.D. Mahlman, and R.W. Sinclair, 1983: Eliassen-Palm
    diagnostics of wave-mean flow interaction in the GFDL "SKYHI"
    general circulation model. J. Atmos. Sci., 40, 2768-2784.
Edmon, H.J., B.J. Hoskins, and M.E. McIntyre, 1980: Eliassen-Palm cross
    sections for the troposphere. J. Atmos. Sci., 37, 2600-2616 (See
    also Corrigendum, 1981, J. Atmos. Sci., 38, 1115.
Fels, S.B., J.D. Mahlman, M.D. Schwarzkopf, and R.W. Sinclair, 1980:
    Stratospheric sensitivity to perturbations in ozone and carbon
    dioxide: Radiative and dynamical response. J. Atmos. Sci., 37,
    2265-2297.
Fels, S.B., 1985: Radiative-dynamical interactions in the middle
    atmosphere. Issues in Atmospheric and Oceanic Modeling, Part A,
    Climate Dynamics, S. Manabe (Ed.), Advances in Geophysics, 28,
    277-300.
Hamilton, K.P., 1982: Stratospheric Circulation Statistics. NCAR Tech.
    Note TN-191-STR. National Center for Atmospheric Research,
    Boulder, Colorado, 174 pp.
Held, I.M., and B.J. Hoskins, 1985: Large-scale eddies and the general
    circulation of the troposphere. Issues in Atmospheric and Ocean
    Modeling, Part A, Climate Dynamics, S. Manabe (Ed.), Advances in
    Geophysics, 28, 3-31.
Mahlman, J.D. and L.J. Umscheid, 1984: Dynamics of the middle
    atmosphere: Successes and problems of the GFDL "SKYHI" general
    circulation model. Dynamics of the Middle Atmosphere, J.R. Holton
    and T. Matsuno (Eds.) Advances in Earth and Planetary Sciences,
    D. Reidel Publishing Co., 501-525.
Matsuno, T, 1971: A dynamical model of the stratospheric sudden
    warming. J. Atmos. Sci., 28, 1479-1494.
McIntyre, M.E., and T.N. Palmer, 1983: Breaking planetary waves in the
    stratosphere. Nature, 305, 593-600.
Miyahara, S, Y. Hayashi, and J.D. Mahlman, 1986: Interactions between
    gravity waves and planetary-scale flow simulated by the GFDL
    "SKYHI" general circulation model. J. Atmos. Sci., 43, 1844-1861.

DYNAMICS AND TRACER TRANSPORT IN THE MIDDLE ATMOSPHERE: AN OVERVIEW
OF SOME RECENT DEVELOPMENTS

M.E. McIntyre
Department of Applied Mathematics and Theoretical Physics
Cambridge CB3 9EW
U.K.

ABSTRACT. Our present understanding of dynamics and tracer transport in
the middle atmosphere, and of the concepts needed for modelling them, is
discussed in the light of recent developments. Included are comments on
wave, mean-flow interaction theory and its relationship to the
generalized concept of wave breaking, on the gravity wave "turbulent
Prandtl number" and associated questions connected with wave-turbulence
inhomogeneity, on the possibility of focusing and self-tuning resonant
effects in the wintertime planetary-scale Rossby wave field, on the
downward influence of wave dissipation regions on diabatic circulations
and stratosphere-troposphere mass exchange rates, on the diabatic
impermeability of isentropic surfaces to (Rossby-Ertel) potential
vorticity and the new view this gives of the zonally asymmetric diabatic
circulation, and on some implications for the representation of dynamical
feedback processes in low-order models of the middle atmosphere,
including models with artificial lower boundary conditions.

1. INTRODUCTION

It would be inappropriate for me to review in detail for this audience
many of the better known, and very notable, recent advances in our
understanding of middle atmospheric dynamics and global-scale tracer
transport. Many of these advances, both observational and theoretical,
are already well documented in a number of excellent reviews. One
immediately thinks for instance of the relevant chapters in the
monumental Ozone Assessment Report No. 16 (WMO 1985, hereafter WMO_3).
Equally noteworthy are the three reviews in the middle-atmosphere section
of the recent Smagorinsky Festschrift (Saltzman and Manabe 1985), several
more in the US-Japan Seminar Proceedings (Holton and Matsuno 1984), and
the review by Fritts (1984) of the gravity-wave aspects, which lucidly
summarizes the leading theoretical and observational facts and their
implications for the global circulation. Several more reviews on a range
of observational and theoretical topics are to appear in a forthcoming
special middle-atmospheric issue of Phil. Trans. Roy. Soc. London.
    It is clear that much of this progress has been made possible not

G. Visconti and R. Garcia (eds.), Transport Processes in the Middle Atmosphere, 267–296.
© 1987 by D. Reidel Publishing Company.

only by the availability of data from a variety of ingenious terrestrial, airborne and space-based observing systems, and by developments in computer technology and numerical modelling, but also by an increasingly fruitful interaction between the different specialist disciplines in terms of theoretical thinking and scientific detective work. A basic example is our awareness of the nature and significance of the global-scale diabatic circulation of the middle atmosphere, including the long-term control of stratosphere-troposphere mass exchange rates. Our picture of this circulation now seems very clear from a combination of classical tracer observations, theoretical and numerical modelling studies, and recent global-scale satellite observations of the distributions of long-lived tracers. The early ideas of pioneers like Brewer and Dobson have been strikingly vindicated, and we are now in the process of refining the picture and making it more quantitative (e.g. Fels 1987; Geller and Wu 1987; Beagley and Harwood 1987; Remsberg 1987, all in this volume). The diabatic circulation is believed to be of great importance for tropospheric as well as for middle-atmospheric chemistry; the distribution of chemical constituents in turn has repercussions for radiative heating and hence for dynamical regimes, as well as climate trends; and the dynamics in turn exerts a controlling influence on the diabatic circulation. It exerts this control in ways on which we have a far better conceptual grip today than I think most of us had ten years ago; and further clarifications are still in the pipeline. The whole picture includes important information about the departure of temperatures from their radiatively determined values, including departures in the lower stratosphere where it is a hopelessly ill-conditioned problem to deduce this information from temperature observations and radiative calculations alone.

The need to consider the interplay between radiation, chemistry and dynamics on an interdisciplinary basis has, of course, been further underlined by the recent discovery of the Antarctic ozone hole by members of the British Antarctic Survey. It is looking more and more as if all three aspects will prove crucial, in different ways, to a sound understanding of what is happening over Antarctica. Such an understanding will in turn teach us a great deal more about the middle atmosphere as a whole.

What I should like to do here is to give a personal commentary on some of the more fluid-dynamical aspects of our current picture of middle-atmospheric behaviour, touching in particular upon some aspects which my collaborators and I have had the opportunity to study in recent years, and here and there looking beyond what is well established today. Among other things, mention will be made of some very recent modelling work done at Cambridge, which gives a first glimpse of what the stratosphere might look like at resolutions finer than 1° latitude, and which may have some bearing on the Antarctic ozone-hole problem.

2. A LIST OF KEY IDEAS

The ideas which seem important for understanding the middle-atmospheric circulation might be summarized succinctly as follows. The

list inevitably contains oversimplifications, some of which will be
commented on in the succeeding sections, as well as some outright
omissions. It represents a set of hypotheses, together with some purely
theoretical principles, which seem sufficiently relevant to be useful in
our present state of knowledge. It begins with the best known and most
widely accepted ideas, and proceeds to some others that may be less so,
but which I think may prove important. For the sake of brevity,
literature citations, including those to other contibutions to this
volume, will be postponed to the succeeding sections.

1. Mean temperatures $\bar{T}$ are held away from their radiatively-determined
values $\bar{T}_r$ by an effective mean zonal force $\bar{\mathcal{F}}$ attributable to eddy
motions. Similarly, $\bar{\mathcal{F}}$ controls the diabatic circulation including
stratosphere-troposphere mass exchange rates.

2. Wave propagation mechanisms play a crucial role in shaping $\bar{\mathcal{F}}$,
especially the gravity-wave and Rossby-wave propagation mechanisms.

3. Propagating waves may transport angular momentum between, and tracers
within, the sites of wave generation and dissipation. It is the angular
momentum transport, or radiation stress, between such sites that gives
rise to $\bar{\mathcal{F}}$. (An approximate but practically useful measure of this stress
is given by the Eliassen-Palm flux.)

4. The waves that contribute to $\bar{\mathcal{F}}$ in the middle atmosphere are generated
predominantly in the troposphere and dissipated predominantly in the
middle atmosphere. (One might call this the Charney-Drazin-Hines
hypothesis.)

5. There can be a two-way interaction between waves and mean state: mean
changes induced by the waves may in turn affect wave generation,
propagation, and dissipation. (For the atmosphere the classic example is
the Lindzen-Holton theory of the quasi-biennial oscillation.)

6. Under the conditions typically met with in the atmosphere, wave
dissipation often involves wave "breaking", in a certain generalized
sense which is directly and simply related to the breakdown of the
nonacceleration theorem of wave, mean-flow interaction theory.

7. Some cases of wave generation, e.g. nonlinear "envelope radiation" of
gravity waves from groups of Kelvin-Helmholtz billows, or the analogous
generation of planetary-scale Rossby waves by tropospheric storm-track
activity, may also involve breaking in the same technical sense.

8. On a phenomenological level, breaking takes a great variety of forms,
but a typical consequence is the generation of a certain amount of
"turbulence". In particular, breaking gravity waves tend to generate
ordinary three-dimensional turbulence, and breaking Rossby waves quasi-
two-dimensional or "geostrophic" turbulence.

9. Another typical consequence of breaking, as understood here, is some

degree of irreversible rearrangement or redistribution of the materially
conserved quantities whose large-scale gradients are involved in the
dynamics of wave propagation. In particular, when gravity waves break,
potential temperature tends to be rearranged vertically, and when Rossby
waves break, potential vorticity tends to be rearranged isentropically.

Here "potential vorticity" means (Rossby-) Ertel potential vorticity,
hereafter "PV". Isentropic distributions, rearrangements, and fluxes of
PV will sometimes be referred to for brevity as IPV distributions,
rearrangements, and fluxes.

10. The conditions required to prove the nonacceleration theorem are, in
essence, conditions ensuring that there can be no irreversible
rearrangement of, or other irreversible change in, the distributions of
potential temperature and PV.

11. This and many other dynamical insights about atmospheric motion
depend on the following, far more general principle. It is familiar in
one form or another to most theoretical dynamicists albeit not always
spelt out explicitly. In its most general form it is sometimes called
the PV "invertibility principle". Briefly it says that if one knows how
the PV is arranged on each isentropic surface, and the potential
temperature on the bottom boundary, then one knows everything about the
dynamics apart from any gravity, inertio-gravity, or tropical Kelvin
waves that may be present. A more precise statement is as follows. If
one knows the mass under each isentropic surface, or some other suitable
quantity defining a reference static stability, then isentropic
distributions of PV, together with the distribution of potential
temperature on the bottom boundary, contain all the information needed to
deduce the wind, pressure, temperature and density fields to the extent
that, and to the accuracy with which, the motion can be regarded as
"balanced". The latter idea involves some subtlety, but roughly speaking
"balanced" means balanced hydrostatically and also geostrophically, or
cyclostrophically, or in some higher-order sense to the extent that
higher accuracy is possible. If the tropics are included then balance,
for the purposes of PV inversion, also entails an absence of tropical
Kelvin waves. The mean flows we usually deal with are all balanced, in

11'. Thus the nonacceleration theorem is really just a corollary of the
invertibility principle. It says, in essence, that if PV and potential
temperature distributions are not changed irreversibly, then the mean
flow will not be changed irreversibly. This version of the
nonacceleration theorem discounts as reversible, or temporary, any mean-
flow change attributable to waves that are not breaking or otherwise
dissipating, such as when for example the wintertime polar vortex is
displaced off the pole and back again by a planetary-scale, wave-1
disturbance in such a way that there is no irreversible PV rearrangement
or other irreversible, e.g. diabatic, change.

12. The ways in which PV distributions can change are restricted, even
in the presence of diabatic heating and frictional or other forces, by

two fundamental theorems. They are undoubtedly important because diabatic heating, and gravity-wave drag, are known to be significant in practice. First, as far as the middle atmosphere is concerned, PV is indestructible. Net creation or destruction of PV can occur only where an isentropic surface meets a boundary, as in the troposphere. Second, isentropic surfaces are impermeable to PV. This is true even when mass is crossing those surfaces diabatically. It means, for instance, that eddy fluxes of PV are always exactly isentropic.

13. Theoretical evidence has accumulated to the effect that Rossby wave "surf zones", meaning regions where breaking is important, can be quite strong partial reflectors or backscatterers of Rossby waves. In this respect they differ greatly from ordinary ocean beach surf zones, which are excellent absorbers of the incoming surface gravity waves which give rise to the surf. The relevant theory, which includes some rigorously proven theorems, again makes use of the ideas of PV rearrangability and PV invertibility. There is no comparable body of theory for gravity waves, but idealized model calculations suggest that some backscattering is likely especially in mesospheric, "steep-beach" conditions where vertical rearrangement (mixing) of potential temperature and material tracers should be relatively strong and sudden.

14. To the extent that Rossby-wave surf zones can act as partial reflectors, their evolution could give rise to focusing and self-tuning resonant effects in the wintertime stratosphere. These might be significant factors in the apparently variable responsiveness of the middle atmosphere and its general circulation to events in the troposphere, and will need to be considered when constructing two-dimensional or other low-order models which attempt to incorporate dynamical feedbacks on the general circulation.

15. Some of the PV rearrangement due to Rossby-wave breaking in the winter hemisphere can be described as "erosion" of high-PV air from the edge of the main polar vortex, into the surrounding surf zone. The polar vortex, when defined to coincide with the main region of high PV on each isentropic surface, is a material entity on which Rossby waves may propagate, and on the edge of which they tend to break, causing the erosion.

16. The wave absorptivity of the surrounding surf zone is related to the rate of erosion of the main vortex. Although the surf zone may often be a strong reflector it need not, on average, be a perfect one. This fact complicates assessments of point 14.

17. The word "erosion" seems apt since, when one looks at the process in detail, it is found to have a strikingly one-sided quality. Unless planetary-scale Rossby wave amplitudes become exceedingly large, close to disrupting the main polar vortex altogether as in a northern major warming, high-PV air tends to be advected into the surf zone but hardly any surf-zone air into the surviving part of the vortex. If more realistic modelling studies confirm this, it may prove important for the

Antarctic problem. It suggests that, especially in the lower
stratosphere in late winter where disturbances are weaker and both PV and
potential temperature better conserved, the Antarctic polar vortex could
be acting as a kind of "containment vessel", chemically isolated from its
surroundings, within which anomalous chemistry can take place. It is
isolated horizontally by the IPV gradients at its edge and the one-
sidedness of the erosion process, and vertically by the static stability
and the weakness of the Antarctic diabatic circulation in late winter.

18. A further point about diabatic circulations may be worth some
emphasis. In contrast with the predominantly upward sense of wave
propagation envisaged by the Charney-Drazin-Hines hypothesis, the control
by the resulting force $\tilde{\mathcal{F}}$ upon the diabatic circulation is exerted
downwards, in the long-time average. This is a consequence inter alia of
the hydrostatic decrease in mass density with height together with the
limitations on the growth of wave amplitude with height imposed by
breaking and other dissipative processes. Long-term stratosphere-
troposphere mass exchange rates are probably controlled by $\tilde{\mathcal{F}}$ values in
the first two or three scale heights of the stratosphere.

Point 18, like point 14, evidently has implications for the construction
of low-order, middle-atmosphere-only models with artificial lower
boundary conditions.

3. THE RADIATIVE SPRING AND THE PRINCIPLE OF DOWNWARD CONTROL

Points 1 and 18 can usefully be taken together. The important idea
expressed by 1 can be traced back at least as far as Dickinson (1969,
p. 80), and there are excellent discussions in several of the reviews
mentioned earlier. It may appear strange at first sight, or even
paradoxical, that a zonally directed force should control a temperature
difference. But it is less surprising when one remembers how tightly
horizontal temperature differences are coupled to dynamics by the
constraints of hydrostatic and cyclostrophic balance under terrestrial
conditions, most tightly of all on the global scale.
      An essential part of the picture is what Stephen Fels aptly calls
the "radiative spring" (Fels 1985, 1987). It is this that makes it
useful not to treat the mean diabatic heating rate $\tilde{\mathcal{Q}}$ on the same footing
as the mean force $\tilde{\mathcal{F}}$, even though they appear in the mean equations in a
similar-looking way. The point is that the dependence of $\tilde{\mathcal{Q}}$ upon changes
in the general circulation is both simpler, and probably stronger, than
that of $\tilde{\mathcal{F}}$. Broadly speaking the sign of $\tilde{\mathcal{Q}}$ tends to be such as to pull
the atmosphere towards its radiatively determined temperature T_r.
Exceptions to this statement tend to be local and temporary. I say
"towards" advisedly, since there is little reason to suppose that even if
all zonal asymmetries could somehow be suppressed the atmosphere would
actually be able to reach temperature T_r everywhere. Attempts to
construct balanced zonal wind fields compatible with T_r tend to
produce either solutions unstable to various modes of zonally symmetric
overturning, or no solution at all. In particular, balance is impossible

with realistic summer hemispheric T_r (Shine 1987), unless we admit
an implausibly strong cyclonic circumpolar vortex at the earth's surface.
Some related points have been made by Tung (1986).

One of the simplifications, perhaps oversimplifications, made in §2
lies in thinking of $\mathcal{F}$ as being zonally directed, and zonally symmetric.
A more refined picture might for example distinguish the $\mathcal{F}$ due to gravity
waves from that due to planetary waves, and take account of various kinds
of planetary-gravity-wave interaction. As suggested by point 11' above,
not all such interactions need be of the familiar kind suggested by the
breakdown of the nonacceleration theorem. I shall return to this
question in §§4 and 6. Another point, which however need not be dwelt on
here since it is widely appreciated today, is that it is simplistic to
talk, as I have been doing, about "the" mean as if the word "mean" had a
self-evident, unique significance. There is an excellent discussion of
the most useful definitions of mean circulation in WMO_3 chapter 6; see
also Andrews (1987), Beagley and Harwood (1987), Remsberg (1987) and
Tung (1987) elsewhere in this volume.

Point 18, the principle of downward control, can be quickly
appreciated if we forget about refinements for the moment and accept the
simple picture envisaged in point 1. The essential assumptions are a
balanced zonal mean state, and statistical steadiness since we are
thinking about long-term means, and trends. Working either in log-
pressure or in log-isentropic coordinates (Holton 1975, 1986), consider
mass exchange across a given "level" z_0, which might for instance
correspond to an isentropic surface such as the 400K or 450K surface in
the lower stratosphere. Take $\mathcal{F} = \mathcal{F}(y, z)$ to be the effective zonal force
per unit mass at the level z_0, where z is the vertical, and y the
latitudinal coordinate defined as latitude ϕ times the earth's radius.
Then an approximate steady-state angular momentum balance of the form
$-f\bar{v}^* = \mathcal{F}$, and a steady-state mass continuity equation of the form

$$\rho_s(z)(\cos\phi)^{-1} \partial(\bar{v}^* \cos\phi)/\partial y + \partial(\rho_s(z)\bar{w}^*)/\partial z = 0 \qquad (1)$$

produces a steady-state vertical velocity $\bar{w}_0^*$ at z_0 given by

$$\bar{w}_0^*(y, z_0) = -\frac{1}{\rho_s(z_0)\cos\phi} \frac{\partial}{\partial y} \left\{ \cos\phi \int_{z_0}^{\infty} \rho_s(z) f^{-1} \mathcal{F}(y, z) \, dz \right\} \qquad (2)$$

together with a corresponding departure of the temperature field from its
radiatively determined values. Here $\bar{v}^*$ and $\bar{w}^*$ are the
steady-state residual, isentropic-diabatic, or other appropriate measures
of the zonal-mean meridional mass circulation (WMO,1985,chaps. 6, 12; Tung
1982, 1987), $\rho_s(z)$ is an appropriate standard density, decreasing
exponentially upwards on scale heights of the order of 7 km, and f is the
Coriolis parameter.

Observationally reasonable orders of magnitude indicate that the
integral in eq. (2) is strongly convergent at its upper limit. This is
to be expected from the constraints on the order of magnitude of $\mathcal{F}$
imposed by the various wave-breaking phenomena, implying that the main

control over $\bar{w}_0^*(y, z_0)$ (and, by implication, over departures of temperatures from their radiatively determined values) comes from $\mathcal{F}$ values in the first few scale heights above z_0. This is the downward control principle.

This long-term control depends on the fact that the Earth is a rapidly rotating planet; and the control is not, of course, exerted locally in the tropics. There, (2) is ill-conditioned, and evidently makes no sense at all unless the force $\mathcal{F}$ goes to zero faster than f as y $\rightarrow$ 0; and in any case $f\bar{v}^*$ is no longer likely to be a useful approximation to a full description of the advection of absolute angular momentum $M(y, z)$ by the mean circulation. In some band of tropical latitudes within which isentropic surfaces touch surfaces of constant M, any nonzero force $\mathcal{F}$ per unit mass must either give rise to a nonzero Eulerian acceleration $\partial\bar{u}/\partial t$, or to vertical shear $\partial\bar{u}/\partial z$ such that $\bar{w}^*$ $\partial\bar{u}/\partial z$ balances $\mathcal{F}$, or to both. In the real stratosphere $\bar{w}^*\partial/\partial z$ tends not to be large enough (besides usually having the wrong sign), and it is $\partial\bar{u}/\partial t$ that tends to respond (Andrews and McIntyre 1976, §6). A nonzero Eulerian acceleration $\partial\bar{u}/\partial t$ cannot persist in the long-time average; and as everyone knows, the tropical lower stratosphere solves this "problem" in a particularly interesting way (point 5). It arranges for $\mathcal{F}$, and $\partial\bar{u}/\partial t$ in response, to change sign every 13 months or so through the well known wave-mean interaction discovered by Lindzen and Holton. According to the Lindzen-Holton theory, the long-time-average $\mathcal{F}$ that should be used in (2) or its refinements is, indeed, zero in the tropics. The waves involved are thought to be upward-propagating tropical Kelvin and Rossby-gravity waves, dissipated diabatically and probably by breaking as well. There could be a significant contribution also from breaking planetary-scale Rossby waves of extratropical origin (for adumbrations of this latter idea see Dickinson 1968, p. 1001 and Andrews and McIntyre 1976, p. 2045, also §7 below). Dunkerton (1983) presents some modelling evidence to show that extratropical Rossby waves and tropical Kelvin waves could suffice by themselves, and Hamilton (1987) discusses further possibilities.

Of course whatever happens locally, the tropics cannot escape the overall, global-scale, long-term control expressed by eq. (2). This is ensured by global mass continuity.

The foregoing discussion has presumed that we may regard $\underset{\sim}{\mathcal{F}}$ as given. This is equivalent to the assumption of "fixed dynamical heating" [sic]. Along with many others I find the resulting view of the diabatic circulation very useful and insightful, even though it is undoubtedly another oversimplification. As hinted in points 8, 14 and 16, the feedback from the general circulation onto $\mathcal{F}$ is ill-understood, and likely to be complicated; the hope is that it will not prove too strong.

In addition, one often makes a tacit but less crucial presumption that $\bar{T}_r$ may be regarded as as given. Together with the presumption about $\underset{\sim}{\mathcal{F}}$ this is certainly a much better, and more robust, idealization than the old idea of regarding $\bar{Z}$ as given. But it cannot be taken too literally since it leaves out of account the feedback onto $\bar{T}_r$ from the diabatic circulation, and from wave breaking, through the redistribution of radiatively active trace chemicals such as ozone. One interesting possibility is illustrated by the negative feedback mechanism

pointed out by Farman et al. (1985), between $\bar{T}_r$, diabatic descent
in the Antarctic lower stratosphere, and ozone photochemistry. Such a
mechanism could well act by itself for a limited time, to some extent
independently of the long-term dynamical control, and might well be
important for understanding seasonal evolution.

4. THE ABILITY OF WAVES TO TRANSPORT MOMENTUM AND ANGULAR MOMENTUM, AND THE PSEUDOMOMENTUM RULE

Point 3 of §2 implies that waves can transport momentum and angular
momentum over distances limited only by how far the waves can propagate
before being dissipated. In this and other respects, the momentum
transport brought about by waves is very different from that brought
about by ordinary turbulence. This is one reason why it may be
important, when looking at observational data showing disturbances to a
fluid medium, to try to determine whether those disturbances are wavelike
or turbulent (or some combination of both, as is typical of breaking
waves). Recognition of the distinction has been helpful in the
elucidation of previously mysterious phenomena, such as the so-called
"negative viscosity" often observed in large-scale atmospheric eddy
motions, again using the principles of point 3.
        Wave-induced momentum transport and mean flow generation have been
studied for many different kinds of waves in fluids, both theoretically
and in the laboratory. One of the most striking laboratory
demonstrations is the celebrated experiment of Plumb and McEwan (1978).
This clearly demonstrated the kind of wave-mean interaction envisaged by
the Lindzen-Holton theory (point 5), and led to a widespread acceptance
that the interaction is not just a theoreticians' plaything, but
something that can take place in real fluids. If one is prepared to be
slightly less ambitious than Plumb and McEwan, then a laboratory
demonstration of wave-induced momentum transport and mean flow generation
is actually very easy. You can do it in the kitchen sink, with no
special apparatus. Run some water into the sink, and sprinkle a little
powder such as ordinary household flour on to the surface of the water.
Dip your fingers into the water, and make your hand oscillate rapidly,
say five or six times per second, so as to radiate short surface waves in
two opposite directions, as suggested in Fig. 1. If the wave amplitude
is large enough, a strong mean flow will quickly appear, streaming away
from the wave source in the directions of strongest radiation, as
suggested in the figure. Momentum is being transferred from the wave
source to the place where the waves are dissipating. This experiment can
easily be (and was) turned into a lecture demonstration, by performing it
in a glass-bottomed dish on an overhead projector. A more refined
version (the one actually carried out in the lecture) uses a horizontally
oriented, vertically oscillating circular cylinder as the wave source.
This tends to give a clearer, simpler mean streaming pattern, and also
gives an unequivocal demonstration that the mean flow observed is
predominantly wave-induced, and not boundary-layer streaming from the
surface of the wavemaker. The latter streaming has been intensively
studied by fluid dynamicists, and is well known to be in the opposite

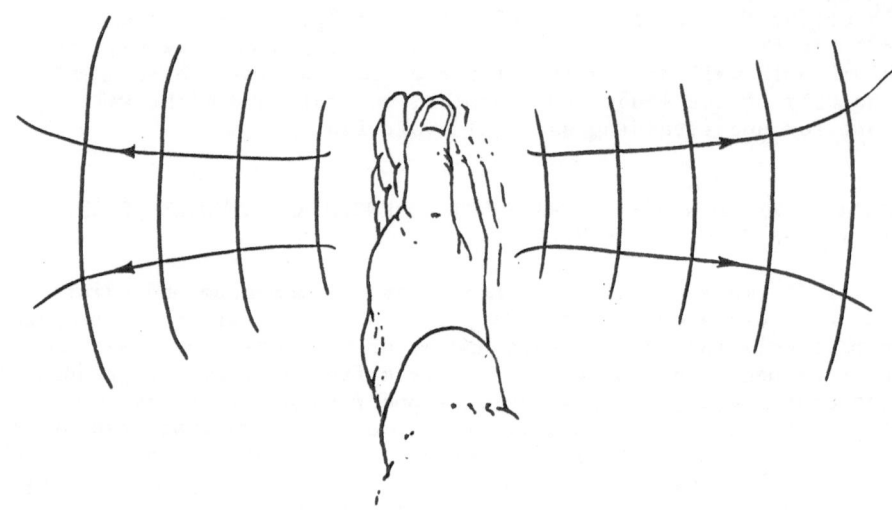

Figure 1.   An experiment on surface gravity waves, demonstrating wave-induced momentum transport and mean flow generation with no special apparatus.

sense when the cylinder oscillates vertically.

The mean flow in this example depends mainly on the dissipation of the waves (the dissipation being enhanced by the presence of the flour and other impurities floating on the surface of the water). The regions of wave generation, propagation and dissipation are hardly separated from each other in this configuration, but you can also play games with larger, concave wavemakers which focus the waves onto a more distant region. Sure enough, that is where the mean flow appears. The longshore current generated by the breaking of obliquely incident ocean-beach waves, a phenomenon well known to coastal engineering experts, is a larger scale version of essentially the same phenomenon. In this case the distance between the wave generation and dissipation sites, over which momentum transfer takes place, may be many thousands of kilometres. Another classic example is the "acoustic streaming" observed when a powerful beam of ultrasound dissipates in air. Fundamentally it is again the same phenomenon. In the acoustic case the wave-induced mean flow is sometimes called the "quartz wind", since some of the earliest devices for generating high-intensity ultrasound used quartz piezoelectric crystals.

In examples like those mentioned so far, the amounts of momentum and angular momentum transported between the sites of wave generation and dissipation are usually the same as if the waves were propagating in a vacuum, and possessed momentum equal to their pseudomomentum. I like to call this the "pseudomomentum rule", if only as a reminder that waves in material media do not always possess an actual well-defined momentum, and that even in cases where they do it may well be irrelevant, as Léon Brillouin pointed out long ago, to the actual transport of momentum.

[I have discussed this point elsewhere (1981); there is some intriguing history going back to Rayleigh and Poynting.] Pseudomomentum, unlike momentum, is a well defined general wave property. Its significance for our purposes lies the properties of its flux, or transport. This often equals the relevant part of the radiation stress or effective wave-induced flux of momentum (Andrews and McIntyre 1978a §8; 1978b §5.2; McIntyre 1981 §5). The pseudomomentum may be defined approximately as wave-action times wavenumber, or intrinsic wave-energy divided by phase speed; precise definitions are given in Andrews and McIntyre (1978a,b), Dunkerton (1980) and, in a more general version of the concept, by McIntyre and Shepherd (1987).

The pseudomomentum rule has wide but not universal applicability. For our purposes, a fairly safe assumption is that the rule is applicable in the aforementioned "interactions of the familiar kind suggested by the breakdown of the nonacceleration theorem" (§3). A precise version of this statement is provided by a general result in Andrews and McIntyre (1978a, Theorem I, p. 621; see also p. 627), and the relation between the nonacceleration theorem and the concept of wave breaking is fully discussed in McIntyre and Palmer (1985, & refs.).

The Lindzen-Holton mechanism for the quasi-biennial oscillation may be described as a feedback oscillation due to two-way interaction between waves and mean state, in which $\bar{\mathcal{F}}$ is given by the pseudomomentum rule, and in which wave dissipation plays an essential role in violating the nonacceleration theorem. It is worth mentioning, by way of contrast, that other kinds of wave-mean feedback oscillation are known for which the pseudomomentum rule and the nonacceleration theorem are irrelevant, and wave dissipation is not involved. Indeed, one such phenomenon can be made the subject of another very interesting laboratory demonstration. It occurs when acoustic waves are excited in a vessel containing a liquid with a free surface (Huntley 1977, & refs.). A large glass laboratory beaker holding four or more litres of water will suffice; more quantitative versions of the experiment have been done in rigidly-enclosed cavities. When the system is driven acoustically just above one of its resonant frequencies, a standing gravity oscillation of the free surface builds up visibly and spontaneously. The gravity oscillation has frequency far lower than the acoustic driving frequency. The sound field sees, in effect, a mean state varying slowly in time, causing the sound field itself to vary, with a certain phase lag, as conditions approach and recede from resonance. The slowly varying mean forces due to the sound waves are able to do work on the gravity oscillations and cause them to grow exponentially. (This can cause troublesome sloshing in rocket fuel tanks.) The phenomenon is one of a broad class of nondissipative interactions which can be regarded equally well as wave-wave interactions or wave-mean interactions. Another example is the interesting non-classical gravity-wave, inertia-wave interaction discovered by Broutman and Young (1986). I suspect that this class of phenomena might include nondissipative planetary-gravity interactions that have not yet been studied at all, but which might yet prove important in the middle atmosphere. Indeed, Theorem I of Andrews and McIntyre (loc. cit.) predicts the existence of nondissipative, zonally asymmetric contributions to $\bar{\mathcal{F}}$ which do not conform to the pseudomomentum rule, and whose order of magnitude is not obviously negligible. This point will be touched on again in §6.

5. THE MESOSPHERE AS AN ILLUSTRATION OF POINT 3 of §2, AND THE "TURBULENT PRANDTL NUMBER" PROBLEM

It is now widely accepted that in order to understand the zonal mean mesospheric circulation, even qualitatively, it is necessary to invoke the transport of angular momentum by gravity waves arriving from below (in accordance with the pseudomomentum rule). There is a consensus among experts on satellite data, radiation and photochemistry that T departs greatly from T_r in the upper mesosphere, around 80km, by amounts which are typically a large fraction of $\pm 100°K$ at solstice. The direct observational evidence on mesospheric gravity waves seems consistent in order-of-magnitude terms with the hypothesis that the breaking of high frequency waves, with periods predominantly less than an hour or so, can account for the required $\mathcal{F}$ (which normalized as force per unit mass has enormous peak values of order $10^2 ms^{-1} day^{-1}$. Some of the observations now include direct measurements of the wave-induced Reynolds stress by radar techniques, although sampling is sparse, with only two sites worldwide. For these high-frequency gravity waves the Reynolds stress is an excellent estimate of the radiation stress, whose divergence gives $\mathcal{F}$. (It is an overestimate for waves nearer the inertial frequency.) For details, and history, the reader is referred to the reviews mentioned earlier and to the papers in this volume by Andrews (1987), Brasseur and Hitchman (1987), Chanin and Hauchecorne (1987), Fritts (1987), Hamilton (1987), and Vincent (1987).

Mesospheric gravity waves may well be a case in which both aspects of point 3 have definitely observable consequences. There is evidence that chemical tracers are being transported within, as well as angular momentum into or out of, the region of strongest wave breaking. For instance recent observational and modelling work has indicated that vertical mixing is needed to account simultaneously for the observed seasonal variation of the 557.7nm green-line atomic oxygen airglow around 100km, and of the 1.27μm near-infrared O_2 airglow around 80km, related to O_3 concentration (e.g. Garcia and Solomon 1985, & refs; Brasseur and Hitchman, op. cit., & refs.). In order to account for what is observed, it seems necessary to hypothesize that water vapour is transported upwards to levels around 80km at the same time as atomic oxygen is transported downwards to the same levels, from 100km. This points strongly to vertical mixing rather than mean ascent or descent. Large transport implies weak airglow, and this is what is observed around the solstices, at the same time as the greatest values of $\mathcal{F}$ are required to account for $T - T_r$. A corresponding seasonal variation in gravity-wave activity has been directly confirmed by radar and lidar observations (e.g. Chanin and Hauchecorne, op. cit; Vincent, op. cit.), and is expected theoretically from the "filtering effect" of the seasonal variation of mean winds between troposphere and mesosphere through which the gravity waves are believed to propagate (e.g. Lindzen 1981).

The argument for simultaneous upward and downward transport is extremely telling in itself, and the whole picture seems to fit together remarkably well, on a qualitative level, and is in accordance with the general expectations summarized in point 3. A model simulation by Garcia and Solomon (op. cit.) provides a useful qualitative check on the mutual consistency of the hypothesized dynamics, radiation and photochemistry. Regarding the dynamical aspects, however, there is at least one very severe difficulty in the way of a more quantitative test. The difficulty is of a kind that seems generic to problems involving turbulent fluid motion, especially turbulence having the extreme spatial inhomogeneity and intermittency that seems characteristic of large bodies of fluid like the atmosphere; and it illustrates how careful one must be in applying ideas from homogeneous turbulence theory and eddy diffusion theory. Another example will be mentioned in §8.

Most mesospheric models, including that of Garcia and Solomon, follow Lindzen (1981) in postulating that the vertical mixing can be parametrized in terms of vertical eddy diffusion coefficients K_{zz} which are (a) spatially homogeneous over a gravity wavelength, (b) equal in value for potential temperature and tracers ("turbulent Schmidt number unity"), (c) equal to the effective eddy viscosity that contributes to wave dissipation ("turbulent Prandtl number unity"), and (d) chosen so that (a)-(c) imply just enough wave dissipation to keep wave amplitudes a not too far above some notional value for incipient breaking, say a = 1 with an appropriate nondimensionalization (the "saturation hypothesis"). [One may take for instance a = (horizontal disturbance velocity)/(horizontal intrinsic phase speed), or (vertical parcel displacement)/(vertical radian wavelength).] It is also assumed (e) that the same K_{zz} applies to the mean state as to the waves.

These assumptions have the virtue of simplicity, indeed they are the simplest that seem reasonable at first sight; and they represent a great advance on the idea of a pre-existing eddy viscosity independent of the incident waves. In particular, as Fritts (1984) clearly explains, the saturation hypothesis automatically takes care of the problem of so-called "critical level" absorption, since the significance of a critical level in a real fluid is simply that breaking amplitudes must be exceeded before the waves reach it, if other dissipation mechanisms have not intervened. However, other aspects are less satisfactory. Chao and Schoeberl (1984), Fritts and Dunkerton (1985), and Coy and Fritts (1987) have pointed out that the K_{zz} required for chemical modelling, measuring the net irreversible vertical rearrangement of the mean state (discounting temporary effects from reversible undulations) often tends to be very much less, by a factor μ say, than the notional spatially uniform K_{zz} calculated from assumptions (d) and (e). Moreover -- and this is the real difficulty -- μ appears to be an exceedingly sensitive function of a − 1, the degree of supersaturation of the breaking wave. I have confirmed this in an independent calculation (unpublished), which indicates that μ varies monotonically from values of the order of 10^{-2} for a = 1.05, to values exceeding unity for a several times the critical value a = 1. Likewise, in the well-known formula

$$K_{zz} = \beta\epsilon/N^2 \quad (N = \text{mean buoyancy or Brunt-Väisälä frequency}), \quad (3)$$

sometimes used to estimate K_{zz} when the turbulent dissipation rate
ϵ has been estimated from observations, the numerical coefficient β is
extremely sensitive to $a - 1$. My model calculation gives β values
ranging monotonically from 10^{-2} or less for $a = 1.05$, through
10^{-1} for a around 1.5, to $\frac{1}{4}$ for $a = 3$ (which already seems an
implausibly large degree of supersaturation). It might therefore be a
mistake to think of the coefficient β as a constant, notwithstanding
estimates from laboratory experiments. It appears that that β, which
can be thought of as a measure of the partitioning between turbulent
viscous dissipation and potential-energy gain due to irreversible
vertical rearrangement, depends very much on how the turbulence is
generated. Some new laboratory experiments which should bear on these
questions are being planned by Dr P.F. Linden at Cambridge.

It could therefore be that the evidence for the vertical mixing of
tracers in the mesosphere amounts to an indication that gravity-wave
amplitudes are strongly supersaturated there. For what it is worth, this
is consistent with the observation that vertical wavelengths increase
with height, and in the mesosphere become comparable to 2π times the
density scale height. It is reasonable that, for the higher frequency
waves, the degree of supersaturation should increase as the wave-breaking
process becomes more 'sudden', like a steeply-sloping as opposed to a
shallowly-sloping ocean beach. This is because the timescale for the
onset of unstable convective overturning when a exceeds 1 is finite,
and may be comparable to the radian wave period for the higher frequency
waves. An increase in vertical wavelength with height is predicted by
the saturation hypothesis applied to a continuous wave spectrum (e.g.
Fritts 1987); with a plausible spectrum the longer waves penetrate higher
before breaking. The sensitive dependence of K_{zz} upon $a - 1$
implies that direct observational information about the degree of
supersaturation, and about the actual vertical mixing of tracers, will be
at a premium.

6. THE SPONTANEOUS BREAKDOWN OF BALANCE AND THE RESULTING GRAVITY-WAVE
    EMISSION: A SIGNIFICANT CONTRIBUTION TO $\bar{\mathcal{F}}$?

How correct is the Charney-Drazin-Hines hypothesis (point 4)? Few of us
would doubt that it is sufficiently correct to be extremely useful in our
present state of knowledge. It has always been plausible that the
denser, lower layers of the atmosphere should affect the more tenuous,
higher layers to a much greater extent than vice versa, on the principle
that the elephant wags the tail. Wave propagation mechanisms are
available to mediate the process because of the fact that, at all
altitudes in the middle atmosphere, vertical gradients of potential
temperature and isentropic gradients of PV are available for the waves to
propagate on. For Rossby waves (Charney-Drazin), the hypothesis still
seems to provide the most plausible explanation of the striking
difference between the observed planetary-scale summer and winter
stratospheric circulations. For gravity waves (Hines), the hypothesis is
well supported by considerations like those of the last section, and by
direct observations of gravity waves, including Rayleigh-scattering lidar

observations (Chanin and Hauchecorne 1987). This lidar technique has
recently detected many examples of what appear to be gravity waves with
upward group velocities occupying the entire altitude range visible to
the technique, 30km to 80km, and with vertical wavelengths Λ in the range
5 to 15km. The vertical wavelengths and likely intrinsic frequencies
seem large enough, on the whole, for infrared-radiative dissipation to be
of secondary importance (e.g. Fels 1985); and consistent with this the
waves are often observed to reach saturation amplitudes especially for Λ
$\gtrsim$ 10km.

However, fluid-dynamical experience cautions us against too
completely unqualified an acceptance of the hypothesis, at least in its
"strong" form that envisages all significant wave generation to be, say,
at tropopause jet-stream levels or lower, and the vertical sense of all
significant propagation to be upward in the middle atmosphere. In
principle, nonlinear fluid motion can backscatter, re-radiate, or even
spontaneously generate waves out of (to use an English colloquialism)
thin air. There is no reason, in principle, why such spontaneous wave
emission should not take place at any altitude. Indeed, it is already an
accepted idea for tropopause jet-stream level, and it might be a mistake
to suppose that this is the only case in point.

For gravity waves, spontaneous emission is also of theoretical
interest as signalling one of the ways in which the balance, PV
invertibility, and nonacceleration concepts may break down (points 11,
11' of §2; see also the review by Hoskins et al., 1985). It is possible
that spontaneous wave emission might have systematic effects on the
angular momentum budget and the distribution of $\bar{\mathfrak{F}}$. An assessment of this
possibility involves some fascinating, ill-understood, and largely
unexplored aspects of wave-mean and wave-wave interaction theory, in
terms of fundamental concepts as well as detailed application, as hinted
at the end of §4. For all these reasons it is a topic of recent and
current research.

It may be worth recalling briefly the two known types of spontaneous
emission. Perhaps the best known example is the nonlinear radiation of
gravity waves from Kelvin-Helmholtz (KH) billows. Such radiation can
occur both on the scale of the KH instability (in some circumstances,
McIntyre and Weissman 1978), and also, probably more commonly and more
powerfully, on the amplitude-modulation scale, i.e. the scale
characteristic of whole groups of billows (Fritts 1982). This is one
example of the "envelope radiation" mentioned in point 7 of §2.
Planetary-scale Rossby radiation from tropospheric storm tracks (e.g.
Opsteegh and Vernekar 1982) is another example of essentially the same
thing. In the case of gravity waves at least, there is a second and
entirely different type of spontaneous emission, having nothing to do
with any instability. It may be called (spontaneous) "geostrophic
adjustment", although this is a misnomer to the extent that the balance
towards which the adjustment is considered to take place, and whose
breakdown marks the emission of gravity waves, may be much more accurate
than than geostrophic. The theoretical evidence to date indicates that,
in contrast with KH envelope radiation, neither wave breaking nor other
dissipative processes play any essential role.

Although spontaneous geostrophic adjustment is far from well

understood, there is little doubt about its physical reality. The
physical principle, and the fact that instability is not involved, was
recently given a clear demonstration within the context of a low-order
atmospheric model by Lorenz and Krishnamurthy (1987). It may be worth
mentioning also that there are close cross-disciplinary analogues in
aerodynamic sound generation, in which sound is emitted by turbulent or
other unsteady vortical motions, and the relevant adjustment process is
adjustment towards elastostatic balance. Such aeroacoustic problems have
been thoroughly studied both theoretically and experimentally (e.g.
Crighton 1981). For the atmosphere there is observational evidence of
what appears to be spontaneous emission, with horizontal wavelengths (~
100km) suggestive perhaps of geostrophic adjustment rather than envelope
KH, from certain "jet-streak" events involving fast, unsteady advection
of PV anomalies on the isentropic surfaces that intersect the upper
troposphere and lower stratosphere. Uccellini and Koch (1987) summarize
thirteen case studies. It seems possible that similar phenomena might
occur also in the wintertime middle or upper stratosphere. The
aeroacoustic analogues suggest that the radiation might be significantly
concentrated in the forward direction relative to the jet stream, with a
corresponding phase-velocity bias, and hence that it could be potentially
significant for the angular momentum budget. It could for instance give
rise to a systematic westward contribution to $\mathcal{F}$ in the stratosphere, and
a complementary eastward contribution in the lower thermosphere.

     It might be thought that such processes would be too weak to matter
much in the stratosphere, or indeed the troposphere; but I don't think
anyone really knows. Both the Lorenz-Krishnamurthy study and the
aeroacoustic analogues indicate that strength of the emission process
increases very steeply as local Rossby and Froude numbers increase toward
values of order unity, suggesting the process might well be under-
represented in general circulation models which do not resolve jet-streak
scales accurately. A quantitative study must await two developments:
first, the wider availability of extremely high resolution middle-
atmospheric models, and second, for diagnostic purposes, developments in
the theory of wave-activity conservation relations which can describe the
angular pseudomomentum exchanges involved in finite-amplitude planetary-
gravity wave interactions. The latter developments are now within sight,
as a result of the recent discovery of the connection between the
Eliassen-Palm theorem and the fundamental Hamiltonian structure of
analytical mechanics (McIntyre and Shepherd 1987). This connection
provides a key to the required generalizations.

     It should not be thought that spontaneous gravity-wave emission is
the only type of nondissipative planetary-gravity wave interaction. The
considerations at the end of §4 suggest that there could also be
significant nondissipative interactions between planetary-scale Rossby
waves and pre-existing gravity waves. This, again, awaits exploration.

7. ROSSBY-WAVE SURF ZONES AND THE WAVE-TURBULENCE JIGSAW PUZZLE

The propagation mechanism to which Rossby waves owe their existence, and
the Charney-Drazin hypothesis its physical basis, is a very strange one

in which the important material displacements occur sideways, on
isentropic surfaces. The propagation mechanism, or restoring mechanism,
which acts to restore such displacements towards zero, depends on the
existence of an isentropic gradient of PV. One strange property is the
weakness of the mechanism at small scales. In the limit of small
wavelength it is weaker than any of the other restoring mechanisms giving
rise to wave propagation in the atmosphere, as evidenced by the well
known scale dependence of the Rossby dispersion relation. The mechanism
is easiest to understand with the help of isentropic maps of PV, and the
PV invertibility principle, as explained for instance in §§6a,c of the
review by Hoskins et al. (1985).

The detailed way in which Rossby waves "break" is almost as strange
as the way in which they propagate, bearing no more resemblance to the
way in which gravity waves break than two-dimensional turbulence does to
three-dimensional turbulence. Some fluid dynamicists understandably
hesitate to call the former "turbulence" at all. The differences are
related not only to the quasi-two-dimensionality but also to the weakness
of the restoring mechanism at small scales. Nevertheless, breaking, as
envisaged here, has exactly the same fundamental significance for wave,
mean-flow interaction in the Rossby case as it does in the gravity case:
as indicated in point 6 of §2 it is one of the ways in which the
nonacceleration theorem can be violated. Precise definitions, and a full
discussion, have been given elsewhere (McIntyre and Palmer 1984, 1985).

Fig. 2, taken from a recent series of high-resolution barotropic
numerical experiments done at Cambridge by M.N. Juckes, illustrates the
kind of small scale, two-dimensionally turbulent motion that can be
expected to result from the breaking of planetary-scale Rossby waves.
The model experiment also illustrates very clearly the generation of mean
flows by wave breaking. The figure shows the PV and wind fields in the
aftermath of a large planetary-scale Rossby-wave breaking event, in some
respects like those observed in the real northern wintertime
stratosphere. Notice the strong westward flow in the tropics. This
develops during the early stages of wave breaking (for more detail see a
forthcoming paper by Juckes and myself, 1987), and results from the
transport of (negative) angular momentum into the region of wave
breaking, exactly as expected from the violation of the nonacceleration
theorem and fundamentally analogous, in this respect, to the effect of
gravity waves on the mesosphere, and to other such situations such as the
generation of longshore currents by obliquely incident ocean-beach waves.

It should be noted that the model is forced, quasi-topographically,
with zonal harmonic wave 1 only, to imitate a tropospherically generated
planetary scale disturbance incident from below, together with the
accompanying radiation stress that is responsible for supplying the
negative angular momentum. This prescribed wave-1 forcing builds up
during the first four days and dies down again between days 12 and 16.
Fig. 2 is for day 17. The initial condition is taken to be a zonally
symmetric state, with a strong, cyclonic polar vortex somewhat like that
in the real wintertime middle or upper stratosphere. Note that on linear
Rossby-wave theory the chaotic, smaller-scale motions evident in Fig. 2
would be predicted to be absent; they arise solely from the highly
nonlinear wave-breaking process. We may aptly speak of a Rossby-wave

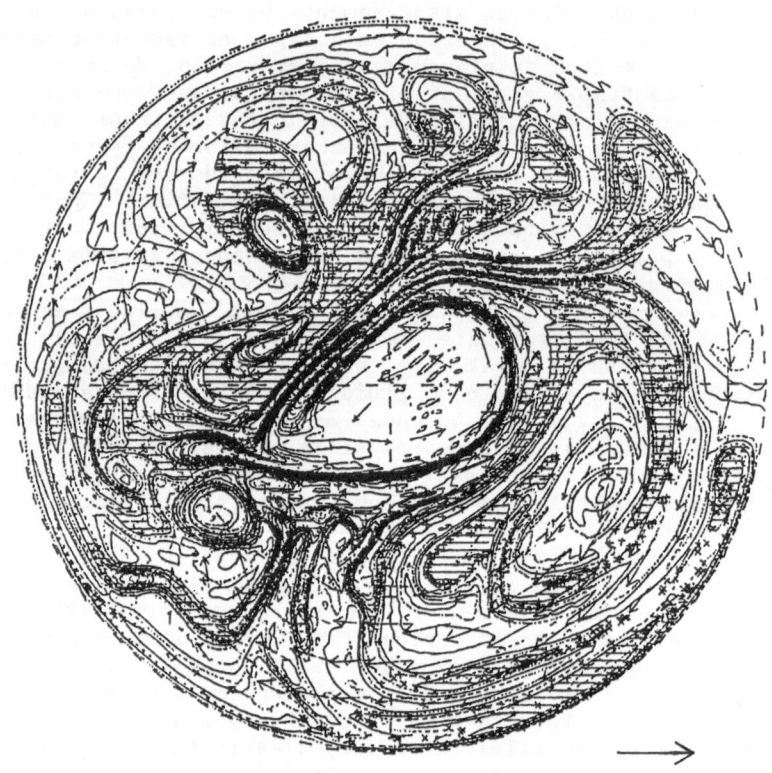

Figure 2. Polar stereographic map of a high-resolution hemispherical
barotropic model of the middle or upper stratosphere, driven by a pure
wave-1 quasi-topographic forcing from a zonally symmetric initial state,
after 17 days of integration. The numerical model is based on those
developed at the University of Reading. The resolution is triangular
truncation T159, roughly equivalent to a grid size between $\frac{1}{2}°$ and $1°$
latitude. The model's artificial dissipation is a Laplacian cubed, and
PV is materially conserved to good accuracy down to scales of the order
of a degree of latitude (so that there is no attempt to represent real-
stratospheric diabatic effects). The contours show PV at equal intervals
of 0.07575, in units of the maximum planetary vorticity 2Ω, ranging from
zero at the equator to 1.36 units within the main polar vortex. The
hatching corresponds to PV values lying between 0.4545 and 0.6060 units.
The arrows show wind velocity and are located with their tails at the
relevant points. The arrow at bottom left corresponds to $100ms^{-1}$.
Further details are given in a forthcoming paper (Juckes and McIntyre
1987).

"surf zone" surrounding the (core of the) main polar vortex, which latter is the conspicuous central region. This has uniformly high PV values corresponding to about 1.36 times the maximum planetary vorticity 2Ω.

Being only a barotropic model (and in this case a nondivergent one), the numerical model does not successfully simulate the real middle stratosphere in all respects. (Nor do shallow-water versions which have been tried.) One point on which the model is clearly unrealistic is that the surf zone in the real stratosphere does not normally, if ever, extend as far as the equator (e.g. Leovy et al. 1985), in the way it does in the model experiment. If the real surf zone did extend as far as that, zonal winds in the real tropics would probably be more like those in the model experiment, i.e. westward and enormously stronger than they are observed to be in the real stratosphere. Conversely, it would take only a very weak, or very sporadic, penetration of extratropical Rossby-wave breaking into the deep tropics to contribute significantly to the relatively small zonal acceleration $\partial\bar{u}/\partial t$ observed there during the westward-acceleration phase of the quasi-biennial oscillation (§3).

On the other hand, the close juxtaposition and sharply differentiated character of the main vortex and the surrounding surf one, and the steep PV gradients at the edge of the main vortex, seem likely to be realistic features. Coarse-grain IPV maps derived from satellite IR data seem consistent with these features, to within their own spatial resolution (e.g. Clough et al. 1985). That resolution is considerably lower than Fig. 2, and so, for instance, such coarse-grain maps may not clearly represent the sharpness of the edge of the main vortex, which is a typical "erosion" feature (points 15 and 17 of §2). There are other tracer measurements, notably aerosol measurements in the lower stratosphere (e.g. McCormick et al. 1983; see also Kent et al. 1985), which do not have global coverage but do indicate that the real polar vortex has a sharp edge like the model one in Fig. 2; and it is worth recalling that the beautiful aircraft observations by Danielsen (1968), of radioactive tracers from different bomb tests, pointed long ago to the same conclusion for tropopause jets.

To the extent that a large part of the pole-to-equator PV difference is concentrated near the edge of the main vortex, that vortex might be expected to act as a waveguide for any upward Rossby-wave propagation that takes place in accordance with the Charney-Drazin hypothesis (point 15). This raises the question of how good, or how leaky, a waveguide it might actually be. The question cannot be settled using concepts like ray theory and refractive index, since virtually all the approximations associated with those concepts break down in the surf zone. However, the leakiness of the waveguide must be presumed to depend somehow on the properties of the surrounding surf zone, which properties are bound to depend in turn on the arrival and breaking of the waves themselves (point 16). Such nonlinear feedback processes seem typical of the highly inhomogeneous, highly nonlinear "wave-turbulence jigsaw puzzle" with which the atmosphere as a whole confronts us. Supporting theoretical evidence will be discussed in the next section, and will suggest that the properties of the presumed waveguide might be quite variable, an idea that may account for some of the observed variability of the Eliassen-Palm planetary wave fluxes. In the next section we shall see moreover that this situation, like the one described in §5, is another case of inhomogeneous turbulent motion in which the notion of eddy viscosity must be used, if at all, with extreme caution.

8.   FOCUSING, SELF-TUNING RESONANCE, AND ARTIFICIAL LOWER BOUNDARIES

The wave-turbulence interaction just alluded to could have significant implications for any of the dynamical feedbacks on $\bar{\mathcal{F}}$ that involve planetary-scale Rossby waves. This is all the more so when it is realized that, if the main polar vortex can behave as a waveguide of possibly variable characteristics (e.g. sometimes focusing upward propagating waves into the high-altitude polar cap, and sometimes letting them defocus or leak equatorwards), then it can probably also, from time to time, act as a self-tuning resonant cavity for Rossby waves, no doubt a somewhat leaky, dissipative one. A likely tuning mechanism is simply the erosion of the vortex by the surf zone, a process which can change the size of the vortex, as is observed to occur, in fact, in high-resolution model experiments like that of Fig. 2.

   So far there are two lines of theoretical modelling evidence that bear on these questions. First, self-focusing, and self-tuning resonance, both definitely seem to occur in idealized models of the Matsuno-Holton type with artificial lower boundary conditions (Dunkerton et al. 1981; Hsu and McIntyre 1987). These idealized models are spectrally truncated in the zonal direction, and cannot represent small-scale details like those illustrated in Fig. 2. Nevertheless, until the corresponding experiments can be carried out at high resolution we must take this evidence fairly seriously. There is theoretical evidence in support of the idea that zonally truncated models can capture part of the wave-turbulence feedback that alters the reflectivity or absorptivity of the surf zone, and hence the leakiness of the waveguide or cavity (Haynes and McIntyre 1987a). Of course the artificial lower boundary implies that the model's resonances will differ from any that the real atmosphere might have, but that does not alter the implication that if such processes can enter the model's behaviour at all, then with different tunings they could also be playing a role in the real middle atmosphere.

   The second line of evidence has emerged from recent work on Rossby-wave critical-layer theory. Although this theory assumes a restricted parameter regime somewhat remote from the regime of Fig. 2, and likewise remote from the regimes in which we are usually interested in the real stratosphere, there are some important insights that carry over to other regimes, some of them provably so. In the latter category is a rigorous theorem which clarifies the general circumstances under which a surf zone can sustain absorption or reflection of Rossby-wave activity (Killworth and McIntyre 1985). Although the technical details of the proof are not trivial -- it involved finding a nonlinear version of the generalized Eliassen-Palm theorem -- its essence is simply that in order to absorb Rossby-wave activity one has to keep rearranging PV isentropically. This can be sustained, in a situation like that of Fig. 2, only as long as one keeps eroding fresh high-PV air from the polar vortex, or as long as some external agency, such diabatic heating, maintains an IPV contrast across the surf zone. Additional discussion is given in §5 of my 1982 review.

In particular, these considerations show that the absorptivity has virtually no connection with the magnitude of the eddy viscosity within the surf zone, contrary to earlier expectations from a naive application of the classical theory of "viscous critical layers". The mistake was, again, to forget about the spatial inhomogeneity of the naturally occurring turbulence; on closer analysis it turned out that the absorptivity of a classical viscous critical layer depends more on the viscosity just outside, rather than that within, the critical layer, since the essential effect was to restore the PV contrast diffusively from outside.

Under mid-stratospheric conditions, restoration of IPV contrasts appears to be too slow to prevent surf zones becoming quite reflective on average. It is this reflectivity that gives rise to the twin possibilities of focusing on the one hand, and resonant cavities, whose tuning can vary as surf zones develop in extent, on the other (point 14). The expected tuning is broadly analogous to the weakly nonlinear self-tuning studied in an idealized model by Plumb (1981), albeit mechanistically quite different, being dependent on wave breaking and therefore categorizable as a strongly nonlinear rather than a weakly nonlinear effect. Self-tuning and focusing could occur simultaneously, as in the examples of Dunkerton et al. (1981) and Hsu and McIntyre (1987); a reflective surf zone might guide waves upward into the polar regions of the upper stratosphere, while moving polewards until a resonant cavity is formed. Even an extremely leaky cavity would be enough to produce significant effects; for example a factor of 2 in planetary-wave amplitude can easily make the difference between whether or not a major stratospheric warming occurs.

The relationship between PV rearrangement and surf-zone absorptivity raises a question about what might be called the self-damping of resonant growth (point 16). If amplitudes were to grow fast enough, strong erosion of PV from the main vortex would be expected, implying less reflection and leading to slower growth. Haynes (1985) has studied an idealized model designed to test inter alia whether this effect is important. He found that for reasonable PV configurations the effect seems to be rather slight. It may be that the greatest difficulty will lie in explaining not why stratospheric planetary-wave amplitudes are as variable as they are, but why resonances are not sharper and amplitudes a great deal more variable.

It is possible that part of the answer may lie in the fact that, as I pointed out in §6 of my 1982 review, we still do not really understand one of the key components of the Charney-Drazin hypothesis, namely the tropospheric source which is presumed to excite stratospheric planetary waves. The answer one obtains from any resonance theory depends on the nature of the forcing function one specifies. The real tropospheric source, assuming such a concept is justified at all, must be related to geographic and orographic factors, but can be expected also to be highly nonlinear and related to the phenomenon of envelope radiation (point 7; §6), and thence to the problems of nonlinear baroclinic instability and tropospheric blocking, and ultimately to the existence of surface potential-temperature gradients at the bottom of the atmosphere (Hoskins et al. 1985, §§6d,e).

9. PV AS A TRACER (?), AND THE SPONGE-ANTISPONGE PICTURE OF
    DIABATIC CIRCULATIONS

    In what sense is PV indestructible in the middle atmosphere, and
unable to cross isentropic surfaces even when mass is crossing those
surfaces diabatically (point 12 of §2)? The answer is, in the sense
dictated by the analogy between PV and chemical·tracers. In this
analogy, which is increasingly being made use of in middle-atmospheric
studies, the PV itself corresponds of course to the tracer mixing ratio,
since it is materially conserved in the absence of diabatic heating and
frictional or external forces, just as tracer mixing ratio is materially
conserved in the absence of photochemical sources or sinks.
    What then should we mean by the total amount of PV contained in a
given part of the atmosphere? Again, it seems logical and useful to
agree that this too should mean the same thing as it does for a chemical
tracer; indeed, it is not clear what "total amount" could otherwise mean,
since use of the term carries a presumption that we are talking about a
additive, extensive, conservable quantity. The total amount of ozone in
a given mass of air is the mixing ratio times the mass element (or mixing
ratio times air density times volume element) integrated over the whole
air mass. The total amount of PV is the PV times the mass element,
integrated similarly.
    With these conventions, the two theorems stated under point 12 are
easy to prove in a few lines of algebra. An equivalent statement is
that, even in the presence of diabatic heating and frictional or external
forces, the total amount of PV in a volume like that shown in Fig. 3 can
change only as a result of what happens at its sides ∂V_s. Details

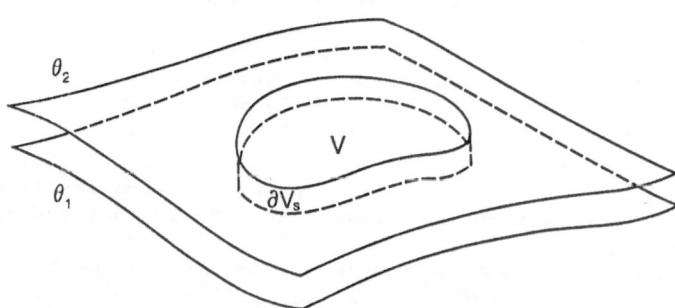

Figure 3. An air mass V bounded above and below by two isentropic
surfaces. The isentropic surfaces are impermeable to PV, which can
therefore enter or escape only through the sides ∂V_s. After Haynes
and McIntyre (1987b).

are given in Haynes and McIntyre (1987b, & refs., hereafter HM), along
with a full discussion of the relationship with various vorticity and
generalized PV concepts. It should be noted that the results depend on
no approximations whatever, and are a direct consequence of the way in
which the PV is constructed mathematically. This means for instance that
they apply not only to the (unobservably fine-grained) exact PV, but also
to any "coarse-grain" PV estimate constructed by substituting
observational data into Rossby's or Ertel's formulae, representing inter
alia the circulation around all resolvable circuits.

It should also be noted, incidentally, that there is no
impermeability theorem for the so-called quasi-geostrophic potential
vorticity. Quasi-geostrophic potential vorticity is very different from
PV, for present purposes, because of its quite different relationship
with adiabatic vertical advection.

An immediate corollary of the impermeability theorem is that the PV
cannot be used as a tracer for diagnosing diabatic circulations.
However, the peculiar properties of PV have advantages as well as
disadvantages. In particular, they have made us realize that not only is
there an important simplifying principle waiting to be exploited in our
thinking about isentropic rearrangement of PV and PV budgets (namely that
fluxes of PV are always exactly isentropic, even when mass is crossing
isentropic surfaces diabatically), but that there is a new, interesting,
and possibly practically useful way of thinking about and evaluating
diabatic circulations and strat-trop exchange rates. Dr Haynes and I are
hoping that this will actually lead to better observational estimates of
diabatic-circulation intensities, even in the lower stratosphere, by, in
effect, exploiting the information in the wind field as well as the
temperature field, together with the expectation that the motion will
usually be balanced.

One of the ways in which PV values within a volume V like that shown
in Fig. 3 can change is by the dilution or concentration of PV, where
again the words "dilution" and "concentration" mean exactly what they
mean for a chemical tracer. If mass enters V diabatically across its
bounding isentropic surfaces, but (necessarily) no PV, then the PV
already within V must be diluted, and PV values will decrease on average.
Similarly, PV will be concentrated if mass leaves V diabatically. If we
ignore isentropic fluxes of PV due to gravity-wave drag, then it can be
shown scale-analytically that, in typical extratropical, synoptic-scale
parameter regimes, dilution and concentration are the dominant effects.
Under these conditions, diabatically attenuating cyclonic PV anomalies
can be thought of, following HM, as "sponges", taking up mass, and
anticylonic ones as "antisponges", expelling mass.

This "sponge-antisponge" idea is no more than a paraphrase of some
simple mathematical facts about PV, but it seemed to us illuminating in
that it is perhaps the most direct way of seeing the relation between
isentropic eddy fluxes of PV on the one hand, and diabatic circulations
on the other (e.g. WMO, 1985, chapter 6). Also, the picture does not
really depend on zonal averaging, and it may give us a genuinely useful
alternative to thinking in terms of the global angular momentum balance.
Continuing to ignore gravity-wave drag, and continuing to follow HM,
consider a layer lying between two given isentropic surfaces. Suppose

that there is a persistent southward eddy flux of PV in middle latitudes,
associated with repeated advection of cyclonic anomalies into low
latitudes and of anticyclonic anomalies into high latitudes, such as
occurs in planetary-wave breaking. If we assume that there is a
statistical tendency for such anomalies to be attenuated diabatically,
then there will be a concomitant tendency for the isentropic layer to
take up mass in low latitudes and expel it in high latitudes. The
tendency will presumably exist whether or not the PV anomalies are also
undergoing quasi-two-dimensional turbulent mixing along the layer. The
implied mass circulation is the diabatic circulation relevant to chemical
tracer transport problems. Again, details are given in HM. The
circulation has to close downwards -- a stronger version of the principle
of downward control, point 18, since neither time averaging nor zonal
averaging is now needed.

10. SOME IMPLICATIONS FOR CHEMICAL MODELLING

Let us return briefly to Fig. 2. Four points are of interest from the
viewpoint of chemical modelling.
       First, as already noted, the (core of the) main polar vortex is a
material entity, or chemically isolated air mass, bounded by very steep
PV gradients. The vortex has considerable ability to withstand total
disruption (owing to the Rossby-wave restoring mechanism associated with
the PV contrast), but its material tends to be eroded into the
surrounding surf zone. The erosion is found to be almost always a
remarkably one-sided process, with almost no surf-zone material finding
its way into the vortex. The model experiment was originally set up with
the northern middle stratosphere in mind, and there was no attempt to
mimic Antarctic lower-stratospheric conditions, but we suspect that the
same kind of fluid dynamics will apply. Further tests of the vortex
isolation hypothesis under conditions more closely relevant to the
Antarctic problem are planned, including high-resolution baroclinic
simulations, in view of the possibly crucial significance of the
isolation question for Antarctic chemistry.
       Second, it is noteworthy that some of the small-scale PV features in
the model surf zone are behaving like passive tracers, being pulled out
into thin streaks by the large-scale straining field, while others
exhibit vortex rollup, indicating some local dynamical control. The
resulting small vortices exhibit much less resistance to disruption than
the main vortex, but still some resistance, suggesting for instance that
quasi-horizontal diffusion coefficients for PV and long-lived chemical
constituents might differ noticeably [cf. assumption (b) in §5, and the
assumptions used in the modelling approach described by Tung (1987),
elsewhere in this volume].
       Third, one may extrapolate from these experiments to get a rough
estimate of the 'mixdown time' taken to bring chemical reagents into
molecular contact in the real, baroclinic stratosphere, so that reactions
may begin. On the assumption that certain ideas from geostrophic
turbulence theory apply, at least as regards order-of-magnitude
estimates, the mixdown timescale $\gtrsim$ a month in the mid-stratospheric

surf zone. This is comparable to some chemical timescales, suggesting the possibility of significant effects on the chemistry that would be missed by a simple diffusion parametrization. For example, the competition between different catalytic cycles might have different effects depending on the order in which reagents came together. Some further details are given in Juckes and McIntyre (1987).

Fourth, the not inconsiderable success of quasi-two-dimensional chemical models prompts one to wonder whether we have yet got the best mileage out of them. For instance, as the Antarctic ozone-hole problem has reminded us, averaging around latitude circles is unlikely to be the optimal way of validating such a model against observational data. It would presumably be better to interpret the models' predictions, and further develop the models themselves, in terms of concepts such as "inside the main vortex", "outside the main vortex", etc, where "main vortex" is understood in the Lagrangian sense already alluded to. Some of the relevant ideas have been discussed most recently by Butchart (1987), elsewhere in this volume.

11. EPILOGUE: THE ERASMUS DARWIN PRINCIPLE

According to Littlewood (1953), "Erasmus Darwin had a theory that once in a while one should perform a damn-fool experiment. It almost always fails, but when it does come off is terrific."

One of the most intensely exciting moments of my life was the moment when I first saw some of the isentropic, mid-stratospheric maps of potential vorticity that were produced in the early 1980s at the UK Meteorological Office, using satellite radiometer data (channels 25 and 26 of the SSU on board Tiros-N). The maps showed, against all expectations concerning data quality, what seemed to be a distorted view of a breaking planetary wave on a huge scale -- a "breaker" the mere tip of which was the size of the United States of America. It was a case of the blind being made to see — specifically, the formerly blind theoretical dynamicist catching his first actual glimpse of what was going on in the real stratosphere, in a way that could be directly related to theoretical fundamentals.

It was the "experiment" of trying to produce these maps in the first place that was important, and this is something for which I can claim no credit whatever. The attempt qualifies as a damn-fool experiment because it was thought (and with some justification) that satellite data could not possibly be good enough to permit a direct observational estimate of such a highly differentiated quantity as PV -- indeed, I had only just been repeating this conventional "wisdom" myself, in my 1982 review, when I talked somewhat glibly about the "near-impossibility of drawing isentropic maps of potential vorticity from even the best data analyses and thus seeing directly what is going on". Subsequent events have proved, to everyone's delight I think, and certainly to the benefit of atmospheric science, that the satellite data, although not without their share of problems, are in some very important ways great deal better than had been thought. We now have case studies where the data quality has been subject to considerable cross-checking (e.g. Clough et al., 1985), and which have given us considerable confidence about the observed phenomena quite independently of theoretical expectations. The whole experience has greatly sharpened our appreciation of the amount of dynamical information contained in the various datasets, and I am sure it must have been rewarding for the many people whose ingenuity and hard work made the acquisition and processing of such data possible in the first place. On the theoretical side, we are beginning to understand somewhat better how the differential and integral conservation properties of PV (§9 above), and the properties of PV inversion operators (e.g. Hoskins et al. 1985), lend some respectability to the idea of a "coarse-grain" PV map and its place within a description of the "resolvable dynamics". The whole chain of events has been a powerful impetus to bringing theory and observation closer together — and little of it could have been planned in advance. I sometimes wonder what Erasmus Darwin would have to say about the difference between the way things are planned, and the way they actually happen.

Acknowledgements. I should like to thank Prof. Visconti, the international Organizing Committee, and the Director of the Ettore Majorana Centre for Scientific Culture, for inviting me to participate in this interesting meeting in such congenial surroundings. I should also like to thank the many colleagues who have helped me towards some understanding of the middle atmosphere, especially those with whom I have had the pleasure of collaborating at various times over the past decade or more. A.H. McIntyre helped with Fig. 1, and M.N. Juckes kindly supplied Fig. 2.

REFERENCES

Andrews, D.G., 1987: Transport mechanisms in the middle atmosphere: An introductory survey. (In this volume).

Andrews, D.G. and McIntyre, M.E., 1976: Planetary waves in horizontal and vertical shear: The generalized Eliassen-Palm relation and the mean zonal acceleration. J.Atmos.Sci. 33, 2031-2048.

Andrews, D.G. and McIntyre, M.E., 1978a: An exact theory of nonlinear waves on a Lagrangian-mean flow. J.Fluid Mech. 89, 609-646. (See also McIntyre, 1980, for some important caveats.)

Andrews, D.G. and McIntyre, M.E., 1978b: On wave-action and its relatives. J.Fluid Mech. 89, 647-664. Corrigendum 95, 796 (1979).

Austin, J., 1987: Evidence for planetary wave breaking from air parcel trajectories. (In this volume).

Beagley, S.R, and Harwood, R.S., 1987: The residual circulation: Inter-hemispheric differences and heating and eddy components. (In this volume).

Bollobàs, B. (ed.), 1986: Littlewood's Miscellany. Cambridge, 200pp. (p. 194.)

Brasseur, G. and Hitchman, M., 1987: The effect of gravity waves on the distribution of trace species in the middle atmosphere. (In this volume).

Broutman, D. and Young, W.R., 1986: On the interaction of small-scale oceanic internal waves with near-inertial waves. J. Fluid Mech. 166, 341-358.

Butchart, N., 1987: Evidence for planetary wave breaking from satellite data: the relative roles of diabatic effects and irreversible mixing. (In this volume).

Chanin, M.L. and Hauchecorne, A., 1987: Lidar sounding of the structure and dynamics of the middle atmosphere: a review of recent results relevant to transport processes. (In this volume).

Chao, W.C. and Schoeberl, M.R., 1984: On the linear approximation of gravity wave saturation in the mesosphere. J. Atmos. Sci. 41, 1893-1898.

Clough, S.A., Grahame, N.S. and O'Neill, A., 1985: Potential vorticity in the stratosphere derived using data from satellites. Quart. J. R. Meteor. Soc., 111, 335-358. See also Austin (1987) and O'Neill (1987).

Coy, L. and Fritts. D.C., 1987: Gravity waves heat fluxes: a Lagrangian approach. J. Atmos. Sci., to appear

Crighton, D.G., 1981: Acoustics as a branch of fluid mechanics. J. Fluid Mech. 106, 261-298.

Danielsen, E.F., 1968: Stratospheric-tropospheric exchange based on radioactivity, ozone and potential vorticity. J. Atmos. Sci. 25, 502-518.

Dickinson, R.E., 1968: Planetary Rossby waves propagating through weak westerly wind wave guides. J. Atmos. Sci. 25, 984-1002.

Dickinson, R.E., 1969: Theory of planetary wave-zonal flow interaction. J. Atmos. Sci., 26, 73-81.

Dunkerton, T.J., 1980: A Lagrangian mean theory of wave, mean-flow interaction with applications to nonacceleration and its breakdown. Revs. Geophys. Space Phys. 18, 387-400.

Dunkerton, T.J., 1983: Laterally propagating waves in the easterly phase of the quasi-biennial oscillation. Atmos.-Ocean 21, 55-68.

Dunkerton, T.J., Hsu, C.-P. and McIntyre, M.E., 1981: Some Eulerian and Lagrangian diagnostics for a model stratospheric warming. J. Atmos. Sci. 38, 819-843.

Farman, J.C., Murgatroyd, R.J., Silnickas, A.M. and Thrush, B.A., 1985: Ozone photochemistry in the antarctic stratosphere in summer. Quart.J.Roy.Meteorol.Soc. 111, 1013-1025.

Farman, J.C., Gardiner, B.G. and Shanklin J.D., 1985: Large losses of total ozone in Antarctica reveal seasonal ClO_x / NO_x interactions. Nature, 315, 207-210.

Fels S.B., 1985: Radiative-dynamical interactions in the middle atmosphere. In Saltzman and Manabe (1985).

Fels, S.B., 1987: Response of the middle atmosphere to changing O_3 and CO_2 - A speculative tutorial. (In this volume).

Fritts, D.C., 1982: Shear excitation of atmspheric gravity waves. J. Atmos. Sci. 39, 1936-1952.

Fritts, D.C., 1984: Gravity wave saturation in the middle atmosphere: a review of theory and observations. Revs. Geophys. Space Phys. 22, 275-308.

Fritts, D.C., 1987: Recent progress in gravity wave saturation studies. (In this volume).

Fritts, D.C. and Dunkerton, T.J., 1985: Fluxes of heat and constituents due to convectively unstable gravity waves. J. Atmos. Sci. 42, 549-556.

Garcia, R.R. and Solomon, S., 1985: The effect of breaking gravity waves on the dynamics and chemical composition of the mesosphere and lower thermosphere. J. Geophys. Res. 90 D, 3850-3868.

Geller, M.A. and Wu, M.-F., 1987: Troposphere-stratosphere general circulation statistics. In Visconti, 1987 (this volume).

Hamilton, K., 1987: A review of observations of the quasi-biennial and semiannual oscillations of wind and temperature in the tropical middle atmosphere. (In this volume).

Haynes, P.H., 1985: A new model of resonance in the winter stratosphere. Handbook for MAP, vol. 18: Extended abstracts of papers presented at the 1984 MAP Symposium, Kyoto. S. Kato, ed., 126-131. Available from SCOSTEP secretariat, University of Illinois, 1406 W. Green St, Urbana, Ill. 61801, U.S.A.

Haynes, P.H. and McIntyre, M.E., 1987a: On the representation of Rossby-wave critical layers and wave breaking in zonally truncated models. J. Atmos. Sci, to appear

Haynes, P.H. and McIntyre, M.E., 1987b: On the evolution of vorticity and potential vorticity in the presence of diabatic heating and frictional or other forces. J. Atmos. Sci. 44, 828-841.

Holton, J.R., 1982: The role of gravity wave induced drag and diffusion in the momentum budget of the mesosphere. J. Atmos. Sci. 39, 791-799.

Holton, J.R., 1983: The influence of gravity wave breaking on the general circulation of the middle atmosphere. J. Atmos. Sci. 40, 2497-2507.

Holton, J.R., 1975: The dynamic meteorology of the stratosphere and mesosphere, Boston, Massachusetts, American Meteorological Society (Meteorol. Monogr. no. 37), 218pp.

Holton, J.R., 1986: Meridional distribution of stratospheric trace constituents. J. Atmos. Sci. 43, 1238-1242.

Holton, J.R. and Matsuno, T., 1984: Dynamics of the middle atmosphere, pp. 333-351, Terrapub/Reidel, Tokyo.

Hoskins, B.J., McIntyre, M.E. and Robertson, A.W., 1985: On the use and significance of isentropic potential-vorticity maps. Quart. J. Roy. Meteorol. Soc. 111, 877-946. See also 113, 402-404.

Hoskins, B.J., McIntyre, M.E. and Robertson, A.W., 1987: Reply to comments by J.S.A. Green. Quart. J. Roy. Meteorol. Soc. 113, 402-404.

Hsu, C.-P.F. and McIntyre, M.E., 1987: Evidence for self-tuning resonant cavity behavior in model stratospheric warmings with an artificial lower boundary condition. J. Atmos. Sci, to be submitted

Huntley, I., 1977: Spatial resonance of a liquid-filled vibrating beaker. J. Fluid Mech. 80, 81-97.

Juckes, M.N and McIntyre, M.E., 1987: A high resolution, one-layer model of breaking planetary waves in the stratosphere. Nature, submitted.

Kent, G.S., Trepte, C.R., Farrukh, U.O. and McCormick M.P., 1985: Variation in the stratospheric aerosol associated with the north cyclonic polar vortex as measured by the SAM II satellite sensor. J. Atmos. Sci., 42, 1536-1551.

Killworth, P.D. and M.E. McIntyre, 1985: Do Rossby-wave critical layers absorb, reflect or over-reflect? J. Fluid Mech. 161, 449-492.

Leovy, C.B., Sun, C.-R., Hitchman, M.H., Remsberg, E.E., Russell, J.M., Gordley, L.L., Gille, J.C., Lyjak, L.V., 1985: Transport of ozone in the middle stratosphere: evidence for planetary wave breaking. J. Atmos. Sci., 42, 230-244.

Lindzen, R.S., 1981: Turbulence and stress owing to gravity wave and tidal breakdown. J. Geophys. Res. 86, 9707-9714.

Littlewood, J.E., 1953: A Mathematician's Miscellany. Methuen, 136pp. (Pp. 43, 20; also Bollobàs 1986).

Lorenz, E.N. and Krishnamurthy, V., 1987: On the non-existence of a slow manifold. J. Atmos. Sci, submitted.

McCormick, M.P., Trepte, C.R. and Kent, G.S., 1983: Spatial changes in the stratospheric aerosol associated with the north polar vortex. Geophys. Res. Lett. 10, 941-944.

McIntyre, M.E. and Weissman, M.A., 1978: On radiating instabilities and resonant overreflection. J.Atmos.Sci. 35, 1190-1198.

McIntyre, M.E., 1980: Towards a Lagrangian-mean description of stratospheric circulations and chemical transports. Phil. Trans. Roy. Soc. A 296, 129-148. [Special Middle Atmosphere issue.]

McIntyre, M.E., 1981: On the "wave momentum" myth. J. Fluid Mech. 106, 331-347. [On p.339, 6th line from bottom, read "wavelength" for "length of the wavetrain"; also note that the term "quasimomentum" is a well established and acceptable substitute for "pseudomomentum".]

McIntyre, M.E., 1982: How well do we understand the dynamics of stratospheric warmings? J. Meteorol. Soc. Japan 60, 37-65.

McIntyre, M.E. and Palmer, T.N., 1984: The "surf zone" in the stratosphere. J. Atm. Terr. Phys., 46, 825-849.

McIntyre, M.E. and Palmer, T.N., 1985: A note on the general concept of wave breaking for planetary and gravity waves. Pure Appl.Geophys. 123, 964-975.

McIntyre, M.E. and Shepherd, T.G 1987: An exact local conservation theorem for finite-amplitude disturbances to nonparallel shear flows, with remarks on Hamiltonian structure and on Arnol'd's stability theorems. J.Fluid Mech., in the press.

O'Neill, A. and Pope, V.D., 1987: The seasonal evolution of the stratosphere in the northern hemisphere. (In this volume).

Opsteegh, J.D., and A.D. Vernekar, 1982: A simulation of the January standing wave pattern including the effects of transient eddies. J.Atmos.Sci., 39, 734-744.

Plumb, R.A., 1981: Instability of the distorted polar night vortex: a theory of stratospheric warmings. J. Atmos. Sci., 38, 2514-2531.

Plumb, R.A. and McEwan, A.D., 1978: The instability of a forced standing wave in a viscous stratified fluid: a laboratory analogue of the quasi-biennial oscillation. J. Atmos. Sci. 35, 1827-1839.

Remsberg, E.E., 1987: Analysis of the mean meridional circulation using satellite data. (In this volume).

Saltzman, B. and Manabe, S., 1985:   Issues in Atmospheric and Oceanic
     Modelling.   Part A: Climate Dynamics. Advances in Geophysics 28A.
     Academic Press, Orlando, Florida, 591pp.
Shine, K.P., 1987:   The middle atmosphere in the absence of dynamical
     heat fluxes. Q.J.Roy.Meteorol.Soc. 113, in press.
Tung, K.K., 1982:   On the two-dimensional transport of stratospheric
     trace gases in isentropic coordinates. J. Atmos. Sci., 39,
     2330-2355.
Tung, K.K., 1986:   Nongeostrophic theory of zonally-averaged circulation.
     Part 1: formulation. J. Atmos. Sci., 43, 2600-2618.
Tung, K.K., 1987:   A coupled model of zonally averaged dynamics,
     radiation and chemistry. In Visconti, 1987 (this volume).
Uccellini, L.W. and Koch, S.E., 1987:   The synoptic setting and possible
     energy sources for mesoscale wave disturbances. Mon. Wea. Rev., in
     press.
Vincent, R.A., 1987:   Radar observations of gravity waves in the
     mesosphere. (In this volume).
Visconti, G., 1987:   Transport Processes in the Middle Atmosphere (Proc.
     NATO Workshop held in November 1986 at Erice, Sicily. Dordrecht,
     Reidel (this volume).
WMO, 1985:   Atmospheric ozone 1985: Assessment of our understanding of
     the processes controlling its present distribution and change.
     Global Ozone Research and Monitoring Report No. 16.   In 3 volumes,
     1095pp. + 86pp. refs. Available from: Global Ozone Research and
     Monitoring Project, World Meteorological Organization, Case Postale
     5, CH 1211, Geneva 20, Switzerland.

EVIDENCE OF PLANETARY WAVE BREAKING IN THE STRATOSPHERE
USING A PHOTOCHEMICAL MODEL ALONG AIR PARCEL TRAJECTORIES

John Austin
Meteorological Office
London Road
Bracknell
Berkshire, U.K. RG12 2SZ

ABSTRACT. Ten day air parcel trajectories are computed from an
approximate material line (a potential vorticity contour) close to the
edge of the circumpolar vortex on the 850K potential temperature
surface for 27 January 1979. Back trajectories are also calculated
from the Aleutian high present on this day. The results show varied
behaviour and along selected trajectories a photochemical model is
integrated. For those parcels just peeling from the vortex the model
calculations agree reasonably well with near coincident measurements
of O_3, H_2O, HNO_3 and NO_2 from the LIMS instrument. In contrast, for
those parcels transported to the tropics, model O_3 values are lower
than observed while model HNO_3 values are higher than observed. For
those parcels ending in the Aleutian high the model HNO_3 discrepancy
has the opposite sign. It is suggested that the model discrepancies
are due to irreversible mixing taking place in preferred regions,
consistent with the concept of planetary wave breaking.

1.   INTRODUCTION

The dynamical concept of planetary wave breaking in the stratosphere
was first introduced by McIntyre (1982) and developed subsequently by
McIntyre and Palmer (1983, 1984). For example, during late January
1979, air parcels starting in the stratospheric polar vortex peeled
out of the vortex and ended up in a region termed the 'surf' zone by
analogy with the situation of breaking water waves on a shoreline. The
situation is best observed with the aid of maps of isentropic
potential vorticity (IPV maps) which show the phenomenon as developing
from a small wavenumber disturbance (eg wavenumber 1) to higher
wavenumbers and finally is thought to lead to irreversible mixing. The
ubiquitous nature of the event is suggested by the study of Clough et
al. (1985) which show similar dynamical behaviour, again using IPV
maps for the 'Canadian warming' of December 1981. The cascade to

297

G. Visconti and R. Garcia (eds.), Transport Processes in the Middle Atmosphere, 297–312.
© *1987 by D. Reidel Publishing Company.*

lower wavenumbers was thought by McIntyre and Palmer (1984) to be similar in many respects to that occurring in a non-linear Rossby wave critical layer Smith et al., (1984) and Smith (1983) have found important wave-wave effects during January-February 1979 although nonlinear interactions have also been found during quiescent periods (Austin and Palmer, 1984). Wave breaking events can have a major effect on the transport of trace constituents and were also believed to be responsible for the preconditioning of the flow prior to the stratospheric warming of February 1979 (Butchart et al., 1982, Palmer and Hsu, 1983). Rood (1986) cautions that the concept of wave breaking groups different dynamical events together and suggests that a more solid theoretical basis is needed before stratospheric transport can be properly understood. In response McIntyre and Palmer (1986) have defended the wave breaking viewpoint and while admitting that the concept may not be indispensable to understanding the dynamics and chemistry of the middle atmosphere, it has nonetheless advanced our understanding.

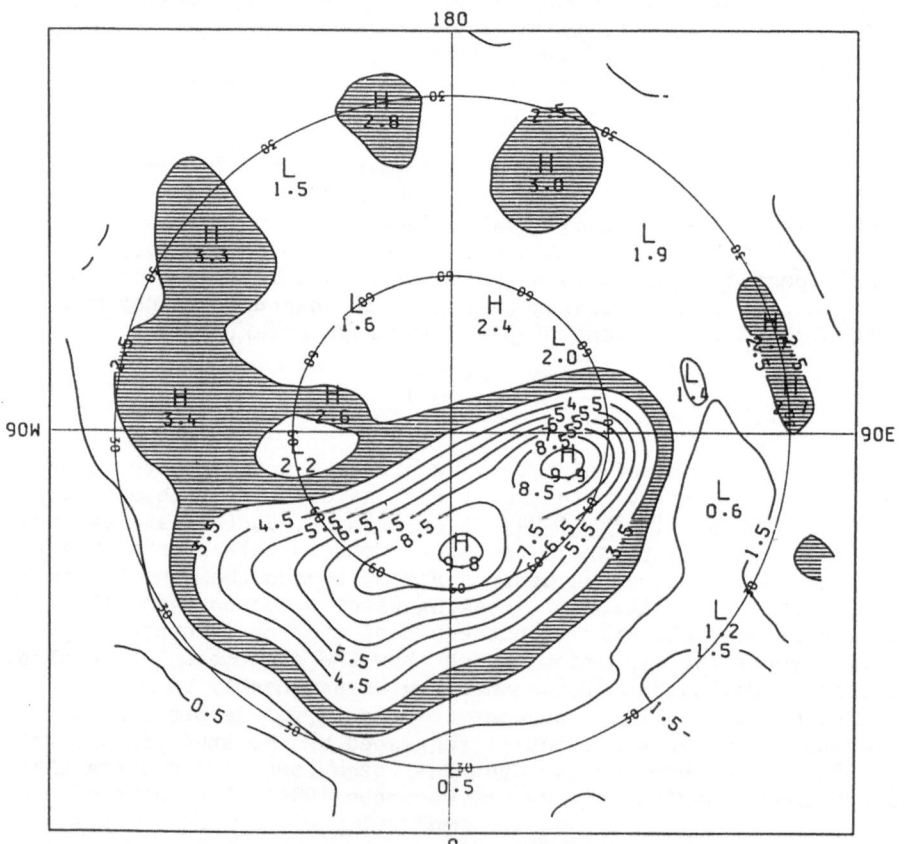

Figure 1.   Isentropic potential vorticity from SSU data in units of
10^{-4} Km^2 kg^{-1}s^{-1}. 27 January 1979 00 GMT.

Using data from the LIMS (Limb Infrared Monitor of the Stratosphere) instrument (Gille and Russell, 1984), Leovy et al. (1985) found regions of irreversible transport of ozone, consistent with planetary wave breaking. Using an area diagnostic for the size of the vortex, Butchart and Remsberg (1986) concluded from the shrinking of the vortex that irreversible mixing was happening almost continuously throughout the winter. This appeared in both potential vorticity and LIMS O_3 and H_2O data but was not very clear from the HNO_3 measurements. For HNO_3 the approach of Leovy et al. (1985) breaks down because of the neglect of photochemistry which can be important even at 30 km where the photochemical timescale is almost comparable to the advective timescale (Brasseur and Solomon, 1984). Also, the method of Leovy et al. (1985) has limitations in that it is not a quantitative technique. The quantitative method of Butchart and Remsberg (1986) could be extended to take account of photochemistry but an alternative method of studying regions of irreversible mixing is with the use of air parcel trajectories. McIntyre and Palmer (1984) have already shown the value of trajectories in clarifying aspects of wave breaking during late January 1979 and recently Dunkerton and Delisi (1986) used trajectories to complement their study of potential vorticity during January and February 1979. Planetary wave breaking appeared to be present during most of the 1978-1979 winter although it was most prominent in the late January period (Butchart and Remsberg, 1986). In this work a photochemical model is integrated along air parcel trajectories (Austin and Tuck, 1985; Austin et al., 1987) to try to clarify the importance of the physical processes occurring in late January 1979. Moreover, by adopting this approach it is possible to take into account the effects of photochemistry which are expected to be particularly important for HNO_3.

2. TRAJECTORY CALCULATIONS

Figure 1 shows the IPV distribution at θ = 850K on 27 January 1979, calculated from Stratospheric Sounding Unit (SSU) data (Clough et al., 1985) with values between 2.5 and 3.5 x 10^{-4} Km^2 kg^{-1} s^{-1} shaded. Note the blobs of high IPV air showing the classical wave breaking picture (McIntyre and Palmer, 1983, 1984). Following Dunkerton and Delisi (1986) trajectories were computed (using SSU data) both forwards and backwards from the IPV contour 2.5, surrounding the vortex. Back trajectories were also computed from the Aleutian high present on 27 January. The technique is that of Austin and Tuck (1985) which has important differences from the geostrophic and isentropic method of Dunkerton and Delisi. Firstly, a radiation scheme is employed to compute cross isentropic flow. Secondly, a gradient wind approximation is employed with a correction for isallobaric effects. These refinements mean that in principle the systematic errors in the trajectory technique are much smaller than in the method of Dunkerton and Delisi (1986), for example. The differences between the trajectory methods, although arguably small in most cases relative to

Table 1
List of trajectory coordinates: all 27 January 1979 0 GMT

Trajectory No.		Lat.	Long.	Description Forwards	Backwards
F1 and	B1	33N	223E	parcel → tropics	vortex peeling
F2	B2	44N	225E	parcels → Aleutian high	....
F3	B3	45N	242E	....	....
F4	B4	55N	254E	....	....
F5	B5	67N	248E	....	comes from Aleutian high
F6	B6	81N	248E	....	comes from vortex edge
F7	B7	81N	140E	parcel → tropics	....
F8	B8	69N	119E	....	joins vortex
F9	B9	56N	110E	stays on vortex edge	....
F10	B10	49N	93E	....	....
F11	B11	49N	78E	....	comes from vortex edge
F12	B12	48N	55E	....	....
F13	B13	47N	37E	....	....
F14	B14	45N	20E	....	....
F15	B15	42N	5E	....	....
F16	B16	37N	353E	parcel → mid latitudes	....
F17	B17	32N	343E	....	joins vortex
F18	B18	34N	332E	....	comes from vortex edge
F19	B19	36N	332E	....	....
F20	B20	32N	309E	....	....
F21	B21	34N	296F	....	....
F22	B22	39N	283E	....	....
F23	B23	32N	271E	....	....
F24	B24	29N	259E	stays on vortex edge	....
F25	B25	33N	248E	parcel → tropics	vortex peeling
F26	B26	28N	237E	....	....
F27	B27	23N	227E	....	....
	B28	50N	210E	completes 1 circuit of vortex	
	B29	50N	230E	completes $^1/_2$	
	B30	50N	250E	completes 1	
	B31	60N	210E	completes $^3/_4$	
	B32	60N	230E	joins Aleutian high from quadrant 0-90E	
	B33	60N	250E	completes $^3/_4$ circuit of vortex	
	B34	70N	210E	joins Aleutian high from quadrant 90-180E	
	B35	70N	230E	....	
	B36	70N	250E	....	

Notation: prefix F refers to forward trajectories prefix B back
trajectories

expected trajectory errors (Austin, 1986) are nonetheless important in
the current work because random errors in the trajectories, which are
a large part of the total error, may be reduced by averaging together
a number of <u>independent</u> trajectories. This will be an important use
of the trajectories in the photochemical model calculations, described
later.

The trajectories are described briefly in Table 1 and in the
following, prefix F refers to forward trajectories and prefix B to
back trajectories. This study is restricted to 3 categories of
trajectories, and in a more extensive paper (Austin and Butchart,
1987), the other 4 categories are also included.

Figure 2 shows the trajectories (B1, B2, B3, B4, B26, B27) which
go around the vortex and just start to peel away from the vortex but
do not quite enter the surf zone. Trajectory B2 is illustrated in
bold with markers every day along the trajectory. Figure 3 shows
those trajectories (F1, F7, F8, F25, F26, F27) which started on the
IPV contour 2.5 and have gone into the tropics with apparently
turbulent motions. Trajectory F8 is marked in bold. Trajectories F1,
F25, F26 and F27 are to the south of the Aleutian high whereas

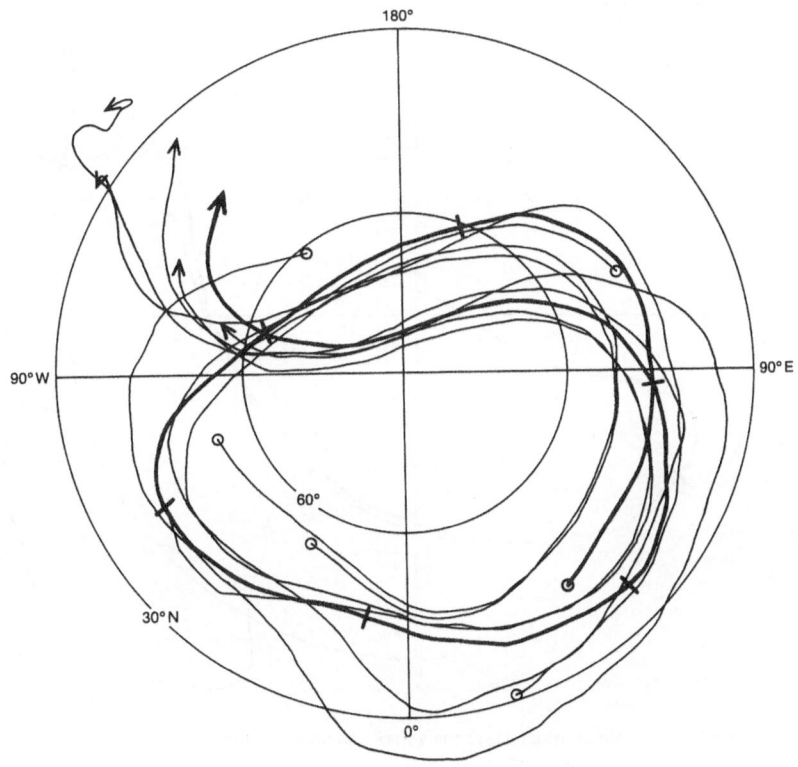

Figure 2.   Trajectories which peel from the vortex. Trajectory B2 is marked in bold.

trajectories F7 and F8 peel from the vortex on the other side of the
hemisphere before going to the tropics. Wave breaking also appears to
be occurring at both sides of the vortex in idealised model
integrations by O'Neill and Pope (1987). Figure 4 shows the back
trajectories from the Aleutian high with trajectory B34 marked in
bold. These show a wide variety of behaviour consistent with the
notion that the anticyclone is in the surf zone or mixed region.

        LIMS profile data were obtained for all points which were within
5 great circle degrees of the trajectory position as discussed in more
detail in Austin et al. (1987), hereafter APPTZ. There were typically
20 such 'near coincidences' per trajectory. Note that the absence of
exact coincidences could cause some of the apparent scatter in the
data (see Section 3) where large gradients in the species exist, for
example on the edge of the vortex.

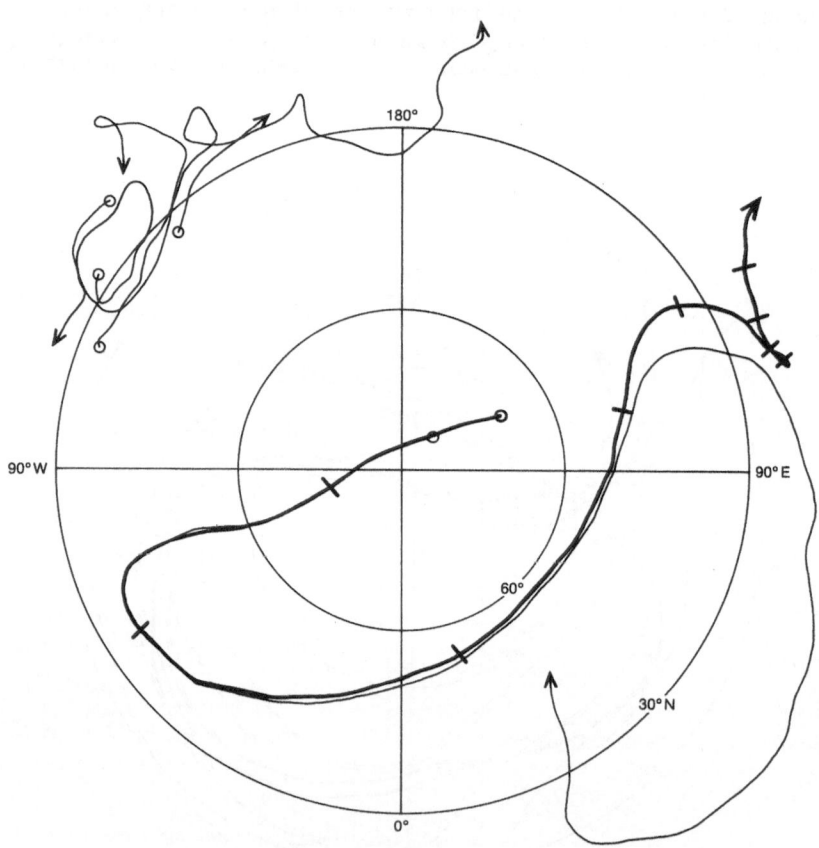

Figure 3.   Trajectories which leave the vortex and go to the tropics.
            Trajectory F8 is marked in bold.

3. PHOTOCHEMICAL MODEL CALCULATIONS

In this work the photochemical model described by APPTZ was integrated along the trajectories of the 3 categories described in the previous section and compared with the LIMS near coincidence data. The model contains the full complexity of the photochemistry, excluding bromine and chlorofluorocarbon reactions which are unimportant for this problem, and incorporates 77 reactions and 32 species. Solar fluxes for the photo-dissociation coefficients are computed using a random band model and spherical geometry is employed to allow for the solar flux at solar zenith angles greater than 90°. The model timestep is 30s during the daytime, 60s at night and 1s at dawn and dusk. Reaction rates are taken from DeMore et al. (1985).

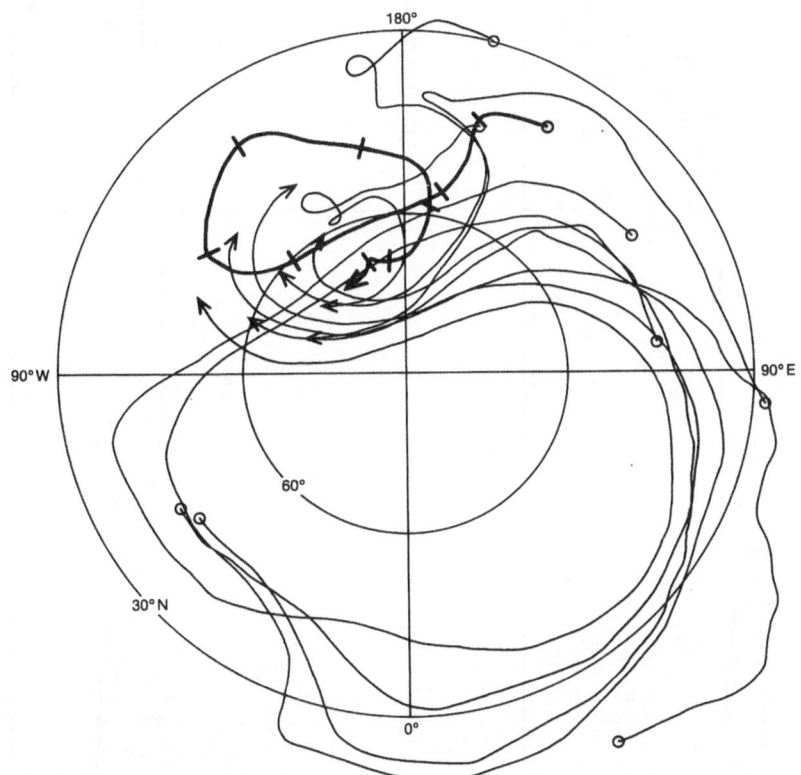

Figure 4. Trajectories which end in the Aleutian high. Trajectory B34 is marked in bold.

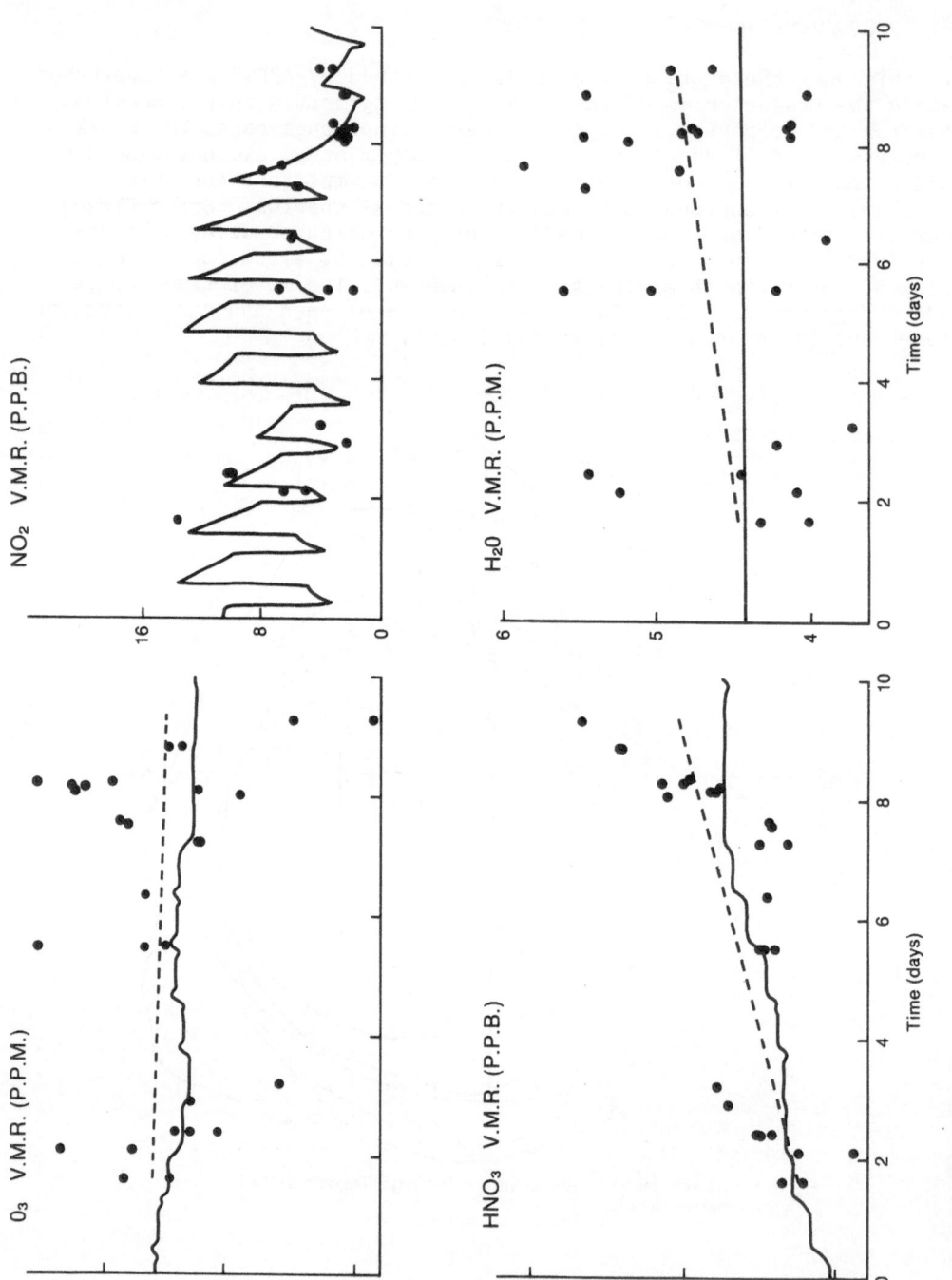

Figure 5.   Chemical model run for trajectory B2.

Table 2

Rates of change of O_3, H_2O (ppm/day) and HNO_3 (ppb/day) in the data and model (rate of change of $H_2O \approx 0$ in the model)

Trajectory	O_3 Data	Model	HNO_3 Data	Model	H_2O Data
(a) Vortex peeling					
B1	−0.034±0.067	−0.036	0.149±0.075	0.122	0.044±0.090
B2	−0.015±0.117	−0.030	0.205±0.107	0.119	0.051±0.107
B3	0.090±0.080	−0.016	0.177±0.055	0.117	0.028±0.081
B4	−0.163±0.121	−0.035	0.261±0.090	0.204	0.024±0.052
B26	0.076±0.126	−0.015	0.167±0.093	0.096	0.023±0.077
B27	0.037±0.067	0.044	0.064±0.058	0.122	0.042±0.072
Mean	0.009±0.035	−0.015	0.153±0.030	0.130	0.032±0.030
$1/_\sigma 2$ weighting					(2.1σ)
(b) Air parcels transported to the tropics					
F1	0.175±0.301	0.120	−0.009±0.130	0.130	−0.001±0.073
F7	0.143±0.084	0.058	−0.087±0.062	0.062	0.030±0.058
F8	0.089±0.091	0.053	−0.037±0.087	0.115	−0.033±0.073
F25	0.069±0.098	0.064	0.037±0.065	0.196	−0.087±0.077
F26	0.147±0.107	0.077	0.060±0.063	0.205	−0.064±0.081
F27	0.302±0.173	0.097	0.017±0.094	0.174	−0.113±0.092
Mean	0.127±0.045	0.078	−0.002±0.031	0.147	−0.032±0.030
$1/_\sigma 2$ weighting		(2.2σ)		(10σ)	(2.1σ)
(c) Parcels arrive at Aleutian high					
B28	0.085±0.103	−0.016	0.109±0.057	0.125	0.029±0.064
B29	−0.302±0.192	−0.108	0.373±0.120	0.154	−0.008±0.072
B30	0.019±0.101	−0.015	0.191±0.062	0.120	0.023±0.052
B31	0.081±0.150	−0.137	0.391±0.057	0.117	0.037±0.077
B32	−0.072±0.086	−0.065	0.279±0.078	0.195	0.053±0.043
B33	−0.241±0.090	−0.071	0.415±0.107	0.199	0.032±0.047
B34	−0.096±0.107	−0.097	0.376±0.060	0.165	0.003±0.042
B35	−0.020±0.042	−0.104	0.457±0.050	0.128	0.052±0.046
B36	−0.090±0.082	−0.082	0.596±0.118	0.201	0.065±0.066
Mean	−0.051±0.027	−0.077	0.330±0.023	0.156	0.032±0.018
$1/_\sigma 2$ weighting			(15σ)		(3.7σ)

3.1. Air parcels peeling from the vortex

Figure 5 shows the results obtained from the chemical model (solid line) for the trajectory B2, illustrated in bold in Figure 2, together with LIMS near coincidences (circles). Linear regression lines have been fitted to the LIMS values for O_3, H_2O and HNO_3 and are shown as pecked lines in the diagram. Both model and data for O_3 agree quite well. For HNO_3 the model and data both increase but the model indicates a less rapid increase than the data. The H_2O data indicate an increase but it is not statistically significant. Model NO_2 agrees very well with the data, in agreement with the study of Solomon et al. (1986), particularly during the extended period of darkness at about day 8.

For O_3, H_2O and HNO3, the mean rates of change along the trajectories have been determined both from the model and from the regression lines through the data and the results for all the air parcels staying on the edge of the vortex are presented in Table 2a. 95% confidence intervals are given for the data and are computed from linear regression statistics (see, e.g., Bentat and Piersol, 1971) so conceptually include random errors and systematic errors excluding bias errors (see APPTZ for a discussion of LIMS errors in the context of trajectory modelling). Model results which disagree with the data are underlined. Similar behaviour of the model is noted for all the trajectories in this category. The mean for the category shows agreement between model and data for both O_3 and HNO_3 but there is a slight discrepancy in H_2O.

3.2. Air parcels transported to the tropics

Figure 6 shows the results obtained from the chemical model together with LIMS observations for the trajectory F8, illustrated in bold in Figure 3. The data show an increase in O_3 and the model, although increasing O_3 slightly, has a smaller rate of change than the data. For HNO_3 the model shows a substantial increase whereas the data decrease. There is also a decrease in H_2O although this is not statistically significant. Note the anticorrelation in the model discrepancies for O_3 and HNO_3 and the correlation in the discrepancies for HNO_3 and H_2O. The NO_2 model values are in quite good agreement with the data. Similar behaviour may also be noted for the other trajectories in this category (see Table 2b). In the mean the model discrepancies are significant for all three species O_3, HNO_3 and H_2O.

3.3. Air parcels arriving at the Aleutian high

The 9 trajectories in this category are illustrated in Figure 4 and the chemical model results for trajectory B34 are illustrated in Figure 7. The data show a substantial increase in HNO_3 and a large decrease in O_3. In comparison, the model change in HNO_3 is much smaller although the model change in O_3 is similar to the data.The agreement between model and data for NO_2 is again very good,

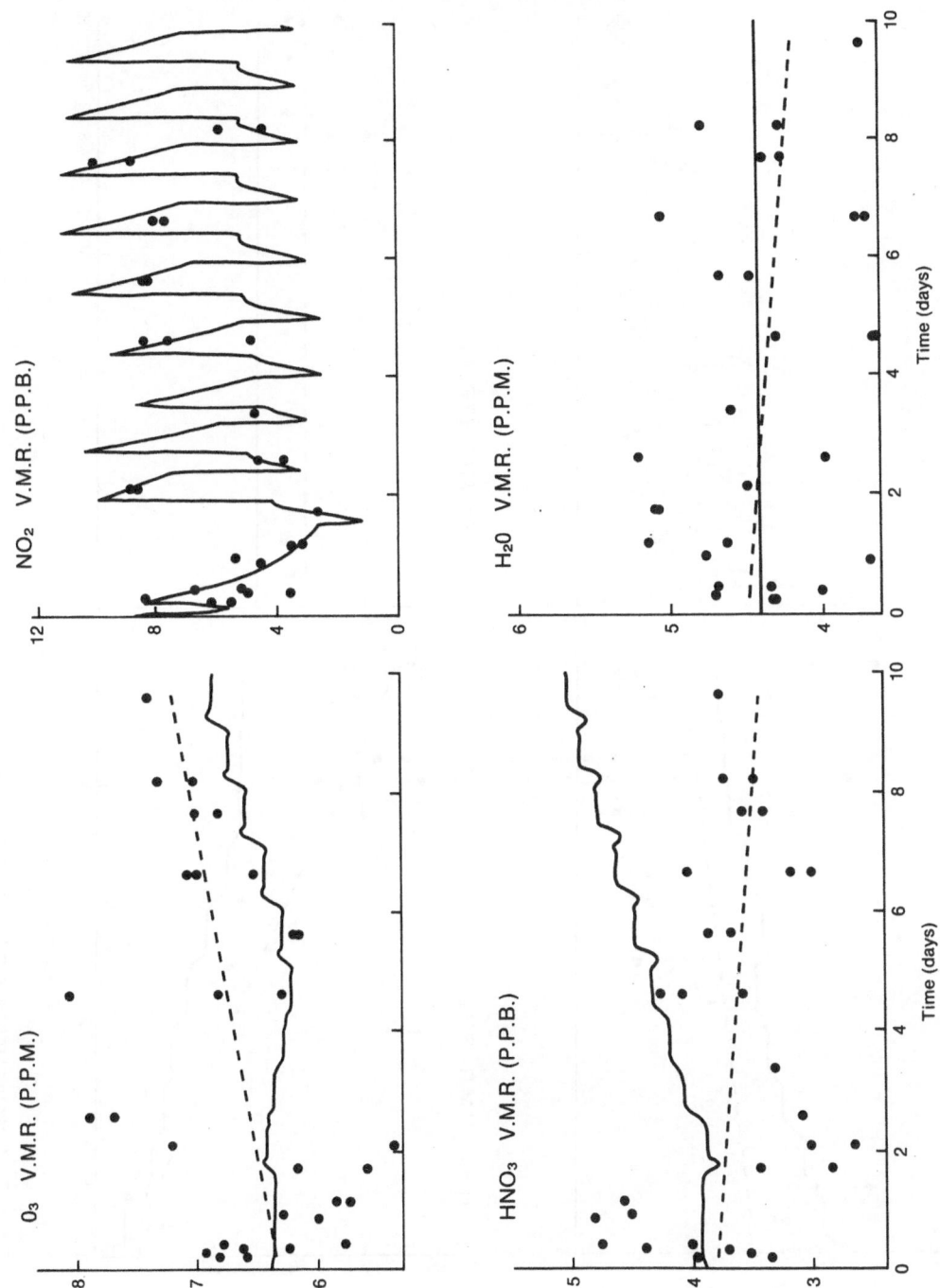

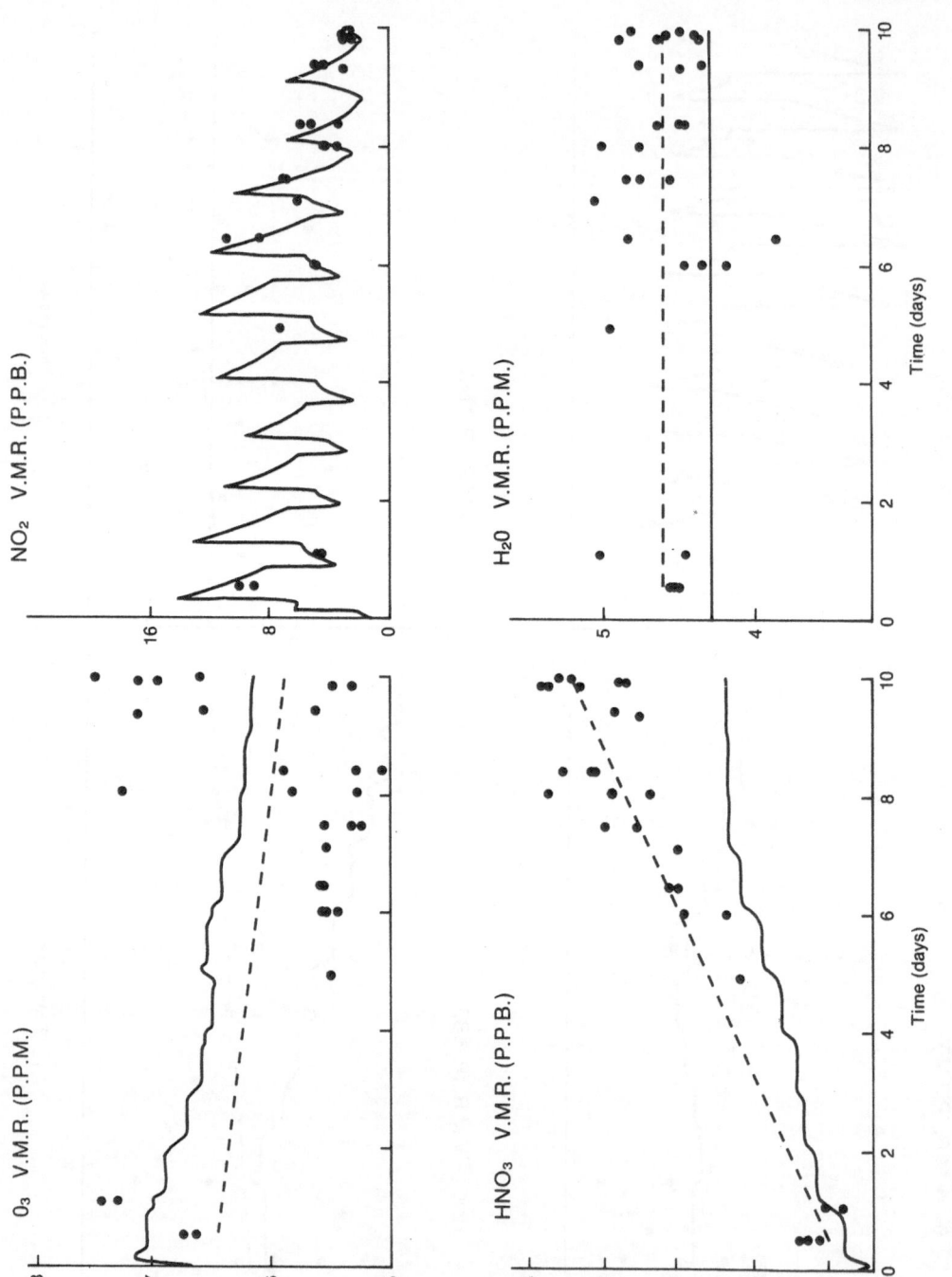

Figure 7.   Chemical model run for trajectory B34.

particularly in the extended period of darkness in the last few days
of the trajectory. The varied nature of the trajectories (Figure 4)
gives rise to varied results both in the chemical model and in the
data, as summarised in Table 2c. Trajectories B29 and B33 have an
anticorrelation between the model discrepancies for HNO_3 and O_3.
Trajectories B31 and B35 give a correlation between the discrepancies.
For the remaining trajectories the results are quite variable,
although in all but trajectory B28 HNO_3 increases at a much faster
rate in the data than in the model. In the mean the model rate of
change of HNO_3 is less than half that observed and H_2O also shows a
significant increase in the data while the difference between model
and data for O_3 is very close to the 95% confidence limit. Thus there
is strong evidence for the presence of unaccounted for
non-conservative processes acting along these trajectories.

4. DISCUSSION

The results of the previous section show strong evidence of
non-conservative processes (in addition to photochemistry) taking
place in preferred regions. It is suggested that these
non-conservative processes can be attributed to irreversible mixing.
However, the results depend critically on the errors associated with
the trajectory calculations themselves. By comparing LIMS and SSU
trajectories Austin (1986) gave an estimate of the absolute errors in
the trajectories and in 10 days these can be quite substantial. The
results of Austin (1986) suggested that SSU trajectory errors may be
smaller than LIMS errors during disturbed conditions because
horizontal resolution is generally more important than vertical
resolution. Consequently, SSU trajectories were used in this paper.
However, that does not preclude the fact that some trajectories may be
substantially in error and that choosing the trajectory categories by
appearance may have inadvertently collected together those
trajectories which have the same type of error. This may have
invalidated the crucial argument that trajectory grouping is used to
reduce the net random error. In practice rigorous defence of the
trajectory accuracy is impossible. However evidence that the
trajectories are adequate comes from the LIMS species which, in
general, show smooth changes whereas large trajectory displacement
errors would be expected to occur in a jerky fashion (see eg Austin,
1986) leading to sudden changes in species mixing ratios. Secondly,
trajectories from both LIMS and SSU data show similar physical
characteristics even though the precise details are different for
individual trajectories. Finally, trajectories with similar
characteristics occur in groups eg F2-F6, F7-F8, F9-F15 etc., showing
that small displacements in the initial positions do not appreciably
affect the results.
     Another area for concern is the performance of the chemical
model. APPTZ showed quite good agreement between the LIMS data and the
model although only 2 trajectories were studied at the 850K level.
The APPTZ results indicated a possible model bias relative to LIMS

data of order 5% in O_3 and 25% in HNO_3, which could produce a spurious rate of change of up to 0.06 ppm/day for O_3 and up to 0.1ppb/day for HNO_3. The precise cause of this bias is uncertain; it may even be a reflection of the bias error in the data itself. A solution to this dilemma is perhaps to consider only those situations where consistent behaviour is noted in the discrepancies between model and data. To do this it may be noted that air from the vortex contains a high concentration of H_2O and HNO_3 and a low concentration of O_3 and air from the tropics has the opposite characteristics. Thus air parcels which start on the edge of the vortex and mix in the tropics (category b) see an increase in O_3 and a decrease in HNO_3 and H_2O relative to a chemical model.

The trajectories arriving at the Aleutian high, category (c), show varied behaviour but typically start in low latitudes and thus transport high O_3 and low HNO_3 into the region. However by 27 January a clear maximum in O_3 is established in the Aleutian high so the model discrepancy for O_3 is not significant. However, the model discrepancies for H_2O and HNO_3 are such that substantial mixing must be occurring in the region surrounding the Aleutian high. Those trajectories which peel from the vortex (category a) would not be expected to show significant discrepancies between model and data because the air parcels have not yet reached the region of apparent turbulence in the 'surf zone'and indeed O_3 and HNO_3 are approximately conserved when allowance is made for photochemistry. However the H_2O change is just significant at the 95% level so it is possible that there is a small systematic error remaining in the trajectories. It is of interest to note that APPTZ found HNO_3 rates of change to be much greater in the model than in the data and it is possible that further analysis may reveal the cause of these differences - for example the calculation of odd hydrogen in the model.

ACKNOWLEDGEMENTS

I would especially like to thank N Butchart for his critical review of the paper and I would also like to thank R L Jones, A O'Neill, V D Pope and E E Remsberg for their many helpful suggestions for improvements to the paper.

REFERENCES

Austin, J., 1986: Comparison of stratospheric air parcel trajectories calculated from SSU and LIMS satellite data. J. Geophys. Res. **91** 7837-7851.
Austin, J., and N. Butchart, 1987: Evidence of planetary wave breaking in the stratosphere from Lagrangian chemical model calculations. In preparation.

Austin, J., R.C. Pallister, J.A. Pyle, A.F. Tuck and A.M. Zavody, 1987: Photochemical model comparisons with LIMS observations in a stratospheric trajectory coordinate system. Quart. J. Roy. Meteor. Soc., 113, 361-392.

Austin, J., and T.N. Palmer, 1984: The importance of nonlinear wave processes in a quiescent winter stratosphere. Quart. J. Roy. Meteor. Soc., 110, 289-301.

Austin, J., and A.F. Tuck, 1985: The calculation of stratospheric air parcel trajectories using satellite data. Quart. Roy. Meteor. Soc. 111, 279-307.

Bendat, J.S., and A.G. Piersol, 1971: Random Data: Analysis and Measurement Procedures. Wiley.

Brasseur, G., and S. Solomon, 1984: Aeronomy of the Middle Atmosphere, Reidel.

Butchart, N., S.A. Clough, T.N. Palmer and P.J. Trevelyan, 1982: Simulations of an observed stratospheric warming with quasigeostrophic refractive index as a model diagnostic. Quart. J. Roy. Meteor. Soc., 108, 475-502.

Butchart, N., and E.E. Remsberg, 1986: The Area of the stratospheric polar vortex as a diagnostic for tracer transport on an isentropic surface. J. Atmos. Sci., 43, 1319-1339.

Clough S.A., N.S. Grahame and A. O'Neill, 1985: Potential vorticity in the stratosphere derived using data from satellites. Quart. J. Roy. Meteor. Soc., 111, 335-358.

DeMore, W.B., J.J. Margitan, M.J. Molina, R.T. Watson, D.M. Golden, R.F. Hampson, M.J. Kurylo, C.J. Howard and A.R. Ravishankara, 1985: Chemical kinetics and photochemical data for use in stratospheric modelling. Evaluation number 7, JPL, Pasadena.

Dunkerton, T.J. and D.P. Delisi, 1986: Evolution of potential vorticity in the winter stratosphere of January-February 1979. J. Geophys. Res., 91, 1199-1208.

Gille, J.C., and J.M. Russell III, 1984: The Limb Infrared Monitor of the Stratosphere: Experiment description, performance and results. J. Geophys. Res. 89, 5125-5140.

Leovy, C.B., C-R. Sun, M.H. Hitchman, E.E. Remsberg, J.M. Russell III, L.L. Gordley, J.C. Gille and L.V. Lyjak, 1985: Transport of ozone in the middle stratosphere: evidence for planetary wave breaking. J. Atmos. Sci., 42, 230-244.

McIntyre, M.E., 1982: How well do we understand the dynamics of stratospheric warmings? J. Met. Soc. Japan, 60, 37-65.

McIntyre, M.E., and T.N. Palmer, 1983: Breaking planetary waves in the stratosphere. Nature, 305, 593-600.

McIntyre, M.E., and T.N. Palmer, 1984: The 'surf zone' in the stratosphere. J. Atmos. Terr. Phys., 46, 825-849.

McIntyre, M.E., and T.N. Palmer, 1986: A note on the general concept of wave breaking for Rossby and gravity waves. Pageoph., 123, 964-975.

O'Neill, A., and V.D. Pope, 1987: Linear and nonlinear perturbations in the stratosphere. In preparation.

Palmer, T.N., and C-P.F. Hsu, 1983: Stratospheric sudden coolings and
    the role of nonlinear wave interactions in preconditioning the
    circumpolar flow. J. Atmos. Sci., **40**, 909-928.
Rood, R.B., 1986: A critical analysis of the concept of planetary
    wave breaking. Pageoph., **123**, 733-755.
Smith, A.K., 1983: Observation of wave-wave interactions in the
    stratosphere. J. Atmos. Sci., **40**, 2484-2496.
Smith, A.K., J.C. Gille and L.V. Lyjak, 1984: Wave-wave interactions
    in the stratosphere: observations during quiet and active
    wintertime periods. J. Atmos. Sci., **41**, 363-373.
Solomon, S., J.M. Russell III and L.L. Gordley, 1986: Observations of
    the diurnal variation of nitrogen dioxide in the stratosphere. J.
    Geophys. Res., **91**, 5455-5464.

THE PRODUCTION OF TEMPORAL VARIABILITY IN TRACE CONSTITUENT CONCENTRATIONS

J. R. Holton
Atmospheric Sciences, AK-40
University of Washington
Seattle, WA, 98195
USA

ABSTRACT. The production of tracer variability by inertia-gravity waves is contrasted with that due to quasi-geostrophic motions. High frequency variability in ozone in the lower stratosphere is often highly correlated with fluctuations in potential temperature, and can be interpreted as evidence for vertical displacements of material surfaces by inertia gravity waves. Examples are shown from recent aircraft measurements.

Tracer variability in the summer stratosphere, as deduced from balloon based measurements of long-lived tracers, is not associated with significant potential temperature oscillations. Such variability may be generated by linearly polarized normal mode oscillations (in which parcel trajectories nearly coincide with the mean isentropes), or may be the result of "frozen in" variance generated by irreversible parcel displacements occurring during the breakdown of the polar vortex in the springtime final warming.

1. INTRODUCTION

Individual vertical profiles of long-lived tracers measured by balloon and horizontal profiles measured by research aircraft generally exhibit substantial deviations from the temporal and spatial mean profiles. Such variability is usually attributed to "transport", but the nature of the transport processes responsible for the variability has not always been appreciated. Observations of the variance of a single tracer are in general difficult to interpret. It is sometimes not even possible to deduce whether observed variability reflects real atmospheric variability, or is simply the result of measurement error. Simultaneous observations of several tracers (both chemical and dynamical) can, however, often be used to infer the probable source of observed variance. This paper compares and contrasts the tracer variability reported in high altitude aircraft observations during the Stratosphere-Troposphere Exchange Project (STEP) with that observed in balloon soundings by Ehhalt et al. (1983). Such a comparison suggests that in the former case mesoscale inertia-gravity waves may account for

G. Visconti and R. Garcia (eds.), Transport Processes in the Middle Atmosphere, 313–326.

much of the observed variance, while in the latter case it is necessary
to invoke variance production by planetary waves as a major source of
variability.

2.   EQUIVALENT DISPLACEMENT HEIGHT (EDH)

Ehhalt et al. (1983) introduced the concept of an equivalent displacement height (EDH) as a normalized measure of the variability of a
vertically stratified tracer. As shown in figure 1, the EDH is defined
as

$$\Delta = \sigma/(\partial M/\partial z) \tag{1}$$

Here, M is the tracer mixing ratio and σ, the standard deviation of M,
can be interpreted as the distance through which the mean tracer profile
must be displaced vertically in order to account for the observed variance of a conservative tracer. For a hypothetical tracer whose surfaces
of constant mixing ratio coincide with surfaces of constant altitude,
variability can only be produced by <u>vertical</u> displacements and the EDH
is then simply the average amplitude of the vertical displacements generated by various types of motion systems. For long-lived chemical

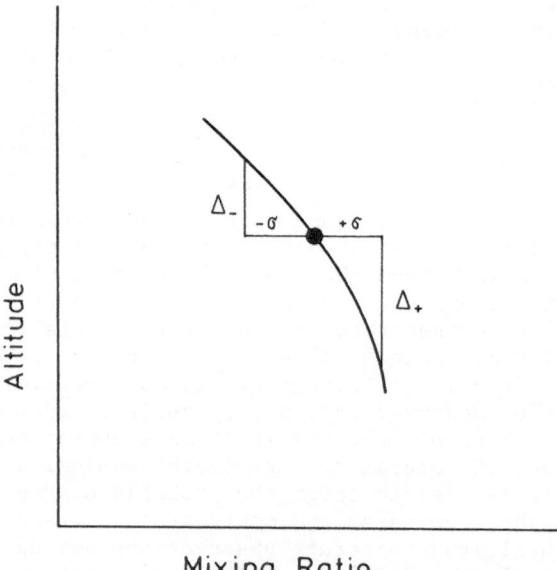

Fig. 1.   Geometry of the equivalent displacement height, Δ.   From Ehhalt
et al., 1985.

tracers of tropospheric origin, however, surfaces of constant mixing
ratio tend to slope poleward and downward relative to surfaces of constant height, and also to surfaces of constant potential temperature
(see figure 2). For such tracers Ehhalt et al. recognized that the
tracer variability characterized by the EDH need not be produced by
vertical displacements, but can be produced by quasi-horizontal parcel
displacements. For a meridional displacement, δy, the EDH can then be
represented as follows:

$$\Delta \equiv \delta y \ (dz/dy)_M \tag{2}$$

where $(dz/dy)_M$ is the slope of the mean tracer mixing ratio surface in
the meridional plane. The role of horizontal parcel displacements in
tracer variance production has been emphasized by Mahlman et al. (1986),
who suggested that the EDH should be thought of as a "scaled displacement length". Either vertical or horizontal displacement may dominate
in the variance production for a given situation depending on the ratio
of the slope of parcel displacements to the slope of the tracer mixing
ratio surface. For high frequency motions, such as inertia gravity
waves, this ratio tends to be large and vertical displacements are

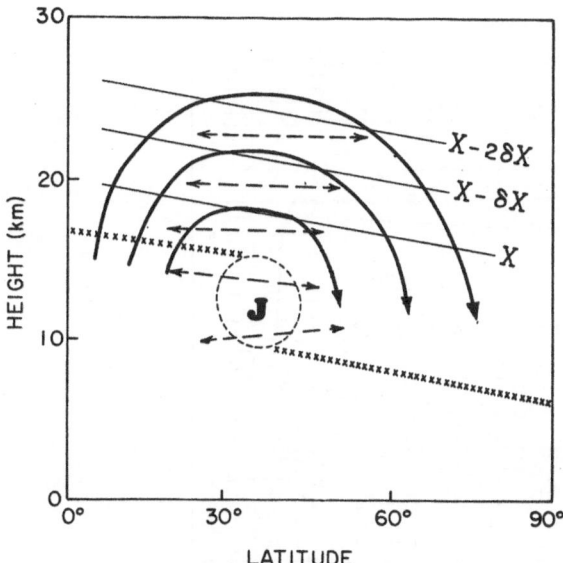

Fig. 2. Schematic illustration of tracer mixing ratio surface slopes in
the meridional plane relative to parcel displacements by quasi-isentropic
planetary waves (dashed arrows). Heavy arrows show the sense of the
diabatic mean meridional flow. Crosses mark the position of the tropopause.

expected to dominate in the production of tracer variance. For quasi-geostrophic planetary scale motions the ratio of vertical to horizontal parcel displacements depends on the baroclinity of the system. For the deep vertical scales characteristic of stratospheric waves, however, it is the horizontal displacements that should dominate in production of tracer variance.

3. MESOSCALE TRACER VARIABILITY

During April of 1984 the NASA stratosphere-troposphere exchange project (STEP) conducted a series of flights of the NASA U-2 research aircraft with the goal of investigating the transfer of air between the stratosphere and the troposphere in the vicinity of the subtropical jet-stream. On each flight in this campaign the U-2 measured several trace constituents (ozone, water vapor, condensation nuclei) and meteorological parameters (Temperature, pressure, horizontal winds) with high frequency response sensors. Horizontal flight paths of several hundred kilometers length were flown orthogonal to the direction of the jet-stream at several levels in the lower stratosphere above the jet core (see figure 3). In addition, vertical profiles were obtained during aircraft climb and descent.

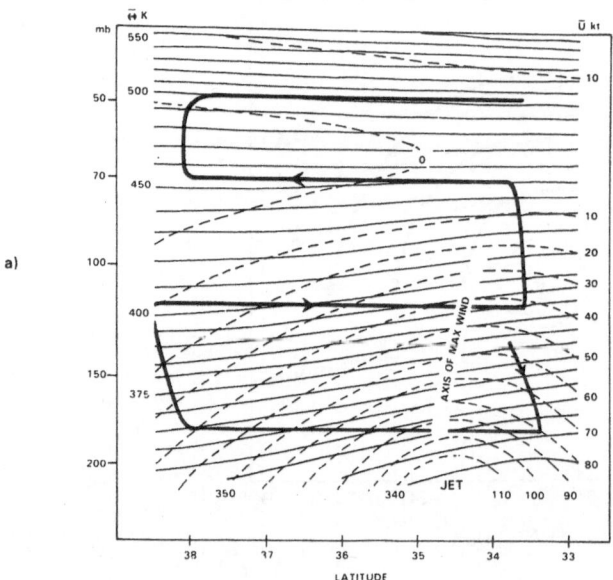

Fig. 3. Vertical cross-sections of low-pass filtered potential temperature (thin solid lines) and mean zonal wind (dashed lines; shown in knots) for the U-2 flight of April 20, 1984 (flight track shown by thick solid line). From WMO (1986).

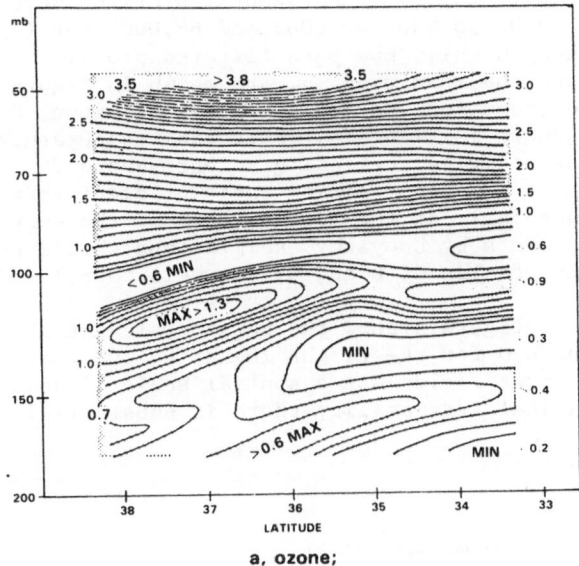

a, ozone;

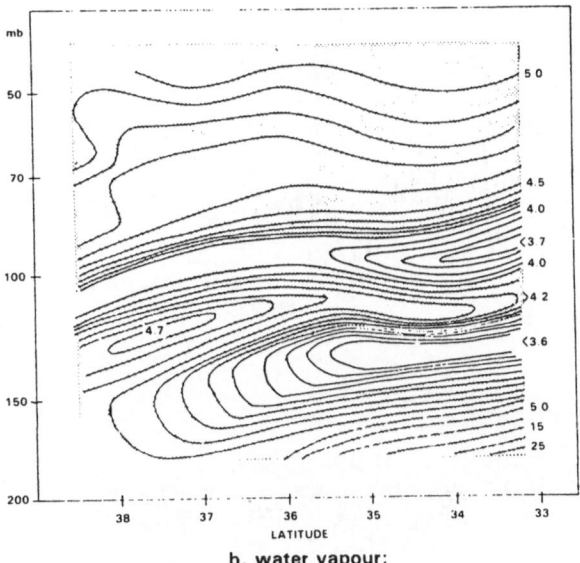

b, water vapour;

Fig. 4. Vertical cross-sections from low pass filtered data showing distributions of (a) O_3 mixing ratio and (b) H_2O mixing ratio, both in ppmv. From WMO (1986).

On the flight of April 20, 1984, horizontal profiles were obtained
at flight levels of 41,000, 50,000, 60,000, and 68,000 feet (approximately 180, 120, 70, and 50 mb). Low-pass filtered profiles of potential temperature, θ, and the mean wind, u, are shown in figure 3. The
corresponding cross-sections for the mixing ratios of ozone, O_3, and
water vapor, H_2O, are shown in figure 4. Potential temperature and
water vapor have strong positive vertical gradients above the 100 mb
level. Water vapor, however, has a mixing ratio gradient that is rather
weak and irregular, especially at the northern end of the section.
Thus, vertical displacements by buoyancy oscillations should produce
variability in which the fields of θ and O_3 are well correlated but H_2O
is less so.

The three panels of figure 5 show the horizontal distributions of
the fields of θ, O_3 and H_2O for the flight leg at 60,000 feet. The primary variation in all three fields has a spatial scale of about 100 km.
Superposed on this mesoscale variability there is substantial varia-

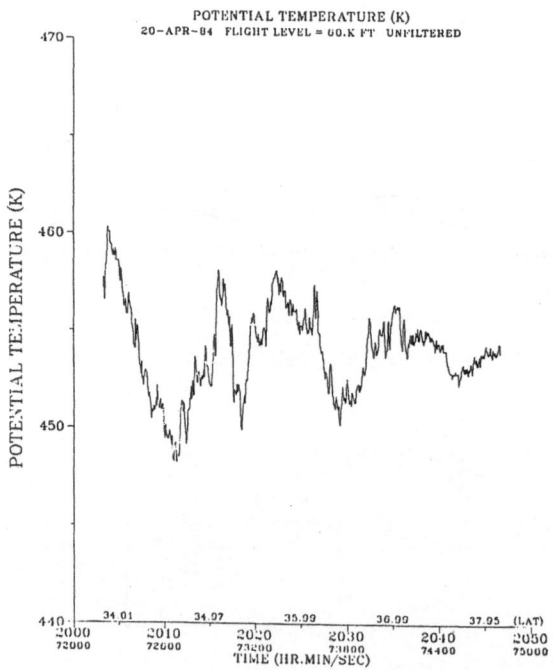

Fig. 5. Observations of potential temperature (a), ozone mixing ratio
(b), and water vapor mixing ratio (c) for the flight leg at 60 kft.
(70 mb) on April 20, 1984. (Courtesy of E. F. Danielsen).

bility at smaller scales. Inspection of figure 5 suggests that there is
a substantial positive correlation between the θ and O_3 fields, while
neither is strongly correlated with the H_2O field.

Since both ozone and potential temperature increase rapidly with
height in the neighborhood of 70 mb (see figure 3), it is plausible to
assume that the correlated behavior in these fields is caused by vertical oscillations associated with inertia-gravity oscillations. To provide some support for this hypothesis, we note that at the 70 mb level
the variance in potential temperature is about 3 K while the vertical
gradient is about 20 K/km; the variance in ozone mixing ratio is about
150 ppmv while the vertical gradient is about 1.0×10^{-3} ppm/km. Substituting these values into (1) we find that in both cases the observed
mesoscale tracer variability can be accounted for by vertical oscillations of amplitude 150 m. As mentioned in section 2, the tracer variance is generated primarily by vertical particle displacements for
motions in which the slope of the parcel trajectories exceeds the mean
slope of constant mixing ratio surfaces. Danielsen (personal communication) has suggested that the variability shown in figure 5 in the 3-22
minute period range is most likely produced by parcel displacements
associated with upward propagating inertia gravity waves with vertical
wavelengths of about 5 km.

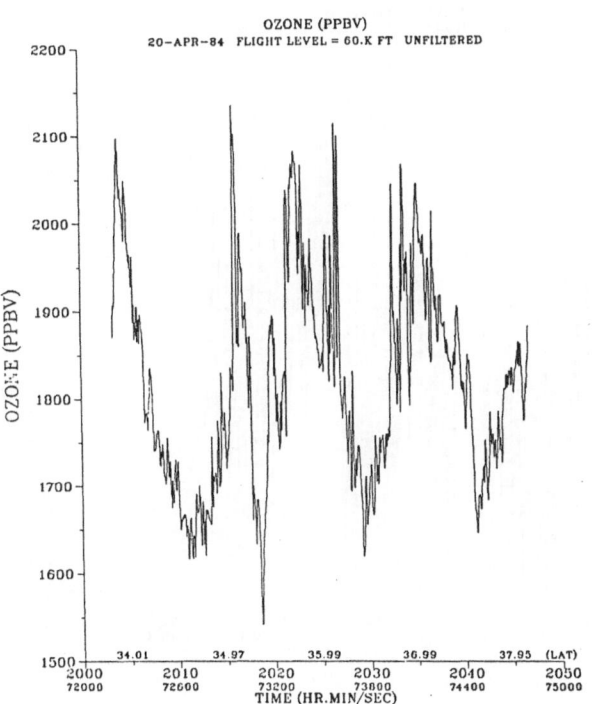

Fig. 5b

For such waves, particle orbits would resemble ellipses in a tilted
plane (see e.g., Gill, 1982, figure 8.4), and would be capable of producing tracer variability for either vertically or horizontally stratified conservative tracers.

For the approximate horizontal scale of 100 km and vertical scale
of 5 km that seem to characterize the observed variability, the slopes
of parcel oscillations will be about 0.05, which is much greater than
slopes typical for ozone mixing ratio surfaces or constant potential
temperature surfaces.

Because it has a weak vertical gradient, the water vapor distribution at 70 mb is not strongly affected by vertical parcel displacements.
Thus, it is not surprising that the water vapor variability is only
weakly correlated with the potential temperature or ozone variability.
Other data collected on this same flight indicated, however, that water
vapor does have significant horizontal gradients in this region. These
observations are consistent with the global water vapor observations
from the LIMS experiment, which show that in the zonal mean there is a
strong meridional gradient and very little vertical gradient in water
vapor mixing ratios near 70 mb in midlatitudes (Remsberg et al., 1984).
Thus, it is possible that the observed variability at the 60,000 ft.

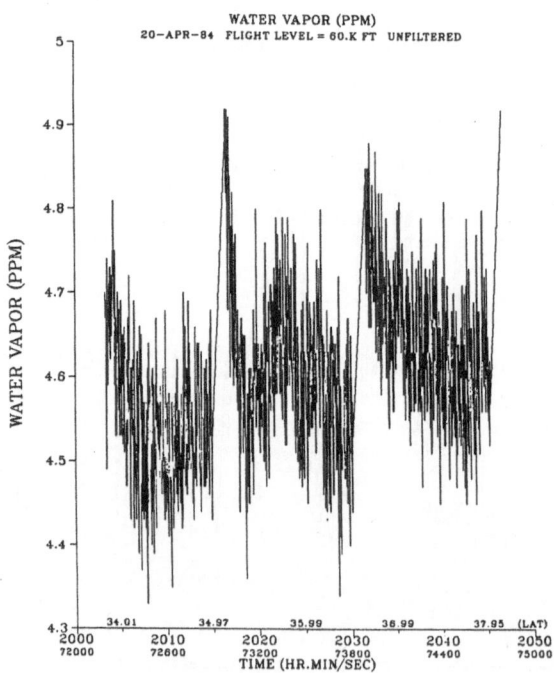

Fig. 5c

level is due to mesoscale horizontal deformation fields providing differential parcel motions in the horizontal plane in the presence of
horizontal gradients. Although there is not much evidence of correlation between the observed horizontal wind variability and the water
vapor variations at 60,000 ft., it is interesting to note that the
variations of water vapor at 60,000 and 68,000 ft. are quite similar in
form. At 68,000 ft. there is a significant correlation between the
water vapor variability and the shear vorticity $(-\partial u/\partial y)$. This suggests that horizontal deformation fields have acted to concentrate
existing gradients of both water vapor and zonal wind.

The limited results presented here do not, of course, conclusively
prove that observed mesoscale variability in ozone and potential temperature can generally be accounted for in terms of inertia-gravity
waves. In reality the flow is seldom dominated by monochromatic waves.
Furthermore, tracer variability is related directly to parcel displacements, rather than perturbation velocity fields. The former can presumably become quite complex when waves of different frequencies are
superposed. Nevertheless for the STEP flight of April 20, 1984 production of tracer variance by inertia gravity waves provides the best
explanation for the behavior of the vertically stratified tracers above
the 100 mb level.

4. LARGE SCALE TRACER VARIABILITY IN THE SUMMER STRATOSPHERE

Ehhalt et al. (1983) originally developed the equivalent displacement height concept in an effort to demonstrate certain universal behavior of the variability in long-lived trace gases for which a number of
vertical profiles were available from balloon measurements in the summer
stratosphere over Southern France. They found that the EDHs obtained
for CH_4, N_2O, $CFCl_3$, and CF_2Cl_2 were all quite similar in magnitude and
vertical variation, with a range of 1 - 2 km in the lower and middle
stratosphere. The EDH derived from ozonesonde observations showed similar behavior; but the EDH for potential temperature was an order of magnitude smaller (i.e., similar in magnitude to that found in the STEP
observations discussed above).

Because of the discrepancy between the trace gas EDHs and the EDH
for potential temperature, it is clear that the variability measured by
Ehhalt et al. (1983) can not be attributed to vertical parcel displacements associated with waves in the summer stratosphere. Rather, the
variance must be due to quasi-horizontal parcel displacements. For tracer mixing ratio slopes typical of those associated with the long-lived
gases measured by Ehhalt et al., horizontal parcel displacements of
order 1000 km or greater would be required to produce the observed EDHs.

It is not clear at present whether the amplitude of wave activity
in the summer is large enough to provide meridional parcel displacements
of the required magnitude. The vertically propagating planetary waves
that dominate the eddy motion field in the winter are blocked from propagation into the summer stratosphere (Charney and Drazin, 1961); synoptic scale waves decay very rapidly above the tropopause at all seasons.
The large scale eddy variance in the summer stratosphere is apparently

due mostly to certain normal mode oscillations, particularly the 5 day
wave and the 16 day wave both of which have zonal wavenumber one structure (Speth and Madden, 1983). Such modes are equivalent barotropic;
they have parcel displacements that are nearly parallel to the mean
isentropes. Thus, they should produce variance in those long lived
tracers that have gradients on isentropic surfaces, but should not produce variance in the potential temperature field. Although this is in
qualitative agreement with the large differences between the EDHs for
chemical tracers and the EDH for potential temperature that were reported by Ehhalt et al., the magnitudes of parcel displacements likely
to be associated with these modes appear to be too small to account for
the observed tracer variances.

The meridional parcel displacements produced by the 5 and 16 day
wavenumber one oscillations can be estimated from linear wave theory.
From the geostrophic wind approximation,

$$v' = g \ f^{-1} \partial Z'/\partial x \tag{3}$$

where f is the Coriolis parameter, g is gravity, and Z' is the disturbance geopotential height. But v' is related to the parcel displacement
η' as

$$v' = d\eta'/dt \cong \hat{U}\partial\eta'/\partial x \tag{4}$$

where $\hat{U}$ is the Doppler shifted mean zonal wind. Combining (3) and (4)
yields

$$|\eta'| \approx |g \ f^{-1}Z'\hat{U}^{-1}| \tag{5}$$

The analysis of Speth and Madden (1983) can be used to estimate the geopotential height disturbance amplitudes of the 5 day and 16 day waves at
40 N latitude in the summer. They are approximately 8 and 10 m, respectively. For a mean zonal wind of -12 m s^{-1} (typical of midlatitude 10
mb conditions during summer) the corresponding Doppler shifted mean
winds are about 50 m s^{-1} for the 5 day wave and 8 m s^{-1} for the 16 day
wave. Application of (3) then yields meridional displacement amplitudes
of approximately 16 km and 125 km for the 5 and 16 day waves, respectively. Thus, it appears that the observed planetary scale normal modes
do not have sufficient amplitude to produce the observed tracer variances in the summer stratosphere.

There remains the possibility that medium scale or synoptic scale
disturbances might have sufficient amplitude even in summer to account
for the observed tracer variances. The quality of stratospheric analyses at present is not sufficient to either support or refute this possibility. However, there is indirect evidence from the behavior of the
aerosol layer produced by the eruption of El Chichon, which suggests
that meridional parcel displacements in the summer stratosphere must be
very small at midlatitudes. The April 1982 eruption of El Chichon produced dramatic increase in the aerosol loading concentrated in a zonal
ring at about the 24 km level and in a latitude band extending from the
equator to 30 N. During the entire summer period there was virtually no

penetration of the aerosol poleward of 30 N; there was also very little vertical spread of the layer (Shibata et al., 1984). This behavior of the El Chichon aerosol is consistent with very weak eddy activity and near radiative equilibrium conditions in the summer stratosphere. Thus, we conclude that present evidence does not support the hypothesis that the tracer variances observed by Ehhalt et al. are generated by quasi-adiabatic eddy motions in the summer stratosphere. Nevertheless we must recognize that reversible meridional displacements of order 500 to 1000 km cannot be ruled out by the data.

An alternative explanation for the large tracer EDHs and small potential temperature EDH of the summer stratosphere has been suggested by Hess and Holton (1985). They noted that large amplitude irreversible tracer transport and mixing occurs over hemispheric scales during the breakdown of the polar vortex in the springtime "final warming". This process generates very large variance in vertically and meridionally stratifed tracers and also in the potential temperature field. Since the breakdown is dominated by planetary scale motions it is likely that the primary variance produced is also on the planetary scale (order 10,000 km). Following the final warming, radiative processes rather quickly damp out the potential temperature variance and establish a summer thermal regime in which the isentropes are nearly zonally symmetric with a very weak temperature increase from equator to pole. Easterlies are established throughout the stratosphere and the circulation becomes nearly zonally symmetric. Under such conditions the tracer variance that was generated during the final warming becomes "frozen" into the mean flow and is passively advected by the symmetric zonal mean flow. In the absence of vertical wind shears such variance will decay only on the slow timescale of the chemical dissipation processes. For trace gases such as nitrous oxide and methane the chemical timescales in the lower stratosphere exceed 100 days. Vertical shears will tend to stretch the blobs of anomalous mixing ratio horizontally and reduce the vertical scale so that small scale mixing processes can destroy the variance. For synoptic scale (1000 km) blobs the observed shears in the summer stratosphere (order $1 \text{ ms}^{-1}\text{km}^{-1}$) should eliminate the variance within a couple of weeks. But planetary scale blobs should be able to retain their identities for more than 100 days. Thus, the tracer variance may last for an entire summer season.

Hess and Holton (1985) presented a preliminary test of the frozen in variance hypothesis based on a highly truncated model of the stratosphere in which the evolutions of the EDHs for N_2O and θ were computed following a simulated final warming. As shown in figure 6 the potential temperature EDH in the model quickly relaxed towards zero following the switch from westerly to easterly flow in the stratosphere. The EDH for the long-lived tracer, however, remained at an elevated level for more than 90 days following the final warming.

The model used by Hess and Holton was too simplified to provide definite results. However, studies of the behavior of simulated N_2O in a stratospheric version of a global climate model (the NCAR Community Climate Model) are currently in progress in an attempt to better test the "frozen in" variance hypothesis.

ZØNAL WIND (M/SEC)

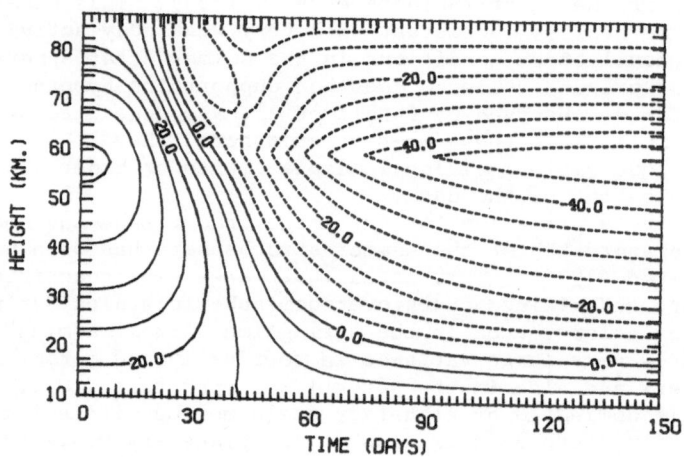

DISPLACEMENT AMPLITUDES AT 25 KM

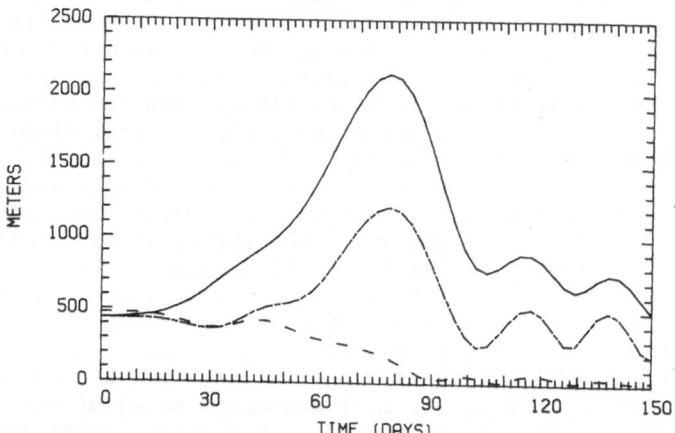

Fig. 6. Time-height cross-section of the mean zonal wind (upper panel) and time variation of the displacement amplitudes at 25 km (lower panel) in the Hess-Holton model. Tracer displacement amplitudes are shown as solid lines, potential temperature displacement amplitudes as dashed lines, and vertical parcel displacement amplitude as long-short dashes. From Hess and Holton (1985).

5. CONCLUSION

In this paper we have discussed the interpretation of observations of tracer variance in the stratosphere. We have shown that small scale, high frequency variance of vertically stratified tracers can be accounted for by vertical displacements due to inertia-gravity wave oscillations. But for tracers with small vertical stratification (e.g., H_2O at 70 mb) horizontal displacements are apparently important in producing the observed variance, even on the scale of 100 km.

Variances measured in the summer stratosphere by Ehhalt et al. (1983) can not be associated with inertia-gravity waves since the EDH calculated from the potential temperature profiles is an order of magnitude less than those calculated for long-lived chemical tracers, and the amplitude of the EDH is much greater than could plausibly be generated by gravity waves. Quasi-horizontal eddy motions in the summer stratosphere, on the other hand, appear to have insufficient amplitude to account for the 1-2 km range of EDH values calculated from the observations. Thus, it is possible that the tracer variances observed in the summer stratosphere do not reflect the active "sloshing" motion of summertime eddies, but rather, represent patches of tracer variance generated during the breakdown of the winter vortex which are passively advected in the summer easterlies as frozen in variance.

ACKNOWLEDGMENTS. This work was supported by the National Aeronautics and Space Administration, NASA Grant NAGW-662. I am indebted to Drs. Alan Plumb and Ed Danielsen for helpful discussions on various aspects of this work. I would also like to thank Dr. Danielsen for giving me access to the STEP data prior to publication.

REFERENCES

Charney, J. G. and P. G. Drazin, 1961: Propagation of planetary-scale disturbances from the lower into the upper atmosphere. J. Geophys. Res., 66, 83-109.

Ehhalt, D. G., E. P. Roth, and U. Schmidt, 1983: On the temporal variance of stratospheric gas concentrations. J. Atmos. Chem., 1, 27-51.

Gill, A. E., 1982: Atmosphere-Ocean Dynamics. Academic Press, New York, 682 pp.

Hess, P. G., and J. R. Holton, 1985: The origin of temporal variance in long-lived trace constituents in the summer stratosphere. J. Atmos. Sci., 42, 1455-1463.

Mahlman, J. D., H. Levy II, and W. J. Moxim, 1986: Three-dimensional tracer structure and behavior as simulated in two ozone precursor experiments. J. Geophys. Res., 91, 2687-2707.

Remsberg, E. E., J. M. Russell III, L. L. Gordley, J. C. Gille, and P. L. Bailey, 1984: Implications of the stratospheric water vapor distribution as determined from the NIMBUS-7 LIMS experiment. J. Atmos. Sci., 41, 2934-2945.

Speth, P., and R. A. Madden, 1983: Space-time spectral analyses of Northern Hemisphere geopotential heights. J. Atmos. Sci., 40, 1086-1100.

Shibata, T., M. Fujiwara, and M. Hirono, 1984: El Chichon volcanic cloud in the stratosphere: lidar observation at Fukuoka and numerical simulation. J. Atmos. Terr. Phys., 46, 1121-1146.

WMO, 1986: Ozone Assessment Report, 1986. Atmospheric ozone, 1985: Assessment of our understanding of the processes controlling its global distribution and change. Global Ozone Research and Monitoring Report No. 16. WMO, Geneva.

PARCEL DISPERSION IN STRATOSPHERIC MODELS

R. Alan Plumb
CSIRO Division of Atmospheric Research
Aspendale 3195
Australia

ABSTRACT. Investigations of the parcel dispersion rates in stratospheric
models are reviewed. These studies have helped to identify the dynamical
circumstances which give rise to sustained dispersion and which are
therefore of central importance in the transport of trace constituents.
Some of these studies, for example, have confirmed the fundamental
difference between the wavy but orderly trajectories in the wintertime
westerly vortex and the more choatic behaviour in the "surf zone".
Stratospheric models used in these investigations have limitations in
their representation of the stratospheric circulation and therefore
results of such studies need to be treated with caution in such
quantitative applications as diffusion coefficients for 2D transport
models.

1. INTRODUCTION

Constituent transport is, in essence, an elementary process which,
for conserved tracers, can be expressed by the simple Lagrangian
conservation law

$$q[\underset{\sim}{x}(t),t] = q[\underset{\sim}{x}(0),0] \qquad (1)$$

where q is the constituent mixing ratio within a given air parcel and
$\underset{\sim}{x}(t)$ is the location of the parcel at time t. Despite this simplicity,
we are still far short of a quantitative understanding of atmospheric
transport processes, even though profound conceptual advances have been
forthcoming during the last decade. The basic difficulty is, of course,
the need for Lagrangian information on atmospheric motions - $\underset{\sim}{x}(t)$ is not
directly available from the routine observational network! There are
means of calculating air parcel trajectories from wind analyses but, in
the stratosphere, errors in the analyses and approximations used in the
calculation of wind velocity limit the validity of trajectory
calculations to about 6 days' duration (Austin and Tuck, 1985).
Therefore, while trajectory analyses of the observed atmosphere can be
very useful in individual case studies (e.g., McIntyre and Palmer,1984),
they are of limited use in revealing the climatological characteristics
of long-term transport.

G. Visconti and R. Garcia (eds.), Transport Processes in the Middle Atmosphere, 327–342.
© *1987 by D. Reidel Publishing Company.*

Apart from the obvious conceptual need for a soundly-based perspective on transport processes, there is the more quantitative requirement of a representation of transport in reduced-dimension constituent models. Such models incorporate some kind of parameterization of transport processes; in the two-dimensional (zonally-averaged) case, the zonally averaged tracer flux $\underset{\sim}{F}$ is usually written as

$$F_i = K_{ij} \frac{\partial q}{\partial x_j} \tag{2}$$

where K is a 2 2 matrix of transport coefficients. For small-amplitude eddies and a conserved tracer, it has the form:

$$\begin{pmatrix} K_{yy} & K_{yz} \\ \\ K_{zy} & K_{zz} \end{pmatrix} = \begin{pmatrix} \frac{1}{2} \frac{\partial}{\partial t}(\overline{\eta^2}) & \frac{1}{2} \frac{\partial}{\partial t}(\overline{\eta\zeta}) \\ \\ \frac{1}{2} \frac{\partial}{\partial t}(\overline{\eta\zeta}) & \frac{1}{2} \frac{\partial}{\partial t}(\overline{\zeta^2}) \end{pmatrix} + \begin{pmatrix} 0 & \frac{1}{2}(\overline{v^\ell\zeta}-\overline{w^\ell\eta}) \\ \\ \frac{1}{2}(\overline{w^\ell\eta}-\overline{v^\ell\zeta}) & 0 \end{pmatrix} \tag{3}$$

(e.g., Plumb 1979) where (η,ζ) and (v^ℓ,w^ℓ) are the (y,z) components of the eddy displacements and Lagrangian perturbation velocities respectively in the generalized Lagrangian-mean theory of Andrews and McIntyre (1978). The first, symmetric, matrix on the right of (3) represents diffusion of constituent in a zonal-mean sense arising from parcel dispersion. The second, antisymmetric matrix corresponds to a correction to the Eulerian mean circulation in the presence of eddies, manifesting the now-well-known fact (e.g., Dunkerton,1978) that the Eulerian mean circulation is not a correct description of the advective component of transport.

What are required to determine $\underset{\sim}{K}$, therefore, are elementary, kinematic, statistics of the eddy motion. Again, the required information is Lagrangian, and is therefore neither directly obtainable from observations, nor calculable from available circulation statistics such as velocity covariances (although some attempts have been made to do so). For this reason, many studies of stratospheric transport processes have relied on data obtained from numerical models of various degrees of complexity. The models' circulation statistics and (implicitly) transport characteristics all fall short of reality to some extent and therefore the applicability of the quantitative results obtained in this way is limited by the errors in the model representation of the real world. However, given the demands on data quality of trajectory analyses, the models' attribute of being dynamically self-consistent is a crucial one, and model-based analyses have provided the basis for much, though by no means all, of our understanding of the nature of stratospheric transport processes. It is essential, however, that their quantitative aspects be interpreted with caution and with due consideration to the characteristics of the model on which the analyses are based.

In what follows, attempts to describe and to quantify stratospheric transport processes on the basis of model circulations are reviewed. The approaches have been of three types: trajectory analyses in mechanistic models of stratospheric warmings; estimates of climatological parcel dispersion rates in three-dimensional circulation models; and determination of the transport coefficients by inversion of the flux-gradient relation (2) using eddy fluxes generated by a three-dimensional (GCM-based) transport model.

2.PARCEL TRAJECTORIES IN MECHANISTIC MODELS OF STRATOSPHERIC WARMINGS

Two of the earlier significant studies of parcel trajectories in stratospheric models were those of Hsu (1980,1981), using a highly truncated (in zonal wavenumber) model of the winter middle atmosphere. In the first of these, parcel trajectories were described in a wave - mean flow model of a wave-2 warming. We focus on results from the second study; the experiment discussed here is the simulation of a wave-1 warming in a model retaining zonal wavenumbers 0, 1 and 2. The model was initialized with a purely zonal mean state approximating a northern hemisphere winter climatology, assumed to be in radiative equilibrium; a planetary wave of zonal wavenumber 1 was forced at the lower boundary. The amplification of this wave produced a major warming, with the high-latitude westerlies being reversed above 30 km after about 22 days of the integration. The dynamics of this behaviour are discussed in Hsu's paper and will not be further addressed here.

Parcel displacements (in fact the displacement and distortion of a material tube) in this experiment were examined by tracking large numbers of marked parcels initially located around a line of latitude at a given height. Horizontal and meridional projections of the location of such rings just before and just after the mature stage of the warming are shown here for two initial locations; (60 N, 32 km) in Fig 1 and (30 N, 30.8 km) in Fig 2. The high-latitude ring exhibits relatively moderate distortions in the presence of the planetary waves. It is clear in the meridional projection that the parcels are on average moving downward, in accord with our understanding of the heat budget of polar regions at such times (e.g., Dunkerton et al., 1981); however, this alone is not an adequate description of the parcel excursions as there has been a substantial dispersion of the air parcels about their descending centre of mass, with parcels being spread over 45 degrees of latitude and about 6 km in height. This dispersion is even more evident in the low-latitude ring, which by day 18 has become severely distorted as low-latitude air is drawn poleward around a developing anticyclone centred at about 50N near 180 degrees longitude (behaviour which has been described in the actual stratosphere by McIntyre and Palmer[1983,1984] and identified as an manifestation of planetary wave breaking). During the mature stage of the event, the flow becomes more complex and the marked ring becomes highly and irregurlarly distorted; by day 26, the ring has been scattered over most of the hemisphere. It should be noted that this later behaviour appears to depend to some extent on the nonlinearity of the dynamic model; in a similar but "linear" experiment (i.e., one with identical forcing but in which only

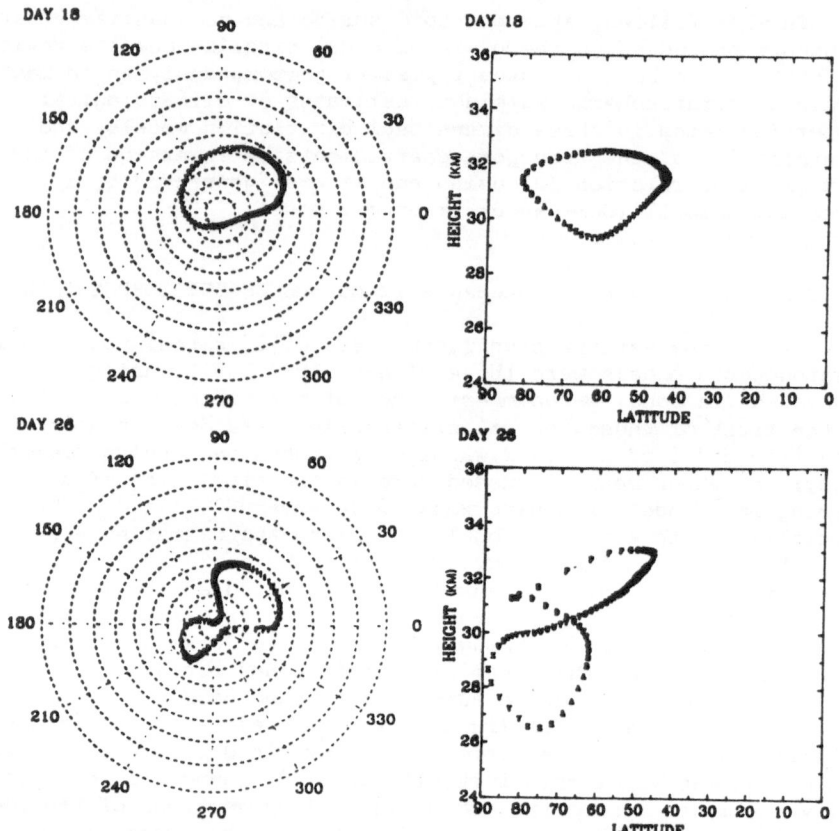

<u>Fig.1</u>: Position of ring of marked parcels initially located at latitude 60 N, height 32 km, in the mature stage of a model warming. Horizontal and meridional projections on the indicated days. [After Hsu(1981)].

zonal wavenumbers 0 and 1 were retained), the early stage of trajectories being wrapped around the anticyclone was sustained, without (at day 26) the the more chaotic behaviour evident in Fig. 2. However, this may merely reflect the less rapid development of the warming in the linear experiment.

Kohno(1984) described a similar study of parcel displacements during wave-1 warming events, using a quasi-linear model (i.e., truncated to retain only zonal wavenumbers 0 and 1). The experiment differed in two significant ways from that of Hsu just described. First, the warming event studied was a minor one, in the sense that the high latitude westerlies were weakened but at no time reversed. Second, the planetary wave forcing was a transient pulse of 25 days duration so that the mean state recovered after the warming event to its initial configuration. Therefore marked parcels could be followed until long after the event and it was thus possible to separate those aspects of material displacements which were permanent from those which were purely transient. As will become clear, the distinction is an important one.

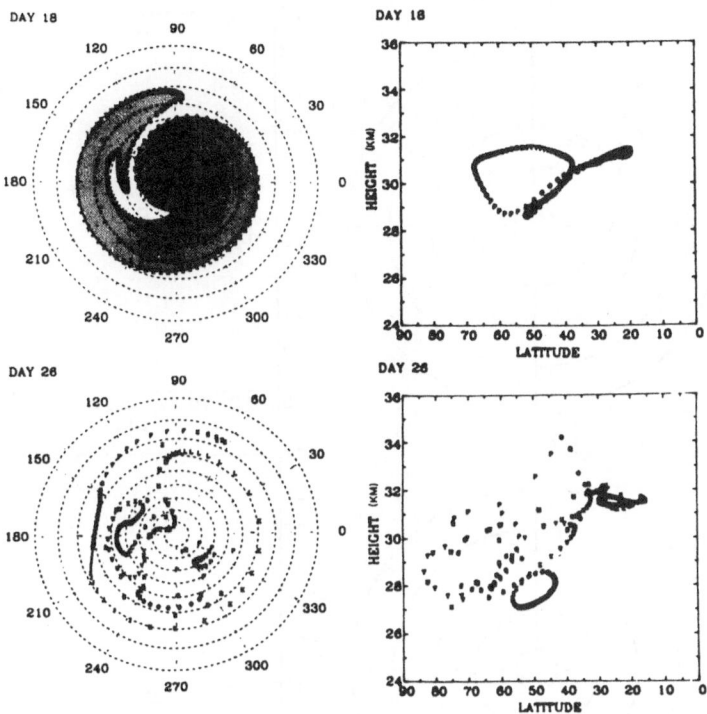

Fig.2: As Fig.1, but for ring of parcels initially located at latitude 30 N, height 30.8 km.

The evolution of rings of marked parcels initially located around latitude circles at (60 N, 25 km) and (30 N, 30 km) are shown in Figs 3 and 4 respectively. Until the mature stage of the warming event (days 20-25) the characteristics of the parcel behaviour are qualitatively as found in Hsu's study. In low latitudes, this similarity is sustained throughout the course of Kohno's integration, with a residual (weak) poleward-upward displacement of the centre of mass of the parcels and a strong permanent dispersion. This dispersion is in a poleward-downward/equatorward-upward plane which, consistent with the arguments of Tung(1984), appears to approximate the plane of the mean isentropes. In high latitudes, however, the dispersion occurring prior to the mature stage is afterward reversed so that, while there remains a large downward displacement, very little net dispersion is evident (i.e., the ring of parcels returns to an almost circular configuration after the decay of the wave pulse). Thus the high-latitude dispersion is essentially a temporary response to the transient wave pulse, indicative of conservative wave dynamics there (Andrews and McIntyre,1978). In

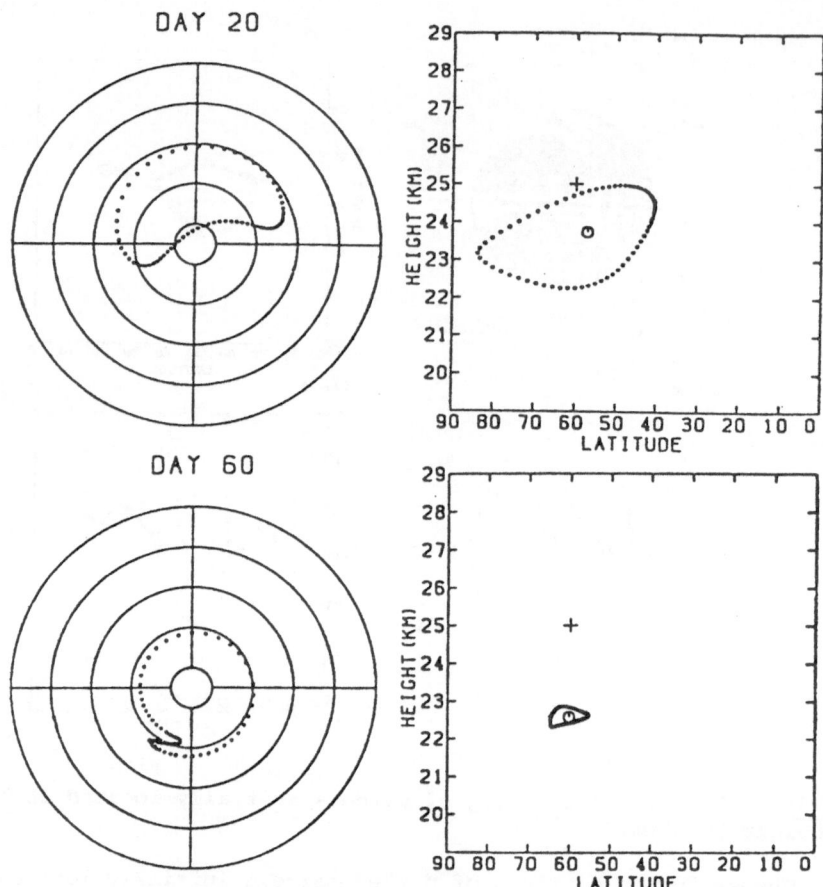

Fig.3: Horizontal and meridional projections of a set of marked parcels initially distributed uniformly around the 60 N latitude circle at 25 km altitude. The symbols "+" and "o" indicate the location of the center of mass of the parcels on day 0 and on the indicated day, respectively. [After Kohno, 1984].

contrast, the sustained dispersion evident in lower latitudes where the wave pulse, as measured by Eulerian statistics such as geopotential or velocity amplitude, is also a transient one is evidence that the wave dynamics are not conservative in those regions. The morphology of the material lines traced out by the parcel rings in both studies suggests that the reason for this is that proposed by McIntyre and Palmer (1983,1984), i.e., the irreversible cascade to small scales associated with the deformation field in the vicinity of the quasi-stationary anticyclonic eddy.

The results of Kohno's study clearly illustrate a distinction between the kinematics of transport in the "surf zone" and the main vortex during a minor warming. Within the vortex, even though the wave amplitudes may become large, the dynamics remains essentially linear

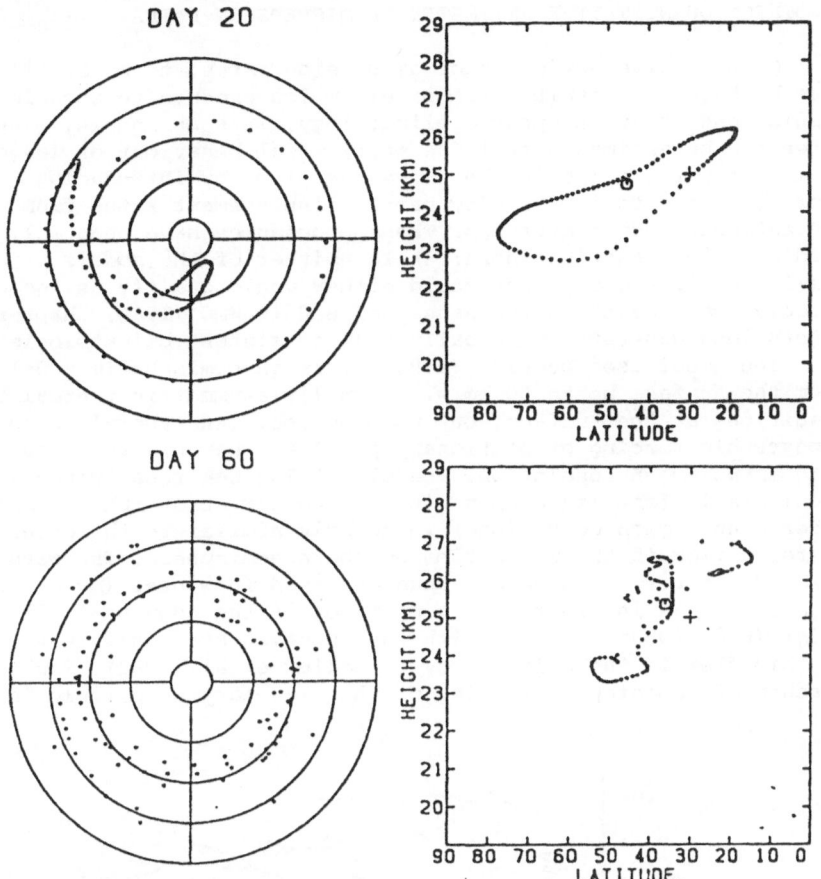

Fig.4: As Fig.3, but for initial latitude 30 N and altitude 25 km. [After Kohno, 1984].

(because of the strong background flow), i.e., parcel displacements can still be characterized as oscillations about a mean zonal motion. Within the weak mean flow of the surf zone, however, trajectories take on a different character in the region of closed eddies where the deformation field of even a simple flow can produce complex and sustained net displacements.

It should be emphasized that Kohno's result of little net displacement in high latitudes over the course of a minor warming cannot be taken to imply that the same would be true for a major warming. If the vortex breaks down completely (as in Hsu's study) then the "surf zone" may extend all the way to the pole. Indeed, the irreversibilty or otherwise of parcel displacements in high latitudes may be the fundamental dynamical distinction between "major" and "minor" events.

3. DIRECT QUANTITATIVE ESTIMATES OF DISPERSION RATES

Quantitative estimates of dispersion rates, to be at all useful, must be based on results from models which produce reasonable simulations of stratospheric climatology. In such models, dispersion rates can be estimated from the statistical behaviour of large numbers of marked parcels (as in the above cases) or by introducing approximations to derive eddy parcel displacement rates from conventional eddy statistics. These procedures have been followed in two studies to be described here; while neither of the models used has a realistic climatology, nor would either would qualify as "general circulation models" in the sense defined in WMO (1986, Chapter 6), they nonetheless generate dynamically self-consistent climatologies.

The model used by Kida (1983a,b) is an hemispheric model extending from the surface up to 50 km with zonally-asymmetric thermal boundary conditions at the surface, but no orography (and therefore no topographic forcing of stationary planetary waves); the model was run with annual-mean insolation. The climatological zonal winds in this model are in fact (in common with some other stratospheric models run under annual-mean conditions) reasonably similar to the observed westerly flow in the wintertime northern hemisphere. The wave activity, however, is very much weaker than observed in winter (by an order of magnitude), as is clear in a comparison of the eddy heat flux in Kida's model (Kida 1983a, Fig A9) with the observed stationary wave component of this flux in the climatology of Geller et al.(1983). Presumably the absence of topographic forcing of the planetary wave field in the model

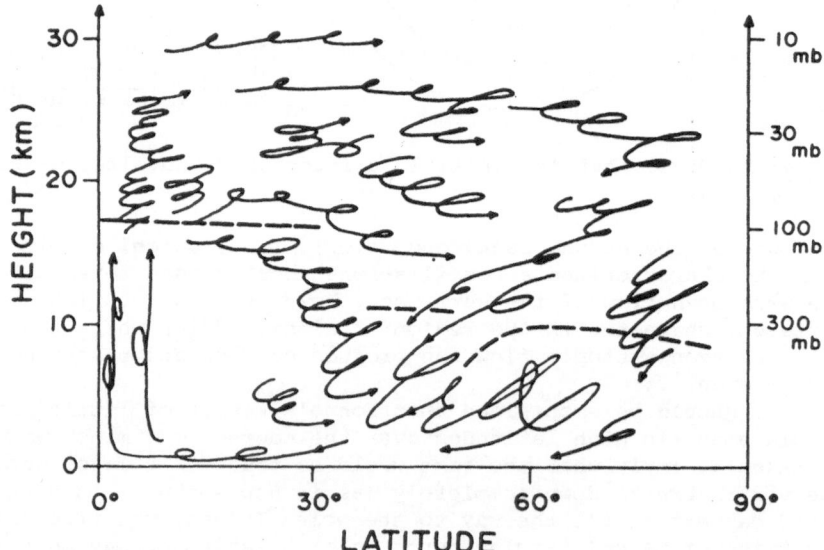

Fig.5: Schematic meridional projection of the trajectories of air parcel motions in the model of Kida(1983a). The broken lines indicate the approximate location of the tropopause.

is a major factor in this discrepancy. The mean meridional circulation
is also much weaker than observed in winter; this is implicit in a
comparison of heating rates in the model (typically 0.2K/day: see Fig 4
of Kida 1983a) and in the lower stratosphere where, according to Kiehl
and Solomon (1986), typical winter high-latitude heating rates are
around 2K/day. Since the diabatic vertical velocity is $W_D = Q/S$, where Q
is proportional to diabatic heating and S is the static stability, and
the static stability in the model stratosphere is not too different from
that in the real world, it follows from this comparison that the
diabatic circulation in the model is about an order of magnitude weaker
than in the winter stratosphere. This is consistent with the weakness of
the model eddy field as, ultimately, it is the eddies that drive the
extra-tropical mean meridional circulation (see WMO, 1986, Chapter 6,
and references therein).

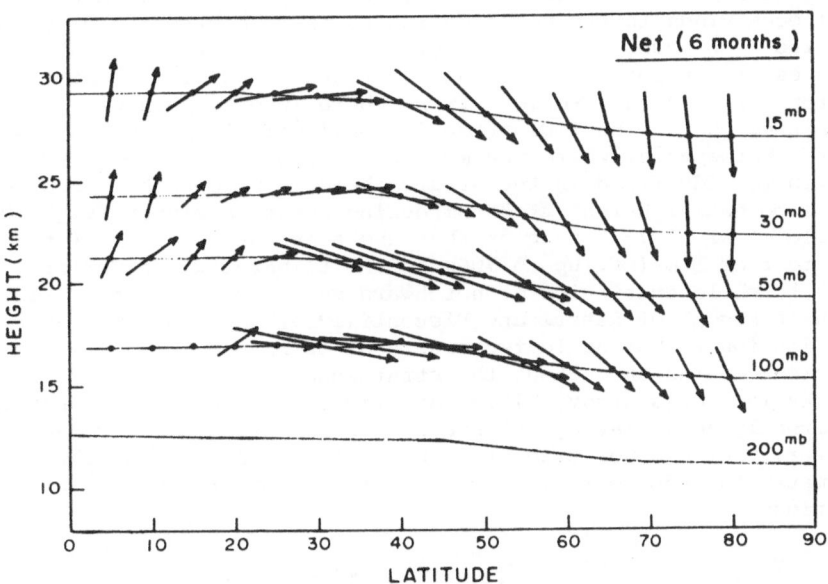

Fig.6: Advective air mass motion in the model of Kida. Vectors are scaled
to show displacement over a 6 month interval. [After Kida, 1983a].

    Air parcel motions were revealed in this model by tracking large
numbers of marked parcels. A schematic view of trajectories thus obtained
is shown in Figure 5. The wiggles in the trajectories are indicative of
dispersion about a mean position in the meridional plane, while the net
drift of the arrows corresponds to a mean motion. In a sense, this motion
is just (by definition) the Lagrangian mean circulation. However, the

Lagrangian mean motion is not in general nondivergent, the divergent
component being associated with spatial inhomogeneities in parcel
dispersion (Andrews and McIntyre, 1978). Kida defined a nondivergent
"advective mass flux" which is time-reversible and calculated this
quantity by running trajectories both backward and forward in time. The
circulation thus revealed is shown in Figure 6, with upward motion across
the tropopause confined to the tropics, poleward flow within the
stratosphere in a single hemispheric cell, and downward motion into the
troposphere occurring mostly, but not exclusively, through the
mid-latitude tropopause gap.

On the basis of the dispersion of trajectories about this mean
motion, Kida estimated horizontal diffusion rates, using (3), to be
typically $K_{yy} \cong 3 \ 10^5 \ m^2 s^{-1}$ in the model stratosphere. Using this value,
he assessed the relative importance of horizontal diffusion and vertical
advection for constituents with meridional structure characteristic of
stratospheric tracers; overall, the two components of transport were
found to be comparable.

A somewhat different approach to the quantification of dispersion
rates was taken by Pitari and Visconti (1985) using a global,
quasigeostrophic model of Cunnold et al. (1975). The climatological
zonal mean winds in the model during northern hemisphere winter are
compared with observations in Fig. 6 of Cunnold et al.(1975). The major
features of the observed flow are reproduced but with significant
discrepancies in the tropical troposphere (where, perhaps, it is not
surprising that a quasi-geostrophic model will fail) and in the winter
lower stratosphere, where the mean westerlies in the model are weak. The
stationary eddy field in the stratosphere is also weak: synoptic maps of
the geopotential height at 30 km during northern winter (Fig. 10 of
Cunnold et al.[1975]) suggest that the stationary wave amplitude is of
the order of 10m (cf. up to 700m in the climatology of Geller et al.
[1983]), while statistics of northward velocity variance at equinox
shown in Fig. 1 of Pitari and Visconti(1985) show the stationary eddy
contribution to be one to two orders of magnitude less than that due to
transient eddies throughout the stratosphere.

As in Kida's study, Pitari and Visconti calculated dispersion rates
by directly accumulating the statistics of parcel displacements.
However, they did not track the absolute location of parcels, but
estimated the eddy parcel displacements by integrating the prediction
equations

$$\left(\frac{\partial}{\partial t} + \bar{u} \cdot \nabla\right) \begin{pmatrix} \eta \\ \zeta \end{pmatrix} = \begin{pmatrix} v' \\ w' \end{pmatrix} \tag{4}$$

which is a small-amplitude approximation to the generalized Lagrangian-
mean definition of these quantities. Thus, the displacements were
evaluated over a three-month season and the dispersion rates then
determined from (3). Results of this procedure for K_{yy} and K_{zz} at
solstice are shown in Figure 7. Typical values found for the horizontal
dispersion rate were $1-3 \times 10^5 m^2 s^{-1}$ in the winter stratosphere and somewhat
smaller values in summer except in a broad region at about
20 km, where values as large as $1 \ 10^6 m^2 s^{-1}$ were found (though there are

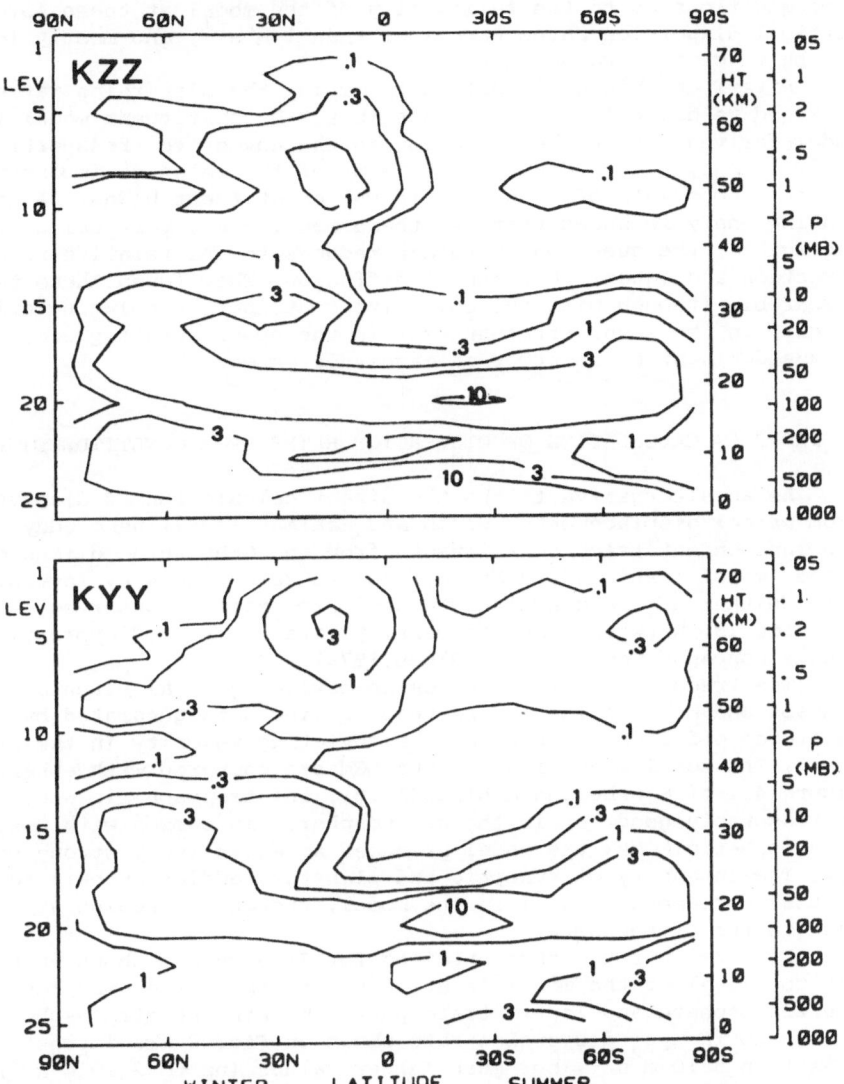

Fig.7: Diffusion coefficients K_{zz} ($10^{-1}m^2s^{-1}$) and K_{yy} ($10^5m^2s^{-1}$) at solstice as determined from a three-dimensional model by Pitari and Visconti (1985).

some questions as to the reliability of the model at these levels). Vertical dispersion rates were less than $1 \text{ m}^2\text{s}^{-1}$, and mostly less than one-third of this value.

Pitari and Visconti went on to employ the dispersion rates thus obtained as diffusion coefficients in a 2-D constituent model (using the model-derived residual circulation as the advective transport). Amongst other things, they addressed the issue of the relative importance of the various components of transport in the constituent budget. A simple scaling analysis shows that, at these magnitudes, vertical diffusion is negligible; the question therefore reduces to the relative roles of advection and (quasi-)horizontal diffusion. They found these two to be comparable (though with the advective contribution being numerically larger) in the lower stratosphere; in the upper stratosphere, advection became dominant by an order of magnitude or more.

4. INDIRECT CALCULATION OF DISPERSION RATES FROM CONSTITUENT FLUXES

As an alternative to the the direct calculation of dispersion rates from parcel displacements, Plumb and Mahlman (1987) used eddy fluxes of two independent trace constituents from the GFDL three-dimensional tracer model (Mahlman and Moxim,1978) to determine K by inversion of (2). Neither of the constituents used was exactly conserved, but the relation (3) between K and dispersion rates is a good approximation for weakly nonconserved tracers (Plumb,1979).

The tracer model is based on an earlier general circulation model (Manabe and Mahlman,1976); one year of wind data generated by this GCM is stored off-line and used as the advecting velocity in the tracer model. The zonal mean winds in the GCM are compared with observations in Figure 4.1 of Manabe and Mahlman(1976). The tropospheric wind structure is reasonably good but in the stratosphere, in common with most other stratospheric GCMs, the model produces an excessively strong polar night jet. The intensity of synoptic and planetary eddies appears to be simulated reasonably well in the model, though the eddies are rather weak in the troposphere.

Dispersion rates thus obtained for January are shown in Figure 8. The top level of the model is at 30 km, so this study was restricted to the troposphere and lower stratosphere. Within the stratosphere, values of the horizontal dispersion rate shown in Fig. 8 are in most places less than $3 \ 10^5 \text{m}^2\text{s}^{-1}$, but much larger values (up to $2 \ 10^6 \text{m}^2\text{s}^{-1}$) are found in a band located in the winter subtropics and extending down across the tropical upper troposphere. As shown on the figure, this band coincides closely with the zero-wind line. For this and other reasons (including the morphology of tracer mixing ratio isopleths in the region of the model's Aleutian anticyclone) Plumb and Mahlman concluded that this band of intense mixing corresponds with the stratospheric "surf zone" identified by McIntyre and Palmer (1983,1984), where planetary wave breaking is occurring in the weak mean flow. In these model results, this phenomenon appears to be the primary mechanism for parcel dispersion in the winter hemisphere (and, of course, very little is occurring in summer). In mid-latitudes, the diffusion was primarily confined locally

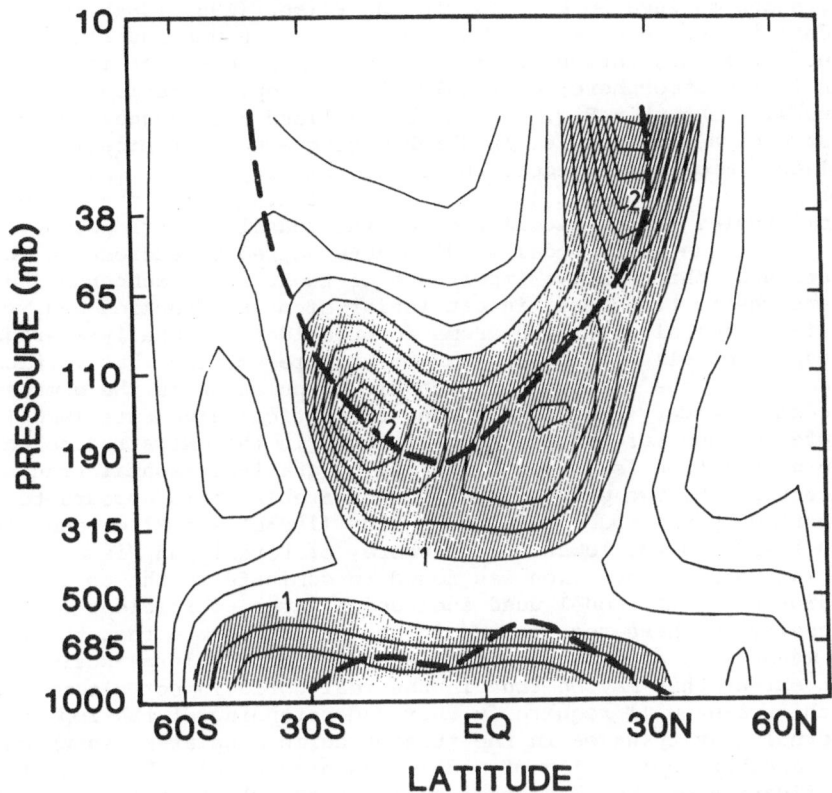

Fig.8: Horizontal diffusion coefficients ($10^6 m^2 s^{-1}$) derived for January from the GFDL 3D tracer model. Dashed line is locus of zero monthly-mean zonal wind. [After Plumb and Mahlman,1987].

to a single plane which in fact was found to be a little steeper than the mean isentropic surfaces.

Plumb and Mahlman further determined the "transport circulation" as the net advective circulation [i.e., incorporating the antisymmetric component of K in (3)]. This is apparently the same as the "advective mass flux" which Kida evaluated in his model by other means. It was found that the transport circulation is well approximated by the residual circulation in the model stratosphere, but not in the troposphere.

5. DISCUSSION

The studies discussed here, when taken in conjuntion with analyses of the real atmosphere, have helped to highlight the kinematics of sustained parcel dispersion and hence to identify those processes responsible for large-scale tranpsort. The "surf zone - main vortex"

separation delineated by McIntyre and Palmer (1983,1984) is particularly
evident in the results of Kohno(1984) and of Plumb and Mahlman(1987).
Indeed, this separation is, if anything, more distinct in the models
than in the atmosphere; certainly the subtropical region of strong
dispersion shown in Fig. 8 is more localized than suggested by analyses
of stratospheric tracers, in which mixing appears typically to extend
well into middle latitudes. While this discrepancy is indicative of
shortcomings of the models, it does not detract from the usefulness of
model studies in the elucidation of the underlying processes.

One issue which model studies have helped to address is the
question of the relative importance of the various components of
stratospheric transport, in particular of mean advection and
quasi-horizontal diffusion. Some formulations of zonally-averaged
transport models have been based on the presumption that vertical
advection by the residual or diabatic circulation is the dominant
process. However, there are strong theoretical arguments (WMO [1986],
Chapter 6, and references therein) that, in the extratropics where the
mean circulation is driven primarily by quasi-horizontal transport
processes, the two effects must be comparable. This appears to be
supported by the model results of Kida (1983a) and Plumb and Mahlman
(1987) and, in the lower stratosphere, of Pitari and Visconti (1985). In
the last study, advection was found to dominate in the upper
stratosphere; the model used included a large coefficient of vertical
viscosity in these regions and it appears therefore that the driving of
the mean circulation there is independent of the large-scale eddies. To
what extent this may be true in the real world is not clear; resolution
of this point will require further understanding of the importance of
internal gravity waves in the stratospheric angular momentum budget.

Another application of model-generated results has been the
verification of the flux-gradient relation (2) which forms the basis of
2D transport models. Plumb and Mahlman constructed such a model using
their transport coefficients derived from the GCM. Comparison
experiments were then run in both 2D and 3D models; it was found that
the 2D model performed very well in reproducing the zonally-averaged
results from the 3D model. The validity of this kind of application of
model-generated dispersion rates is largely independent of the accuracy
with which the GCM simulates the real world.

Model shortcomings do, however, place limitations on the confidence
with which the quantitative results of model studies can be applied in
2D transport models. The errors are not just quantitative; as noted
above, the models appear to err in the location of the surf zone and,
moreover, such dramatic events as major warmings are not simulated in
most general circulation models. Nevertheless, model studies do give
self-consistent estimates of dispersion rates and, in the current
absence of quantitative estimates of dispersion in the real atmosphere,
these at least constitute valuable constraints on the choice of
transport coefficients for use in 2D transport models.

What are required, of course, are quantitative calculations of
dispersion rates in the real atmosphere. The difficulty of achieving
this on the basis of trajectory analyses was noted in the Introduction.
There have been recent attempts to make estimates by following the
procedure of inverting the flux-gradient relation for quasigeostrophic

potential vorticity (Newman et al. [1986]) but there are some
difficulties in the interpretation of results, including uncertainty as
to what extent this quantity can be regarded as a conserved tracer.
There will for the forseeable future be a continuing dependence on model
analyses. Recent improvements in stratospheric GCMs have brought about
susbtantial advances in the quality of simulation of the stratospheric
circulation and it may be anticipated that analyses of their transport
properties, when taken in conjunction with what information observations
can provide, will bring us much closer to accurate simulations of
stratospheric transport.

REFERENCES

Andrews, D.G., and M.E. McIntyre (1978): An exact theory of nonlinear
waves on a Lagrangian mean flow. J.Fluid Mech., 89, 609-646.

Austin, J., and A.F. Tuck (1985): The calculation of stratospheric air
parcel trajectories using satellite data. Quart.J.Roy.Meteor.Soc., 111,
279-307.

Cunnold, D., F. Alyea, N. Phillips and R. Prinn (1975): A
three-dimensional dynamical-chemical model of atmospheric ozone.
J.Atmos.Sci., 32, 170-194.

Dunkerton,T. (1978): On the mean meridional mass motions of the
stratosphere and mesosphere. J.Atmos.Sci., 35, 2325-2333.

Dunkerton, T., C.P.F. Hsu and M.E. McIntyre (1981): Some Eulerian and
Lagrangian diagnostics for a model stratospheric warming. J.Atmos.Sci.,
38, 819-843.

Geller, M.A., M.-F. Wu and M.E. Gelman (1983): Troposphere-stratosphere
(surface-55km) monthly winter general circulation statistics for the
northern hemisphere - four year averages. J.Atmos.Sci., 40, 1334-1352.

Hsu, C.-P. (1980): Air parcel motions during a numerically simulated
stratospheric warming. J.Atmos.Sci., 37, 2768-2792.

Hsu, C.-P. (1981): A numerical study of the role of wave-wave
interactions during sudden stratospheric warmings. J.Atmos.Sci., 38,
189-214.

Kida, H. (1983a): General circulation of air parcels and transport
characteristics from a hemispheric GCM, Part 1. A determination of
advective mass flow in the lower stratosphere. J.Meteor.Soc.Japan, 61,
171-188.

Kida, H. (1983b): General circulation of air parcels and transport
characteristics from a hemispheric GCM, Part 2. Very long-term motions
of air parcels in the troposphere and stratosphere. J.Meteor.Soc.Japan,
61, 510-523.

Kiehl, J.T., and S. Solomon (1986): On the radiative balance of the stratosphere. J.Atmos.Sci., 43, 1525-1534.

Kohno, J. (1984): Stratospheric ozone transport due to transient large-amplitude planetary waves. J.Meteor.Soc.Japan, 62, 413-439.

Mahlman, J.D., and W.J. Moxim (1978): Tracer simulation using a global general circulation model: results from a midlatitude instantaneous source experiment. J.Atmos.Sci., 35, 1340-1374.

Manabe, S., and J.D Mahlman (1976): Simulation of seasonal and inter-hemispheric variations in the stratospheric circulation. J.Atmos.Sci., 33, 2185-2217.

McIntyre, M.E., and T.N. Palmer (1983): Breaking planetary waves in the stratosphere. Nature, 305, 593-600.

McIntyre, M.E., and T.N. Palmer (1984): The "surf zone" in the stratosphere. J.Atmos.Terr.Phys., 46, 825-850.

Newman, P.A., M.R. Schoeberl and R.A. Plumb (1986): Horizontal mixing coefficients for two-dimensional chemical models calculated from National Meteorological Center data. J.Geophys.Res., 91, 7919-7924.

Pitari, G., and G. Visconti (1985): Two-dimensional tracer transport: derivation of residual mean circulation and eddy transport tensor from a 3D model data set. J.Geophys.Res., 90, 8019-8032.

Plumb, R.A. (1979): Eddy fluxes of conserved quantities by small-amplitude waves. J.Atmos.Sci., 36, 1699-1704.

Plumb, R.A., and J.D. Mahlman (1987): The zonally-averaged transport characteristics of the GFDL general circulation/transport model. J.Atmos.Sci., 44, 298-327.

Tung, K.-K. (1984): Modeling of tracer transport in the middle atmosphere. In Dynamics of the Middle Atmosphere, J.R.Holton and T.Matsuno, eds., Terrapub, Tokyo, pp 417-444.

WMO (1986): Atmospheric Ozone 1985. World Meteorological Organization Global Ozone Research and Monitoring Project - Report no. 16. WMO, Geneva.

DIFFUSION COEFFICIENTS CALCULATED FROM SATELLITE DATA

Lawrence V. Lyjak
National Center for Atmospheric Research
P. O. Box 3000
Boulder, CO   80307

ABSTRACT. Lagrangian parcel statistics are used to estimate the horizontal and vertical diffusion coefficients, K_{yy} and K_{zz}, for an inert tracer in the wintertime stratosphere. Ensembles of air parcels are placed on latitude circles and the subsequent parcel trajectories are calculated using winds derived from LIMS satellite data. The diffusion coefficients are then obtained by calculating the time rate of change of the mean squared deviation of the parcels from their center of mass. The calculational procedure for middle and high latitude ensembles is modified to remove the unrealistically large initial dispersion resulting from placing parcels on latitude circles. During quiet periods, when the vortex is nearly axisymmetric, $K_{yy} \leq 7.5 \times 10^5$ m^2/s (typically 1-5 $\times$ 10^5 m^2/s) and $K_{zz} \approx .1$ m^2/s throughout the middle and high latitude stratosphere. For this same domain during a period that included a major warming, K_{yy} varied between 7.5 $\times$ 10^5 and 3.75 $\times$ 10^6 m^2/s throughout the same region, with K_{zz} varying between .3 and 1.4 m^2/s.

1.   INTRODUCTION

The most sophisticated approach to numerically simulating trace constituent transport in the stratosphere is provided by general circulation models (GCMs). However 2-dimensional transport models offer some advantages over their 3-dimensional counterparts. They are simpler and therefore less expensive to develop. They are also less costly to run, allowing the inclusion of more complex chemistry than is currently practical in a GCM.

In the 2-dimensional formalism it is assumed that transports by a mean advective circulation plus a diffusive component can be used to parameterize the 3-dimensional transport. The diffusive portion of the transport, which for an inert tracer is due solely to air parcel dispersion, is usually parameterized in terms of a flux gradient relation. When the effects of diffusion are parameterized in this way, the zonal mean continuity equation for an inert tracer μ assumes the form:

$$\frac{\partial \overline{\mu}}{\partial t} + \overline{v}\frac{\partial \overline{\mu}}{\partial y} + \overline{w}\frac{\partial \overline{\mu}}{\partial z} - \nabla \cdot (\mathbf{K}\nabla\overline{\mu}) = 0 \tag{1}$$

G. Visconti and R. Garcia (eds.), Transport Processes in the Middle Atmosphere, 343–352.
© *1987 by D. Reidel Publishing Company.*

where (y,z) denote the meridional and vertical coordinates [z = -H ln(p/1000), where H is a scale height (typically 7 km) and p is pressure in millibars], $(\bar{v}, \bar{w})$ are the meridional and vertical components of a mean meridional circulation, and **K** is a diffusion tensor having the form:

$$\mathbf{K} = \begin{pmatrix} K_{yy} & K_{yz} \\ K_{zy} & K_{zz} \end{pmatrix} \tag{2}$$

Several types of mean circulations have been used with (1) and the elements of **K** should be consistent with the one chosen. The traditional Eulerian-mean formulation of 2-dimensional transport fails to properly separate the purely advective and diffusive components of the transport (Plumb, 1979). Mean circulations accomplishing this separation and which are more representative of the long term net parcel motion include the residual (Andrews and McIntyre, 1976, 1978a; Boyd,1976), diabatic (Dunkerton, 1978), and transport (Plumb and Mahlman, 1987) circulations. (Plumb and Mahlman have argued that the transport circulation is equivalent to Kida's (1983) "advective mass flux".)

For chemically inert species advected by small amplitude, adiabatic waves, diffusion coefficients that are consistent with both the transport and residual circulations may be expressed as (Holton, 1981):

$$K_{yy} = \frac{1}{2}\frac{\partial}{\partial t}(\overline{\eta'^2})$$

$$K_{yz} = K_{zy} = \frac{1}{2}\frac{\partial}{\partial t}(\overline{\eta'\varsigma'})$$

$$K_{zz} = \frac{1}{2}\frac{\partial}{\partial t}(\overline{\varsigma'^2}) \tag{3}$$

where η' and ς' denote meridional and vertical Lagrangian parcel displacements (as defined by Andrews and McIntyre, 1978b). Since wave activity on which air parcel dispersion depends varies greatly in the real atmosphere, the elements of **K** should be functions of both latitude and altitude.

Kida (1983) determined K_{yy} and K_{zz} for the stratosphere by applying (3) to parcel trajectories obtained from a GCM. The K_{yy} and K_{yz} computations of Newman et al. (1986) were based on a flux gradient relation applied to NMC derived quasi-geostrophic potential vorticity. Plumb and Mahlman (1987) applied a similar procedure to calculate K_{yy} from GCM generated data. In this paper estimates of K_{yy} and K_{zz} for the winter stratosphere are obtained by applying Kida's (1983) procedure to parcel trajectories calculated from winds derived from LIMS satellite data.

Section 2 discusses the computational details and practical problems involved in applying this method to atmospheric data. Results are presented in Section 3 for a quiet period, an active period, and for a time period that included a major warming. Conclusions are discussed in Section 4.

2.   CALCULATIONS

The procedure used to calculate K_{yy} and K_{zz} is a modified version of the one used by Kida (1983). The calculations involve initializing an ensemble of air parcels, calculating the subsequent parcel trajectories, and then using a discrete form of (3) to obtain K_{yy} and K_{zz}.

Kida (1983), beginning his model simulation from a zonally symmetric state, chose the natural initialization procedure of placing each ensemble on a latitude circle equally spaced in longitude. In the present study two different initialization procedures were used. The first procedure was identical to Kida's, even though the atmosphere is in general not in a state of zonal symmetry. This procedure was useful as a simple exploratory first experiment. However, when waves are present streamlines do not coincide with latitude circles and defining the initial ensembles on latitude circles results in an unrealistically large initial amount of parcel dispersion.

The second initialization procedure placed the ensemble of parcels on isentropic geostrophic streamlines. It has been shown by Andrews and McIntyre (1976) and Boyd (1976) that steady, conservative waves of small amplitude do not force changes in the zonal mean state. Under such conditions (and assuming there are no non-conservative zonal-mean forcings present) the transport circulation is zero, and, by an argument based on momentum balance considerations given by Plumb and Mahlman (1987), the diffusion should also be zero. Diffusion coefficient calculations based on the streamline initialization method would in this case correctly show that no air parcel dispersion is occuring, while those based on the latitude circle initialization would not. Initializing ensembles along streamlines is therefore the most physically defensible of the two initialization procedures as it does not suffer from the unrealistically large initial dispersion associated with the latitude circle method.

The parcel ensembles for both initialization methods were placed on 13 pressure levels ranging between 50 and .4 mb. For the latitude circle initialization procedure the latitudes used ranged between the equator and 80 N, with a 10 degree latitude spacing, plus 85 N. Less confidence can be placed in locating geostrophic streamlines at lower latitudes where height gradients are weaker. Therefore, parcel ensembles for the streamline initialization method were placed only at middle and high latitudes.

The second step of the calculations was determining the subsequent trajectories for each parcel of an ensemble. The 3-dimensional wind field required for this calculation was calculated once daily and was derived using LIMS temperatures (Gille et al., 1984). The temperatures, available as Fourier coefficients up to wavenumber 6, were used to calculate thicknesses. Isobaric heights were obtained by integrating the thicknesses, using 50 mb FGGE height fields as the tie-on height (50 mb NMC heights were used for November 1978). Heights poleward of 84 N (the northern limit of the LIMS data) were obtained by extrapolation from 80 and 84 N. The mean and eddy components of the zonal wind, as well as the eddy component of the meridional wind, were calculated from the geostrophic wind equation. The zonal mean meridional velocity and both the mean and eddy components of the vertical wind were calculated from the thermodynamic and continuity equations in the same manner as in Smith and Lyjak (1985). The net radiative heating rates used in the thermodynamic equation were calculated as described in Gille and Lyjak (1986). The parcel trajectories were calculated using:

$$X(t_2) = X(t_1) + V(X(t_1), t_1)\Delta t \tag{4}$$

where Δt is the time step, $X(t)$ is the 3-dimensional position vector of the parcel at time t, and $V(X(t),t)$ is the velocity vector at position X and time t. V was determined by linearly interpolating in space and time from once daily grids. The time step was 15 minutes and each ensemble consisted of 120 parcels. Initial exploratory computations determined that using smaller time steps or more parcels per ensemble had no appreciable effect on the computed diffusion coefficients. Integrations were performed for about 15 days. This integration time length was considered long enough to estimate the mean wave effect on parcel dispersion.

The final step of the calculations was to calculate K_{yy} and K_{zz} using a discrete form of (3). Letting N denote the number of parcels in an ensemble, (y_i,z_i) the latitude and altitude of the ith parcel of the ensemble, and $(\bar{y}, \bar{z})$ the mean latitude and altitude of the ensemble, (3) may be written as:

$$K_{yy} = \frac{1}{2}\frac{\partial}{\partial t}\left(\sum_{i=1}^{N}\frac{(y_i - \bar{y})^2}{N}\right)$$

$$K_{zz} = \frac{1}{2}\frac{\partial}{\partial t}\left(\sum_{i=1}^{N}\frac{(z_i - \bar{z})^2}{N}\right) \qquad (5)$$

where we have let $\eta' = y_i - \bar{y}$ and $\varsigma' = z_i - \bar{z}$. For both the latitude circle and streamline initializations, time derivatives in (5) were computed using values of $\overline{\eta'^2}$ (or $\overline{\varsigma'^2}$) from the first and last days of the integrations.

As already discussed, initializing the ensembles to lie along streamlines is more appropriate from a physical viewpoint than initializing along latitude circles. However from a computational viewpoint, it is difficult using the streamline method to achieve uniform latitudinal coverage over the Northern Hemisphere. Therefore the K_{yy} computational procedure was modified so that values of $\overline{\eta'^2}$ based on the latitude circle initialization could be used to approximate results obtained by initializing along streamlines. (Computations of K_{zz} are not nearly as sensitive to the initialization method selected as are those for K_{yy}. Therefore K_{zz} was computed by applying the procedure as already described to ensembles coinciding with latitude circles.) The temporal variation of $\overline{\eta'^2}$ for a middle to high latitude ensemble initialized along a latitude circle is shown in Fig. 1. This curve is typical of a period when the eddies are relatively steady. As the parcels are transported from their initial locations along a latitude circle into the configuration of the streamlines, the curve rapidly rises from a value of zero on January 1 (this data point is not shown in the plot) to a value of 2.3×10^{12} m^2 on January 3. Thereafter the curve oscillates in a relatively narrow range. Based on results such as this, it was concluded that for middle to high latitude ensembles the first local minimum of $\overline{\eta'^2}$ occurring after day 1 of the integration would provide a more appropriate initial value of $\overline{\eta'^2}$ for use in (5).

A comparison of K_{yy} values obtained using the latitude circle, the modified latitude circle, and the streamline initialization methods is shown in Fig. 2 with results plotted as functions of latitude. As expected, the largest values are associated with the unmodified latitude circle method. While not duplicated for every pressure level and time period, the close agreement seen here between the modified latitude circle and streamline methods is fairly typical.

An initial value of $\overline{\eta'^2}$ other than its value on day 1 of the integration was used in the K_{yy} calculations for middle and high latitude ensembles. However, a

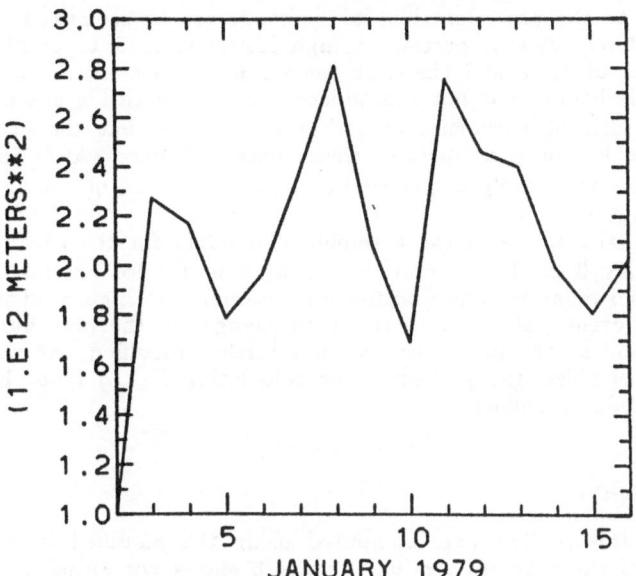

Figure 1. Variation of $\overline{\eta'^2}$ initializing along latitude circles at 7 mb, 50°N, from January 2 through January 15, 1979.

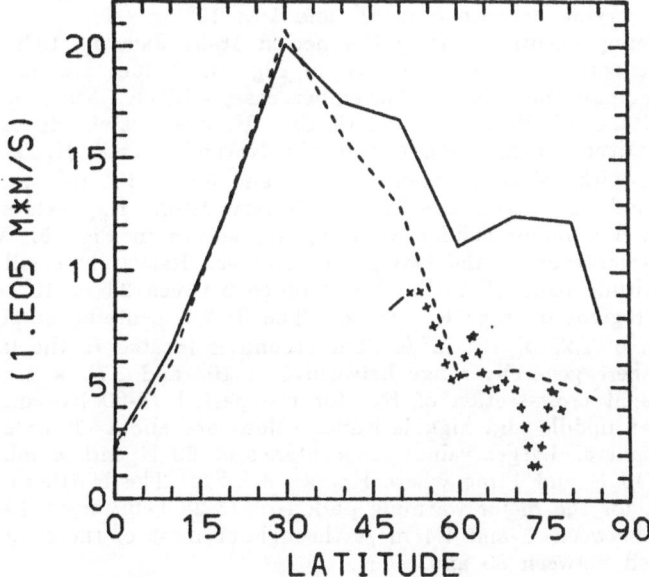

Figure 2. K_{yy} for February 15–March 1, 1979 and 10 mb using unmodified latitude circle (solid), modified latitude circle (dashed), and streamline (plus signs) initialization procedures.

different modification to the procedure had to be made for lower latitude ensembles. When waves are relatively steady, parcels at high latitudes tend to oscillate about their ensemble's center of mass and therefore remain in the vicinity of their original latitude. However, this behavior is not reproduced for lower latitude ensembles. Time series of $\overline{\eta\prime^2}$ for low latitude ensembles (not shown) indicate that the parcels tend on average to continually move away from their center of mass. After integrating for 15 days the parcels are widely dispersed in latitude and the $\overline{\eta\prime^2}$ value associated with the ensemble at this time is no longer indicative of the amount of dispersion occurring at the original latitude of the ensemble. Therefore for the end value of $\overline{\eta\prime^2}$, the value of $\overline{\eta\prime^2}$ on day 8 of the integration was used in (5) for ensembles in lower latitudes. Of course a similar situation occurs for ensembles at higher latitudes during highly transient wave events. However in this case changes in the procedure would be much more difficult and so the procedure was not further modified. As was the case for higher latitude ensembles, the procedure for calculating K_{zz} did not have to be modified for low latitude ensembles.

3.   RESULTS

Results presented in this section were computed using the modified latitude circle procedure discussed in the previous section. Figure 3 shows computed values of K_{yy} for the Northern Hemisphere stratosphere during the relatively quiet period of 15-30 November 1978. Mid-latitude values generally decrease with height and range between 7.5×10^5 m²/s at 50 mb to 2.5×10^5 m²/s between 3 and 1 mb. High latitude K_{yy} values in the middle and upper stratosphere show little variation with height or latitude with values near 2.5×10^5 m²/s . At low latitudes, K_{yy} increases with decreasing latitude, varying between 5×10^5 and 1×10^6 m²/s.

A minor warming occurred during the period 16-31 January 1979 and K_{yy} values for this more active period are shown in Fig. 4. Values are now generally a factor of 2 larger than those of the November case, with $K_{yy} \approx 5 \times 10^5$ m²/s throughout much of the middle and high latitudes. However values do exceed 7.5×10^5 m²/s at high latitudes near 5 mb. As in the November case, K_{yy}'s tend to be larger in the tropics with values between 5×10^5 and 3.5×10^6 m²/s.

The final period considered was 15-28 February 1979. K_{yy} values for this period which included a major sudden warming are shown in Fig. 5. Values are now considerably larger than in the two previous cases. Relatively small values are restricted to the latitude band 55 to 75 N and range between 2.5×10^5 and 5×10^5 m²/s. There are 2 regions of large K_{yy} values. The first is centered at 16 mb and 84 N with a maximum of 1.25×10^6 m²/s. The second is located in the latitude band of 15-40 N. Values here generally range between 1×10^6 and 3.75×10^6 m²/s.

A latitude-height cross-section of K_{zz} for the period 15-30 November 1978 is shown in Fig. 6. At middle and high latitudes values are about .1 m²/s throughout most of the stratosphere. Larger values are centered at 50 N and .4 mb where $K_{zz} \approx .7$ m²/s and at 10 N and 1 mb where $K_{zz} \approx .8$ m²/s. The Northern Hemispheric cross-section of K_{zz} for the major warming period of 15-28 February 1979 (not shown) shows K_{zz} varying between .3 and 1.4 m²/s throughout most of the stratosphere with largest values located between 60 and 80 N.

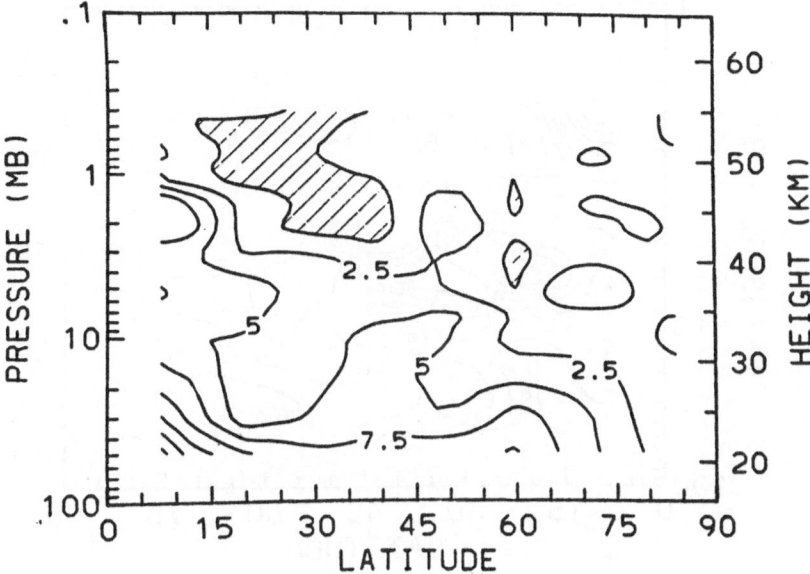

Figure 3. Latitude-height cross-section of K_{yy} for 15-30 November, 1978. Units are 10^5 m^2/s.

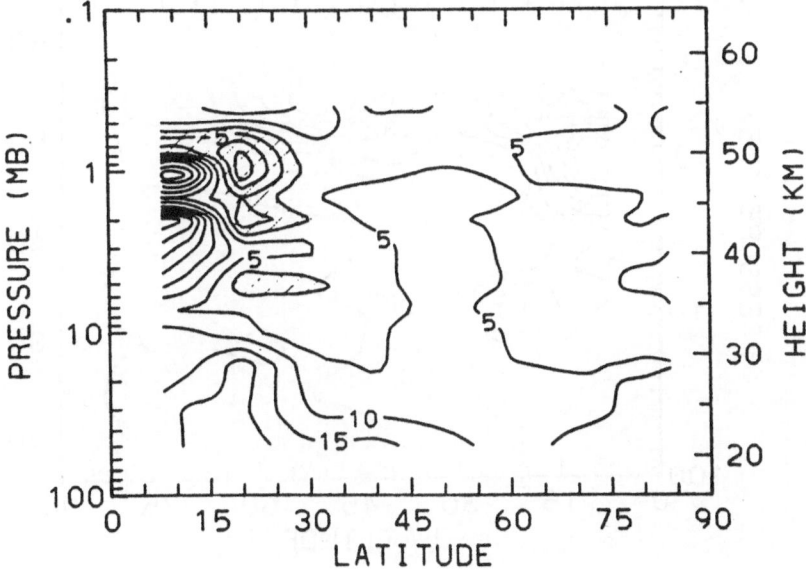

Figure 4. Latitude-height cross-section of K_{yy} for 16-31 January, 1979. Units are 10^5 m^2/s.

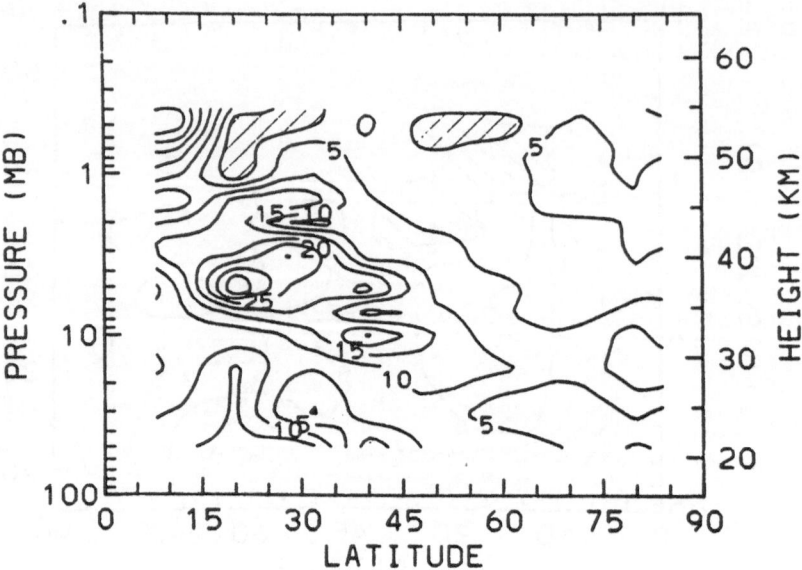

Figure 5. Latitude-height cross-section of K_{yy} for 15-28 February 1979. Units are 10^5 m^2/s.

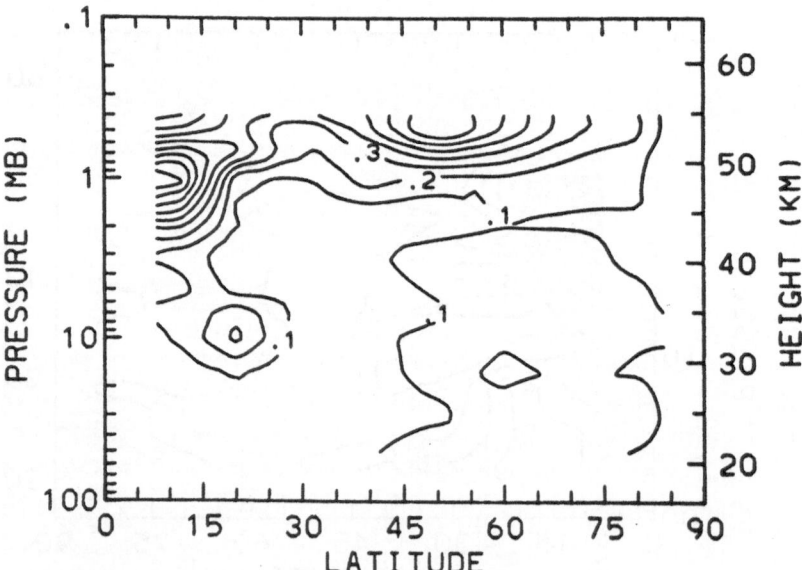

Figure 6. Latitude-height cross-section of K_{zz} for 15-30 November 1978. Units are m^2/s.

4.   SUMMARY

Estimates of horizontal and vertical diffusion coefficients have been made for the winter stratosphere using Lagrangian parcel statistics. The sense of the temporal variation of K_{yy} and K_{zz} is consistent with the relative amount of wave activity present for the three periods considered here. Smallest values of K_{yy} were associated with the relatively quiet period of late November 1978 and were generally $< 5 \times 10^5$ m^2/s. For the active period of late January 1979 K_{yy} values ranged between 5×10^5 and 1×10^6 m^2/s. During the sudden warming period of late February 1979 mid to high latitude statospheric values reached maximum values of 2.5×10^6 m^2/s. The large K_{yy} values estimated for sudden warming periods, while of dynamical interest, are not representative of K_{yy} values averaged over the winter season and this is generally what is required in a 2-dimensional model.

Based on the K_{yy} values calculated for the November and January cases, $K_{yy} \approx 5 \times 10^5$ m^2/s was selected as a representative value for the winter stratosphere. This value is slightly larger than Kida's value of 3×10^5 m^2/s for the lower stratosphere. However, it is considerably smaller than the K_{yy} values computed by Newman et al. (1986). They estimated K_{yy} for much of the winter stratosphere at middle latitudes to be between 1×10^6 and 2×10^6 m^2/s. For the middle and high latitude stratosphere Plumb and Mahlman (1987) obtained estimates of $K_{yy} \leq 5 \times 10^5$ m^2/s.

Low and middle stratosphere K_{zz} computations for late November 1978 gave an estimate of $K_{zz} \approx .1$ m^2/s. This value was also obtained by Kida (1983) for the lower stratosphere. However, for the major warming period of late February 1979 estimates of K_{zz} calculated in this paper ranged between .3 and 1.4 m^2/s.

Finally, it should be recalled that the expressions used to calculate K_{yy} and K_{zz} were derived under the assumption that the waves were conservative and of small amplitude. This is clearly not the case during the late January and late February 1979 periods.

Acknowledgments. I wish to thank Rolando Garcia, John Gille, and Anne Smith for several helpful discussions. This work was supported in part by the National Aeronautics and Space Administration through grants W15439 and W16215. The National Center for Atmospheric Research is sponsored by the National Science Foundation.

REFERENCES

Andrews, D. G., and M. E. McIntyre, 1976: Planetary waves in horizontal and vertical shear: The generalized Eliassen-Palm relation and the mean zonal acceleration. J. Atmos. Sci., 33, 2031-2048.

Andrews, D. G., and M. E. McIntyre, 1978a: Generalized Eliassen-Palm and Charney-Drazin theorems for waves on axisymmetric flows in compressible atmospheres. J. Atmos. Sci., 35, 175-185.

Andrews, D. G., and M. E. McIntyre, 1978b: An exact theory of nonlinear waves on a Lagrangian-mean flow. J. Fluid Mech., 89, 609-646.

Boyd, J., 1976: The noninteraction of waves with the zonally-averaged flow on a spherical earth and the interrelationships of eddy fluxes of energy, heat and momentum. J. Atmos. Sci., 33, 2285-2291.

Dunkerton, T., 1978: On the mean meridional motions of the stratosphere and mesosphere. J. Atmos. Sci., 35, 2325-2333.

Gille, J. C., J. M. Russell III, P. L. Bailey, L. L. Gordley, E. E. Remsberg, J. H. Lienesch, W. G. Planet, F. B. House, L. V. Lyjak, and S. A. Beck, 1984: Validation of temperature retrievals obtained by the Limb Infrared Monitor of the Stratosphere (LIMS) experiment on NIMBUS 7. *J. Geophys. Res., 89,* 5147-5160.

Gille, J. C., and L. V. Lyjak, 1986: Radiative heating and cooling rates in the middle atmosphere. *J. Atmos. Sci., 43,* 2215-2229.

Holton, J. R., 1981: An advective model for two-dimensional transport of stratospheric trace species. *J. Geophys. Res., 86,* 11989-11994.

Kida, H., 1983: General correlation of air parcels and transport characteristics derived from a hemispheric GCM: I. A determination of advective mass flow in the lower stratosphere. *J. Meteor. Soc. Japan, 61,* 171-187.

Newman, P. A., M. R. Schoeberl, and R. A. Plumb, 1986: Horizontal mixing coefficients for two-dimensional chemical models calculated from National Meteorological Center data. *J. Geophys. Res., 91,* 7919-7924.

Plumb, R. A., 1979: Eddy fluxes of conservative quantities by small-amplitude waves. *J. Atmos. Sci., 36,* 1699-1721.

Plumb, R. A., and J. D. Mahlman, 1987: The zonally-averaged transport characteristics of the GFDL general circulation/transport model. *J. Atmos. Sci., 44,* 298-327.

Smith, A. K., and L. V. Lyjak, 1985: An observational estimate of gravity wave drag from the momentum balance in the middle atmosphere. *J. Geophys. Res., 90,* 2233-2241.

VACILLATIONS INDUCED BY INTERFERENCE OF STATIONARY AND TRAVELING PLANETARY WAVES

Murry L. Salby*
Department of Astrophysical, Planetary,
  and Atmospheric Sciences
University of Colorado
Campus Box 391
Boulder, CO 80309
USA

Rolando R. Garcia*
National Center for
  Atmospheric Research
P.O. Box 3000
Boulder, CO 80307
USA

ABSTRACT. The interference pattern created when a traveling planetary wave propagates over a stationary forced wave is explored. Interference leads to a modulation of all the transport properties of the stationary wave even if the traveling wave is barotropic, as is typical in the troposphere and lower stratosphere. In so doing, the steady uniform stream of wave activity associated with the stationary wave is organized into a series of capsules or wavepackets which propagate with time. Consequently, the signature emerges locally as a series of *bursts* of wave activity. Rising values of Eliassen-Palm flux at the leading edge of these wavepackets and opposite behavior at the trailing edge induce an alternating mean flow response which captures several aspects of the observed behavior of quasi-periodic disturbances. Synoptic patterns of geopotential and Ertel potential vorticity also resemble behavior observed during disturbed conditions.

1. Introduction

Madden (1975) has suggested that interference between traveling and stationary planetary waves may be responsible for the fluctuations in eddy fluxes of heat and momentum observed in the stratosphere. The interference mechanism has also been invoked in connection with the upward propagation of wave amplitude vacillations (Madden, 1977). These observations have led to the speculation that this process might be important in large-scale stratospheric disturbances.

Of all the traveling waves relevant to the problem, atmospheric normal modes have received much of the attention. Part of the reason is that they are preferred in the response to random or broad-band forcing (Salby, 1984) and capture a significant fraction of the planetary scale variance (e.g., Lindzen et al., 1984). Among the normal modes documented, the second symmetric mode of wavenumber one—the 16-day wave—has captured the attention of investigators because of its large amplitude and recurrent appearance. It is a significant climatological feature of Northern Hemisphere winter statistics (Madden, 1978; 1983; Rinne and Sarkenan, 1985), its variance being concentrated between 12 and 21 days. In transient episodes such as that of January, 1979, the 16-day wave appears capable of dwarfing the forced wave over the entire troposphere and lower stratosphere (Madden and Labitzke, 1981). At upper levels, the instantaneous wavefield observed during January, 1979, constitutes one of the most disturbed configurations on record (Fig. 1).

In this paper we examine in detail the interference between forced stationary waves and traveling planetary waves. The signature of the process is explored in the wave field, in the mean flow, and in combined synoptic fields. We examine the behavior starting

---

  Center for Atmospheric Theory and Analysis, University of Colorado, Boulder, CO 80309-0391

G. Visconti and R. Garcia (eds.), Transport Processes in the Middle Atmosphere, 353–367.

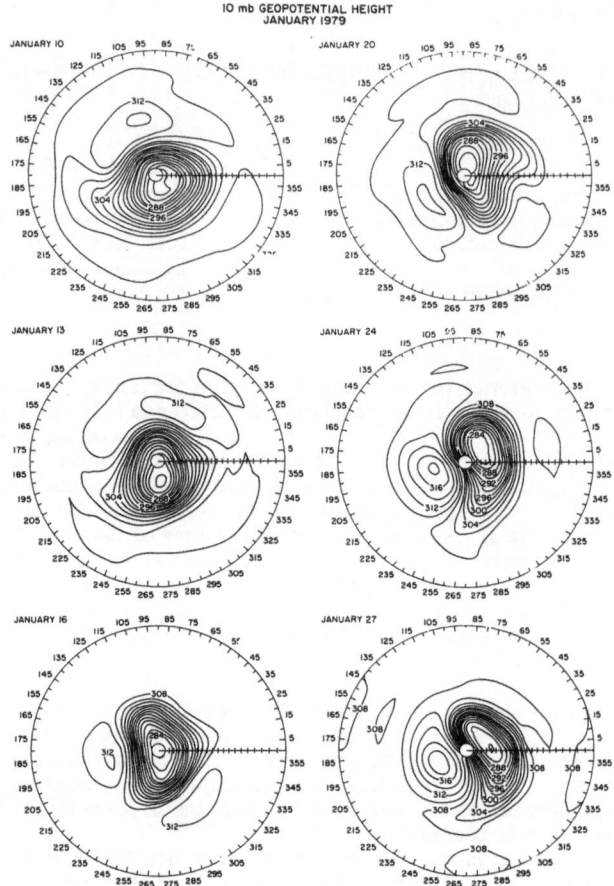

Figure 1 Sequence of N.H. 10 mb geopotential height maps during January, 1979.

from simple considerations such as local wave amplitude, progressing to higher order quantities such as the EP flux. We conclude by deriving the interference signature in synoptic fields of geopotential and Ertel potential vorticity. The latter is of interest because maps of this quantity derived from satellite data suggest the presence of nonlinear advection (McIntyre and Palmer, 1983; 1984). It has been proposed that nonlinear "wave breaking" plays a major role in the erosion of the stratospheric polar night vortex.

In section 2 we formulate the necessary equations and derive the interference behavior in general form. In section 3, solutions are obtained for stationary and traveling wave fields calculated numerically from the linearized Primitive Equations. The reaction of the mean flow to fluctuating eddy forcing is derived in section 4, and the synoptic behavior of the geopotential and Ertel potential vorticity fields is examined in sections 5 and 6, respectively. Section 7 summarizes our findings and conclusions.

2. Formulation

We consider an arbitrary unsteady streamfield represented as

$$\psi'(\lambda, \phi, z, t) = \frac{1}{2\pi} \int\limits_{-\infty}^{\infty} d\sigma e^{-i\sigma t} \sum_{m=-\infty}^{\infty} e^{im\lambda} \Psi_m^\sigma(\phi, z) \tag{1}$$

where λ, ϕ, z, and t denote longitude, latitude, height and time respectively, and Ψ_m^σ is the complex amplitude for wavenumber m and angular frequency σ. We examine the flux of wave activity or Eliassen-Palm flux, F, resulting from (1). Because the instantaneous streamfield evolves, so too will the instantaneous EP flux field $F(\phi, z, t)$.

2.1 Disturbances to the Primitive Equations

For disturbances satisfying the linearized Primitive Equations, the latitudinal and vertical components of F are:

$$
\begin{cases}
F_\phi = \bar{\rho} a \cos\phi \left[\bar{u}_z \frac{\overline{v'\theta'}}{\bar{\theta}_z} - \overline{v'u'} \right] & (2.1) \\[3mm]
F_z = \bar{\rho} a \cos\phi \left[\left(f - \frac{(\bar{u}\cos\phi)_\phi}{a\cos\phi} \right) \frac{\overline{v'\theta'}}{\bar{\theta}_z} - \overline{w'u'} \right] & (2.2)
\end{cases}
$$

where $\bar{\rho} = \rho_o e^{-z/H}$, z is log pressure height, ρ_o and H are constant reference density and scale height respectively, u, v, and θ represent zonal and meridional velocities and potential temperature respectively, and overbar denotes zonal average. Additional symbols are in standard notation.

The EP flux can be rewritten in terms of its space-time Fourier components as follows:

$$F(\phi, z, t) = \sum_{m=0}^{\infty} \frac{1}{(2\pi)^2} \int\limits_{-\infty}^{\infty} d\sigma \int\limits_{-\infty}^{\infty} d\sigma' F_m^{\sigma\sigma'}(\phi, z, t) \tag{3}$$

where

$$F_m^{\sigma\sigma'}(\phi, z, t) = 2\bar{\rho} a \cos\phi \left[\begin{matrix} Re\left\{ \left[\bar{u}_z \frac{V_m^\sigma \Theta_m^{\sigma'*}}{\bar{\theta}_z} - V_m^\sigma U_m^{\sigma'*} \right] e^{-i(\sigma-\sigma')t} \right\} \\[3mm] Re\left\{ \left[\left(f - \frac{(\bar{u}\cos\phi)_\phi}{a\cos\phi} \right) \frac{V_m^\sigma \Theta_m^{\sigma'*}}{\bar{\theta}_z} - W_m^\sigma U_m^{\sigma'*} \right] e^{-i(\sigma-\sigma')t} \right\} \end{matrix} \right] \tag{4}$$

The instantaneous EP flux vector may be recognized as a convolution of frequency components of the space-time spectrum. $F_m^{\sigma\sigma'}$ represents the contribution from wavenumber m at frequencies σ and σ' and oscillates as $e^{-i(\sigma-\sigma')t}$, i.e. beating at the difference in frequencies between the components. In particular, the steady or time-mean contributions are derived from spectral components of the same frequency, while fluctuating contributions arise from the cross terms involving different frequencies.

Because the EP flux field is evolving, so too will its divergence and hence the eddy driving of the mean flow. As will be seen shortly, this is due to the spatial modulation of the time-mean wave and migration of the ensuing pattern with time. We explore now the response of the zonal-mean state to this induced transient forcing.

2.2 Reaction of the basic stream

The EP flux represents the essential wave driving of the mean flow. Specifically, its divergence constitutes the sole wave forcing of the (quasi-geostrophic) transformed Eulerian equations governing the zonal-mean state:

$$
\begin{cases}
\frac{\partial \bar{u}}{\partial t} - f\bar{v}^* = \frac{1}{\bar{\rho}a\cos\phi}\nabla \cdot F + \bar{F} & (5.1) \\[2mm]
f\bar{u}_z + \frac{S}{a}\bar{\theta}_\phi = 0 & (5.2) \\[2mm]
\frac{1}{a\cos\phi}\frac{\partial}{\partial\phi}[\cos\phi\,\bar{v}^*] + \frac{1}{\bar{\rho}}\frac{\partial}{\partial z}[\bar{\rho}\bar{w}^*] = 0 & (5.3) \\[2mm]
\frac{\partial\bar{\theta}}{\partial t} + \bar{\theta}_z\bar{w}^* = \bar{Q} & (5.4)
\end{cases}
$$

where $\bar{F}$ and $\bar{Q}$ represent zonal-mean friction and heating, and $\nabla \cdot F$ is the divergence of the EP flux.

The residual mean system (5) may then be consolidated into a single diagnostic equation for the mean meridional streamfunction $\bar{\psi}^*$:

$$
L[\bar{\psi}^*] = \frac{f}{N^2}\frac{\partial}{\partial z}\left[\frac{1}{\bar{\rho}a\cos\phi}\nabla \cdot F\right] + \frac{f}{N^2}\frac{\partial\bar{F}}{\partial z} + \left(\frac{S}{N^2a}\right)\frac{\partial\bar{Q}}{\partial\phi} \tag{6}
$$

We will focus on the transient eddy driving embodied in the first term, specifically that introduced by interference as represented in the cross frequency terms of (4). To this end, we presume the zonal-mean friction and heating are in balance with the time-mean zonal flow and with the time-mean EP flux.

2.3 Diatomic spectrum

In general the convolution (4) involves a continuum of frequencies. However, it is of interest to examine situations where the variance is concentrated about particular frequencies, for instance a red spectrum about zero frequency associated with the quasi-stationary waves and one or more discrete peaks associated with normal modes. However, it should be understood that the ideas we are about to develop carry through regardless of the spectral make-up of the traveling wave.

We consider the simple two-component space-time spectrum representing the combination of a single stationary wave ($\sigma_1 = 0$) and a monochromatic westward traveling wave ($\sigma_2 < 0$), both of wavenumber m. In addition to being of considerable advantage in elucidating the process, this idealization is also of practical merit because of the concentration of variance associated with the stationary and 16-day waves. The frequency convolution (4) reduces to

$$
F(\phi, z, t) = F_o(\phi, z) + F'(\phi, z, t) \tag{7.1}
$$

$$
F_o(\phi, z) = \frac{1}{4}\left(F_m^{\sigma_1\sigma_1} + F_m^{\sigma_2\sigma_2}\right) \tag{7.2}
$$

$$
F'(\phi, z, t) = \frac{1}{4}\left(F_m^{\sigma_1\sigma_2} + F_m^{\sigma_2\sigma_1}\right), \tag{7.3}
$$

i.e., a combination of steady and fluctuating components. The steady component is derived from matching frequencies and the fluctuating component from cross terms. The fluctuating vector field F' may be thought of in terms of amplitude and phase, both of which vary spatially according to the structures of the two wave components.

The steady component F_o is just the sum of the individual time-mean contributions due to components 1 and 2. The fluctuating component F' is new in that it exists only when both waves are simultaneously present. It exists even when the time-mean contribution to F of the traveling wave vanishes, e.g. for a normal mode.

3. Stationary wave and traveling normal mode in realistic mean flow

We examine the interference signature with a stationary wave and an atmospheric normal mode, both derived under realistic conditions. These are obtained numerically with a global, linear Primitive Equations calculation, and assigned amplitudes representative of January 1979. Both waves are generated by surface forcing. A Gaussian mountain, centered at $\phi = 45°$ and having a sigma of 15° latitude is used to force the stationary wave. The procedure described in Salby (1981) is used to locate in frequency the second symmetric normal mode of wavenumber 1 (i.e., the 16-day wave).

Evolution of the instantaneous wave field is shown in Fig. 2. Because the stationary and traveling waves have similar phase behavior in the middle and upper stratosphere, the interference at these levels proceeds with near simultaneity. Below these altitudes, a bulging up from below can be seen. The peak amplitude, occurring at about 40 km, intensifies over the course of the cycle from 650 gpm to in excess of 2100 gpm. The region influenced by the wave expands both upward and equatorward. During this evolution the phase changes from a regular vertical phase tilt, indicating simple upward radiation, to one virtually barotropic at peak amplitude, suggesting vertical trapping. Neither of these characterizations is completely correct, as the refractive properties of the zeroth order flow are unchanged during the cycle. What actually appears to occur is that the barotropic region of the traveling wave below bulges up into the stratosphere, driving the region of phase tilt equatorward and to higher altitudes. Following peak amplification the entire pattern collapses, with the phase tilt being restored to the middle and upper stratosphere.

Evolution of the instantaneous EP flux field and eddy driving of the mean flow are given in Fig. 3. Over much of the stratosphere the time-mean component of eddy driving is dwarfed by the transient component, which peaks at a value nearly four times that of the steady contribution. During the initial phase of the cycle wave activity radiates vertically out of the troposphere above the source region. It is refracted both poleward and equatorward but rapidly attenuates with altitude as a result of dissipation. As the traveling wave constructively interferes with the stationary wave, the wave activity flux swells upward out of the troposphere, penetrating to substantially greater heights. Following peak surge, the EP flux field veers equatorward and subsequently collapses.

The eddy driving of the mean flow has maximum values at upper levels, due to the inverse density effect, and at high latitudes, due to the polar focusing mechanism. Initially there is a region of strong mean flow deceleration (about -70 ms^{-1}day^{-1}) in the middle polar stratosphere. This gradually breaks down and is replaced by an intensifying lobe of mean flow acceleration which migrates up from below. The sequence at middle and upper stratospheric levels proceeds with near simultaneity due to the similarity of phase structures of the stationary and traveling waves at these altitudes. These characteristics describe a vacillation which propagates through and is confined by an envelope.

4. Reaction of the mean flow

As shown in Fig. 3, the instantaneous eddy driving of the mean flow is dominated by the transient contribution at high latitudes. This behavior results from the focusing of wave activity when F is diverted poleward as a result of the interference. As we shall explore in more detail shortly, wave amplitudes are large enough to cast doubt on the validity of linear and quasi-linear descriptions. Nevertheless, it is important to examine how much of the observed behavior of January 1979 can indeed be captured by such lower order descriptions. In so doing, we may provide insight into which facets of the evolution are truly indicative of strong nonlinearity.

The reaction of the mean flow to the fluctuating driving is obtained by solving the global boundary value problem (5) with the aid of the meridional streamfunction equation (6). To solve Eq. (6) the Hough spectral scheme described by Plumb (1982) was modified to apply to nonzero frequency components of the meridional streamfunction and include thermal and mechanical damping. Newtonian cooling and Rayleigh friction

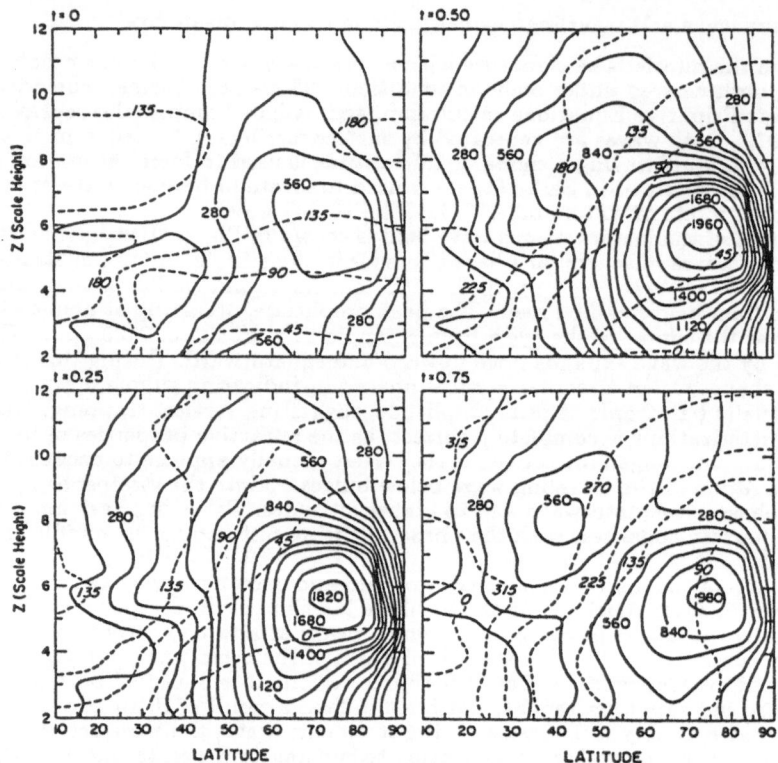

Figure 2 Latitude-height cross-sections of the instantaneous wavefield during
the vacillation. The wavefield is derived from the linearized Primitive
Equations under N.H. winter conditions.

of the same magnitudes as in the wave calculations were prescribed. The problem was
then forced by the eddy driving shown in Fig. 3.

Figure 4 shows the evolution of the mean flow. Initially, polar winds are accelerated
due to the positive EP flux driving occurring shortly before. (From (5.1) it can be inferred that if the balance of the zonal-mean momentum equation is dictated approximately by the zonal wind tendency and the eddy forcing, the zonal flow response should
lag the EP flux driving by roughly 90°.) By one-quarter cycle, a strong easterly flow has
developed almost simultaneously between 6 and 10 scale heights (45 and 70 km). This
subsequently spreads upwards into the mesosphere and is replaced by renewed acceleration in the polar stratosphere which comes up from below. By comparison with Fig. 2
it can be seen that the mean flow reaction is nearly out of phase with the instantaneous
wave field at these altitudes, in agreement with observed variability during winter 1978-
79 (Smith, 1985). Tropical behavior, albeit much less pronounced, is approximately out
of phase with the sequence of events at polar latitudes.

5. Synoptic behavior of geopotential height

We now examine the interference signature in the synoptic field of geopotential height.
We establish here the connection between latitude-height sections of mean flow and wave
fields described previously and the implied behavior in synoptic maps.

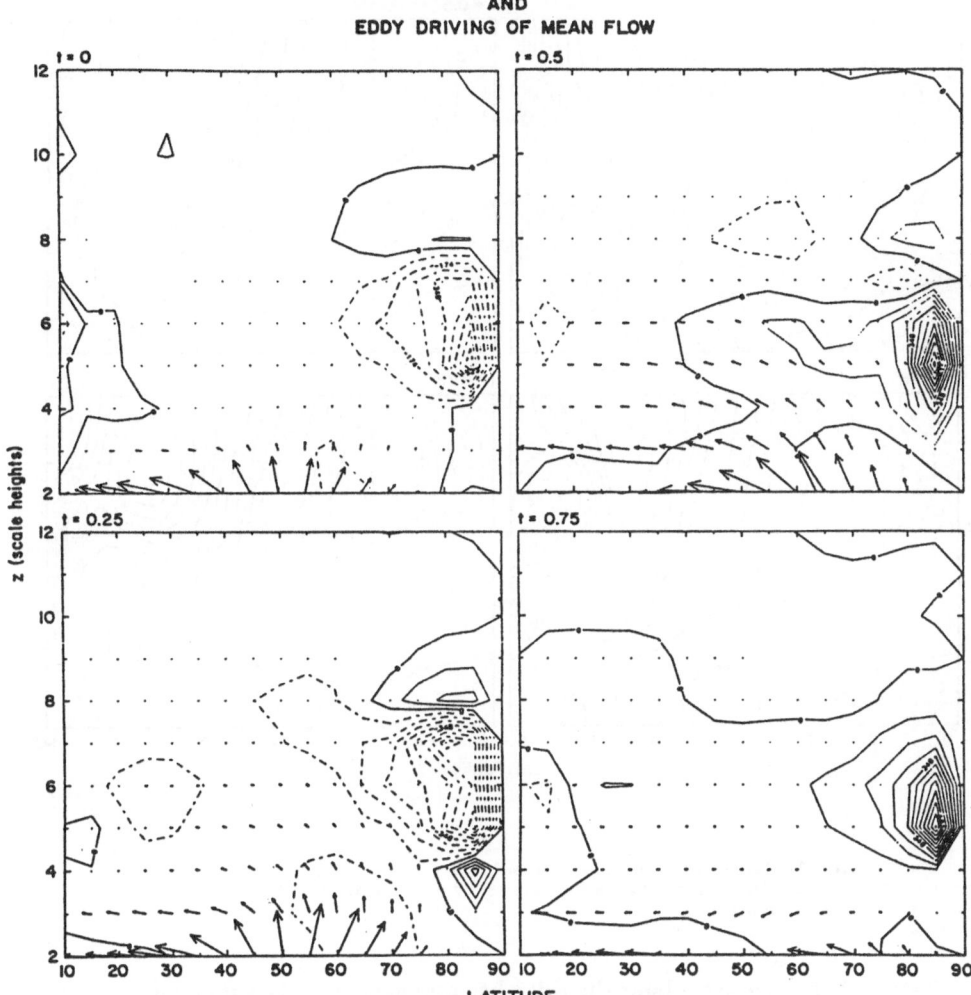

**INSTANTANEOUS EP FLUX
AND
EDDY DRIVING OF MEAN FLOW**

Figure 3 Latitude-height cross-section of the instantaneous EP flux and eddy
forcing of the mean flow for the wavefield shown in Fig. 2.

The evolution of the total geopotential field (zonal mean plus wave) at 4 scale heights
is shown in Fig. 5. The instantaneous synoptic patterns exhibit a rather familiar signature (cf. Fig. 1). The phase of the vacillation during the cycle depends upon the level
chosen. We shall take the initial phase to be at 0.75 cycles and discuss the evolution forward from there. The vortex, centered initially over the pole, is quite intense, having a
maximum depression of 2600 gpm. One-quarter cycle later a ridge of 300 gpm builds
in, and the vortex is simultaneously displaced off the pole and diminished in amplitude

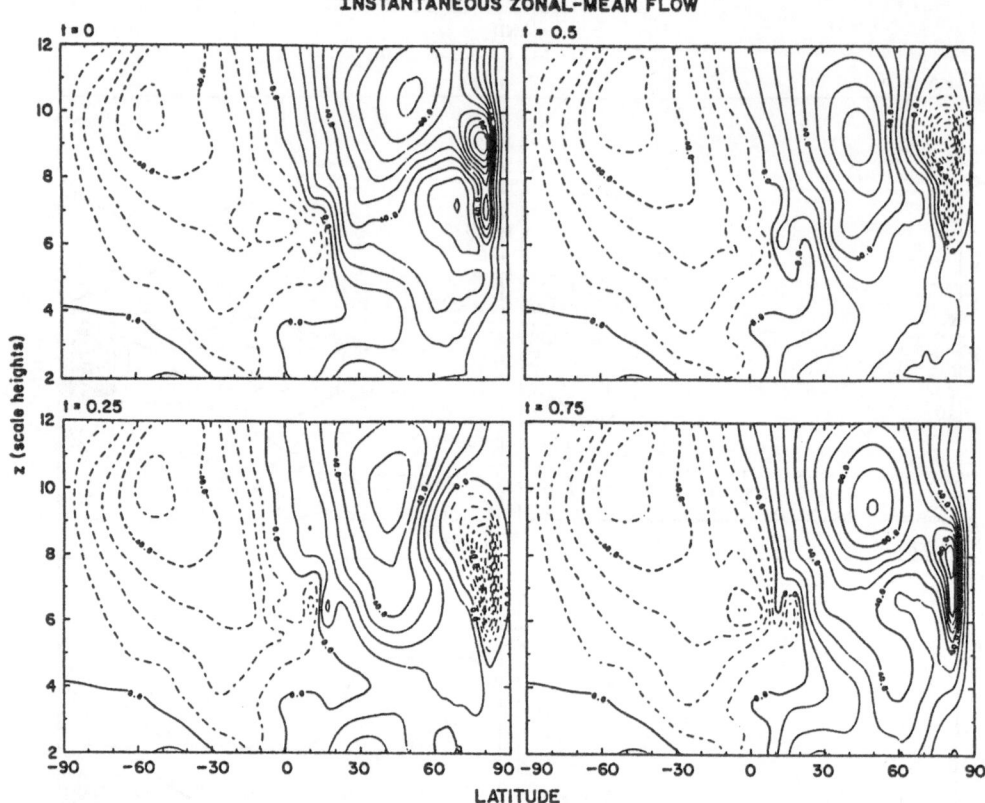

Figure 4 Evolution of the zonal-mean flow during the vacillation cycle. Polar
acceleration and deceleration are nearly out of phase with tropical
variations.

to 1400 gpm. A quarter-cycle later the ridge has propagated westward and amplified to
1000 gpm. At the same time, the vortex has rotated about the pole opposite the ridge
and has become further displaced, its amplitude falling to 1200 gpm. As a result of the
marked intensification of the ridge and simultaneous displacement and collapse of the
vortex, a strong cross-polar flow is established.

Accompanying the retrograde motion is a considerable deformation of regions of high
and low geopotential, each taking on comma shapes and spiraling equatorward about one
another (cf. Fig. 1). As the cycle moves into the final phase, this wobbling about the pole
and distortion continues. The ridge now collapses and the vortex becomes restored to the
pole and reintensifies. During the process the trough axis becomes elongated, spiraling
equatorward about the ridge and ultimately ending up in the tropics. This is a consequence of the lateral phase tilt of the instantaneous wave (Fig. 2).

INSTANTANEOUS GEOPOTENTIAL HEIGHT
(z = 4 scale heights)

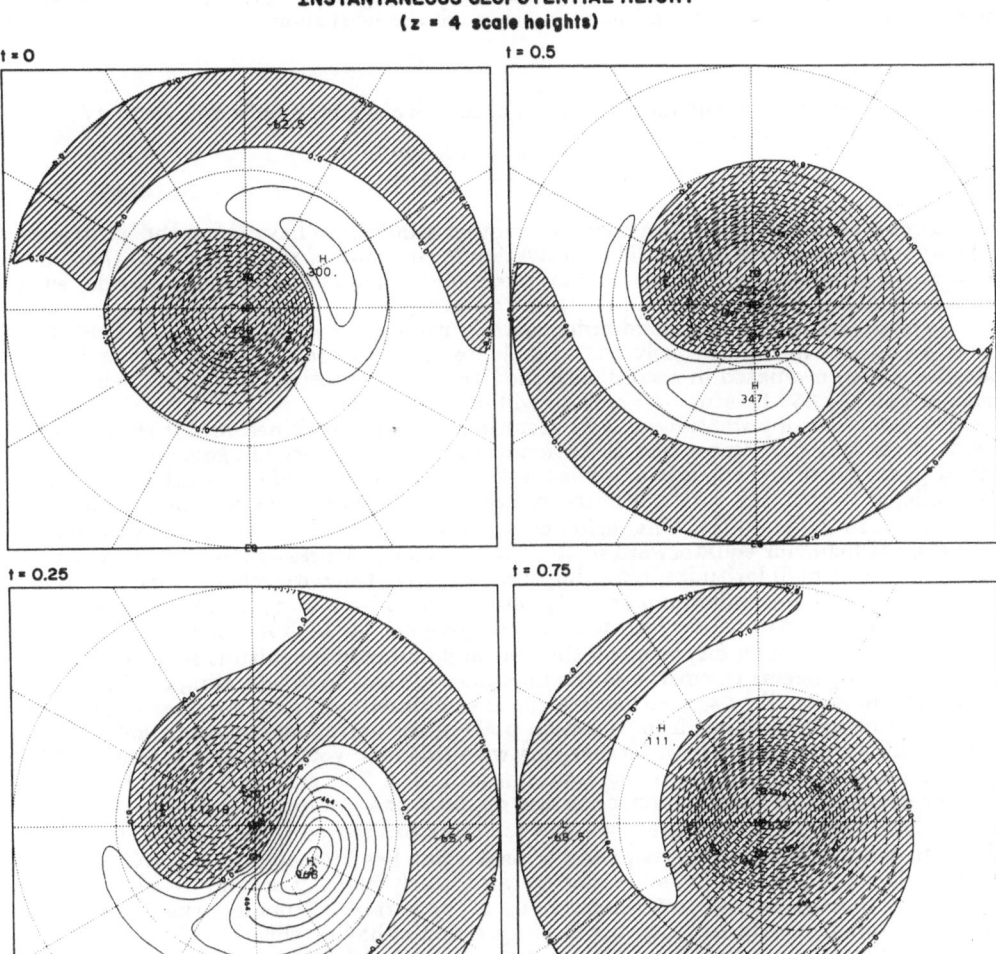

Figure 5 Instantaneous total geopotential height field at 4 scale heights during
the vacillation cycle.

The intensification sequence signals the passage of a capsule or packet of wave activity through the level shown. Because of its equatorward tilt, the eddy geopotential spirals westward with decreasing latitude. At this altitude, the traveling component of the instantaneous wave predominates, resulting in systematic retrogression throughout the cycle. Consequently, the ridge and vortex move continually westward. Together with the equatorward spiraling of ridge and trough axes, the retrogression gives the wavepacket the appearance of a corkscrew propagating through the level shown.

7. Ertel potential vorticity

As the final diagnostic of the interference signature, we examine potential vorticity

$$q = \frac{(f + \mathbf{k} \cdot \nabla \times \mathbf{v}_h) \frac{\partial \theta}{\partial z}}{\rho} \tag{8}$$

on isentropic surfaces where, consistent with other studies, we have taken the isentropes to be approximately horizontal. The motivation for examining this quantity is that under inviscid adiabatic conditions, q is conserved following material elements. Under such conditions, it is therefore a dynamical tracer of transport.

We have evaluated q to second order, namely using the combined synoptic fields to locate a θ surface instantaneously and to evaluate $\mathbf{v}_h$, ρ, and $\partial\theta/\partial z$. All disturbance quantities have been evaluated in accord with the linearized Primitive Equations. Synoptic maps of q on the 850 K isentropic surface will be presented.

The evolution of potential vorticity resulting from the combined field is shown in Fig. 6. Note that the comma-like distortion of the vortex, induced in geopotential by the lateral phase tilt of the wave, is amplified in the higher order field of potential vorticity. Beginning at $t = 0.0$, the vortex is fairly well centered over the pole, attaining a maximum in q of 1.1×10^{-2} (SI units). The region of poleward q gradient is readily distinguished from the flat behavior equatorward of 30°. At $t = 0.25$ a depression in q appears near 30° longitude at high latitudes, reflecting the buildup of the anticyclonic ridge in geopotential (Fig. 5). This q minimum actually can be traced back along an axis around the periphery of the vortex to low latitudes. Accompanying the high latitude depression is the maximum in q being displaced off the pole and falling in magnitude to 0.75×10^{-3}. A tongue of high q begins to emerge at the boundary of the vortex, spiraling clockwise away from its main body. By $t = 0.50$ this tongue is well developed, giving a clear impression of material being drawn out from the main body of the vortex and spiraling anticyclonically about the depression in q. Exchanging with this high vorticity air is a tongue of low q which has intruded into the vortex from low latitudes. The central body of the vortex remains displaced off the pole, retrogressing opposite the q minimum, and has dropped in magnitude to 0.62×10^{-3}. At $t = 0.75$ the vortex has moved back towards the pole, although continuing to retrogress, and the peak value has increased to 1.3×10^{-2}. The outer shell of the high q tonque has now spiraled nearly halfway around the globe, while the inner core has returned to the main body of the vortex. Ultimately the outer shell separates from the main body, giving the impression of having sheared off and left a region of high potential vorticity at low latitudes evident at $t = 0.0$.

Insofar as the material is concerned, detachment in a strict sense cannot occur. Instead of truly separating from the main body of the vortex, the two regions would remain connected by a narrow but nonvanishing filament. No doubt this is the case for the material field. The apparent separation of regions of high potential vorticity is introduced by the narrowing and ultimate collapse of the meridional scale of the tongue below that which can be resolved in this simple calculation. Given greater resolution, i.e., more wavenumbers and frequencies, the q distribution would be capable of remaining intact throughout.

Perhaps equally important is that potential vorticity is not truly conserved. Erosion of q by radiative dissipation may lead to a genuine break in the potential vorticity distribution, even though no such discontinuity occurs in the material field. Similarly, the return of the inner shell of the tongue to the main body of the vortex may also not accurately reflect material displacements. Radiative dissipation can be expected to play an important role here, because relaxation times at this level are of the order of 5-10 days, comparable to the time scale for material advection from polar to tropical regions. In particular, as horizontal gradients of potential vorticity are exaggerated in the advection

INSTANTANEOUS ERTEL POTENTIAL VORTICITY
(θ = 850K)

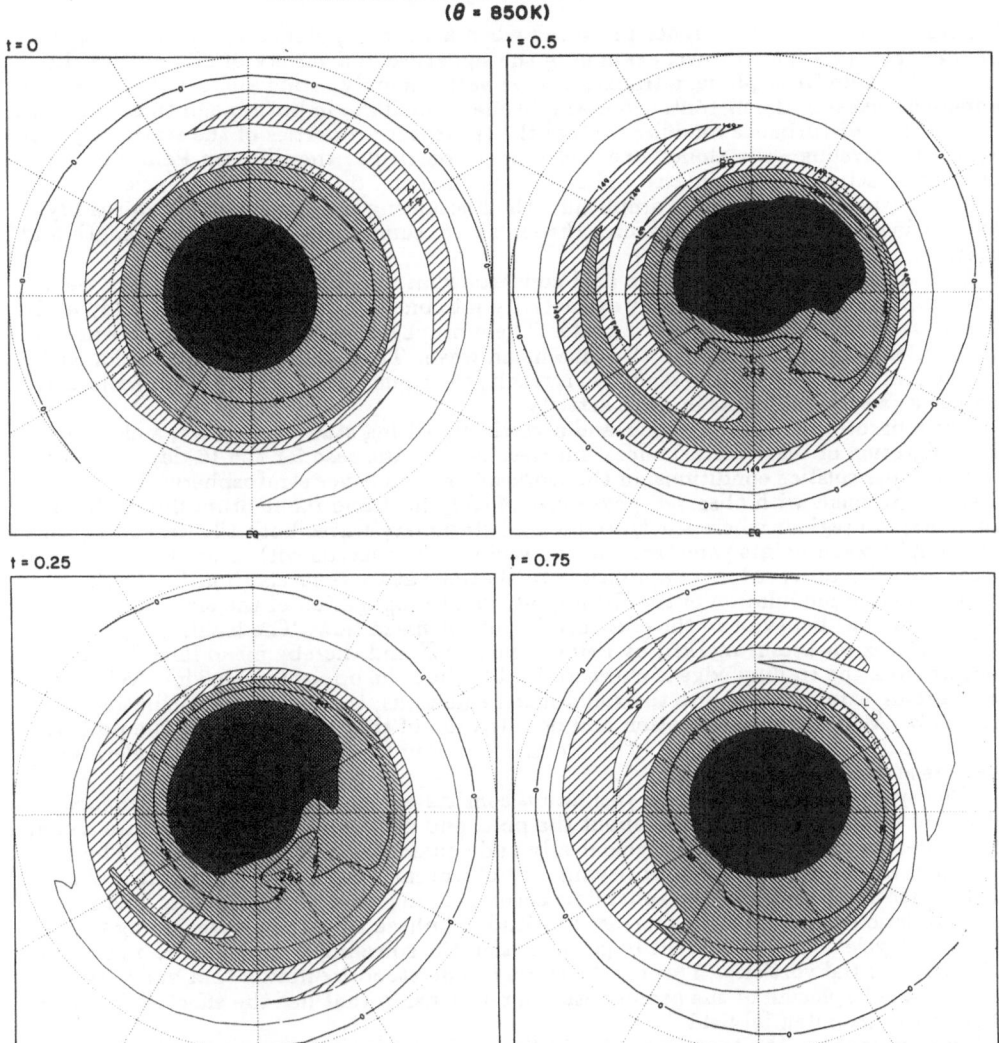

Figure 6 Instantaneous Ertel potential vorticity field on the 850 K isentrope during the vacillation cycle.

process, by thermal wind balance so too are vertical temperature gradients. These would be expected to lead to increased radiative dissipation. Such diminution of q is presumably responsible for the eventual collapse of the trailing segment of the tongue, or *debris*, seen to have spiraled further at one quarter cycle later ($t = 0.25$).

7. Conclusions

We have explored the interference produced when a traveling planetary wave propagates across a stationary forced wave, examining the signature in a variety of diagnostics. In the complex amplitude plane, a transient wave vector simply orbits about the time-mean component, causing the instantaneous amplitude to vacillate in both magnitude and phase. The transient disturbance modulates all of the transport properties of the stationary wave, even if the traveling wave alone fluxes nothing. In particular the Eliassen Palm flux, or flux of wave activity, is modulated in both magnitude and direction. For transient and stationary waves of comparable amplitude, the fluctuating component of F can readily drive the instantaneous EP flux vector through zero, completely altering its strength and direction.

Spatial modulation of the wave field converts what may be, for the stationary wave alone, a simple uniform stream of wave activity into one which is corpuscular, eddy activity localized into capsules or wavepackets. These propagate upward and equatorward in a manner similar to the phase of the time-mean wave. They are attended by a change in both magnitude and direction of F. As a result, the signature realized at a fixed location emerges as a series of *bursts* in wave activity.

For a barotropic traveling wave and a vertically tilting stationary wave, a clear upward migration of these disturbances emerges. Such is the case for the 16-day wave, calculated under solstice conditions, in the troposphere and lower stratosphere. The behavior is in agreement with observed phase lags of eddy heat and momentum fluxes (Madden, 1983). In regions where the traveling and stationary waves both tilt westward, the phase lag between points vanishes, and the vacillation proceeds with near simultaneity, in agreement with observed wave amplitude vacillations at these levels (Smith, 1985).

The spatial modulation of the EP flux field and propagation of the ensuing pattern produces similar behavior in the eddy driving of the mean flow. The leading edge of one of the EP flux packets is marked by rising values of F and thereby mean flow deceleration, whereas the trailing edge is characterized by just the opposite behavior. When this wave forcing is introduced into the zonal-mean equations, the result is a vacillation in the basic flow. Alternating acceleration and deceleration of the zonal flow near the pole was indeed observed during January 1979. Palmer and Hsu (1983) refer to the acceleration phase as stratospheric cooling.

The synoptic signature of the vacillation consists of two basic elements: Displacement and wobbling of the vortex about the pole, and distortion of the vortex into a comma-like shape, its axis spiraling anticyclonically and equatorward about the ridge. The first of these is a consequence of the antisymmetry of wavenumber 1 about the pole and simultaneous deceleration of the zonal flow accompanying wave amplification. The second characteristic results from the increased latitudinal phase tilt of the wave during the interference cycle. Though evident in geopotential, the spiraling is exaggerated in synoptic fields of potential vorticity. This can be simply understood by noting that vorticity is essentially the Laplacian of the geopotential, a relationship that has the effect of *high-pass filtering* the geopotential field.

In combination, the equatorward spiraling and rotation of the pattern about the pole suggest material being drawn out of the vortex. To the extent that potential vorticity is conserved, these motions reflect material displacements. The behavior may also be interpreted in terms of the enstrophy carried aloft by the wavepacket. In particular, eddy potential vorticity on an isentropic surface exhibits a conspicuous and pronounced spiral. If vertical phase tilt is taken into account, the 3-dimensional distribution of eddy enstrophy associated with the wavepacket is in fact helical and localized in the vertical. The spiraling pattern is achieved when this helix of eddy potential vorticity, propagated upwards with the wavepacket, intersects a given θ surface. It is manifest in the combined field initially as distortion of the vortex into a comma-like shape and ultimately as potential vorticity being drawn anticyclonically around the ridge to low latitudes. As potential vorticity spirals equatorward, it becomes elongated and eventually appears to shear off at low latitudes. We have associated the apparent separation of regions of potential vorticity with the limited resolution of our simple calculation and with radiative dissipation of q. Episodes where potential vorticity appears to reattach to the vortex, as has been noted in observations (Clough et al., 1985), occur as well and may be attributed to the same mechanisms.

This alternate interpretation of potential vorticity behavior, viewed in terms of eddy enstrophy being carried upwards with the wavepacket, is perhaps complementary to the interpretation predicated on material motions. Because the eddy potential vorticity field is of zero mean, it contributes no net change to the total hemispheric potential vorticity. Instead, it acts to disturb an existing distribution on a θ surface, in effect redistributing the potential vorticity. It may therefore be viewed as a perturbing influence which rearranges but neither creates nor destroys potential vorticity. To the degree that q is conserved, the same argument can be applied to the material field, in which case the wavepacket may be thought of as inducing advection along the θ surface.

There is an important distinction between the wave breaking and interference descriptions. In an adiabatic interference process, there is no mechanism by which small scales of a *tracer* distribution can be generated, whereas in wave breaking strong nonlinearity presumably leads to a cascade of variance to higher wavenumbers. However, small scales *could* be generated by interference in the potential vorticity distribution through diabatic effects, as we have suggested. Likewise, genuine transport can be accomplished through the induced advection without the intervention of strong nonlinearity. The debris in q remaining after passage of the wavepacket may genuinely reflect a net exchange of material between high and low latitudes. This would follow from diabatic effects which prevent parcels from returning to their original positions.

We have presented calculations based on rather large amplitudes, because they are indicative of a highly documented event (January, 1979). Behavior resulting from smaller amplitudes is similar, only not as dramatic. The morphology which emerges in this simple calculation, with a single wavenumber and only two frequencies, captures the essential character of quasi-periodic disturbances in the stratosphere as reported by Madden and others. It thus supports Madden's speculation that interference might play an important role in disturbed conditions of the stratosphere. Indeed, the fluctuating response of the mean flow, observed for several cycles during winter 1978-79 out of phase with the wave field (Smith, 1985), is reasonably well predicted.

What is remarkable is that even for the large amplitude event of January 1979, one of the strongest on record, the low order formalism is successful in capturing many of the salient aspects of the overall unsteady behavior —whether viewed from the perspective of mean and eddy components or from synoptic fields. At such amplitudes, particularly when the mean flow is strongly decelerated, the mathematical underpinnings of linear and quasi-linear descriptions break down at least locally, i.e., they cannot be justified *a priori*. Behavior where disturbance and mean velocities are comparable, e.g., near a critical surface or in general where the streamfield has closed contours, is most in doubt. The quasi-linear description may be regarded as an asymptotic series solution to the complete problem, neglecting contributions third order and higher in eddy amplitude. A fundamental issue is just how much and which aspects of the actual behavior can be captured by the lowest orders of the complete description. Our results indicate that the gross aspects of the evolution are indeed represented in the low order formalism. In particular, the displacement and wobbling of the vortex and its distortion and equatorward spiraling are all predicted.

Our results *do not* preclude the existence of strong nonlinearity in such episodes. On the contrary, there are suggestions that such behavior might well arise out of amplification and from the synoptic configurations realized during the interference process. For instance when the vortex buckles under distortion, regions of reversed potential vorticity gradient are created. These may be unstable under local considerations analogous to Rayleigh's criterion, or in three dimensions Charney and Stern's (1962). Ensuing behavior might lead to the mixing of potential vorticity and concomitant depletion of the vortex as proposed by McIntyre and Palmer (1983). However, by the same token, such features would be expected to be of smaller dimension than the large-scale pattern from which they evolve. McIntyre and Palmer's interpretation is complicated by the fact that there exist close parallels in the quasi-linear behavior. Elements of potential vorticity, through limited resolution and thermal dissipation, can give the appearance of shearing off at low latitudes and at times reattaching to the main body of the vortex. What is rather strongly indicated by our results is that the gross morphology of the January 1979 warming, specifically the distortion and spiraling of potential vorticity to low latitudes and arguably all that is genuinely resolved in satellite measurements, is not in itself a reflection of strong nonlinearity and eddy mixing.

8. Acknowledgement

The National Center for Atmospheric Research is sponsored by the National Science Foundation.

9. References

Charney, J. and M. Stern, 1962: In the stability of internal baroclinic jets in a rotating atmosphere. *J. Atmos. Sci.*, **19**, 159-172.

Clough. S.A., N.S. Grahame and A. O'Neill, 1985: Potential vorticity in the stratosphere derived using data from satellites. *Quart. J. Royal Meteor. Soc.*, **111**, 335-358.

Lindzen, R.S., D.M. Strauss and D. Katz, 1984: An observational study of large- scale atmospheric Rossby waves during FGGE. *J. Atmos. Sci.*, **41**, 1320- 1335.

Madden, R.A., 1975: Oscillations in the winter stratosphere: Part 2 The role of horizontal eddy heat transport and the interaction of transient and stationary planetary scale waves. *Mon. Wea. Rev*, **102**, 717-729.

Madden. R.A., 1977: Evidence for large-scale regularly propagating waves in a 73-year data set. *Extended Summaries of Contributions*, IAGA/IAMAP, Seattle, WA, International Association for Atmospheric Physics, National Center for Atmospheric Research.

Madden, R.A., 1978: Further evidence of traveling planetary waves. *J. Atmos. Sci.*, **35**, 1605-1618.

Madden, R.A., 1983: The effect of interference of traveling and stationary waves on time variations of the large-scale circulation. *J. Atmos. Sci.* **40**, 1110-1125.

Madden. R.A. and K. Labitzke, 1981: A free Rossby wave in the troposphere and stratosphere during January 1979. *J. Atmos. Sci.*, **86**, 1247-1254.

McIntyre, M.E. and Palmer, 1983: Breaking planetary waves in the stratosphere. *Nature* **305**, 593-600.

McIntyre, M.E. and Palmer, 1984: The surf zone in the stratosphere. *J. Atm. Terr. Phys.*, **46**, 825-850.

Palmer, T.N. and C.-P. Hsu, 1983: Stratospheric sudden coolings and the role of non-linear wave interactions in preconditioning the circumpolar flow. *J. Atmos. Sci.*, **40**, 909-928.

Plumb, R.A., 1982: Zonally-symmetric Hough modes and meridional circulations in the middle atmosphere. *J. Atmos. Sci.*, **39**, 938-991. 645-648.

Rinne, J. and A. Sarkanen, 1985: 500 mb geopotential height spectral peaks in the 1- to 5-week range. *Tellus*, **37**A, 323-335.

Salby, M.L., 1981: Rossby normal modes in nonuniform background configurations. Part I: Simple nonuniformities. *J. Atmos. Sci.*, **38**, 1803-1826.

Salby, M.L., 1984: Transient disturbances in the stratosphere: implications for theory and observing systems. *J. Atmos. Terr. Phys.*, **46**, 1009- 1047.

Smith, A.K., 1985: Wave transience and wave-mean flow interaction caused by the interference of stationary and traveling waves. *J. Atmos. Sci.*, **42**, 529-535.

4. RADIATIVE PROCESSES AND TRANSPORT

RESPONSE OF THE MIDDLE ATMOSPHERE TO CHANGING O_3 AND CO_2
- A SPECULATIVE TUTORIAL

Stephen B. Fels

Geophysical Fluid Dynamics Laboratory / NOAA
Princeton University, Princeton NJ 08542, USA

I) Introduction

The notes that follow are very much in the spirit of the original
lecture - they are pedagogical in intent, and contain almost nothing
which has not already appeared somewhere else in the literature. The
one important exception to this are the remarks on the obstacles which
seem to me to stand in the way of a reliable prediction of the response
of the middle atmosphere to altered composition.

Given the premise that CO_2 and O_3 amounts will change, several
zeroth order questions suggest themselves:

   * How will the zonal mean climatology change?

   * How will the planetary wave structure change?

   * How will various time dependent phenomena such as the SAO, QBO,
     and seasonal cycles change?

In what follows, we shall concentrate on the first question; we shall
see however that the answer to this must ultimately depend on understanding the second. We shall have nothing to say here about the third;
In the original lecture, one brief speculation which may be relevant was
offered.

At the outset, we remind the reader of one trivial but essential fact:

It is only through their radiative effects that CO_2 and O_3 affect the
structure of the middle atmosphere. It is for this reason that any
discussion of altered amounts of these gases must begin with radiative
transfer.

We shall take a fairly pedestrian approach, and begin with a radiation
primer, using a few figures to remind ourselves of the fundamentals of
that subject. This will lead us to the idea of radiatively determined

G. Visconti and R. Garcia (eds.), Transport Processes in the Middle Atmosphere, 371–386.
© 1987 by D. Reidel Publishing Company.

states, and the quantities

$$T_{rad}, \ U_{rad}, \ \tau_{rad}, \ \text{and} \ Q_{dyn}.$$

Next we shall discuss simple radiative models for the response to altered ozone and carbon dioxide, and move on to some possible radiative-dynamical feedbacks. This latter portion will draw on some interesting work done several years ago at NCAR.

II) Radiatively Determined States

The cartoon below illustrates the rudiments of shortwave heating and long wave cooling in the middle atmosphere; The most important things to keep in mind are written in the boxes at the bottom of the figure.

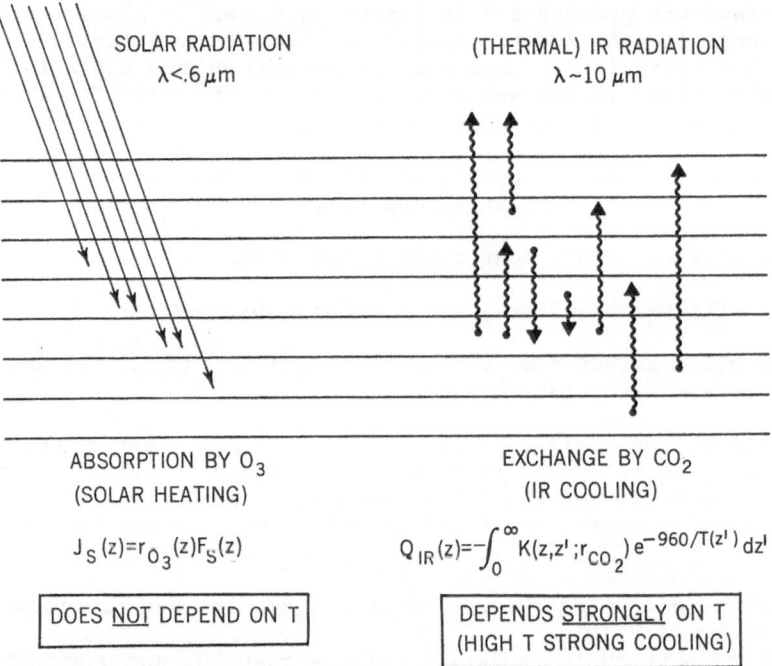

SOLAR RADIATION
$\lambda < .6 \, \mu m$

(THERMAL) IR RADIATION
$\lambda \sim 10 \, \mu m$

ABSORPTION BY O_3
(SOLAR HEATING)

$$J_S(z) = r_{O_3}(z) F_S(z)$$

DOES NOT DEPEND ON T

EXCHANGE BY CO_2
(IR COOLING)

$$Q_{IR}(z) = -\int_0^\infty K(z,z'; r_{CO_2}) e^{-960/T(z')} \, dz'$$

DEPENDS STRONGLY ON T
(HIGH T STRONG COOLING)

Fig.1   The essentials of radiative transfer in the middle atmosphere

The next figure reminds us that radiative transfer usually acts to damp out small perturbations of temperature away from an equilibrium profile. This has been discussed by many authors - see, for example, Fels, (1982). An interesting discussion of a case where radiation amplifies perturbations is given in Held, Pierrehumbert, and Panetta, (1986).

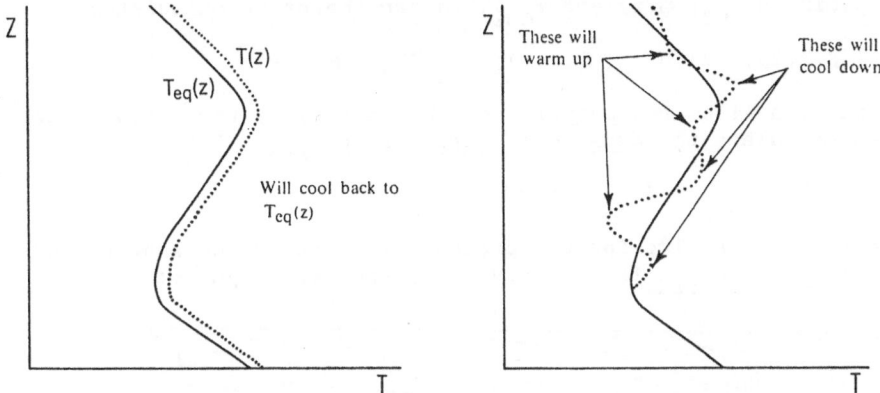

Fig.2 The rudiments of radiative relaxation

It is sometimes possible to treat this non-local damping of
disturbances by radiation as a local process, leading to the idea of a
radiative relaxation time τ_{rad}: this is always true for
perturbations with large vertical scale, and can be usefully
generalized for many sinusoidal waves. The basic ideas are shown below.

Choose a profile $T_o(z)$, and calculate the IR cooling $Q_o(z)$ for it. Now
let $T(z)$ be a perturbation of T_o, and $Q(z)$ the corresponding cooling
rate. Then

$$Q(z) = Q_o(z) + \int_o^\infty K(z,z')(960/T_o^2)e^{-(960/T_o)}(T(z')-T_o(z'))dz' .$$

    If the function K is sharply peaked about $z=z'$, we can pull
everything else out of the integral, and write:

$$Q(z) = Q_o(z) + [T_o(z)-T(z)]/\tau_{rad}$$

    τ_{rad} is called the radiative relaxation time.

With the above in mind, we can now consider the thermodynamic
equation: all of the dynamics is on the left, and all the radiation
on the right.

$$\frac{\partial T}{\partial t} + \underbrace{\vec{u}\cdot\vec{\nabla}T + w\left(\frac{\partial T}{\partial z} + g/c_p\right)}_{-Q_{dyn}} = J_{solar} + Q_{IR}$$

For a specified $r_{O3}(\theta,z)$ and r_{CO2}, we can therefore calculate:

1) $T_{rad}(\theta,z,t)$;   2) $U_{rad}(\theta,z,t)$;   3) $\tau_{rad}(\theta,z,t)$

If we are also given the <u>actual</u> (not the radiative) temperature, we can also calculate the <u>dynamical heating rate</u> Q_{dyn}:

$$0 = Q_{dyn} + J_{solar} + Q_{IR}$$

What do these look like for the present atmosphere? We show a time-marched T_{rad} and U_{rad} for January 15 in the next figures:

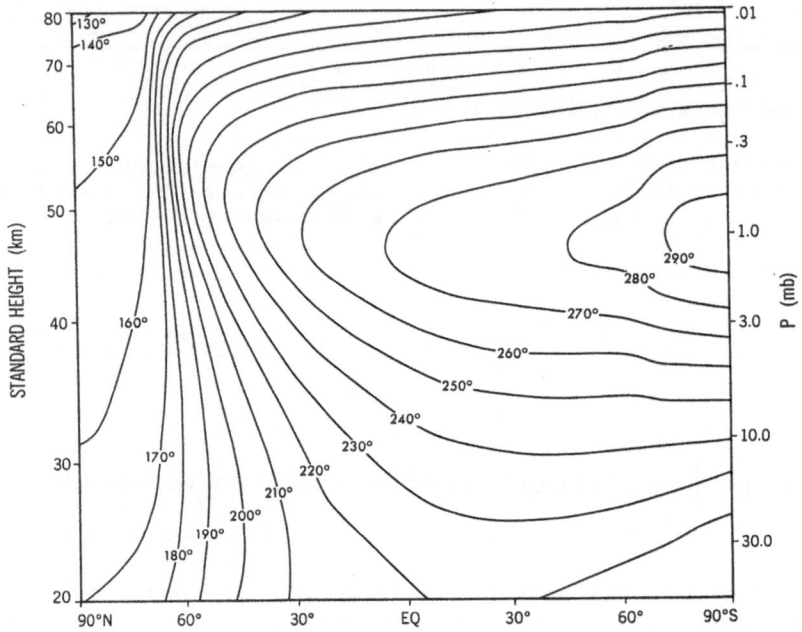

Fig.3 Radiative-convective-photochemical T_{rad} for January 15

A few comments are in order here:

1) The ozone mixing ratio above about 10 mb is not taken from observations; instead it is calculated by means of a rather complicated 1-D photochemical model developed by my collegues S. Liu, J.D. Mahlman, and M.D. Schwarzkopf.

2) The temperatures at the surface are taken from observations, using the data of Oort (1983).

3) There is a convective adjustment which never allows the lapse rate to exceed $-6.5°$/km.

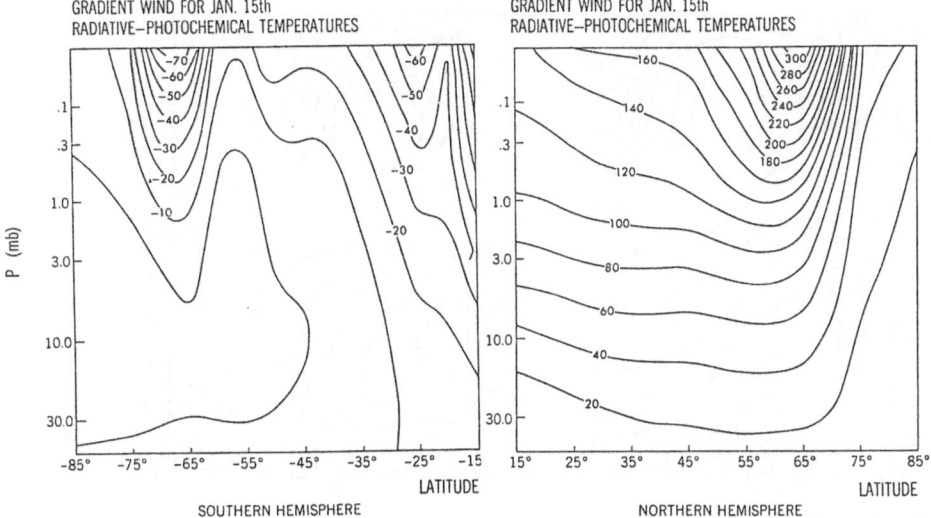

Fig.4   The gradient winds U_{rad} (m/s) for January 15

Upon comparing these results to the observed temperature and zonal winds shown below, (taken from Barnett and Corney, 1985), we immediately recognize that although the tropics and summer hemisphere agrees crudely with the radiative results, the winter polar night is much warmer: 100° at 50 km!   This is of course also reflected in the ridiculously large values of the radiative winter hemisphere zonal winds.

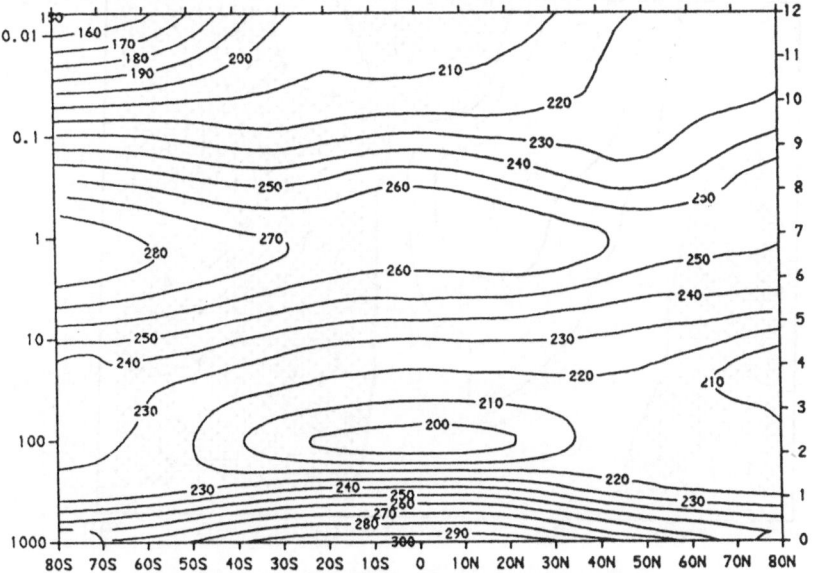

Fig.5   The observed January temperature

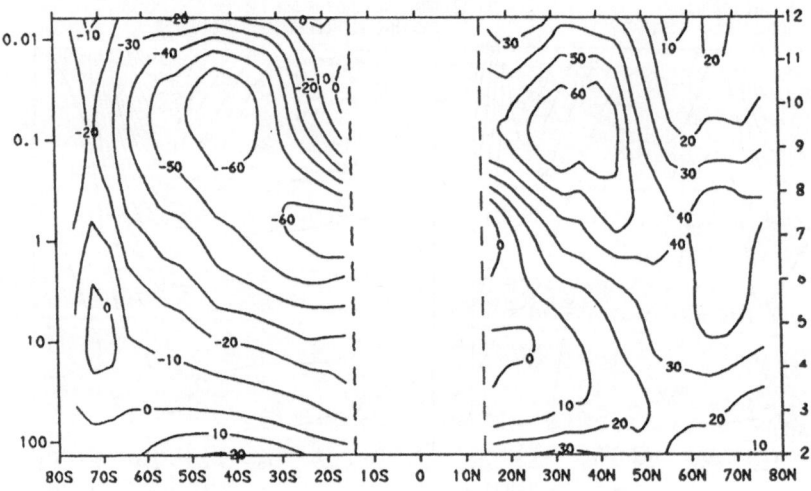

Zonal mean wind (m/s)

Fig.6 The observed January zonal mean zonal wind U

The above is simply telling us that the winter polar night is
undergoing very strong dynamical heating, which is seen very clearly
in the following plot of $J_{solar} + Q_{IR} = -Q_{dyn}$, taken from
Kiehl and Solomon (1986).

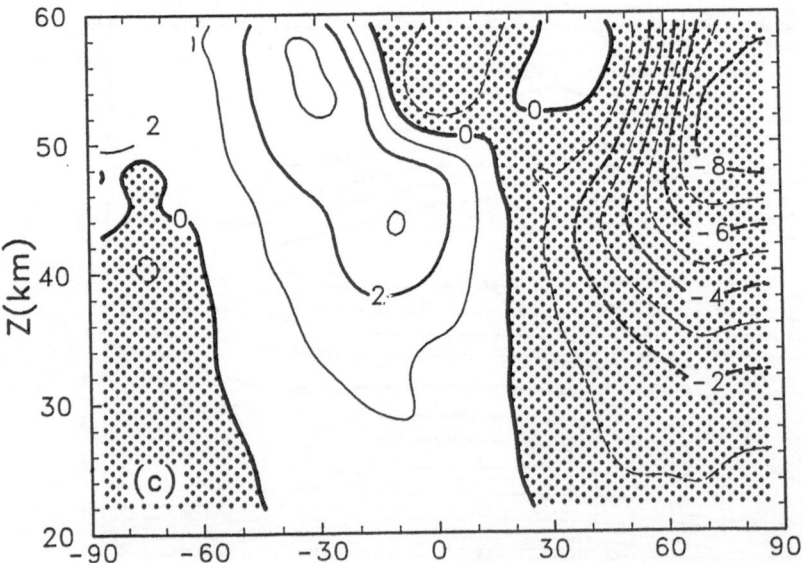

Fig.7 The net radiative heating ($= -Q_{dyn}$) ($^\circ$/day) for January

The last of the relevent quantities to be displayed are the radiative relaxation times τ_{rad}. In the next figures, we show both the "classical" Newtonian cooling values, recently computed by Kiehl and Solomon (1986), and their scale dependent generalizations from Fels (1982).

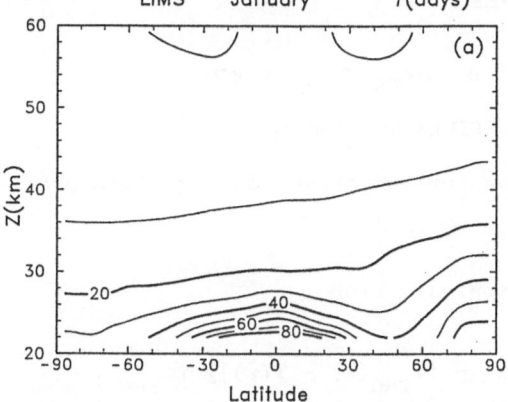

Fig. 8   The radiative relaxation times for disturbances of very large vertical scale, as a function of height and latitude

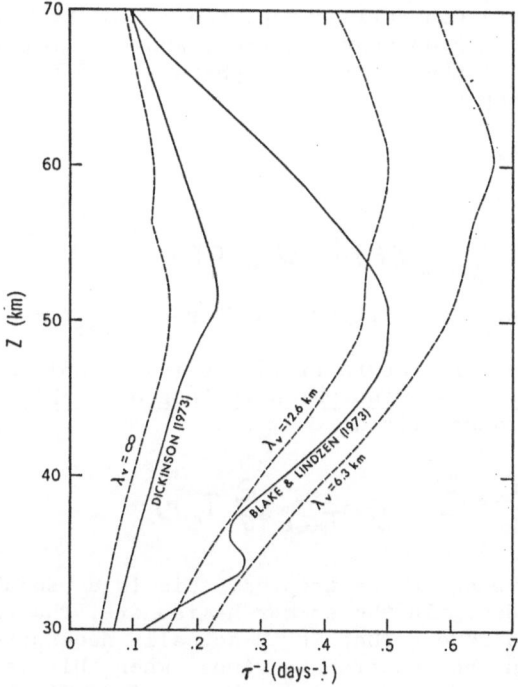

Fig. 9   The scale-dependent radiative damping rates for the sub-tropical middle atmosphere

We can now define several simple models of the middle atmosphere which
entirely neglect dynamical heating.

1) Radiative Equilibrium:

$$J_{solar} + Q_{IR}(T) = 0$$

This DEFINES $T_{rad}(\theta, z)$

If, in the expression for radiative cooling given previously, we let

$T_o(z) = T_{rad}(z)$,

we have the useful approximation

$$Q_{IR}(z) + J_{solar}(z) = [T_{rad}(z) - T(z)]/\tau_{rad}$$

In most parts of the middle atmosphere, the radiative relaxation time
is short compared to the seasonal time scale, so that lag effects due
to thermal inertia are generally unimportant. In the polar night,
however, this is not a good approximation, and a better indication of
the radiative state of the middle atmosphere is had by retaining the
tendency term, leading to

2) Radiative March:

$$\frac{\partial T}{\partial t} = J_{solar}(t) + Q_{IR}(T)$$

This defines a slightly different $T_{rad}(\theta, z)$.

Given this radiative temperature field, it is a simple matter to
calculate U_{rad}, that zonal wind which is in geostrophic balance with
the radiative temperature field:

$$f \frac{\partial U_{rad}}{\partial z} = - g \frac{1}{T_{rad}} \frac{\partial}{\partial y} T_{rad}$$

Although of no relevence in the tropics, this is a useful quantity in
mid and high latitudes. In the summer hemisphere, the radiative
easterlies become so strong that it is actually necessary to use the
gradient rather than the geostrophic wind. When this is done,
however, problems can arise due to breakdown of gradient wind balance
in the mesosphere. (Shine, 1987)

III) Effects of Altered O_3 and/or CO_2

1) Suppose we change O_3; this will primarily effect J_{solar}.
In equilibrium then,

$$0 = \delta J_{solar} + \delta Q_{dyn} + \frac{\delta Q_{IR}}{\delta T} \delta T$$

The simplest plausible model of the response to this is to assume that
the dynamical heating does not change

$$\delta Q_{dyn} = 0 \; ;$$

$$\delta T(z) = - \delta J_{solar} \left(\frac{\delta Q_{IR}}{\delta T} \right)^{-1}$$

$$\delta T(z) = \mathcal{C}_{rad} \, \delta J_{solar}$$

This is called the Fixed Dynamical Heating (FDH) model, (Fels and
Kaplan, 1975, Fels et. al., 1980). Notice that this does not assume
that the original state was one of radiative equilibrium, but only
that the dynamical heating does not change.

2) We can use the same model for the response to altered CO_2; we need
only recognize that in this case, the dependence of Q_{IR} on the CO_2
mixing ratio r must be considered:

$$\frac{\delta Q_{IR}}{\delta r} \delta r + \frac{\delta Q_{IR}}{\delta T} \delta T = 0 \; ;$$

$$\delta T = \mathcal{C}_{rad} \left(\frac{\delta Q_{IR}}{\delta r} \delta r \right)$$

Notice that these calculations do not require any dynamical
information, but only the actual temperature structure of the
unperturbed atmosphere.

When the FDH model is used to predict the middle atmosphere's response

a) a uniform 50% reduction in O_3

b) a uniform doubling of CO_2,

the results (taken from Fels et. al., 1980) are as shown below:

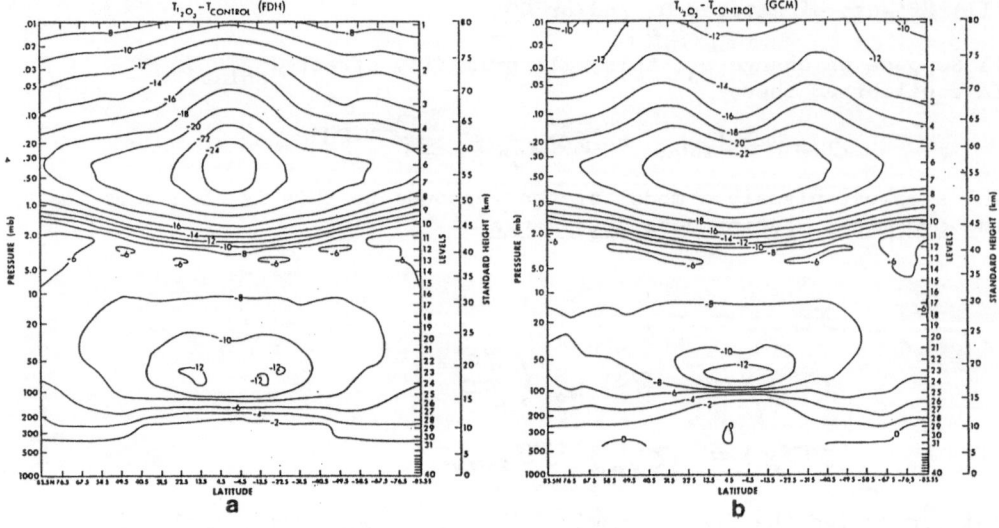

Fig.10   FDH and GCM response to a uniform 50% reduction in O_3

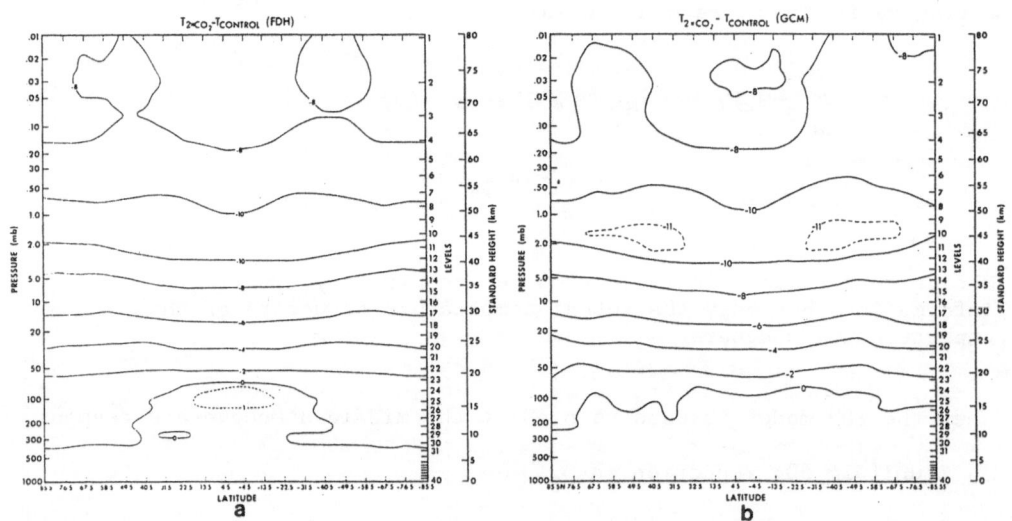

Fig.11   FDH and GCM response to a doubling of CO_2

Adjacent to each of the two FDH results is also shown the response of a GCM with _seasonally_ _averaged_ _insolation_ to these two perturbations. Remarkably enough, this GCM has a climatology which is not terribly different from that observed for the winter hemisphere middle atmosphere We observe that the agreement between the FDH and GCM runs is generally excellent. This is a very misleading result, however, because the temperature structure of the GCM is very close to T_{rad} throughout its entire middle atmosphere, while of course the same is most emphatically not true for the winter middle atmosphere, as discussed previously. _Thus,_ _this_ _GCM_ _run_ _really_ _contains_ _little_ _more_ _information_ _about_ _the_ _probable_ _response_ _of_ _the_ _real_ _middle_ _atmosphere_ _than_ _does_ _the_ _much_ _simpler_ _FDH_ _model._ Now of course, it is possible that the FDH assumptioi is indeed correct - indeed, in the tropics and summer hemisphere, which are close to radiative equilibrium, it probably is. In the winter hemisphere, however, we must have either a theory or a good numerical model of the response of Q_{dyn} to altered O_3 and CO_2. We next look briefly at each of these.

The trouble with most seasonal GCM's is that they do a very poor job at simulating the winter middle atmosphere. For example, the GFDL SKYHI models constructed by Mahlman and collaborators is generally much too radiative at the winter pole, with dynamical heating rates of only one o two degrees at 50 km, rather than the eight or more required (Mahlman an Umscheid, 1983). These models therefore can be expected to respond in a manner similar to a FDH model whose basic temperature state is much too cold at the winter pole. In fairness to Mahlman, I should point out tha this "cold bias" becomes demonstrably less serious as one increase the horizontal resolution, and as has been discussed in his lecture in this series, the 1° horizontal resolution model seems to be very close to observations. Unfortunately, the model takes a long time to run - about 10 hours of CYBER 205 time per model day. It is thus not a toy to be played with frivolously, and I do not expect to use it for altered composition experiments in the next several years.

Now of course, there are other models which apparently do an excellent job of simulating the thermal structure of the stratosphere, and these are often very simple to run. Many of the 2D models, such as that described by K.K. Tung in these lectures fall into this catagory. Why not use them?

The problem with many of these is that in effect, the dynamical heating is put in "by hand". There are many ways of doing this, some of them quite ingenious. Use of eddy transports of heat and momentum deduced fr observations is one way, and of appropriately inserted Rayleigh friction another. While by no means trivial, it is possible to make such models give a reasonable representation for Q_{dyn} in the current middle atmosphere. The real question, though, is how this quantity will (or will not) change when the temperature is altered radiatively. For this purpose, the 2D models would seem to be inadaquate.

IV) <u>Response</u> <u>of</u> <u>Dynamical</u> <u>Heating</u> <u>to</u> <u>Changed</u> <u>Thermal</u> <u>Structure</u>
   (A lesson from NCAR?)

There are several intriguing hints that Q_{dyn} in the polar night
stratosphere may depend very sensitively on other characteristics of the
stratosphere. To my mind, the very existence of sudden warmings is the
most suggestive of these. The large difference between the winter
climatologies of the southern and northern hemispheres is another. Quite
recently, a series of experiments performed by Ramanathan and his
collaborators at NCAR have provided another interesting example
(Ramanathan et. al., 1983).

In an attempt to discover why their model had the same sort of polar cold
bias as mentioned above, these workers made a number of rather subtle
changes in the radiative transfer algorithm used in the GCM, and
discovered that by virtue of changing the radiative parameters, <u>the</u>
<u>dynamical</u> <u>heating</u> <u>also</u> <u>changed</u> <u>significantly</u>. In fact, due to increased
dynamical heating, their polar stratospheric temperatures warmed up
substantially more than would be predicted by the FDH assumption.
Apparently, then, a relatively small change in the radiation "flipped"
the model from a state with little dynamical heating to one with
significant dynamical heating. It is worth pointing out that this effect
seems to depend crucially on the particular dynamical model used. When
the same changes were made in a different NCAR model, no interesting
changes in the dynamical heating occured. These results, while by no
means conclusive, suggest that use of the FDH assumption ought to be
regarded with some caution; in the next section, we will look at a simple
conceptual model which may throw some light on how the FDH model might
fail.

A <u>Way</u> <u>of</u> <u>Thinking</u> <u>about</u> <u>Dynamical</u> <u>Response</u>
(Really only interesting if there are waves in the system)

Let's keep things as simple as possible, and use a zonally averaged,
Boussinesq, f-plane model for the residual circulation of the middle
atmosphere. (The "residual circulation" is a concept that plays a central
role in current thinking about stratospheric dynamics and transport; if
you don't know what it is, read Andrews and McIntyre, 1976)

1) $$-f v^* = \nabla \cdot E$$

2) $$f U_z = -\frac{R}{H} T_y$$

3) $$N^2 w^* = \frac{R}{H} \frac{1}{\tau_{rad}} (T_{rad} - T)$$

4) $$v_y^* + w_z^* = 0$$

In the above, $\nabla \cdot E$ is the famous Eliassen-Palm Flux Divergence. It contains all of the information about the <u>effect</u> of <u>waves</u> <u>on</u> the <u>zonally</u> <u>averaged</u> <u>flow</u>. Notice also that we have put a lot of radiation physics into the thermodynmic equation; the quantities T_{rad} and τ_{rad} have been introduced.

We can boil these down to one fairly simple equation:

$$\frac{\partial}{\partial z} \left(\frac{T - T_{rad}}{\tau_{rad}} \right) = \frac{N^2}{f} \frac{H}{R} \frac{\partial}{\partial y} (\nabla \cdot E)$$

A more useful form of this makes use of the thermal wind equation to write:

$$\frac{f^2}{N^2} \frac{\partial^2}{\partial z^2} \left(\frac{U - U_{rad}}{\tau_{rad}} \right) = \frac{\partial^2}{\partial y^2} (\nabla \cdot E)$$

  1) Radiation determines U_{rad}
  2) Radiation determines τ_{rad}
  3) Radiation partially determines $\nabla \cdot E$.

If we take the horizontal and vertical scales of the forcing to be L and D respectively, we can write

$$U_{rad} - U \sim \frac{N^2 D^2}{f^2 L^2} \tau_{rad} \nabla \cdot E \equiv \tau_{rad} F$$

This is most interesting: it tells us that <u>the</u> <u>effect</u> <u>of</u> <u>the</u> <u>waves</u> <u>(or</u> <u>eddies)</u> <u>is</u> <u>to</u> <u>drive</u> <u>U</u> <u>away</u> <u>from</u> <u>its</u> <u>radiative</u> <u>value</u>;

Moreover, <u>for</u> <u>a</u> <u>given</u> $F_\perp$ <u>the</u> <u>smaller</u> τ_{rad}, <u>the</u> <u>smaller</u> <u>will</u> <u>be</u> <u>the</u> <u>deviation</u> <u>of</u> <u>U</u> <u>from</u> U_{rad}.

For this reason, it is useful to think of a "radiative spring", trying to restore the system to it radiative equilibrium state.

Now we come to the really interesting part: It is possible that F may depend very strongly on U, due to the complicated effect of the mean flow on waves propagating through it. In the sketch below, we illustrate a graphical solution to the above equation. We plot two quantities as a function of U : U_{rad}-U (solid line), and $\tau_{rad}F$ (dashed and dotted lines).

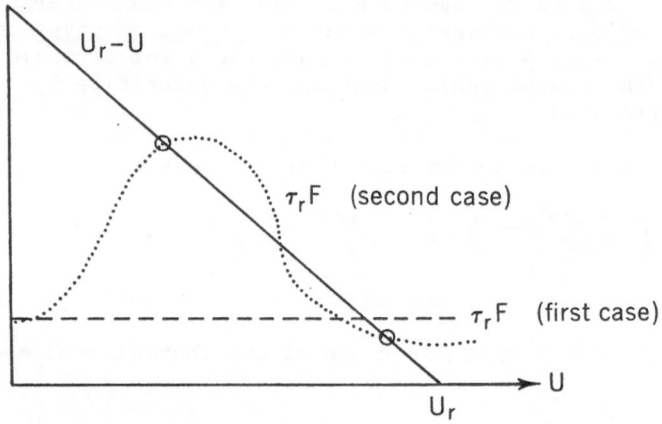

Fig.12   Schmematic illustration of multiple and single equilibrium
         solutions to the radiative-dynamical equilibrium problem

In one case, (the dashed line), $\tau_{rad}F$ is independent of U, and we have
only one solution. In the other case (the dotted line), we have the sort
of dependence which a simple model of a Rossby wave propagating upward
into a westerly flow would imply. Now there are three solutions, of
which two are stable. One of these, the "radiative" solution, has a
large U and small wave forcing, while the other, the "dynamical"
solution, has a small U and large waves.

In this model, then, prediction of the effect of altered composition
requires that we answer three questions:

1) How will U_{rad} change?

2) How will τ_{rad} change?

3) How will $\nabla \cdot E$ change? (for a given U)

In the doubled CO_2 problem, there are some partial answers:

1) U_{rad} will change very little, assuming no important changes in
   temperature gradients in the free troposphere.

2) τ_{rad} will be decreased by about 25% throughout most of the stratosphere. This will <u>increase</u> the strength of the radiative spring.

We do not really know much about how $\nabla \cdot E$ will change; to the extent that
it arises from mechanical mixing and wave breaking, however, we expect it
to be insensitive to radiation.

The figure below shows what might happen for the two different scenarios;

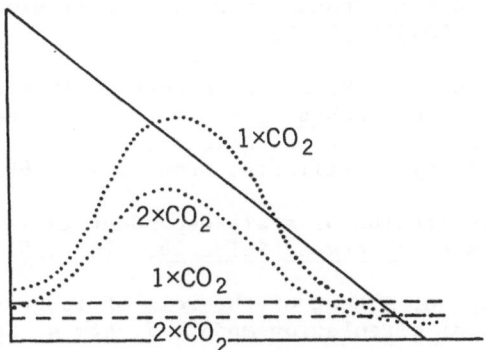

Fig.13   Response of radiative-dynamical equilibria to changes in τ_{rad}, assuming that U_{rad} does not change

In one case, (the dashed line), the solution was radiative originally, and changing τ_{rad} makes very little difference. In the second case (the dotted line), however, we see that for 2X CO_2, we have lost the "wavey" solution, so that the system will revert to the very different radiative one!

There are many grounds on which one could criticize the above, and it is certainly a drastic oversimplification at best. If I had to bet money, my guess would be that the all-important dependence of F on U has some structure, but is much less sharply peaked than our dashed curves. In this case, there would be only one solution, somewhere in between the wavey and radiative ones. Even here, however, we would expect that altering τ_{rad} would result in a significant violation of the FDH assumption, as indicated in the final figure.

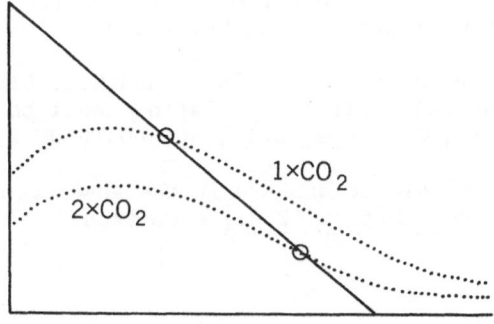

Fig.14   Response of less highly tuned atmosphere to a decrease in τ_{rad}

References

Andrews, D.G., and M.E. McIntyre, Planetary waves in horizontal and vertical shear: Asymptotic theory for equatorial waves in weak shear, J. Atmos. Sci., 33, 2049-2053, 1976.

Barnett, J.J., and M. Corney, Middle atmosphere reference model derived from satellite data, in Handbook for MAP, Vol. 16, edited by K. Labitzke, J.J. Barnett, and B. Edwards, pp. 47-85, SCOSTEP Secretariat, University of Illinois, Urbana, IL, 1985.

Fels, S.B., A parameterization of scale-dependent radiative damping in the middle atmosphere, J. Atmos. Sci., 39, 1141-1152, 1982.

--------, and L.D. Kaplan, A test of the role of longwave radiative transfer in a general circulation model, J. Atmos. Sci., 33, 779-789, 1975.

--------, J.D. Mahlman, M.D. Schwarzkopf, and R.W. Sinclair, Stratospheric sensitivity to perturbations in ozone and carbon dioxide: radiative and dynamical response, J. Atmos. Sci., 37, 2266-2297, 1980.

Held, I.M., R.T. Pierrehumbert, and R.L. Panetta, Dissipative destabilization of external Rossby waves, J. Atmos. Sci., 43, 388-396, 1986.

Kiehl, J.T., and S. Solomon, On the radiative balance of the stratosphere, J. Atmos. Sci., 43, 1525-1534, 1986.

Mahlman, J.D., and L.J. Umscheid, Dynamics of the middle atmosphere: Sucesses and problems of the GFDL "SKYHI" general circulation model, in Dynamics of the Middle Atmosphere, pp. 501-525, Reidel Publishers, Dordrecht, Netherlands, 1984.

Oort, A.H., Global Atmospheric Circulation Statistics, 1958-1973, NOAA Professional Paper 14, U.S. Department of Commerce, National Oceanic and Atmospheric Administration, Rockville, MD., 1983.

Ramanathan, V., E.J. Pitcher, R.C. Malone, and M.L. Blackmon, The response of a spectral general circulation model to improvements in radiative processes, J. Atmos. Sci., 40, 605-630, 1983.

Shine, K., 1987: The middle atmosphere in the absence of dynamical heat fluxes. Quart. J. Roy. Meteor. Soc., submitted.

The residual circulation: Interhemispheric differences and heating and eddy components

S.R. Beagley and R.S. Harwood, University of Edinburgh
Department of Meteorology, JCMB,
King's Buildings, Mayfield Rd.,
Edinburgh EH9 3JZ
SCOTLAND.

Abstract. The residual circulation has been diagnosed using an "omega-equation" technique, which allows it to be split into the components driven by the various factors which are tending to destroy the thermal wind balance of the zonal mean state. This clarifies how the stratosphere is maintained away from zonal-mean radiative equilibrium conditions by the eddy heat and momentum fluxes. The "diabatic" circulation has also been diagnosed. This shows many similarities to the residual circulation, though we find the latter to be larger in the lower stratosphere.

The interhemispheric differences in residual circulation are discussed too. The differences found during months near the equinox are in the sense which is expected to result in the observed interhemispheric differences in total ozone.

1. Scope of the investigation

The introduction of the residual circulation and the related transformed Eulerian mean formalism (Andrews & McIntyre 1976, 1978) has led to a greater understanding of the processes governing the evolution of the zonal mean wind and temperature distributions. The residual circulation is also of importance in studying distributions of trace gases such as ozone, being closely related to the "diabatic" circulation and both being related to a Lagrangian mean or "transport" circulation (Dunkerton 1978). The residual or diabatic circulations have been proposed, or used, with suitable parametrizations for certain eddy effects, as the main basis for calculating constituent transport in 2-D chemical models (eg Pyle & Rogers 1980, Holton 1981, Garcia & Solomon 1983). It is to be anticipated therefore that the interhemispheric differences in such quantities as total ozone, will be clearly associated with interhemispheric differences in the residual circulation. In section 4 below, we present the stratospheric residual circulation for several months for the whole globe, tentatively noting features of the interhemispheric differences which may be related to those in the total ozone distribution.

G. Visconti and R. Garcia (eds.), Transport Processes in the Middle Atmosphere, 387–399.
© 1987 by D. Reidel Publishing Company.

In general the pattern of zonal mean (diabatic) heating and the torque produced by the divergence of Eliassen-Palm flux both disturb the thermal wind balance of the zonal mean state. The residual circulation can be regarded as the response to those disturbing tendencies, which opposes them to restore thermal wind balance. In section 3 below the relative contributions of heating and EP-flux convergence in "driving" the residual circulation for January 1979 are discussed and the total circulation compared with the "diabatic" circulation.

2. Computational details

The zonal mean momentum and thermodynamic equations may be written in the quasi-geostrophic approximation (Lorenz 1960) as

$$[u_t] - f[v_i] = (c^2[v^*u^*])_y/(c^2) \tag{1}$$

$$[\theta_t] + [v_i][\theta]_y + [w_i][\theta]_\eta =$$
$$[Q] - (c[v^*\theta^*])_y/c - (p[w_i^*\theta^*])_\eta/p \tag{2}$$

where $\eta = \ln(p_o/p)$; p = pressure; p_o = 1000 mbar; $w = D\eta/Dt$; subscripts t, y and η denote differentiation with respect to that variable; subscript i denotes "irrotational part"; square brackets and superscript star denote zonal mean and departure therefrom respectively and $c=\cos(\varphi)$ with φ = latitude.

Mass continuity implies the existence of a streamfunction ψ for the zonal-mean meridional circulation

$$[v] = -(p\Psi)_\eta/(pc), \quad [w] = \Psi_y/c \tag{3}$$

To this level of approximation the streamfunction ψ_R for the residual circulation (v_R, w_R) is given by

$$\Psi_R = \Psi + c[v^*\theta^*]/[\theta]_\eta \tag{4}$$

Equations 1 & 2 are thereby transformed to

$$[u_t] - fv_R = ((cF^\varphi)_\varphi/(ac) + (F^\eta)_\eta)/(acp) \tag{5}$$

$$[\theta]_t + v_R[\theta]_y + w_R[\theta]_\eta =$$
$$[Q] - (p([\theta]_y[v^*\theta^*]/[\theta]_\eta + [w^*\theta^*]))_\eta/p \tag{6}$$

where (F^φ, F^η) is the Eliassen Palm flux:

$$F^\varphi = -apc[v^*u^*] \quad \& \quad F^\eta = apcf[v^*\theta^*]/[\theta_\eta]$$

At this level of approximation, the meridional momentum equation leads to the thermal wind equation

$$[\theta]_y = -A[u]_\eta \qquad (7)$$

where $A = f(p_o/p)^K/R$

Eliminating the time derivatives between 5 & 6 using 7 gives a 2nd order linear partial differential equation for ψ_R:

$$([\theta]_\eta \Psi_{Ry}/c)_y - ([\theta]_y (p\Psi_R)_\eta /pc)_y + Af((p\Psi_R)_\eta /pc)_\eta$$

$$= (A/ac)((\nabla.F)/p)_\eta + [Q]_y -$$

$$-(p([\theta]_y [v^* \theta^*]/[\theta]_\eta + [w^* \theta^*])_y)_\eta /p \qquad (8)$$

This equation is elliptic provided the static stability exceeds a certain rather high value, which is the case for all the cases studied here.

Appropriate boundary condition at the poles are $\psi = 0$ ie $\psi_R = 0$. At the lower boundary, $\eta = 0$, $\psi = 0$ giving $\psi_R = \cos \phi [v^* \theta^*]/[\theta_\eta]$. The upper boundary condition is more problematical. Our approach is to impose an artificial boundary condition, viz $\psi_R = 0$ but to place the boundary sufficiently high not to affect the region of interest. Our aim is to diagnose the stratospheric circulation up to 1 mbar. Accordingly the boundary has been set at 0.006 mbar. Trial integrations show that the solution in the stratosphere is insensitive to the value of ψ_R at that height (Beagley 1987).

Equation 8 and similar ones discussed below have been solved on a y-η grid, spaced at 5 degree intervals in latitude and 0.75 in η, by a relaxation technique, with the winds temperatures and EP fluxes based on temperatures deduced from either the SSU or LIMS temperature measurements and tropospheric data supplied by the U.K. Meteorological Office, using the geostrophic and hydrostatic equations as appropriate. As the satellite data do not extend beyond 1 mbar, the E-P fluxes have been smoothly reduced above this level to reach zero at the upper boundary. A sensitivity test has shown that the solution below 1 mbar is very little affected by the values above.

The diabatic heating rates needed for the right hand side of eq. 8 have been calculated using the radiation code of Haigh (1984).

3. Processes forcing the residual circulation

Equation 8 is of the form $L(\psi_R) = R_Q + R_E$ where L is the elliptic operator, R_Q represents the term in heating and R_E represents the terms on the right involving eddy fluxes. Since L is linear, we may write $\psi = \psi_Q + \psi_E$ where ψ_Q and ψ_E satisfy $L(\psi_Q) = R_Q$ and $L(\psi_E) = R_E$. The boundary conditions are satisfied by specifying $\psi_E = 0$ and $\psi_Q = 0$ on all boundaries except that at the bottom boundary $\psi_E = \psi_R$. ψ_Q can be regarded as being forced by diabatic processes and ψ_E by eddies.

(a)

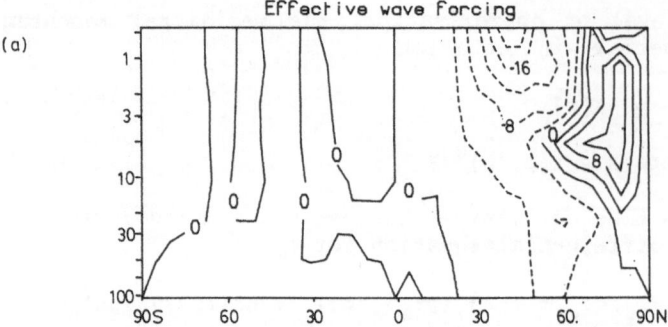

(b)

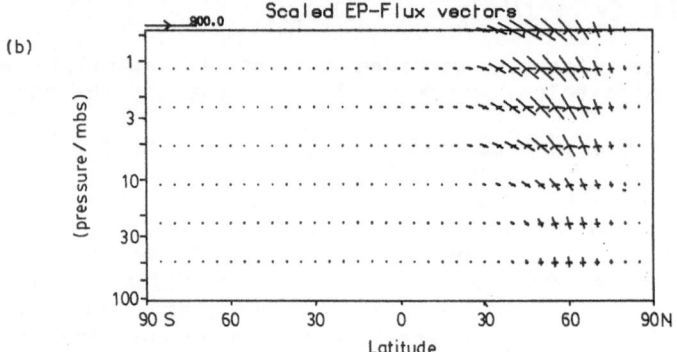

(c)

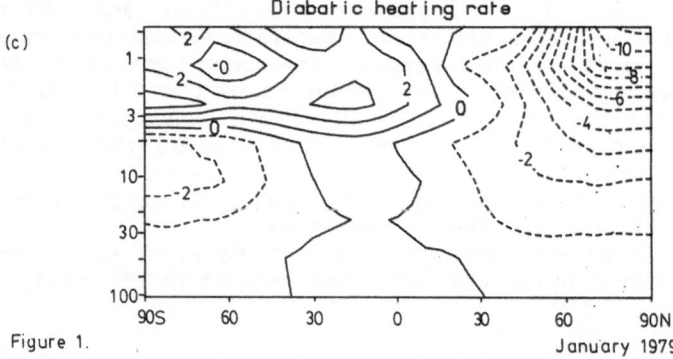

Figure 1.   January 1979

(a) Divergence of E-P flux expressed as a specific zonal force
(equivalent acceleration).The contour interval is 4 x 10^{-5} m
s^{-2}. (b) E-P flux, scaled as F/p. (c) Diabatic heating rate.
The Contour interval is 1 K/day. January 1979.

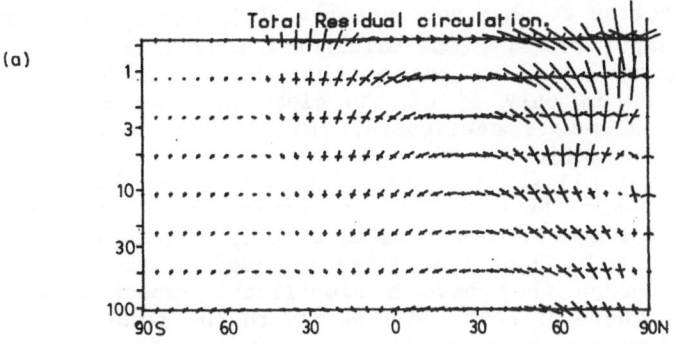

(a)

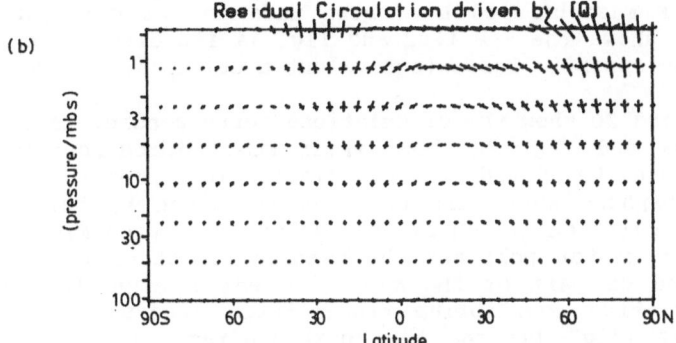

(b)

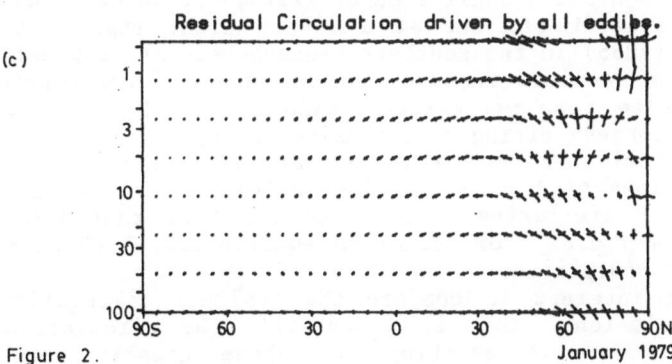

(c)

Figure 2.   January 1979

Components of the residual circulation for january 1979. Arrows show velocity (distance travelled in 10 days). (a) Total circulation. (b) Circulation corresponding to Ψ_Q. (c) Circulation corresponding to Ψ_E.

In this section we show some details of ψ_E and ψ_R for January 1979. The required temperatures, ozone values and eddy fluxes were taken from LIMS data, extrapolated to the south pole as appropriate. As the extrapolated area is only 8% of the globe and the eddies have small amplitudes in the summer hemisphere, this is not thought to introduce significant uncertainties.

Note that not all physical processes have been taken into account. The technique adopted here contains no estimates of the convergence of vertical fluxes of momentum associated with gravity or equatorial waves. The circulations induced by the latter are small (see fig.7 of Gray and Pyle (1986)), though they have a significant impact upon the tracer distributions. These effects add linearly to the right of equation 8.

Fig. 1c shows the diabatic heating rate. It is broadly similar to those presented by Kiehl & Solomon(1986) and Gille & Lyjak (1986) but with some differences. Some of these probably arise because no adjustments for mass-balance have been made in the calculations shown here. Fig. 1b shows the E-P flux and fig. 1a its divergence, expressed in terms of the specific force (equivalent acceleration) produced on the mean flow.

Figs. 2b and 2c show the circulations corresponding to figs. 1c and 1b respectively and Fig. 2a shows their sum. Throughout most of the winter hemisphere the eddy-driven circulation is dominant, especially in the low stratosphere where the circulation driven by heating is very small. The circulation driven by the eddies has a predominately sinking motion near the winter pole and the heating effect of this circulation can be regarded as part of the mechanism which keeps the winter pole above the radiative equilibrium temperature, leading in turn to the heating pattern (fig. 1a) and the corresponding circulation driven by the heating (which also acts in a way to warm the winter pole).

We have diagnosed a small area of rising motion near the north Pole which is reminiscent of, but at a lower height than, that found by Al-Ajmi et al (1985) in the southern hemisphere. It is possible however that this is a spurious feature, as the use of geostrophic winds is known to lead to some misrepresentation of the E-P fluxes (Robinson 1986). The strongest rising motion driven by eddies is in low latitudes of the summer hemisphere with some descent at the summer pole. Thus the general directions of the circulations driven by eddies and by heating have an overall similarity, reflecting the fact that the heating is produced by departures from radiative equilibrium which arise from the eddy induced circulations.

It is of interest to compare the residual circulation with the diabatic circulation. This is essentially the circulation diagnosed from the thermodynamic equation (6) alone usually with the time derivative set to zero and an adjustment made (equivalent to adjusting Q) to ensure mass-consistence (cf Gille et al 1987). Substituting for ψ_D (the diabatic streamfunction) and differentiating with respect to latitude gives an equation for ψ_D which may be compared with that for ψ_R or ψ_Q, namely

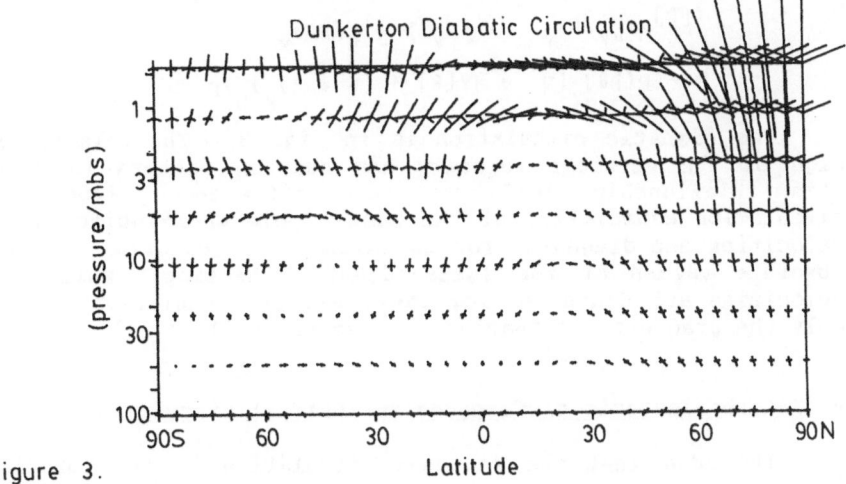

Figure 3.

Diabatic circulation (corresponding to Ψ_D) for January 1979.

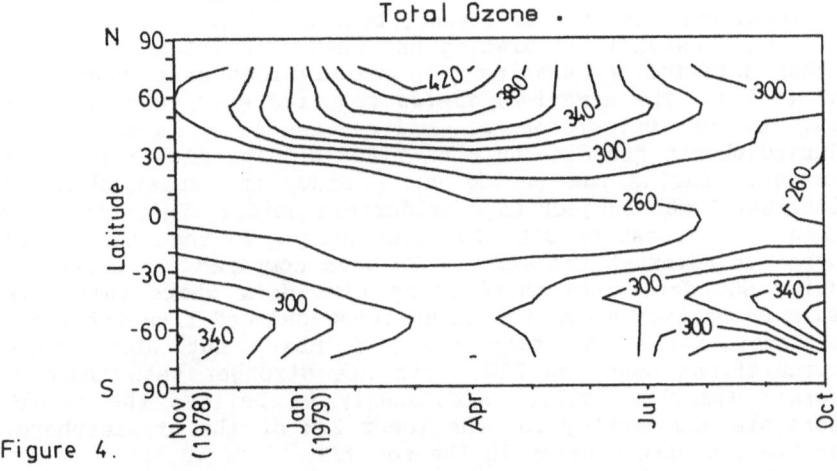

Figure 4.

Total Ozone in Dobson units based on McPeters et al 1984.

$$([\theta]_\eta \Psi_{0y}/c)_y - ([\theta]_y (p\Psi_0)_\eta/pc)_y$$

$$= -(p([\theta]_y [v^* \theta^*]/[\theta]_\eta + [w^* \theta^*])_y)_\eta/p \qquad (9)$$

The diabatic circulation is in fig. 3. The relation between the sign of the vertical velocity and heating pattern is very apparent. This relationship leads to some differences from the residual circulation especially, at the summer pole where quite strong vertical velocities are diagnosed for the diabatic circulation (which depends on absolute values of the (mass-adjusted) heating), while much smaller velocities are diagnosed for the residual circulation (which involves only the gradients of heating -see equation 8).

4. Interhemispheric differences in residual circulation

The idea that the residual circulation is an approximation to a Lagrangian mean circulation (Dunkerton 1978) leads to the expectation that interhemispheric differences in the circulation should be discernible which are related to those in total ozone. Some indirect evidence for this contention comes from the (Eulerian) 2-D modelling experiment of Harwood and Pyle (1977) which produced fairly realistic interhemispheric differences in total ozone although the only differences imposed on the model were through the momentum fluxes. Accordingly we have investigated the interhemispheric differences in residual circulation for 1 year, December 1978 - November 1979.

The residual circulation has been calculated from equation 8. The ozone data for the heating rate calculations were taken from McPeters et al (1984). The dynamical fields required were taken from data supplied by the UK meteorological office based on measurements from the Stratospheric Sounding Unit on TIROS-N. The measurements are global in extent. During the period under study the upper channel of the SSU instrument was subject to considerable noise. This results in a loss of about 30 percent of data for each month, so that the results presented here are somewhat tentative. We have compared the circulations deduced using SSU data with those using LIMS data where these are available. This comparison shows the circulation deduced from the two data sets to be essentially the same below 5 mbar, but above this level the circulations based on LIMS data are stronger than those based on SSU data. (Beagley 1987). Accordingly we believe the results presented here are trustworthy for the lower 2/3 of the stratosphere, but should be treated with caution in the top 1/3.

Figs. 5-7 show a selection of three of the interhemispheric differences. Fig. 5a shows the circulation for December 1978 and fig. 5b that for June 1979. To obtain fig. 5c, the June circulation is reflected about the equator and subtracted from the December circulation. Very little difference is found between the two circulations for these months

Fig. 6 shows February and August 1979 treated in the same way. February 1979 was the occasion of a major warming in the northern

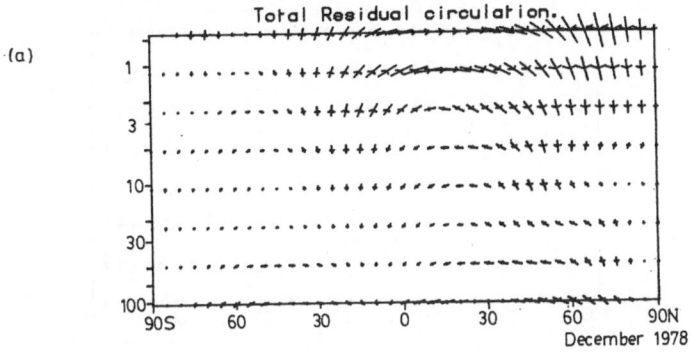

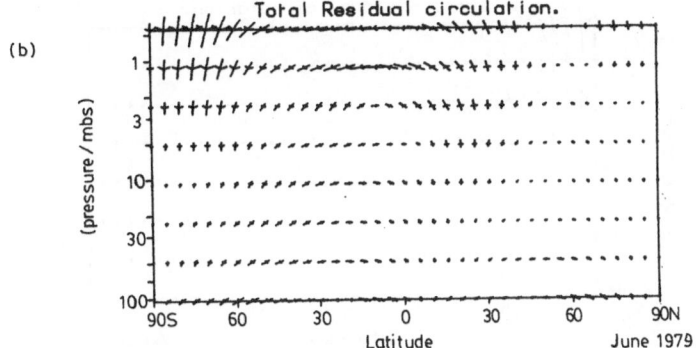

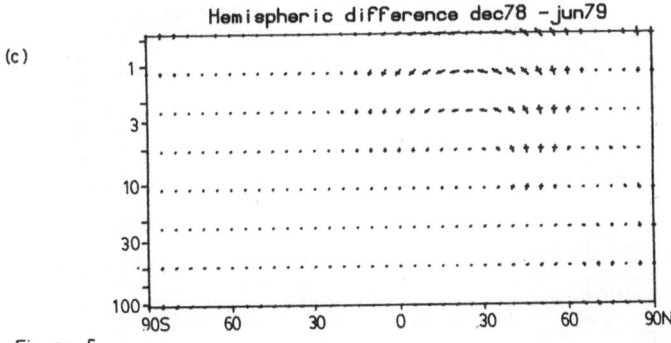

Figure 5.

Residual circulation for: (a) December 1978, (b) June 1979, (c)
The difference between (a) and the reflection of (b) about
the equator. The arrows are scaled as in fig 2.

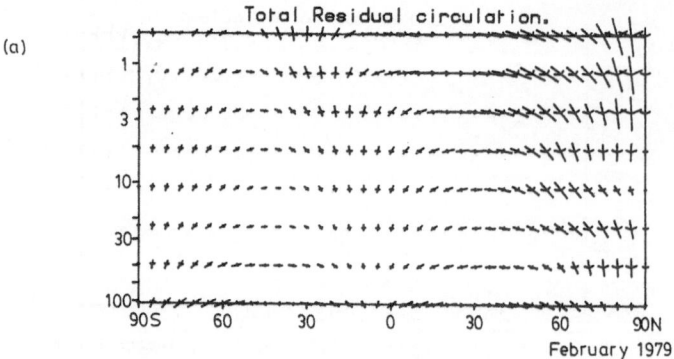

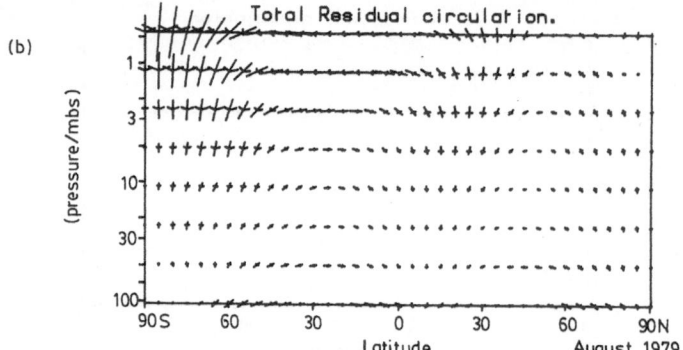

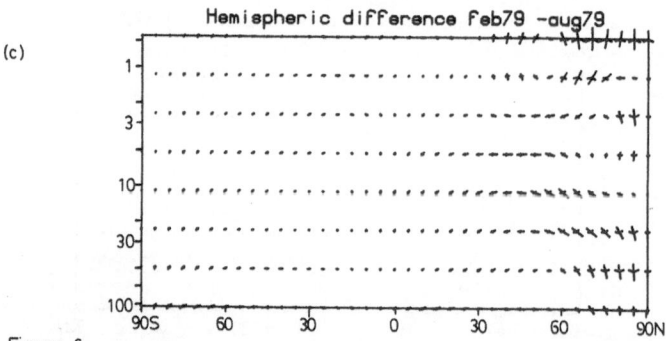

Figure 6.

As fig 5 but for: (a) February 1979, (b) August 1979.

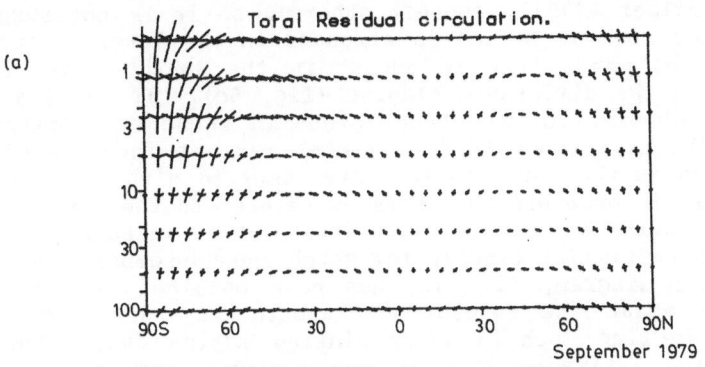

(a)

Total Residual circulation.

September 1979

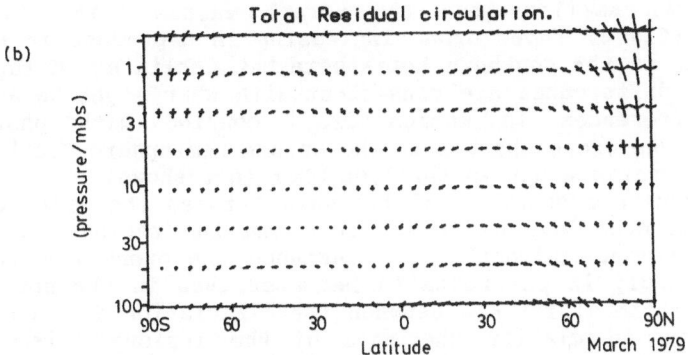

(b)

Total Residual circulation.

(pressure/mbs)

Latitude   March 1979

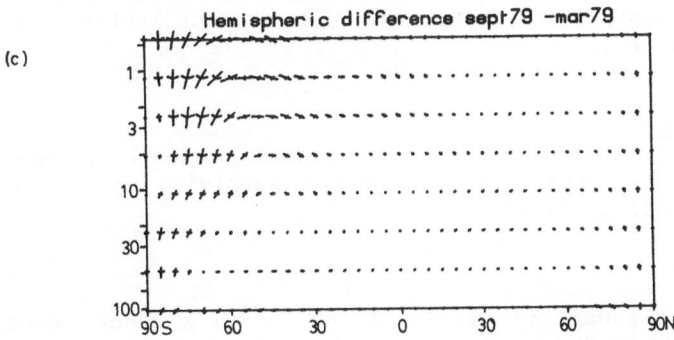

(c)

Hemispheric difference sept79 −mar79

Figure 7.

Similar to fig 5 except: (a) September 1979, (b) March 1979.

Hemisphere (Palmer (1981), amongst others), so it is not surprising to see the winter sinking motion is stronger in the lower stratosphere in February in the north than in August in the south. This is clearly brought out in the difference diagram (fig. 6c). It is of interest to compare the behaviour of the total ozone in these two months (see fig 4). A notable difference is the greater rate of increase of ozone at high latitudes in the north in February compared with that in the south in August. It is probable that this is related to the stronger poleward and downward motion below the middle stratosphere in February.

Fig 7 shows similar figures for March and September. Here, however the difference diagram, fig. 7c, has been obtained by reflecting the March values about the equator and subtracting them from those for September. We find much stronger sinking motion during the southern spring in high latitudes than in the northern spring. The autumnal circulations are fairly similar leading to only small differences, and are in general smaller than the vernal values. The total ozone distribution (fig 4) shows ozone increasing in September in middle and high latitudes of the southern hemisphere but declining in the north in March. These differences are consistent with what might be anticipated from the differences in motion field coupled with photochemical destruction. The ozone changes in the autumn hemisphere are less rapid. Very similar remarks apply to April/October (not shown).

These results suggest a relationship between the interhemispheric differences in behaviour of total ozone and the residual circulation, but they are not unequivocal. For instance the ozone was building up much more strongly in the north in December than in the south in June but we find little difference between the circulation for those months. A possible way to clarify the role of the residual circulation in creating interhemispheric differences in ozone is to use the circulations deduced here in a transformed Eulerian mean 2-D chemical model of stratospheric chemistry. We are currently attempting to arrange such a test.

Acknowledgements

S.R. Beagley was supported by a NERC studentship when this work was performed. We thank Dr. Haigh for supplying us with the code of the radiation calculations.

References

Al-Ajmi,D.N., R.S.Harwood and T. Miles, 1985: A sudden warming in the middle atmosphere of the southern hemisphere. Quart. J. R. Met.Soc. 111,359-389.

Andrews,D.G and M.E.McIntyre, 1976: Planetary waves in horizontal and vertical shear: The generalised Eliassen-Palm relation and the mean zonal acceleration. J.Atmos. Sci., 33 2031-2048

Andrews,D.G. and M.E.McIntyre, 1978: Generalized Eliassen-Palm and Charney-Drazin theorems for axisymmetric flows in compressible atmosphers. J.Atmos.Sci., 35, 175-185.

Beagley,S.R., 1987. Unpublished Ph.D. thesis. Edinburgh University.

Dunkerton,T., 1978: On the mean meridional mass motions of the
    stratosphere and mesosphere. J.Atmos.Sci., 35 2325-2333.
Gille,J.C. and L.V.Lyjak, 1986: radiative heating and cooling rates in
    the middle atmosphere. J.Atmos.Sci., 43,2215-2229
Gille, J.C., L.V.Lyjak and A.K.Smith, 1987:   The global residual mean
    circulation in the middle atmosphere for the northern winter period.
    J.Atmos.Sci., 44, 1437-1452.
Garcia, R.R. and S.Solomon, 1983: A numerical model of the zonally
    averaged dynamical and chemical structure of the middle
    atmosphere. J.Geophys. Res., 88, 1379-1400.
Grey, L.J. and J.A.Pyle, 1986: The semi-annual oscillation and
    equatorial tracer distributions. Quart.J.R.Met.Soc, 112, 387-407.
Haigh, J.D.,1984: Radiative heating in the lower stratosphere and the
    distribution of ozone in a two-dimensional model.
    Quart.J.R.Met.Soc., 110, 167-185.
Harwood, R.S. and J.A.Pyle, 1977:   Studies of the ozone budget using a
    zonal mean circulation model and linearized photochemistry. Quart.
    J.Roy.Meteor.Soc., 103, 319-343.
Holton, J.R.,1981. An advective model for two-dimensional transport of
    stratospheric trace species. J.Geophys.Res., 86, 11989-11994
Kiehl,J.T. and S.Solomon, 1986: On the radiative balance of the
    stratosphere. J.Atmos.Sci., 43, 1525-1534.
Lorenz,E.N., 1960: Energy and numerical weather prediction. Tellus 12,
    364-373.
McPeters,R.D., P.F.Heath and P.K.Bhartia, 1984: Average Ozone profiles
    for 1979 from the Nimbus 7 SBUV instrument. J.Geophys.Res., 89
    D4, 5199-5214.
Palmer,T. 1981: Diagnostic study of a wavenumber-2 stratospheric sudden
    warming in a transformed Eulerian-mean formalism. J.Atmos.Sci.,
    38, 844-855.
Pyle,J.A. and C.F.Rogers, 1980: A modified diabatic circulation model
    for stratospheric trace transport. Nature, 287, 711-714
Robinson,W.A., 1986: On the application of the quasi-geostrophic EP-flux
    to the analysis of stratospheric data. J.Atmos.Sci., 43,
    1017-1023.

ANALYSIS OF THE MEAN MERIDIONAL CIRCULATION USING SATELLITE DATA

Ellis E. Remsberg
Atmospheric Sciences Division
NASA Langley Research Center
Hampton, Virginia 23665-5225

ABSTRACT. Near-global satellite data on temperature, ozone, and water vapor are now being used to calculate both residual mean and diabatic circulation fields in the stratosphere. For the residual circulation, the effects of wave forcing must also be calculated from satellite temperatures. Both calculations require very accurate and preferably consistent (ozone and temperature for same month and year) data sets. The derived circulations can be used to conduct tracer transport simulations which can be compared with the observed tracer behavior. Because the precision of the satellite tracer measurements is very good, this procedure provides a means for evaluating the derived circulation.

1. INTRODUCTION

    Mahlman et al. (1984) gave an excellent review of atmospheric tracer transport studies through 1981 based on data available then. Much insight was derived from studies of radioactive debris and balloon measurements of long-lived tracers (e.g. N_2O and CH_4) in the Northern Hemisphere, although ideas were also developed from careful analyses of rocket, balloon, and aircraft measurements of temperature, ozone, and humidity. Since 1981, there has been a significant advance in information about the middle atmosphere based on satellite data (WMO, 1986). However, it is important to note that the basic ideas about middle atmospheric transport remain unchanged; the satellite data have primarily served to verify and refine those ideas and extend them into the Southern Hemisphere. The possible exception to this is in the tropical middle atmosphere, where atmospheric forcing and dissipation mechanisms have only recently begun to be quantified.
    Table 1 summarizes those satellite data sets which have been described in some detail and made generally available since 1981. Some of the radiance data from earlier Nimbus experiments (SCR and PMR) have also been processed recently to give a global temperature climatology (Barnett and Corney, 1985). Characteristics of the various data sets such as coverage, altitude range, and vertical resolutions (VR) are given by Barnett and Corney (1985) and references therein, and in Clough et al. (1985). Examples of many of the species data are given in Russell et al. (1986) and Keating and Young (1985). The SBUV and TOMS data are being reanalyzed with updated ultraviolet

401

G. Visconti and R. Garcia (eds.), Transport Processes in the Middle Atmosphere, 401–419.
© *1987 by D. Reidel Publishing Company.*

TABLE 1.   SATELLITE DATA SETS

A. Temperature and/or Thickness Fields

Instrument	Comments
Selective Chopper Radiometer (SCR) and Pressure Modulated Radiometer (PMR)	multi-year; excellent coverage; vertical resolution (VR) about 15 km
TIROS Operational Vertical Sounder (TOVS): Stratospheric Sounding Unit (SSU), Microwave Sounding Unit (MSU), High Resolution Infrared Sounder (HIRS)	multi-year; VR 10-15 km; excellent coverage; geopotential thickness; base height and T field required for T profiles
Stratospheric and Mesospheric Sounder (SAMS)	multi-year; 50 S to 70 N; VR about 8 km
Limb Infrared Monitor of the Stratosphere (LIMS)	7 months; 64 S to 84 N; VR about 2.5 km (IPAT); 3.5 km (MAT); includes geopotential height

B. Ozone Profile Data

Instrument	Comments
Solar Backscatter Ultraviolet (SBUV and SBUV 2)	multi-year; VR about 5 km; excellent coverage during sunlight
Solar Mesosphere Explorer (SME)	multi-year; UV and IR measurements; VR about 3.5 km; limited in longitude
LIMS	7 months; day and night; VR about 2.9 km
Stratospheric Aerosol and Gas Experiment (SAGE I)	multi-year; coverage limited each day; VR about 1 km

C. Total Ozone Data

Instrument	Comments
SBUV	multi-year; daytime
TOVS	day and night
Total Ozone Mapping Spectrometer (TOMS)	multi-year; daytime; excellent spatial resolution

D. Other Species Distributions

Instrument	Comments
SAMS (methane and nitrous oxide)	multi-year; mid-stratosphere to lower mesosphere; zonal-mean only; VR about 5 km
SME (nitrogen dioxide, ozone and aerosols)	limited longitudinal coverage; VR about 3.5 km; multi-year
LIMS (nitric acid, water vapor, nitrogen dioxide)	7 months; excellent coverage; VR about 2.9 km for HNO_3, 5 km for H_2O and NO_2; day and night
SAGE I (ozone, nitrogen dioxide, and aerosols)	VR about 1 km; sunrise and sunset data; coverage limited for a given day
Stratospheric Aerosol Measurement, SAM II (aerosols)	VR about 1 km; high latitude only; multi-year

absorption coefficients, and those results should be archived in early
1987.   Finally, data from the SAGE II experiment launched in 1985
(ozone, NO_2, H_2O, aerosols) should also be available in 1987.

The quality of all these data has been assessed and, in general,
they exhibit accuracies, at least for the vertical resolutions of the
satellite results, which are comparable to or better than those
available from in situ sensors or from remote sensors on balloon-borne
platforms. Certainly a key advantage of satellite data, in addition
to their coverage and duration, is their high precision. This means
their distributions can be used to study both wave activity and the
mean circulation.

This paper contains a brief discussion of the use of satellite
data for determinations of the various forms of the mean meridional
circulation. These include (1) the generalized Lagrangian-mean (GLM)
circulation and its approximate form, the transport circulation, (2)
the residual or transformed Eulerian mean circulation, and (3) the
diabatic circulation (see WMO (1986), Chapter 6). This is in no way a
comprehensive discussion, and other papers in this volume should be
consulted (e.g. Andrews; Beagley and Harwood). Following that, the mean
meridional circulation is diagnosed with the aid of tracer-like species, particularly water vapor. Finally, the use of area diagnostics of
tracers (Butchart and Remsberg, 1986) is mentioned as a means to infer
any irreversible trends in stratospheric species distributions.

Estimates of the GLM circulation (Andrews and McIntyre, 1978)
from data entail tracking the center-of-mass of material tubes of
conserved quantities.   This approach defines the total transport,
advection plus dispersion, but becomes impractical when the tubes are
severely deformed or stretched out in the atmosphere. To know whether
this is a problem, one must have access to an accurate, three- dimensional data set with relatively high spatial resolution.   Such data
sets do not exist, although LIMS and SSU (potential vorticity, PV)
provide opportunities for initial studies.   Alternatively, one can
approximate the GLM transport using wind fields and tracing parcel
trajectories for short periods.   This approach is appropriate for
model calculations of transport, but performing similar calculations
using satellite-derived winds is subject to the same data inaccuracies
as are true for a normal Eulerian mean circulation.   Furthermore,
under the assumption that the dispersive transport is due to small-
amplitude, adiabatic eddies, the GLM circulation is equal to the
transformed Eulerian or residual circulation. This being the case, it
is appropriate to consider uncertainties in a residual circulation
derived from data.

2.   THE RESIDUAL CIRCULATIONS

The residual circulation is an alternate or transformed form of
the usual Eulerian circulation (Andrews and McIntyre, 1978).   Under

quasi-geostrophic assumptions (Edmon et al., 1980), the zonal-mean momentum, thermodynamic, and continuity equations in the transformed formulation become

$$\bar{u}_t - f\bar{v}^* = (r_o \cos \phi)^{-1} \ (\nabla \cdot F) \tag{1}$$

$$\bar{\theta}_t + \bar{\theta}_p \bar{\omega}^* = \bar{Q} \tag{2}$$

$$(r_o \cos \phi)^{-1} \ (\bar{v}^* \cos \phi)_\phi \ + \ \bar{\omega}_p^* = 0 \tag{3}$$

where the "residual" meridional and vertical velocities are $\bar{v}^*$ and $\bar{\omega}^*$, respectively, in pressure (p) - latitude (ϕ) coordinates. A straightforward way of obtaining $\bar{\omega}^*$ and $\bar{v}^*$ is to use temperature data and a radiative code to evaluate $\bar{\theta}_t$ and $\bar{Q}$, then deduce $\bar{\omega}^*$ from equation (2). Then $\bar{v}^*$ is obtained from the continuity equation (3). Both the temperature tendency $\bar{\theta}_t$ and net heating $\bar{Q}$ are known fairly well, so this calculation should be reasonably accurate.

An alternate approach to $\bar{v}^*$ is through the momentum equation (1). Using geostrophic or gradient winds and temperature data, the time tendency of the zonal wind $\bar{u}_t$ and the Eliassen-Palm flux divergence $\nabla \cdot F$ can be calculated. Unfortunately, $\nabla \cdot F$ involves several differentiation operations, and uncertainties in such highly derived quantities are a combination of both the measurement and analysis procedures. The term $f\bar{v}^*$ can then be deduced from (1).

In order to gain confidence in the values obtained for $\nabla \cdot F$, Grose (1984) compared $\bar{v}^*$ determined both ways using LIMS data during Northern Hemisphere winter (Fig. 1). The two independent results track each other although the mean transport for brief periods differs by a factor of two. Such results are encouraging nonetheless, given the uncertainties in the data and the number of differentiations required in the analysis. At other locations and during less dynamically active periods, $\bar{v}^*$ can be smaller and subject to even larger errors.

The success of these calculations led to an attempt to include the effects of $\nabla \cdot F$ forcing in the model calculations of mean residual circulations by Al-Ajmi et al. (1985) using SCR data and by Solomon et al. (1986a) using LIMS data. The effects of adding $\nabla \cdot F$ were to "flatten" zonal-mean tracer isolines at high latitudes, bringing more agreement with observations. They found that the winter stratosphere circulation was accelerated, and there were also significant increases in upward velocities in the summer hemisphere Tropics far removed from the regions of direct momentum forcing. Distributions of satellite-derived tracers or global distributions of mass were used to check these findings, but further verification is anticipated along the lines of that reported by Gray and Pyle (1986). Hitchman and Leovy (1986) have inferred the forcing of a meridional circulation by tropical waves using wind fields calculated from LIMS data. They have verified their conclusions by estimating residual circulations using a radiative heating code with LIMS data as input. The high vertical resolution of the satellite data has made it possible to determine the connection between tropical waves and the meridional circulation for the first time. The satellite data coverage has also revealed significant longitudinal asymmetries in that tropical wave forcing.

3. DIABATIC CIRCULATIONS

The residual meridional circulation, $\bar{v}^*$, $\bar{\omega}^*$ is more appealing than the Eulerian-mean circulation because the sense of the circulation is consistent with observed tracer motions. Dunkerton (1978) pointed out that in the absence of wave transience and dissipation and when $\bar{\theta}_t$ in Eq. (2) is small, the residual meridional circulation can be approximated by a diabatic circulation $\bar{v}_D$, $\bar{w}_D$ that responds to the net radiative heating, although the wave forcing that must be included in a residual circulation (Solomon et al., 1986a) has been set to zero in the generation of diabatic streamfunctions. Some effects of waves are implicitly present in $(\bar{v}_D, \bar{w}_D)$ because the monthly changes in the data fields used in the heating calculation are due, in part, to actions of waves. Until recently, accurate diabatic circulations were difficult to obtain because the necessary stratospheric data sets for temperature, ozone, and water vapor were inadequate; an approximate Newtonian radiative relaxation rate was often used. With the advent of the new stratospheric climatologies, more accurate radiative codes were employed as well in order to achieve the desired accuracies in $\bar{v}_D$, $\bar{w}_D$.

Initially, it is important to analyze the net diabatic heating of the entire stratosphere. One might anticipate that a self-consistent data set would be required. Several potentially consistent data sets are contained in Table 1. First, there are the 7 months of LIMS data for temperature, ozone, and water vapor. Because these data do not quite extend to the southern Pole, they may be supplemented with temperatures derived from SSU thicknesses and a base height temperature and with ozone from SBUV or SAGE. Another possible combination is the SAMS (or SSU derived) temperatures plus the SBUV

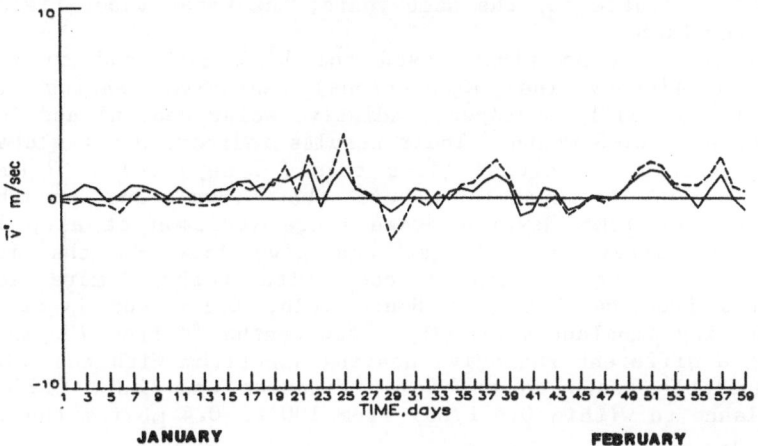

Figure 1. Residual meridional velocity inferred from the momentum equation (---) and from the thermodynamic equation (—) for January-February 1979 at 68°N latitude and 10-mb pressure (after Grose, 1984).

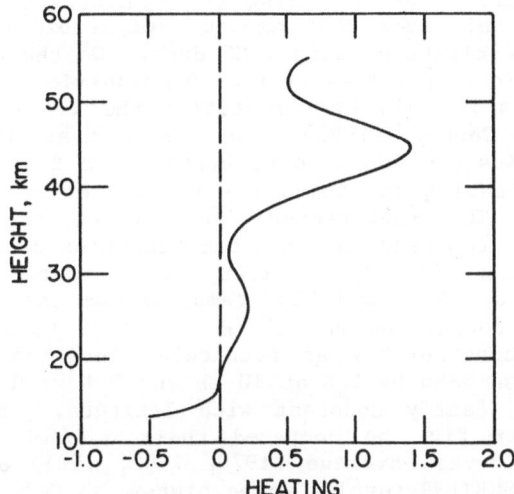

Figure 2. Global and annual average of the net radiative heating (K/day) after Callis et al. (1987).

ozone, both available for the same years; the water vapor field could
be taken from LIMS.

Kiehl and Solomon (1986) used the LIMS data and an accurate
(compared to line-by-line calculations) radiative heating code to
calculate the globally averaged, radiative solar heating and longwave
cooling for each LIMS month. Their results indicate a net global heat
balance to within 25 percent (less than 1 K/day) below 52 km, with
slightly larger discrepancies above that level due possibly to
overestimates of LIMS daytime ozone there (Solomon et al., 1986b).
The local imbalances are larger and give rise to the diabatic
circulation. Using a similar code with perhaps more accurate
temperatures from 64 S to the South Pole, Gille and Lyjak (1986)
achieve heating imbalances of only a few tenths °K from 100 mb to 0.7
mb. Using a different radiative heating algorithm with the LIMS data
and extrapolating data to the Pole, Hitchman and Leovy (1986) find
global balance to within 0.6 K/day from 100 to 0.4 mb for the 7-month
LIMS period.

Callis et al. (1987) derived the profile of global net heating
shown in Fig. 2 using a 12-month "climatology" based on 2 years of SCR
and 3 years of PMR temperatures from Barnett and Corney (1985) and 4
years of ozone data compiled by Keating and Young (1985). Their
heating code was similar to that used by Kiehl and Solomon. They
obtain a near balance, except near 45 km where there is an excess
heating rate of up to 1.4 K/day. It is unlikely that errors in the
radiative codes are the cause of the imbalance that they found at 45
km because their auxiliary results using the 7 months of LIMS data
yielded imbalances of only half that amount, comparable to what was
found by the other investigators using LIMS data. On the other hand,
a systematic increase in temperature over an atmospheric scale height,
for example, of 2 to 4 K would have rectified the 1.4 K/day excess
there. Barnett and Corney (1985) have shown that the PMR/SCR
temperature climatologies are known to no better than 2 K, so some of
the imbalance in net heating in Fig. 2 may be due to the temperature
data uncertainties. The rocketsonde data, which represent the
comparison standard for temperature, have uncertainties of that order.

Grose and Rodgers (1986) reported same year, zonal-mean
differences between SSU, SAMS, and LIMS temperatures that average on
the order of 1 K for comparison periods during 1979 to 1981. That
performance degrades considerably at particular locations; however,
e.g. LIMS is warmer than SAMS by 5 K at 10 mb and 3 K at 1 mb for May
1979—both differences fairly constant with latitude. Barnett and
Corney (1985, in their Fig. 3) compared their SCR/PMR zonal mean
temperatures with the 3-year average (1979, 1980, 1981) of SAMS for
January and found the SCR/PMR results to be higher, by 4 K at the same
pressures, 10 mb and 1 mb, the differences being maintained between
±50° latitude. That comparison is less instructive, however, because
it includes significant interannual variations as well as possible
differences in satellite retrieval methods and/or instrument
resolutions. Thus, both the accuracy and the consistency of the
relevant temperature data can still be major concerns for heating rate

calculations. The LIMS temperature data have been carefully validated against both rocket temperatures and winds, the latter comparison a measure of the accuracy of temperature gradients (Smith and Bailey, 1985; Hitchman and Leovy, 1986). The quality of both results is excellent even during disturbed atmospheric conditions.

Uncertainties in ozone are also important. Gille and Lyjak (1986) report solar heating rates of 10 K/day at 1 mb between ±30° latitude at equinox. Since the zonal mean ozone data at that level are known to no better than 10 percent, this can represent a significant uncertainty in the net heating. It is clear that heating rate calculations for residual mean and diabatic circulations require satellite data that are both accurate and precise.

Several points about data consistency can be made. Callis et al. (1987) compared diabatic streamfunctions determined using monthly averaged LIMS data versus those obtained with the 12-month "climatologies" of ozone and temperature. The same radiative code was used in both calculations. For example, their diabatically diagnosed streamfunctions for March present almost identical two-cell circulations for the two data sets. The circulation changed markedly by May to more of a single cell motion rising in the north and moving into the Southern Hemisphere (Fig. 3). Data are plotted versus potential temperature, where 1000 K is about 33 km and 2000 K is near 50 km. Dashed contours represent clockwise flow in the meridional plane; solid lines give counterclockwise flow. Poleward of 50°N and below 1000 K, the sense of the circulation is opposite'for the two data sets in May, despite the fact that there was no need to extrapolate the LIMS data to the Pole as was the case for the Southern Hemisphere, where the circulations seem to agree. However, it must be emphasized that the apparent circulation differences in the north are actually very small because individual heating terms and the resulting net motions are weak then. Those differences are more likely due to differences in lower stratospheric ozone between the 1979 LIMS result and the SBUV climatology for May. SBUV shows more ozone at higher latitudes leading to larger values of solar heating there. Use of the improved UV absorption coefficients in the SBUV retrievals leads to better agreement between SBUV and LIMS, and once those new SBUV results are archived, the May calculations can be compared again. Figure 3 points out the sensitivity of even the direction of a weak diabatic circulation to the accuracy of the data sets.

Finally, there is the prospect that only temperature and ozone data sets for the same month and year should be used for calculations of net heating rates. Rosenfield and Schoeberl (1986) show significant interannual variability in both satellite T and O_3 in the Southern Hemisphere for 1984 versus 1980 in October, even at low latitudes near 3 mb ($\approx$40 km). Of course, this is the time of final stratospheric warmings in the Southern Hemisphere and when year-to-year variations in wave forcing are larger. They also pointed out that the $\bar{\theta}_t$ term in Eq. (2) is not negligible for the south polar stratosphere during spring. During dynamically disturbed periods when

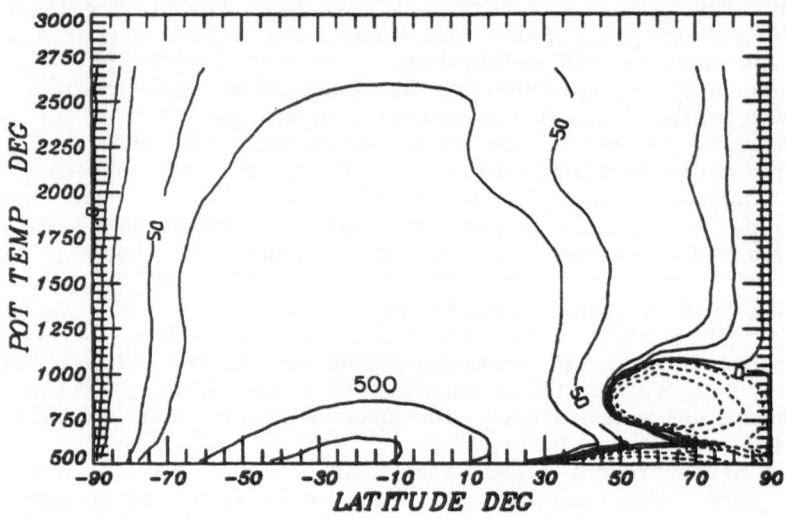

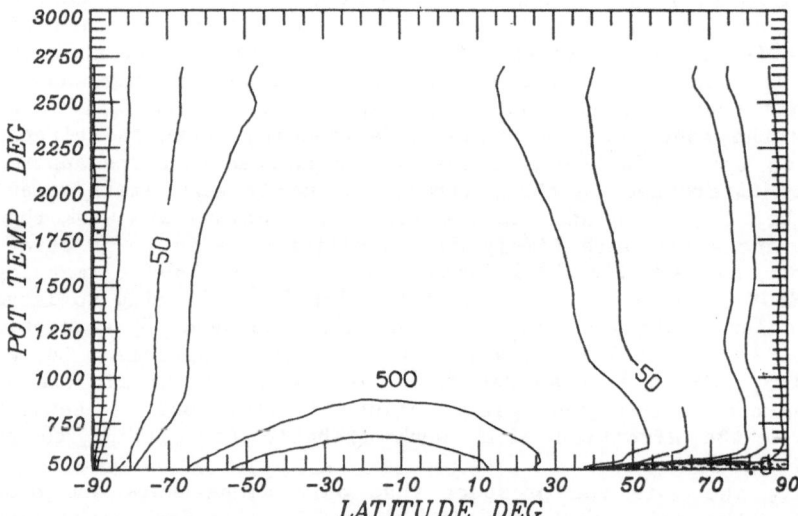

Figure 3. Mass weighted stream functions for diabatic circulation for
May derived (top) from a "climatology" of temperature and ozone and
(bottom) from LIMS data. Values are scaled by 10^{-3} divided by the
radius of the Earth with units of kg/km^2-day and contour intervals 0,
1, 5, 10, 50, 100, 500. Solid contours represent counterclockwise
motion, dotted contours are clockwise, northern latitudes are shown
positive (after Callis et al., 1987).

the vortex is shifted off the Pole, mean circulations based on a zonal mean net heating or ∇•F can be misleading.   In those cases, diabatic effects should be obtained locally.

4. MERIDIONAL CIRCULATIONS FROM TRACERS

The use of tracers to study stratospheric transport depends on their relative inertness to chemical conversion.   In Table 2, the satellite tracer data have been grouped, accordingly, into species that are relatively non-reactive in the upper and lower stratosphere. In the case of the SAMS methane and nitrous oxide, for example, the categorization is also due to the altitudes over which accurate data were obtained.   The strength of species sources and sinks is a factor in the usefulness of species as tracers, as well.   For instance, ozone has its source region in the mid to upper stratosphere and exhibits a strong seasonal dependence at mid latitudes.   Aerosols are often restricted to point and later line sources (volcanic) or to atmospheric layers over extended distances ("background" or dispersed aerosols).   Nitrous oxide and methane have their source in the tropical troposphere, and their lifetimes are very long until they reach the upper stratosphere.   These latter two tracer molecules, along with water vapor, have extended source and sink regions which give rise to the gradual variations in their distributions with altitude and latitude.   In this section, the focus will be on the LIMS water vapor data because its distribution spans the entire stratosphere and is essentially nonreactive throughout.

WMO (1986) contains a discussion of the quality of the seven monthly LIMS water vapor distributions.   It is noted that the descending mode results are more reliable, particularly in the upper stratosphere.   That conclusion is supported by the constancy of total hydrogen in the upper stratosphere using the descending mode water vapor and the SAMS methane (Jones et al., 1986).   Figure 4 shows the zonal-mean distribution of water vapor for May.   The tropical lower stratosphere is a region of minimum water vapor--a hygropause several kilometers above the tropical tropopause.   The mechanisms for the maintenance of that dry air are uncertain, but its pattern is nearly fixed for the entire LIMS period, commensurate with a long turnover time for that stratospheric reservoir.   A small portion of the relatively dry air ascends to the upper stratosphere and branches to higher latitudes.   The March streamfunctions of Callis et al. (1987) led to a distribution similar to that in Fig. 4 (vertical and horizontal movement at 5 mb due to the net circulation is about 6 km and 2000 km, respectively, in 2 months).   During April and May (Fig. 3) above 10 mb, transport to higher northern latitudes is restricted, but toward the south is enhanced, and that effect may be partly responsible for the appearance of the distribution in Fig. 4.

Net circulations in the lower stratosphere extend poleward in both hemispheres, but Fig. 3 for May indicates a stronger net circulation toward the southern high latitudes; water vapor appears to

TABLE 2.   TRACERS AND THEIR DOMAIN FOR TRANSPORT STUDIES

Parameter   Comment

A. Mid and Upper Stratosphere

N_2O, CH_4 (SAMS)   zonal mean only

O_3 (SBUV, LIMS, SAGE)   higher latitudes mainly

H_2O, nighttime NO_2 (LIMS)   four longitudinal waves for
                                                            NO_2

Potential Vorticity, PV   derived quantity
(LIMS, SSU)

B. Lower Stratosphere

H_2O, HNO_3, O_3 (LIMS)   six longitudinal waves

O_3 and aerosols (SAGE)   limited coverage per day

aerosols (SME)   El Chichon dispersion

PV (LIMS and SSU)   derived quantity

NO_y = HNO_3 + NO_2 (LIMS)   appropriate for nighttime
                                                            data only; four longitu-
                                                            dinal waves

C. Total Column

Ozone (TOMS, SBUV)   85 to 90% in stratosphere

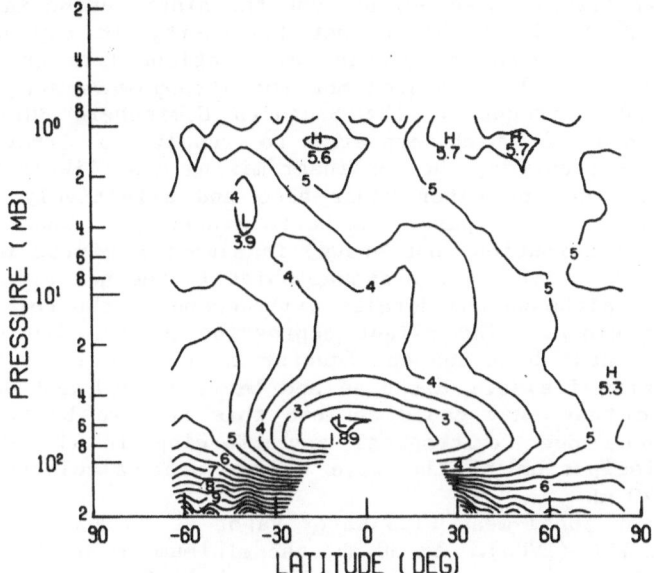

Figure 4.   LIMS zonal-mean water vapor cross section for May 1979 for descending mode data.   Contour interval is 0.5 ppmv.

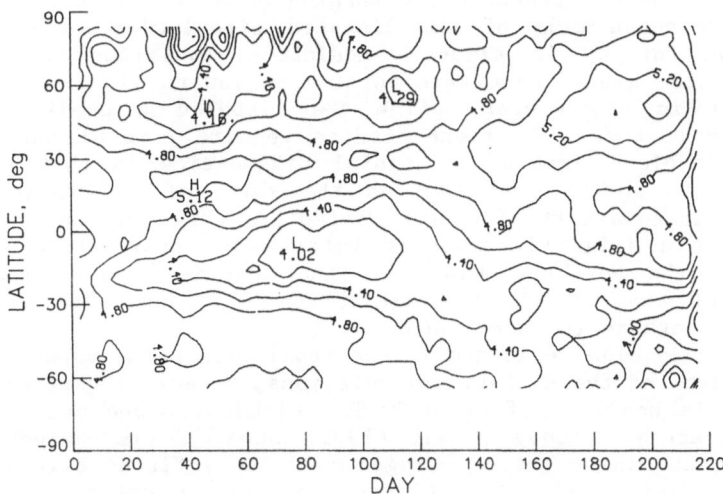

Figure 5.   Time series of zonal-mean, descending mode, LIMS water vapor at 3 mb.   Contour interval is 0.2 ppmv and day 0 is October 25, 1978.

disagree with that trend.   However, because the distribution in Fig. 4 is likely the result of months of net transport, it may be more appropriate to compare with the average circulations from an earlier period.   Leovy et al. (1985) pointed out the strong wave activity in LIMS ozone at mid-latitudes of the Southern Hemisphere during May (late autumn), some of which appeared to result in irreversible mixing.   Where this occurred, the nonlinear mixing most likely reduced the meridional gradient of water vapor also and effectively altered the poleward motions of the purely diabatic transport.   The November 1978 water vapor distribution (not shown) is almost a mirror image of that in May 1979 (WMO, 1986).   A seasonal shift in the net circulation can be postulated, although the details will depend on the role of any weak sources and sinks.   The slight depression of the 4.0 and 4.5 contours in Fig. 4 near 5 mb and the Equator is not predicted by most diabatic circulation fields.   Hitchman and Leovy (1986) and Gray and Pyle (1986) found that they could achieve this feature by including momentum divergence due to tropical wave forcing in the diabatic circulation.   Dissipation of the wave energy was calculated to be important above 20 mb.

Time series of zonal-mean LIMS water vapor at 50 and 10 mb have been presented in WMO (1986).   At 30 mb, the minimum is at the Equator during winter with some variations at high northern latitudes.   Figure 5 for 3 mb shows a minimum of about 4.0 ppmv at day 80 (January 14), indicating an acceleration in the upward net circulation.   This observation is supported by the fact that tropical ozone in the lower stratosphere is a minimum at the same time.   There may be a connection with the minor sudden stratospheric warming event in mid-January, but if so, the response time of the diabatic circulation to the high latitude wave forcing is almost immediate.   Dunkerton and Delisi (1986) documented significant wave forcing at low to mid latitudes at 5 mb earlier in the winter, which they labeled a predisturbance period.   Actual cause and effect studies require careful analysis-- tracer-like species ought to be useful diagnostics in that regard. Progressing into spring, one can see the shift of the water vapor minimum at 3 mb toward the Southern Hemisphere (Fig. 5).   At the same time, there is a slight increase in water vapor near 40 N.   It is probable that the Northern Hemisphere increase at 3 mb is due to a combination of changes in the circulation plus the slow chemical conversion of methane to water vapor.

There is no clear evidence in the zonal-mean water vapor for the descending parts of the diabatic circulations, because the water vapor mixing ratio is nearly uniform at high latitudes; ozone and aerosols are better tracers.   Leovy et al. (1985) showed a time-height cross section of ozone mixing ratio averaged over the latitude belt 76-84 N for the LIMS mission (Fig. 6).   Changes in early December and in late January are fairly reversible and reflect temporary movements of the polar vortex off the Pole.   In the mid-stratosphere in early winter, the ozone contours crossed the zonal mean θ-surfaces in a way that is consistent with the wintertime diabatic cooling rates in that region. Cooling rates of 1.3 K/day at 10 mb and 0.3 K/day at 50 mb were determined. Those rates agree with values reported by Kiehl and Solomon (1986) and Callis et al. (1987). Kent et al. (1985) calculate downward diabatic velocities of 0.1 km/day (cooling of about 1.0 K/day) at 23 km (about 30 mb) from trends in aerosol extinction data in the north polar vortex during autumn 1979.

Russell et al. (1984) examined polar night LIMS NO_2 data in the upper stratosphere and lower mesosphere and deduced downward transport of that species at high latitudes during winter. A more exact estimate from NO_2 data requires knowledge of the conversion rates of NO_2 to N_2O_5, or alternatively, profile measurements of N_2O_5 and HNO_3 in the polar night.

Recent observations of the wintertime Antarctic total ozone minimum by the TOMS instrument and in profile by SAGE and SBUV also imply a significant downward diabatic circulation in the polar vortex (Stolarski et al., 1986; McCormick and Larsen, 1986). Of course, the rapid decreases in ozone in the lower stratosphere during September and October are more likely due to a combination of adiabatic transport and chemical loss processes. A similar, though much less dramatic, LIMS ozone minimum occurs in the Northern Hemisphere. In that case, the diabatic circulation approximates the net transport best during the autumn when wave action is still fairly weak (see Leovy et al., 1985, their Figs. 6 and 7).

The seasonal variation of potential vorticity (PV) on an isentropic surface is an appropriate indicator of changes in the forcing of the extratropical diabatic circulation. Butchart and Remsberg (1986), for example, compared PV derived from LIMS data with that from an annual cycle integration of a zonally symmetric, general circulation model. Time series of the average PV between 20 N and the North Pole were plotted from both data and model at θ = 850 K. The mean PV values reach a maximum in mid-winter in response to the gradual downward diabatic transport at high latitudes. The PV results from LIMS indicate rapid smaller-scale variations in the seasonal trends that are due to wave activity not included in the model study. They verified the PV changes locally by intercomparing with changes in the LIMS water vapor distribution on a θ-surface (see also paper by Butchart, this volume).

5. CONCLUDING REMARKS

Analyses of recent satellite data sets are revealing an amazing consistency in the large-scale variations in stratospheric tracer-like parameters—a testimony to the high precision of the measurements. Several methods for deriving net circulations have been considered. The indirect approach involves combining temperature, ozone, and water vapor data with a fairly detailed radiative code to derive a net heating distribution. Net diabatic and/or residual mean circulations are determined from imbalances in that net heating. Two disadvantages arise. (1) Data sets for temperature and ozone and the attendant calculations for such a study must be very accurate, as well as precise; where circulations are weak, even the sense of the derived circulation can be in error. (2) Diabatic effects due to irreversible wave mixing are not normally factored into the derived stream-functions, leading to faulty interpretations from diabatic circulations, particularly in the lower stratosphere. The inclusion of information about any temperature tendencies gives a more complete net circulation.

A direct method for deducing circulations involves changes in tracer-like parameters. Data accuracy can be sacrificed somewhat, if precision is high, for species such as N_2O, CH_4, H_2O, and total O_3. PV is the exception here because it is a highly derived, dynamical quantity requiring accuracy in the gradients of fields. Still, extra-tropical PV signatures derived locally from several data sets (SSU and LIMS) appear to be of high quality. There are drawbacks to the direct method, too. (1) Sources and sinks of the proposed tracers must be well understood. (2) No information can be obtained about a net circulation where tracer gradients are very weak. The various tracers that are available for study fortunately display distributions of differing gradients, so that one can obtain a more accurate picture of the total circulation by examining trends in several concurrent tracers. One approach is to map one tracer distribution against another--the "modified Lagrangian mean" approach (Butchart and Remsberg, 1986). Three-dimensional distributions of the tracers (e.g. LIMS data) are required. The analyses can also be used to examine non-conservative processes or budgets of more reactive species, such as HNO_3 or NO_2. Finally, an examination of subregions of the stratosphere (upper and lower; tropical and higher latitudes) using satellite tracer data appears feasible for the first time. While three-dimensional models have been used for more detailed studies of transport in the past, analytical and numerical model/data intercomparisons can now be initiated using fields other than just temperature and winds.

6. ACKNOWLEDGEMENTS

I thank Linwood B. Callis for supplying unpublished material in Fig. 3. William L. Grose provided information regarding the calculations of $\bar{v}^*$ from the momentum versus the thermodynamic equation. That material was presented at IAMAP/IAGA in Prague, Czechoslovakia in 1985 by Grose and O'Neill for both the LIMS and SSU data sets.

7. REFERENCES

Al-Ajmi, D. N., R. S. Harwood, and T. Miles, 1985: 'A sudden warming in the middle atmosphere of the southern hemisphere.' Quart. J. Roy. Meteor. Soc., 111, 359-389.

Andrews, D. G., and M. E. McIntyre, 1978: 'Generalized Eliassen-Palm and Charney-Drazin theorems for waves on axisymmetric mean flows in compressible atmospheres.' J. Atmos. Sci., 35, 175-185.

Barnett, J. J., and M. Corney, 1985: 'Temperature data from satellites' (pp. 3-11); 'Middle atmosphere reference model derived from satellite data' (pp. 47-85). MAP Handbook, 16, University of Illinois, Urbana, IL.

Butchart, N., and E. Remsberg, 1986: 'The area of the stratospheric polar vortex as a diagnostic for tracer transport on an isentropic surface.' J. Atmos. Sci., 43, 1319-1339.

Callis, L. B., R. E. Boughner, and J. D. Lambeth, 1987: 'The stratosphere: Climatologies of the radiative heating and cooling rates and the diabatically diagnosed net circulation fields.' J. Geophys. Res., 92, 5585-5608.

Clough, S. A., N. S. Grahame, and A. O'Neill, 1985: 'Potential vorticity in the stratosphere derived using data from satellites.' Quart. J. Roy. Meteor. Soc., 111, 335-358.

Dunkerton, T., 1978: 'On the mean meridional mass motions of the stratosphere and mesosphere.' J. Atmos. Sci., 35, 2325-2333.

Dunkerton, T. J., and D. P. Delisi, 1986: 'Evolution of potential vorticity in the winter stratosphere of January-February 1979.' J. Geophys. Res., 91, 1199-1208.

Edmon, H. J., Jr., B. J. Hoskins, and M. E. McIntyre, 1980: 'Eliassen-Palm cross sections for the troposphere. J. Atmos. Sci., 37, 2600-2626.

Gille, J. C., and L. V. Lyjak, 1986: 'Radiative heating and cooling rates in the middle atmosphere.' J. Atmos. Sci., 43, 2215-2229.

Gray, L., and J. Pyle, 1986: 'The semiannual oscillation and equatorial tracer distributions.' Quart. J. Roy. Meteor. Soc., 112, 387-407.

Grose, W. L., 1984: 'Recent advances in understanding stratospheric dynamics and transport processes: Applications of satellite data to their interpretation.' Adv. Space Res., 4, 19-28.

Grose, W. L., and C. D. Rodgers, 1986: 'Coordinate study of the behavior of the middle atmosphere in winter: Monthly mean comparisons of satellite and radiosonde data and derived quantities' (pp. 79-111). MAP Handbook, 21, University of Illinois, Urbana, IL.

Hitchman, M. H., and C. B. Leovy, 1986: 'Evolution of the zonal mean state in the equatorial middle atmosphere during October 1978 - May 1979.' J. Atmos. Sci., 43, 3159-3176.

Jones, R. L., J. A. Pyle, J. E. Harries, A. M. Zavody, J. M. Russell III, and J. C. Gille, 1986: 'The water vapor budget of the stratosphere studied using LIMS and SAMS satellite data.' Quart. J. Roy. Meteor. Soc., 112, 1127-1143.

Keating, G. M., and D. F. Young, 1985: 'Interim reference ozone model for the middle atmosphere.' MAP Handbook, 16, University of Illinois, Urbana, IL, 205-229.

Kent, G. S., C. R. Trepte, U. O. Farrukh, and M. P. McCormick, 1985: 'Variation of the stratospheric aerosol associated with the north cyclonic polar vortex as measured by the SAM II satellite sensor.' J. Atmos. Sci., 42, 1536-1551.

Kiehl, J. T., and S. Solomon, 1986: 'On the radiative balance of the stratosphere.' J. Atmos. Sci., 43, 1525-1534.

Leovy, C. B., C.-R. Sun, M. H. Hitchman, E. E. Remsberg, J. M. Russell III, L. L. Gordley, J. C. Gille, and L. V. Lyjak, 1985: 'Transport of ozone in the middle stratosphere: Evidence for planetary wave breaking.' J. Atmos. Sci., 42, 230-244.

Mahlman, J. D., D. G. Andrews, H. U. Dutsch, D. L. Hartmann, T. Matsuno, and R. J. Murgatroyd, 1984: 'Transport of trace constituents in the stratosphere.' Dynamics of the Middle Atmosphere, Reidel, 387-416.

McCormick, M. P., and J. C. Larsen, 1986: 'Antarctic springtime measurements of ozone, nitrogen dioxide, and aerosol extinction by SAM II, SAGE, and SAGE II.' Geophys. Res. Lett., 13, 1280-1283.

Rosenfield, J. E., and M. E. Schoeberl, 1986: 'A computation of stratospheric heating rates and the diabatic circulation for the Antarctic spring.' Geophys. Res. Lett., 13, 1339-1342.

Russell, J. M. III, S. Solomon, L. L. Gordley, E. E. Remsberg, and L. B. Callis, 1984: 'The variability of stratospheric and mesospheric NO_2 in the polar winter night observed by LIMS.' J. Geophys. Res., 89, 7267-7275.

Russell, J. M. III, S. Solomon, M. P. McCormick, A. J. Miller, J. J. Barnett, R. L. Jones, and D. W. Rusch, 1986: 'Middle atmosphere composition revealed by satellite observations.' MAP Handbook, 22, University of Illinois, Urbana, IL 302 pp.

Smith, A. K., and P. L. Bailey, 1985: 'Comparison of horizontal winds from the LIMS satellite instrument with rocket measurements.' J. Geophys. Res., 90, 3897–3901.

Solomon, S., J. T. Kiehl, R. R. Garcia, and W. L. Grose, 1986a: 'Tracer transport by the diabatic circulation deduced from satellite observations.' J. Atmos. Sci., 43, 1603–1617.

Solomon, S., J. T. Kiehl, B. J. Kerridge, E. E. Remsberg, and J. M. Russell III, 1986b: 'Evidence for nonlocal thermodynamic equilibrium in the ν3 mode of mesospheric ozone.' J. Geophys. Res., 91, 9865–9876.

Stolarski, R. S., A. J. Kreuger, M. R. Schoeberl, R. D. McPeters, P. A. Newman, and J. C. Alpert, 1986: 'Nimbus 7 satellite measurements of the springtime Antarctic ozone decrease.' Nature, 322, 808–811.

World Meteorological Organization, 1986:   'Atmospheric ozone 1985: Global ozone research and monitoring project.' Report No. 16.

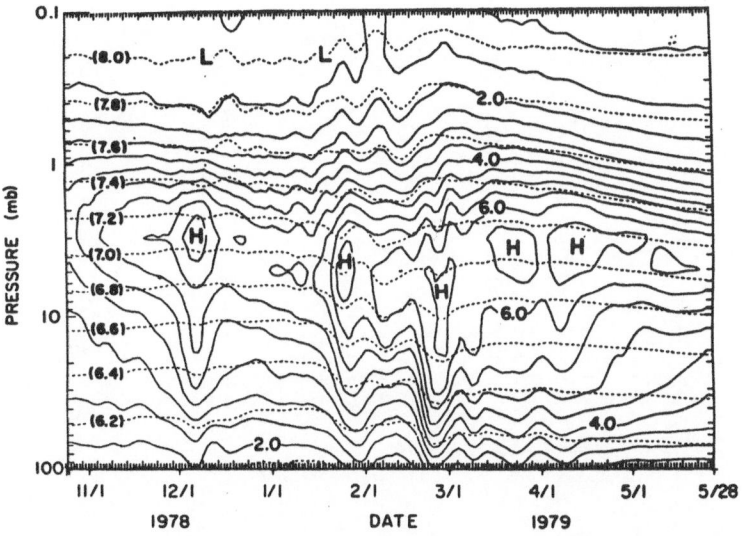

Figure 6.   Time-height cross section of zonal-mean ozone (contour interval is 0.5 ppmv) averaged over the latitude belt 76-84°N for the period October 25, 1978, through May 28, 1979.   Dashed lines are isentropes with values of lnθ indicated (after Leovy et al., 1985).

5. OBSERVATIONAL TECHNIQUES

IRREGULAR AND DIURNAL VARIABILITY IN ASYNOPTIC MEASUREMENTS OF STRATOSPHERIC TRACE SPECIES

Murry L. Salby*
Department of Astrophysical, Planetary,
  and Atmospheric Sciences
Campus Box 391
University of Colorado
Boulder, CO 80309
USA

ABSTRACT. Two classes of variability in stratospheric trace species: (i) dynamically introduced tracer irregularities and (ii) diurnal variations in photochemically active species, are investigated with regard to asynoptic satellite measurements. The fidelity with which the *continuous* behavior can be derived from the *discrete* asynoptic measurements is examined. Irregular tracer variability is described in terms of an advected random process, while diurnal fluctuations are constructed with a solar waveform. In each case, the continuous behavior retrieved is corrupted by aliasing from variability unresolved by the asynoptic measurements. In essence, the discrete observations are *misinterpreted*. Contamination by diurnal variations is particularly grave, because it leads to variance well removed from the Nyquist limits of asynoptic sampling and aliases to time-mean behavior. The availability of contemporaneous measurements from several instruments viewing different regions on the globe (e.g. UARS), may make it possible to alleviate these difficulties by capitalizing on the expanded information content of the collective data set.

1. Introduction

Satellite data have provided an unprecedented view of stratospheric behavior, contributing on both dynamical and chemical fronts. Because of their global and uninterrupted coverage, they can be expected in the future to play an even greater role in advancing our understanding of atmospheric behavior.

Early stages of remote sensing focused on determining basic dynamical fields, such as geopotential which could be built up hydrostatically from retrieved temperatures. However, recent years have witnessed the development of a number of fundamental diagnostics, e.g. Eliassen-Palm flux and transformed Eulerian circulation, which are essential ingredients of mean meridional motion and transport in general. With the recognition of these central dynamical quantities and the growing importance of tracers in elucidating transport, emphasis has shifted from conventional fields, such as streamfunction, to higher order behavior. In particular, attention has been drawn to the divergence of Eliassen-Palm flux, representing the essential mean flow driving by the eddies, and to Ertel potential vorticity, a dynamical tracer.

These quantities are *higher order*, often termed highly derived, because they are related to streamfunction through several spatial derivatives. For this reason, they are of inherently richer structure than low order fields such as geopotential, from which they are currently derived. The more complex behavior reflected in these fields is a consequence of spatial differentiation, which acts to high-pass filter the variability (Båth, 1974). In so doing, the granularity of the fields is increased. There are also solid kinematic reasons (developed below) why tracer distributions should exhibit richer structure than say geopotential. In any event, the more rapid variability inherent in these quantities places a stronger demand on the sampling needed to resolve the behavior.

---

*Center for Atmospheric Theory and Analysis, University of Colorado, Boulder, CO 80309-0391

G. Visconti and R. Garcia (eds.), Transport Processes in the Middle Atmosphere, 423–438.
© 1987 by D. Reidel Publishing Company.

Two important physical mechanisms operating in the stratosphere challenge the limited resolving power of asynoptic sampling. Each is associated with an opposite extreme in photochemical lifetime τ_c, or more accurately its magnitude relative to a dynamical time scale τ_D which reflects the time for material to be advected across mean gradients of long-lived species. Both of these processes can lead to rapid variability in space and/or time which may not be fully captured in asynoptic observations.

1.1. Dynamically introduced irregularity: $\frac{\tau_c}{\tau_D} \rightarrow \infty$

In the limit of long photochemical lifetime (i.e. with respect to the advective time scale), mixing ratio is conserved following material elements. Such a species is said to be a "material tracer", it's value tracking through space along with fluid elements. Because of this property, mixing ratio inside a fluid parcel reflects the element's *initial environment*. Equivalently, a tracer within a fluid parcel may be regarded as having an infinite memory. The behavior is exemplified by ozone in the extratropical lower stratosphere.

The distribution of such a long-lived species is controlled by advective drive. As such, it is subject to irregular reordering, as may be introduced through nonlinear dynamics and parcel dispersion. From a Fourier perspective, this redistribution of material and tracer mixing ratio appears as a "cascade", variance passing from large spatial scales to progressively smaller ones. It is this dynamical property which has important consequences to asynoptic sampling. Specifically, if variance cascades to small enough spatial scales, or if concomitant time scales are sufficiently short, the global field will be *undersampled*.

Tracer mixing ratio variance may be created in a variety of ways. Perhaps the most ubiquitous means is through parcel dispersion accompanying the forced wave critical surface, e.g. in a neighborhood of the zero wind line. During large amplitude wave events the critical region, where perturbation velocities are comparable to the zonal mean, expands and may engulf much of the hemisphere. In either event large material strains, realized in the region of nonlinearity, are capable of generating a complex and irregular redistribution of material and therefore any tracer field. Under such circumstances, the behavior may be more sensibly described in terms of its spatial and temporal statistics, as in the theory of turbulence. As is typical of geophysical signatures, tracer spectra are "red", power eventually falling off monotonically with wavenumber and frequency. The key question regarding asynoptic sampling is: How quickly does variance decay? Alternatively, how much variance lies beyond the scales which can be resolved in the measurements?

1.2. Diurnal variations: $\frac{\tau_c}{\tau_D} \rightarrow 0$

In the limit of short photochemical lifetime, mixing ratio within a fluid element is always in photochemical equilibrium, adjusting continually to its changing environment. Such a species (e.g. within a fluid parcel) may be regarded as having zero memory and reflects *local and instantaneous conditions*. Ozone in the tropical upper stratosphere is an example. Advection is ineffective at redistributing a species in this limit. Consequently, the distribution of such a constituent is controlled by photochemical drive.

Because of the immediate response to photochemical equilibrium, daily variations of solar flux will lead to marked changes in mixing ratio. Although dependent upon local concentration and on the chemical reactions involved, diurnal variations in short-lived constituents are expected to be sizable, mirroring the daily growth and collapse of solar radiation.

Both of these classes of variability challenge the resolving power of asynoptic measurements. Rapid spatial and temporal behavior associated with each may not be fully captured by the finite information content of asynoptic sampling. We explore here the capability of asynoptic observations to represent these two important classes of behavior. In particular, we focus on the ability with which the continuous behavior may be extracted from discrete asynoptic measurements. For each case, known behavior is constructed, sampled asynoptically, and used to derive a variety of diagnostics of the continuous variability, e.g. space-time spectra, synoptic maps, etc. These are then compared with the known continuous behavior, and errors emerging in the process are identified with sampling deficiencies.

2. SAMPLING, RESOLUTION, AND ALIASING

2.1. 1-Dimensional Sampling

Before proceeding to the issue at hand, we review some basic properties of discrete sampling. A 1-dimensional signature ψ, sampled discretely at uniform increments of longitude or time, λ or t (Fig. 1a), may be used to construct the complex Fourier spectrum Ψ (Fig. 1b) over wavenumbers m or frequencies σ. The question arises: How many spectral components can be determined? This will be seen shortly as equivalent to asking: How much of the behavior (e.g. between discrete observations) do we really know?

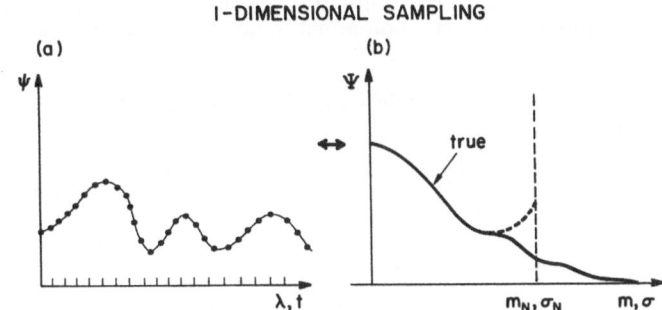

Figure 1 (a) 1-dimensional signature ψ sampled discretely at uniform increments (longitude or time). (b) Fourier spectrum Ψ (real or imaginary component) over wavenumber m or frequency σ. Power lying beyond the Nyquist limit, m_N or σ_N, is unresolved and folds back onto the resolvable components.

A discrete sample contains only a finite number of pieces of information. Consequentl only a limited number of spectral components Ψ may be determined independently. If the spectrum is computed from simple oscillatory behavior, the true spectral component surfaces, but so too do infinitely many spurious entries, known as "aliases". The aliases, appearing as mirror images about the "Nyquist" wavenumber: $m = 1/\lambda$, or frequency: $\sigma = \pi/\Delta t$, are degeneracies which arise out of the limited information content of the sample. They are *indistinguishable*, in the discrete observations, from the true variability. The Nyquist values, corresponding to a complete oscillation across two grid intervals, represent the smallest scales which can be resolved in the measurements. Wavenumbers (frequencies) less than the Nyquist define the "information content" of the discrete sample: those scales genuinely resolvable in the observations.

The Sampling Theorem (Båth, 1974) ensures that the two representations: ψ in physical space and Ψ in Fourier space, are unique and *equivalent* descriptions of the behavior, provided no variance lies beyond the Nyquist limit. That is, if we know one, we know the other. In such case, the variability is said to be "adequately sampled", and the complete behavior is at hand. The continuous behavior may be reconstructed by inverting the spectrum, in effect Fourier interpolating.

If, however, the actual behavior involves scales lying beyond the Nyquist, the calculated spectrum is "aliased" by such variability. Spectral components within the information content, otherwise correctly recoverable, are misinterpreted with components outside the Nyquist which, in any event, are not resolvable. In 1-dimensional sampling, unresolved variance *folds* back across the Nyquist, corrupting scales which are resolvable.

This misinterpretation of behavior is a fundamental property of discrete data. It carries over to the variability in physical space. In attempting to interpolate to the continuous behavior, the discrete observations will be misinterpreted.

The Sampling Theorem tells us under what circumstances the continuous behavior may be inferred from the discrete observations. This result has profound implications for satellite measurements because the data, taken asynoptically (viz. different locations at different times), must somehow be interpolated (in space and time) to synoptic coordinates, giving in effect an instantaneous snapshot of the global field. This requires knowledge of the continuous space-time behavior.

2.2. 2-Dimensional *Synoptic* Sampling

Consider now longitude-time variability on a latitude circle. The behavior, sampled uniformly in space at regular increments of time, defines a rectangular grid in the longitude-time plane (Fig. 2a). As in the 1-dimensional case, a Fourier spectrum may be constructed from the discrete observations, here the "space-time spectrum" over wavenumber and frequency (Fig. 2b). Again, owing to the finite number of pieces of information in the sample, only a limited number of spectral components may be unambiguously determined.

2-DIMENSIONAL SYNOPTIC SAMPLING

Figure 2 (a) 2-dimensional signature sampled asynoptically at uniform increments of longitude and time. (b) Space-time spectrum over wavenumber and frequency. Nyquist wavenumber and frequency, defining the information content, describe a rectangle oriented parallel to the wavenumber frequency axes.

Extension of concepts, such as aliasing and information content, from 1-dimensional sampling is straightforward. Nyquist wavenumber and frequency follow directly from the uniform grid of observations. These again define the resolvable scales (i.e. the information content), here a rectangle in the wavenumber-frequency plane (Fig. 2b). (As in the 1-dimensional case, only half of the spectral domain need be displayed, the other half being simply the complex conjugate.) A 2-dimensional extension of the Sampling Theorem ensures that if no variance lies beyond the Nyquist boundaries, the complete behavior is captured by the discrete sampling.

2.3. Asynoptic Sampling

Consider the behavior on any latitude circle. The sampling made by satellite (Fig. 3) is again discrete, involving ascending and descending traversals of the orbit. Only in this case, the observation points are sampled asynoptically, and the resulting grid is not regular in longitude. As the satellite moves about the latitude circle, the crossings or "nodes" do not return to those of the previous day. Instead, they fall between sample points of earlier cycles. After sufficient time, a dense, but irregular nest of points has been sampled, leaving it unclear which spatial scales are actually resolvable.

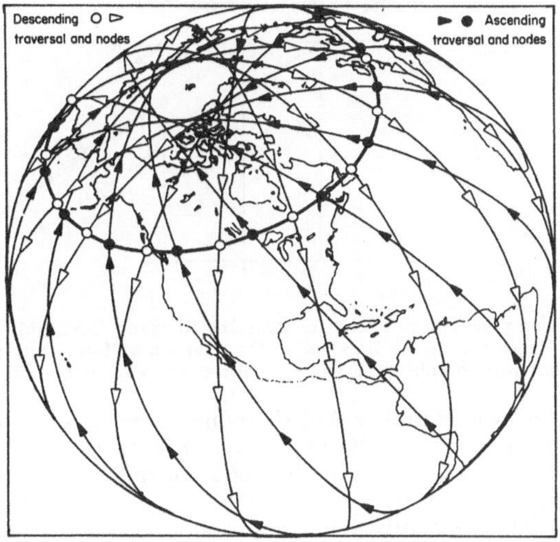

SUB-SATELLITE TRACKS

Figure 3 Sampling geometry of polar orbiting satellite with nadir-viewing instrument. Limb-viewing geometry similar.

The key to eliminating this uncertainty lies in the fact that the observations are made *asynoptically*. Therefore, the spatial resolution cannot be defined independently of the temporal resolution, as is the case in synoptic sampling. Although on the latitude circle asynoptic sampling appears irregular, if plotted in the extended longitude-time plane (Fig. 4), a regular array of sample points again emerges. However, in this case the data have symmetry about *asynoptic coordinates*: s along which ascending and descending data lie, and r orthogonal to this direction. Each is an amalgam of space and time.

In the case of synoptic sampling, the space-time transform could be evaluated directly over the longitude-time rectangle D shown in Fig. 4. However, for asynoptic sampling the data do not lend themselves easily to such evaluation, being distributed irregularly over this region. It can be shown (Salby, 1982a) that the space-time spectrum can be derived equivalently over the asynoptic strip D', which is directly amenable to asynoptic observations.

Construction of the space-time spectrum requires use of the discrete data. As in the case of synoptic sampling, two characteristic separations emerge (Fig. 4): Δs along which the individual data lie, and Δr separating the arrays of ascending and descending data. This second scale also is a fundamental component of the sampling, since the domain is cyclic (e.g. in λ), and therefore each array has periodic images (not shown) spaced Δr units away.

Resolvable scales following from this distribution of data are analogous to synoptic sampling, only rotated into the asynoptic wavenumbers: k_s and k_r (Fig. 5). Each of these is an amalgam of zonal wavenumber and frequency, paralleling the mixed space-time dependence of sample points in physical space. For single node sampling (ascending *or* descending data) and typical orbital characteristics, the Nyquist limits correspond roughly

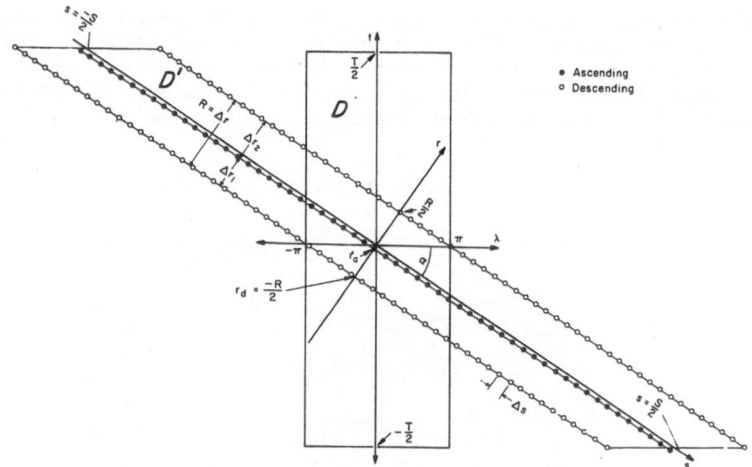

Figure 4 Observation points on a latitude circle, sampled asynoptically, plotted
in the extended longitude-time plane. Data form a uniform grid, lying
along asynoptic coordinates s and r, each a mixture of space and time.

to frequencies of 0.5 cpd and wavenumber 6. In analogous fashion to the 1-dimensional
case, the Asynoptic Sampling Theorem (Salby, 1982a;b), ensures that if no variance lies
beyond this information content, the continuous behavior in physical space, *in particular
at synoptic coordinates*, may be recovered.

An important distinction of satellite sampling is that asynoptic aliasing is *mixed*. Unlike aliasing in synoptic data, unresolved spatial variability in one time scale can corrupt
spatial variability of a completely different time scale. Because of this, steady unresolved
small-scale structure can be misinterpreted with larger-scale variability which is evolving. As an example, plotted on Fig. 5 is a resolvable traveling wavenumber 3 component
and two of its aliases. One is located at wavenumber 10 and 0 frequency, viz. steady, fine
structure. The reason for this degeneracy emerges clearly if one considers the true feature
and its steady alias from the frame of the satellite moving about a latitude circle (Fig. 6).
Although the continuous signatures are readily discerned, at the discrete sample points
the two are completely indistinguishable.

3. IRREGULAR TRACER VARIABILITY

We consider random behavior, in the form of an advected space-time process. The variability may be constructed (Salby, 1987) by randomly superposing local anomalies, representing distinct *blobs* of tracer mixing ratio, according to prescribed statistics (Fig. 7).
Specifically, tracer anomalies of characteristic spatial dimension appear and disappear
from the field with a characteristic lifetime. While present, they are advected about the
globe by a specified zonal flow $\bar{u}$. The tracer variability can be *frozen in the flow* (Hess
and Holton, 1985) by letting the characteristic lifetime tend to infinity.

3.1. Space-Time Spectra

This random behavior appears in Fourier space as a spectrum which is red. If the tracer
anomalies have Gaussian structure and time dependence, their space-time spectrum is
likewise Gaussian (Fig. 8a), with symmetry about the wavenumber-frequency axes. The
random process, advected at angular velocity $\bar{u}$, has a spectrum which is also Gaussian
(Fig. 8b), only rotated into the Doppler shifted line

$$\sigma = \bar{u}m$$

(1)

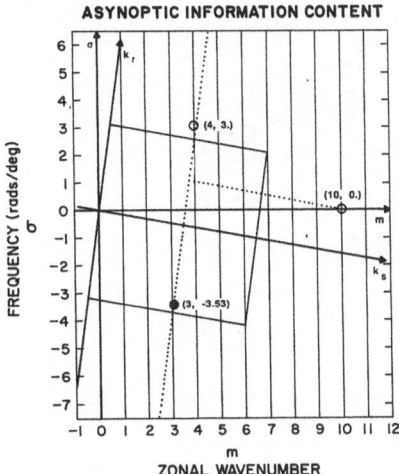

Figure 5 Wavenumbers and frequencies resolvable in asynoptic sampling (information content). Define rectangle rotated into the asynoptic wavenumbers, k_s and k_r, each an amalgam of wavenumber and frequency. Boundaries of rectangle correspond to Nyquist limits in these wavenumbers. Resolvable spectral component (dot) and two of it's aliases (circles) also shown.

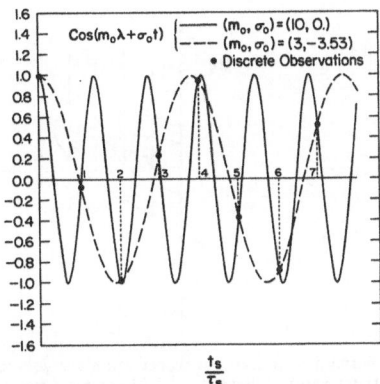

Figure 6 Alias pair, shown in Fig. 5, as seen from reference frame moving about latitude circle with the satellite. Discrete asynoptically sampled versions are indistinguishable.

Collapsing the characteristic spatial or temporal scale makes the variability in physical space more rapid, and therefore results in a spectrum which is *whiter* (variance falling off more slowly) along the line (1) and in frequency. For sufficiently small scales in physical space, variance will spill beyond the Nyquist limits of asynoptic sampling (cf. Fig. 5). Increasing the characteristic scales has just the opposite effect, behavior in physical space being more gradual and the spectrum becoming redder. In a similar fashion, increasing

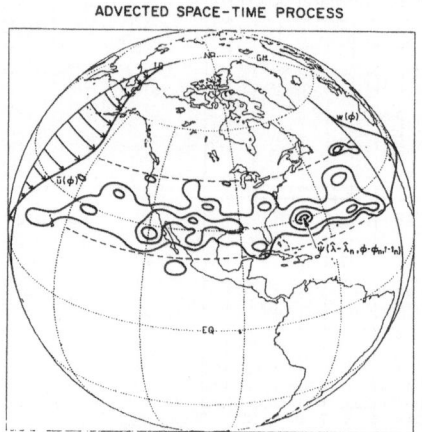

Figure 7 Advected space-time random process, describing creation and advection of tracer mixing ratio anomalies.

the flow $\bar{u}$ rotates the Doppler shifted line (1) to higher frequencies. Consequently, fewer wavenumbers are available within the Nyquist boundaries to represent the variability. For sufficiently rapid advection, power will again spill beyond the scales resolvable in asynoptic measurements.

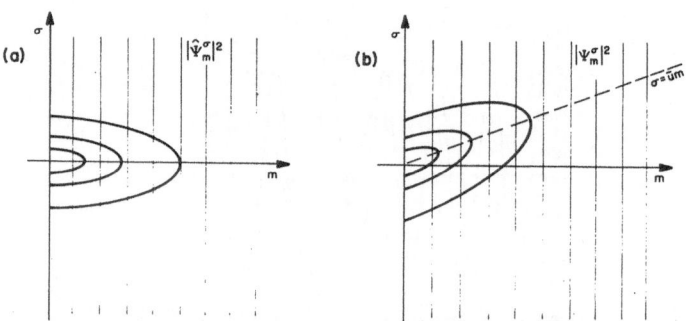

Figure 8 (a) Space-time power spectrum of individual tracer anomaly $\hat{\psi}$, superposed randomly to construct random field ψ. (b) Space-time spectrum of advected random process corresponding to anomalies in (a). Both are red, variance falling off with wavenumber and frequency, the spectrum of the random process being rotated into the Doppler shifted line $\sigma = \bar{u}m$.

We explore now the ability of satellite measurements to resolve this behavior by sampling asynoptically and constructing the space-time power spectrum. Several characteristic space and time scales are considered. We examine first the situation without zonal flow.

Figure 9 shows the space-time power spectrum recovered from anomalies of hemispheric dimension and having a characteristic time scale of 6 days. A very red spectrum emerges, power falling off sharply with wavenumber and frequency, so that all of the variance is captured by the asynoptic information content. In addition to the true power spectrum, alias images removed in frequency can also be seen. However, because the true spectrum falls off to negligible values before reaching the Nyquist boundaries, these spurious images do not influence the resolvable components.

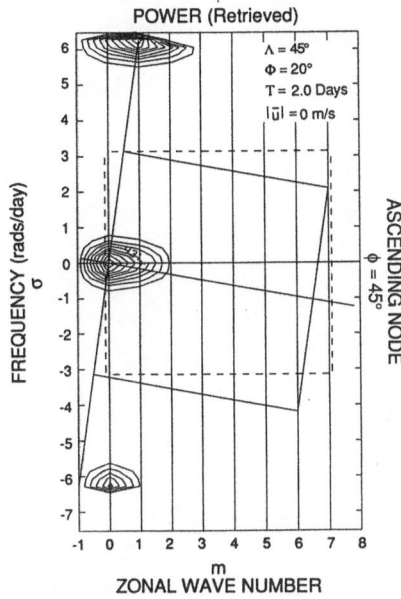

Figure 9   Asynoptically recovered Space-time power spectrum of random tracer behavior. Tracer anomalies of hemispheric scale, having a characteristic lifetime of 6 days, without zonal flow.

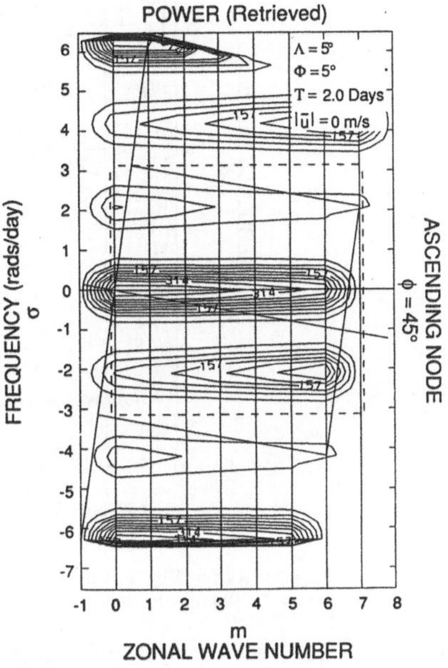

Figure 10   As for Fig. 9, but with tracer anomalies of characteristic zonal scale 15°.

Collapsing the zonal scale to 15°, as may reflect a mature cascade or small-scale variability in the lower stratosphere where synoptic disturbances are important, has the effect of spilling variance beyond the k_s Nyquist (Fig. 10). This results in aliases folding into the region of resolvable scales, corrupting components which would otherwise be correctly interpreted. Although it is nearly steady *spatial* variability which is undersampled, asynoptic aliasing can be seen to lead to spurious *temporal* variability.

Next we collapse the time scale to 1.5 days, holding the zonal dimension at 30°. This reflects the time for a fluid parcel to be advected through the crest of a large amplitude wave, and hence the time for a tracer anomaly to be created. The retrieved spectrum is now aliased by unresolved variance which folds in across the k_r Nyquist limit (Fig. 11).

We now set this random variability in motion by advecting it with zonal velocity $\bar{u}$. For tracer anomalies of characteristic zonal dimension 30° and lifetime of 3 days, advected in westerlies of 60 m/s, the retrieved spectrum (Fig. 12) has only 3 wavenumbers with which to resolve the variability. Consequently, power spills beyond the resolvable limits and can be seen to fold in across the positive k_r Nyquist. Both steady and transient components are affected.

3.2. Synoptic Behavior

The Asynoptic Sampling Theorem ensures that if the variance is captured by the discrete asynoptic measurements, we can interpolate the behavior (in space and time) to synoptic

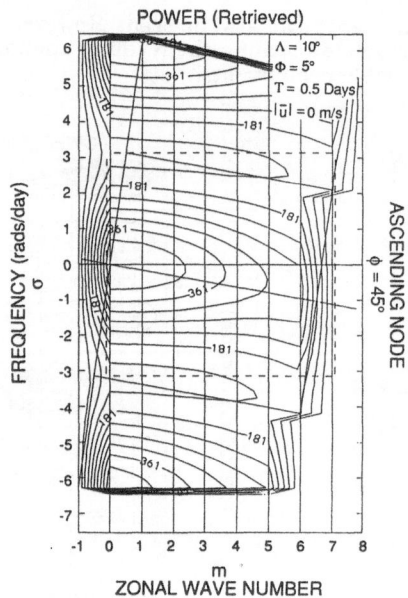

Figure 11 As for Fig. 9, but with tracer anomalies of characteristic zonal scale
30° and characteristic lifetime of 1.5 days.

coordinates and recover synoptic maps at any instant. However, we have seen that under
plausible scales, the behavior is *undersampled*. A significant fraction of the variability is
not captured by the information content of asynoptic measurements. As a result, scales
which would otherwise be correctly recovered, are misrepresented. To the degree that
this is true, synoptic fields interpolated from asynoptic measurements will be in error. As
an alternative, we explore whether we can relax the demands on the data and correctly
recover a weaker diagnostic of the behavior.

3.3. Time-Mean Field

We consider whether the time-mean field can be extracted, even though the information
content is inadequate to resolve the full space-time variability. It turns out (Salby, 1987)
that aliases, problematic in attempting to recover the complete behavior, are eliminated
in time-averaging. Consequently, the true time-mean structure is recovered after sufficient averaging. This result holds even if the variability is undersampled. It follows from
the randomness inherent to this class of behavior, leading to a cancellation of aliases. In
the case where variability is not random, e.g. systematic variations, no such cancellation
occurs.

4. DIURNAL VARIABILITY

Consider now a solar signature, e.g. in some photochemically active constituent, moving
about the globe with the sun and confined to a latitudinal envelope. The signature is presumed to be nonvanishing over only half the globe, viz. on the sunlit portion. For any
latitude circle, the behavior traced out in the longitude-time plane (Fig. 13a) is periodic,
assuming the form of a series of strips slanting westward with time.

In the Fourier plane, the space-time spectrum (Fig. 14) is concentrated along the
Doppler shifted line

$$\sigma = c_o m \tag{2}$$

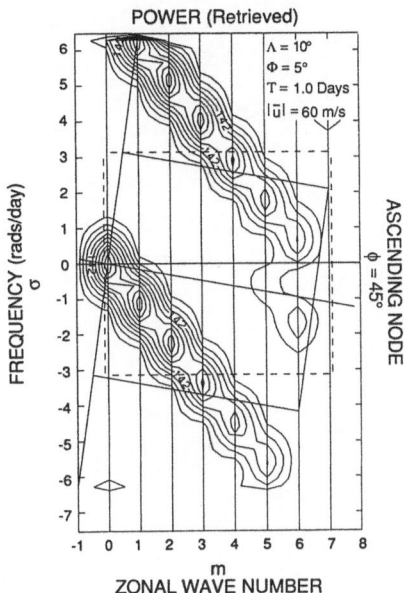

POWER (Retrieved)

$\Lambda = 10°$
$\Phi = 5°$
$T = 1.0$ Days
$|\bar{u}| = 60$ m/s

FREQUENCY (rads/day) σ

ASCENDING NODE $\phi = 45°$

m
ZONAL WAVE NUMBER

Figure 12 As for Fig. 9, but with tracer anomalies of characteristic zonal scale
30°, characteristic lifetime of 3 days, and advected in a 60 m/s westerly zonal flow.

where $c_o = 2\pi$ rads/day. This is analogous the situation of tracer variability being advected by a mean flow. Only here the speed is enormous, so that most of the variability lies outside the limits of asynoptic sampling. Indeed, the wavenumber 1 harmonic is located at 1.0 cpd—beyond the k_r Nyquist limit for single node sampling.

If the solar waveform (Fig. 13b) is a sinusoid, variance along the line (2) is distributed about the corresponding wavenumber, m_o. Truncating the signature to positive values has the effect of smearing the spectrum along the line (2). This introduces additional wavenumbers which lie further removed in frequency, viz. from the resolvable scales.

4.1. Space-Time Spectrum

We explore now this behavior under asynoptic sampling. Only the most favorable situation is presented, namely when the full information content of combined (ascending *and* descending) data is recovered, pushing the k_r Nyquist limit to approximately 1.0 cpd. It should be noted, however, that this expanded information content can be retrieved alias-free only if asymmetries intrinsic to combined sampling are explicitly accounted for (Salby, 1982a).

The space-time power spectrum for a wavenumber 1 sinusoid, truncated to positive values, is shown in Fig. 15. Because the signature is not simple oscillatory, variance is distributed over several wavenumbers, and hence involves higher harmonics (not shown) of the diurnal signal. The wavenumber 0 component lies within the information content, while the wavenumber 1 harmonic coincides with the Nyquist fold. Both of these components turn out to be incorrectly retrieved, significantly aliased by unresolved variability lying beyond the Nyquist limits. In particular, the steady zonal-mean component is aliased by higher even harmonics of the diurnal signature.

The source of this contamination is the high frequency content associated with the rapid phase speed of the solar signature. To demonstrate that the basic morphology is

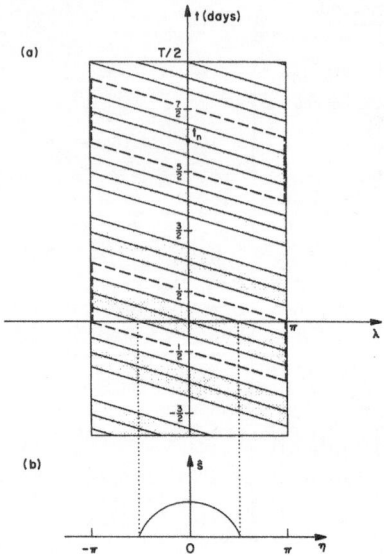

Figure 13 (a) Signature of solar waveform traced out in longitude-time plane. (b) Solar waveform, truncated to positive values, in longitude.

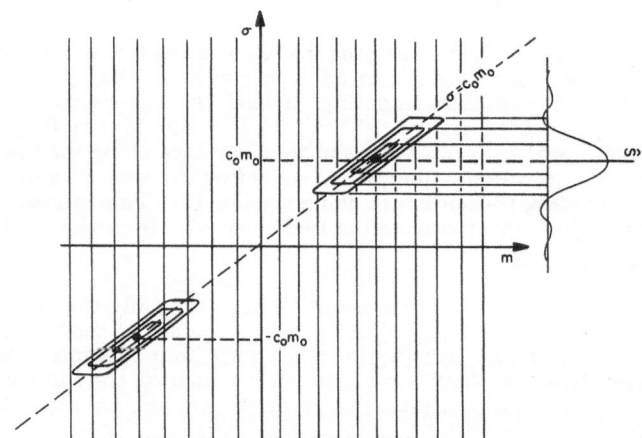

Figure 14 Space-time power spectrum resulting from solar signature shown in Fig. 13. Variance concentrated along Doppler shifted line $\sigma = c_o m_o$; $c_o = 2\pi$ rads/day.

derivable asynoptically, we sample the same signature, only moving at a third of the speed. For this case, the space-time spectrum (Fig. 16) shows the first three harmonics being captured by the information content of combined sampling. In fact, power falls off to

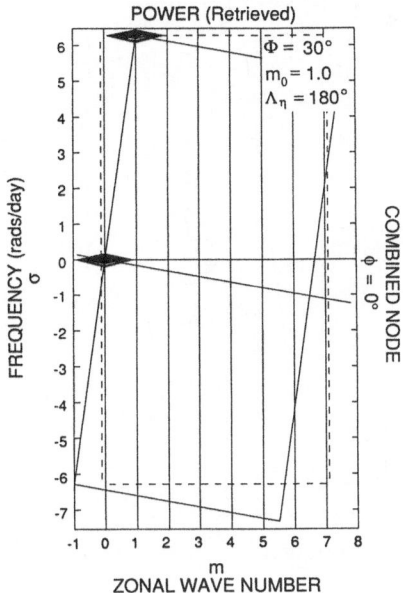

Figure 15 Asynoptically retrieved Space-time power spectrum, of sinusoidal solar waveform, truncated to positive values.

negligible values before reaching the Nyquist limit. Consequently, all of the resolvable scales are faithfully recovered.

4.2. Synoptic Behavior

As was the case for advected random behavior, unresolved variability leads to aliasing of the space-time spectrum. Sampling theory tells us then that the behavior in physical space, e.g. in attempting to interpolate to synoptic coordinates, will be misinterpreted. We move lastly to explore if, as in the case of random variability, we can relax the demands on the the information content and correctly recover the time-mean behavior.

4.3. Time-Mean Field

We saw for random variability that aliasing, which contaminated the full space-time behavior, was eliminated in time averaging. For diurnal variations this is not the case. The systematic relationship between the sampling and the diurnal signature leads to aliasing of time-mean components. Spurious values created in this process are not removed in time averaging.

The true time-mean field, corresponding to the behavior used to construct the spectrum in Fig. 16 and confined to a Gaussian envelope in latitude, is shown in Fig. 17. Time mean structure assumes the form of the envelope itself, because the solar waveform sweeps through repeatedly. The behavior retrieved asynoptically also assumes this form (Fig. 18) However, values recovered are *down by a factor of 3*. The actual numbers retrieved will depend on the relative phasing between the solar waveform and the sampling, e.g. the local time. Nonetheless, because this class of variability is undersampled, even the time-mean structure cannot be faithfully recovered.

In constructing the space-time spectrum, we saw that the true behavior could be recovered if the migrating waveform was made to move more slowly about the globe. This allowed all of the harmonics to be captured by the sampling. The same is true of the

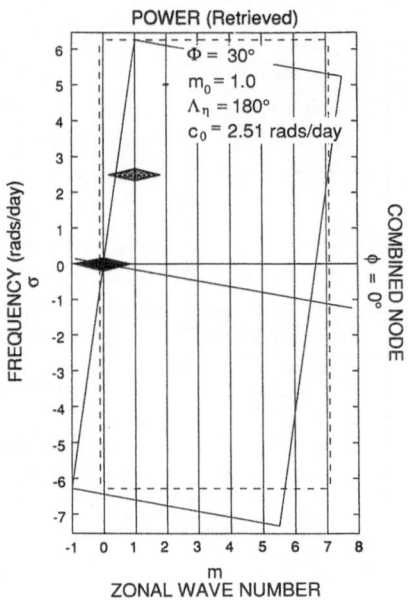

Figure 16 As for Fig. 15, only with the waveform moving about a latitude circle at one third the speed of the sun.

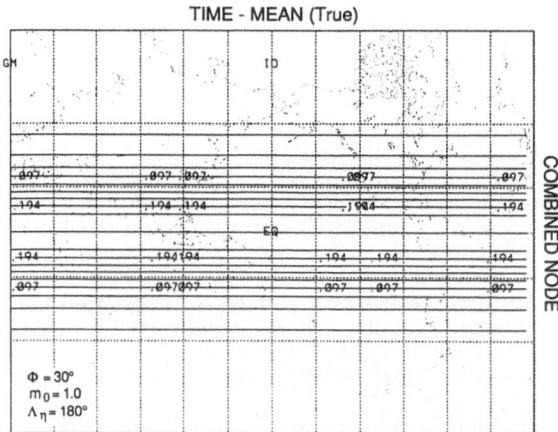

Figure 17 True time-mean field corresponding to diurnal variability indicated in Fig. 15.

time-mean field. Figure 19 shows the time-mean structure recovered when the solar signature is slowed down to one third of its speed. By comparison with Fig. 17, it can be seen that for all practical purposes the behavior is correctly retrieved.

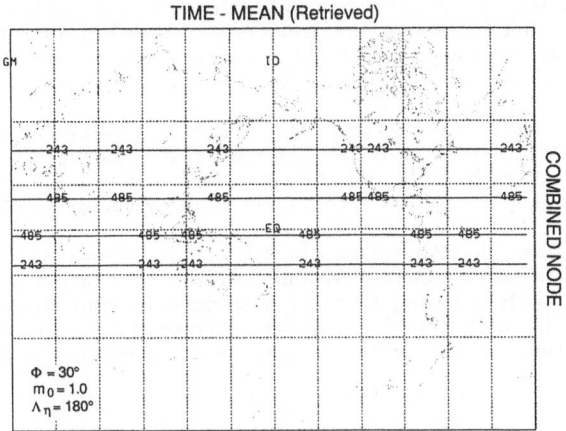

Figure 18 As for Fig. 17 (same contour increment), except retrieved asynoptically. (Contour labels scaled by 10,000.)

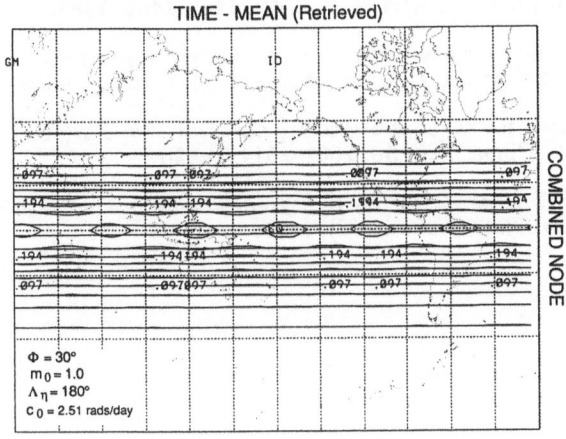

Figure 19 As for Fig. 18, only with the waveform moving about a latitude circle
at one third the speed of the sun.

5. CONCLUSIONS

Under plausible scales, the creation and rapid advection of tracer anomalies and diurnal
variations can each lead to undersampled behavior. Depending on the proportion of unresolved variance, the continuous behavior, e.g. synoptic maps interpolated from asynoptic
measurements, will be misrepresented.

Of the two classes of behavior, diurnal variations pose the most serious challenge,
leading to variance well removed from the Nyquist limits of asynoptic sampling. Unlike
random tracer variability, diurnal variations are systematically related to the sampling
and alias to time-mean behavior. Consequently, spurious contamination is not eliminated
in time averaging. It would be expected to increasingly corrupt the behavior of many

species at higher altitudes, where photochemical lifetimes become short. A small change in the orbital speed moving about latitude circles (e.g. precession) does not obviate this problem, making only a minor adjustment in the maximum resolvable frequency. However, it should permit correct recovery of the time-mean field in a manner analogous to random variability.

The real difficulty posed by these two classes of behavior stems from the fact that they are *both* present. If only diurnal variability existed, it could be readily distinguished from its aliases. However, even in the upper stratosphere and mesosphere, dynamically induced temperature fluctuations introduce a wide range of scales, making such simple interpretations impossible.

The only real means of obviating these difficulties is to increase the resolving power of the data. With contemporaneous observations at different locations on the globe (as may be available from UARS), it may be possible to do so by capitalizing on the expanded information content of the collective data set. The success of such a procedure will hinge on sampling irregularities, inherent to the combined data ensemble, being explicitly accounted for.

6. REFERENCES

Båth, M., *Spectral Analysis in Geophysics*, Elsevier, 563pp, 1974.

Hess, P. and J. Holton, The origin of temporal variance in long-lived trace constituents in the summer stratosphere. *J. Atmos. Sci.*, **42**, 1455–1463, 1985.

Salby, M., Sampling theory for asynoptic satellite observations. Part I: Space-time spectra, resolution, and aliasing, *J. Atmos. Sci.*, **39**, 2577–2600, 1982a.

_____, Sampling theory for asynoptic satellite observations. Part II: Fast Fourier Synoptic Mapping. *J. Atmos. Sci.*, **39**, 2601–2614, 1982b.

_____, Irregular and diurnal variability in asynoptic measurements of stratospheric trace species, *J. Geophys. Res.*, (submitted), 1987.

CONTRIBUTION OF RADAR OBSERVATIONS OF WINDS, WAVES, TURBULENCE AND COMPOSITION TO STUDY TRANSPORT PROCESSES IN THE MIDDLE ATMOSPHERE

J. Röttger [*]
EISCAT Scientific Association
Box 812
S-981 28 Kiruna
Sweden

ABSTRACT.   The measurements of parameters relevant to study transport processes of energy, momentum and mass in the middle atmosphere as well as the neighboring regions of the troposphere and the lower thermosphere, which can be done with mesosphere-stratosphere-troposphere (MST) radars and incoherent scatter radars (ISR), are described. The most essential exchange processes between the troposphere and stratosphere, namely the variation of the tropopause height, synoptic- and meso-scale vertical motions and small-scale eddy or turbulence transport, which can be monitored with MST radars, are summarized. The possibility to study vertical transport within the stratosphere by means of vertical motion and by turbulent diffusion is outlined. The transport and deposition of energy and momentum by atmospheric gravity waves in the middle atmosphere is briefly discussed. A method to deduce the turbulence diffusion coefficient from radar data is presented and applied to mesospheric MST radar observations. A brief outline of ISR observations of the mesopause and lower thermosphere is finally given to stress the importance of the coupling between these upper regions.

1.   INTRODUCTION

Turbulence, precipitation and related dynamical and transport processes in the planetary boundary layer and in clouds have been intensively studied with radars operating at wavelengths of a few centimeters. During the last decade a new kind of radar technique has been developed which uses wavelengths of several meters and allows to detect turbulence and gradient structures in the troposphere, strato-

[*] on leave from Max-Planck-Institut für Aeronomie, D-3411 Katlenburg-Lindau, W.Germany

G. Visconti and R. Garcia (eds.), Transport Processes in the Middle Atmosphere. 439–457.

sphere and mesosphere (e.g., Woodman and Guillen, 1974; and
special issues of Radio Science, Eds.: Gossard and Yeh
(1980), and Liu and Kato (1985)). These radars have become
known as MST (mesosphere-stratosphere-troposphere) radars,
or also as VHF radars because their most common wavelength
of 6 m is in the lower portion of the VHF band of the radio-
wave spectrum. They are considered to be essential tools to
study the dynamics and structure of the middle atmosphere
(e.g., Fritts et al., 1984), and have extensively been used
during the Middle Atmosphere Program -MAP- (see, for instance, the proceedings of the three special MST radar work-
shops held under the auspices of MAP (Bowhill and Edwards,
1983,1984,1986).

Some reasonable results have been obtained with the MST
radars in studying transport processes of passive tracers
in the middle atmosphere. It is attempted in this paper to
summarize the different phenomena which can be qualitatively
(and eventually also more quantitatively) observed with MST
radars. We will only consider mixing, respectively transport
processes in the vertical direction, since these appear to
be appropriately studied by these radars. After briefly out-
lining the radar technique and measurable parameters, we
will discuss a few examples of exchange processes between
and in the different altitude regions of the troposphere,
stratosphere, mesosphere and lower thermosphere. We will
also mention the transport and deposition of energy and
momentum by gravity waves. The discussion of the latter two
altitude regions has also to include a brief outline of the
incoherent scatter radar (ISR) technique. The ground-based
methods, particularly the radar techniques used to study
the middle atmosphere and lower thermosphere are described
in a Handbook for MAP (Vincent, 1984).

2. THE MST AND ISR RADAR TECHNIQUES

The MST radars are used to detect echoes due to scattering
and (partial) reflection from inhomogeneities in the refractive index of the clear air. These inhomogeneities essentially result from humidity variations in the lower and
middle troposphere, from temperature variations in the upper
troposphere and stratosphere, and from electron density
variations in the mesosphere. The latter result from density
variations of the neutral air. The scattered as well as the
reflected radar echo intensity are given by the component
of the spatial spectrum of the variation of the refractive
index whose wavelength is half the radar wavelength. In order to detect and interpret the echoes, it is assumed that
the turbulence causing the refractive index variations is
in the inertial subrange of the Kolmogoroff spectrum. For
3-m scales, this is mostly justified up to mesospheric

heights, and it is consequently reasonable to assume vertical gradient scales of 3 m to be also existent, which
cause partial reflection (Balsley and Gage, 1980, Röttger,
1980).

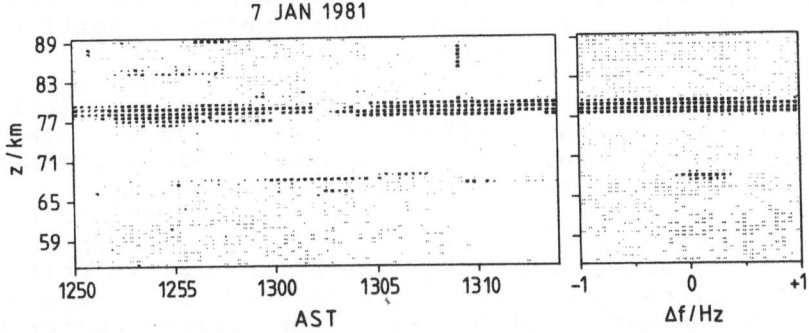

Fig.1 Mesospheric turbulence echoes (left) and their
spectra (right), measured with a VHF/MST radar at the
Arecibo Observatory, Puerto Rico. The frequency
$\Delta f = 1$ Hz corresponds to a velocity of 3.2 m s^{-1}.

The main parameters measured directly with a VHF/MST radar
are the intensity or power, the Doppler spectrum shift and
the Doppler spectrum width. These are measured as function
of time, altitude and antenna beam direction which is mostly
close to the zenith. Figure 1 shows a typical example of a
height-time intensity and a spectra plot of echoes from the
mesosphere. The echo structures are observed in thin laminae
and short-lived blobs (e.g., around 67-68 km altitude) as
well as thick layers (e.g., around 77-79 km). The stratospheric (and tropospheric) echo structures are fairly similar - although often even thinner and much stronger in intensity - to those thin laminae at 67 km altitude. The
spectra plot at the right-hand side of Fig. 1 indicates that
the thin laminae are characterized by a narrow spectrum
width, i.e. small velocity fluctuations, whereas the thick
and intense layers have a wide spectrum resulting from large
turbulent velocity fluctuations. For pure scattering, the
echo intensity is a measure of the intensity of turbulence.
For partial reflection, the echo intensity is a measure of
the steepness and magnitude of the refractive index
gradient. However, care has to be taken when using these
interpretations, since the background profiles, particularly
of humidity in the troposphere, of temperature in the
stratosphere and of electron density in the mesosphere, also
have a substantial influence on the echo intensity. The
Doppler shift is directly proportional to the bulk radial
velocity of the backscatter or reflection region. This

feature is commonly used to deduce the three-dimensional
velocity vector. The Doppler spectrum width is with certain
precautions a good estimate of the turbulent velocity fluctuations, which should be proportional to the turbulence
layer thickness. The temporal variation of intensity is a
measure of the persistency of turbulence or the stability
of the gradient of refractive index, which is particularly
conclusive in the stratosphere. By means of measuring the
intensity at different antenna beam directions, the anisotropy of turbulence and gradient structures is investigated.
The latter two measures allow to also derive the atmospheric
stability. By combining measurements of velocity fluctuations at different beam directions, the covariance of horizontal and vertical velocities, namely the Reynold stress
due to atmospheric waves, is determined. Using three or more
horizontally spaced antennas, the horizontal velocity vector
and the horizontal coherence are deduced. This spaced
antenna technique is also used in the interferometer mode
which allows to investigate the fine structure of turbulence
and gradients. More detailed descriptions how these parameters are measured and which suitable means have to be
applied to deduce relevant conclusions can be found elsewhere (e.g. Hocking, 1983; and papers in Bowhill and
Edwards, 1983,1984,1986, and in Vincent, 1984).

The outlined quantities are typically measurable with altitude resolutions of 150 m to about 1 km in the altitude
range up to 20-30 km in the stratosphere and between about
60 and 90 km in the mesosphere, depending on radar sensitivity and atmospheric conditions, namely the state of turbulence or stability, and electron density (in the mesosphere). The time resolution is adapted to the physical phenomenon of interest and can be as good as some 10 seconds.
The most sensitive radars, namely the Jicamarca-, Arecibo-,
Pokerflat-, MU- and SOUSY-VHF-Radar use antenna apertures
of several 10000 m^2 down to a few 1000 m^2. The MST radars
operate in the monostatic mode with different antenna beams
of 1-7 degrees width directed close to the zenith and use
average transmitter powers of about 100 kW to several kW.

Of particular interest for transport processes is the turbulence occurrence and its intensity. The latter is linked
to the turbulence energy dissipation rate ε. This in turn
determines the structure constants C_x^2 of irregularities in
humidity, temperature and electron density (in the mesosphere), which in total constitute the turbulence refractive
index structure constant C_n^2. This quantity can be deduced
under certain conditions from radar echo power measurements,
which then yield ε. These agree well with estimates from
statistical models (e.g. Warnock et al., 1985). The dissipation rate can also be determined from the Doppler spectrum
width, if instrumental effects are negligible or removed
(e.g. Hocking, 1986). All approaches need the knowledge of

background profiles of temperature etc. The eddy (turbulence) diffusion coefficient K can then directly be deduced
from ε. The average effective vertical eddy diffusion, however, can only be found by knowing ε within the turbulence
layers as well as their spatial distribution and temporal
variability in the atmosphere.

Atmospheric gravity waves transport energy and momentum
from the lower altitudes and can deposit it in the middle
atmosphere. This can be measured with MST radars by observing the altitude variation of vertical velocity fluctuations and the covariance of vertical and horizontal velocity
due to these waves as well as its connection to mean wind
velocity variations. Deposition of energy can be due to wave
saturation by breaking into turbulence which is dissipated
into heat. Deposition of momentum leads to an acceleration-
deceleration of the mean background wind. The latter phenomenon has been successfully studied by Vincent and Reid
(1983), who used the MF partial reflection radar at Adelaide
to implement the method to study momentum transport.

Molecular diffusion starts to dominate turbulent (eddy)
diffusion in the lower thermosphere. This region can efficiently be studied by incoherent scatter radars (ISR). These
radars are commonly applied to study the thermosphere and,
under certain conditions of optimised radar sensitivity and
high electron density, also the ionospheric D-region, i.e.
the mesosphere. The incoherent or Thomson scatter is due to
random thermal fluctuations of free electrons and ions.
These radars mostly operate in the upper VHF and UHF band
at frequencies between 200 MHz and 1200 MHz. Their transmitter power (>1 MW) and antenna gains are substantially
larger than those of the MST radars. Due to viscous subrange
limitations, the ISRs are supposed to be insensitive to turbulence scatter from the mesosphere, except of the Jicamarca
radar which operates at the long wavelength of 6 m. However,
they had efficiently been used to study also turbulence
scatter from the lower altitudes of the stratosphere and
troposphere and the eddy diffusion coefficient was deduced.
The parameters measured with the incoherent scatter radars
in the mesosphere and/or the lower thermosphere are: electron density, electron, ion and neutral temperatures, col-
lision frequency, neutral density, negative ion density as
well as the ion and neutral velocity. These are typically
measured at reasonable time resolutions of several minutes
and height resolutions of some kilometers. Further processes, such as the energy input by Joule and particle heating
and the momentum transfer and particle input from the magne-
tosphere to the lower thermosphere and hence the neutral
atmosphere can be determined. In the context of studying
transport processes they are regarded particularly useful
to investigate the height region of the turbopause in which
the mutual coupling between the lower thermosphere and the

upper mesosphere takes place and which is not accessible by
the common MST radars.

3. TROPOSPHERE-STRATOSPHERE EXCHANGE PROCESSES

The troposphere is the portion of the neutral atmosphere
which is more likely to be convectively unstable than other
altitude regions. Essentially, the vertical transport in the
troposphere is due to the convection processes, particularly
in thunderclouds. Up- and downdrafts transport very efficiently air masses between the bottom and top of the troposphere. Observations with the SOUSY-VHF-Radar during the
passage of a thundercloud showed the echo power, the mean
vertical velocity and the turbulent velocity. Mean upward
velocities were almost 10 m s^{-1} and fluctuating velocities
several m s^{-1}, the cloud top reached the tropopause and
partly regions of upward velocities penetrated to the lower
stratosphere. These first observations unveiled the cap-
abilities of VHF radars to investigate the dynamics of these
convective processes, but have to be done for deducing
qualitative results on entrainment and detrainment of air
masses in and around the thunderclouds as well as on vertical exchange of air masses. Penetrative convection reaching
the lower stratosphere needs to be studied in much more detail since it constitutes transport of tropospheric air masses into the stratosphere besides of being an efficient
source of gravity waves (Röttger, 1980). The vertical transport due to convection is very pronounced in the tropical
regions, which essentially drives the mean global circulation. Since VHF radars exhibit a unique capability to
measure vertical velocities, a continuous operation of a
chain of VHF radars around the globe would comprise a suitable contribution to monitor the mean vertical transport. A
few of such radars are already operated near the equator in
the Pacific (see papers in Bowhill and Edwards, 1986).
    Another process, gaining vertical mixing, is active turbu-
lence generated by shear instability. Pronounced regions of
this clear air turbulence are associated with velocity
shears in jet streams. The mechanism of Kelvin-Helmholtz
instability generating the clear-air turbulence and the corresponding velocity variations was intensively investigated
with VHF radars. It is envisaged that further efforts will
take place, such as the continuous monitoring of vertical
velocity fluctuations with VHF radars, to get an improved
statistical climatology of clear-air turbulence, its connection to synoptic-scale disturbances, and the associated
vertical transport of passive tracers.
    Convergences and divergences in synoptic-scale disturban-
ces result in changes of the flow pattern also in the vertical direction. This apparently occurs around the jet
stream and was measured with VHF radars, and also on larger

scales due to non-horizontal flow of warm and cold air in the frontal systems, occurring in connection with synoptic-scale disturbances. In further VHF radar observations, one evidently has to evaluate more distinctly this kind of synoptic-scale vertical velocity and its impact on large-scale vertical transport. Also potential vorticity is transferred between the troposphere and stratosphere in tropopause breaks, which can be observed with VHF radars (Larsen and Röttger, 1982).

VHF radar observations reveal in general that the troposphere is more turbulent (larger fluctuations of vertical velocity in the troposphere), and wave structures occur more frequently in the more stable stratosphere. These are obviously two regions of different stability, separated by the boundary of the tropopause. The exchange of air masses between the troposphere and the stratosphere is fairly important since it means transport though a region of strongly increasing stability, namely the tropopause. There are basically the following processes responsible for the mass transfer between the stratosphere and troposphere (Reiter, 1975): (1) the seasonal adjustment in the height of the mean tropopause level, (2) organized large-scale motions expressed by the mean meridional circulation, (3) large-scale eddy transport, mainly in jet stream regions, and (4) meso-scale and small-scale eddy transport across the tropopause. All of these processes can be studied with VHF radars. It was estimated that about 40% of the vertical transport is due to the Hadley cell circulation in the tropics, although vertical velocities in mid- and higher latitudes are also non-negligible. One has also to consider that overshooting cumulonimbus towers (penetrative convection) transport tropospheric air into the stratosphere. Approximately 20% of mass exchange is caused by large-scale eddies of synoptic-scale disturbances and associated tropopause breaks, which can representatively be detected with VHF radars. About 10% of mass flux is estimated to be due to the seasonal changes of the tropopause height, which can also be continuously monitored by VHF radars. Although it is assumed that small-scale and meso-scale turbulence diffusion contributes only very insignificantly to the vertical transport, radar investigations resulted in a different conclusion (e.g. Woodman and Rastogi, 1984). There is strong evidence now that gravity-wave motions and turbulence give rise to enhanced diffusion in the middle atmosphere (essentially in the mesosphere), and VHF radars are very suitable tools to study these phenomena. It is, thus, worthwhile to study further if and how much the incoherent wave and turbulence motions, observable with the VHF radars, contribute to vertical transport.

4. TRANSPORT WITHIN THE STRATOSPHERE

Synoptic-scale and meso-scale disturbances in the troposphere can propagate to the stratosphere. Besides of typical
variations of the horizontal velocity also small vertical
velocity variations are expected. These were measured in the
lower stratosphere with MST radar as shown in Fig. 2. The
synoptic-scale disturbance is readily recognized in the
meridional and vertical velocity variations. The latter is
only of a few cm s^{-1} amplitude, but kept consistently its
upward/downward direction during 1-2 days. The average vertical velocity during the 11 days observation period was
less than one cm s^{-1}, but was significantly upward to 18 km
altitude. Passive tracers, such as minor constituents,
should be transported over large horizontal distances due to
the mean zonal circulation and the superimposed synoptic-
scale disturbances. They are also to be transported vertically along the isentrops, as indicated by the vertical
velocity component measured by VHF radar (Fig. 2). Provided
that these disturbances propagate into the stratosphere and
are not exactly periodic, a net vertical transport will
result between the troposphere and stratosphere.

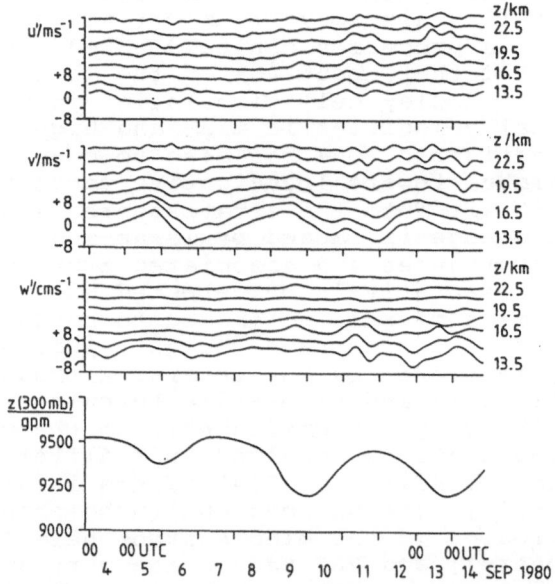

Fig.2 Zonal (u'), meridional (v'), and vertical (w') velocity deviations from the 11-days mean velocities showing the
passage of a synoptic scale wave in the stratosphere. The
lower diagram shows the corresponding altitude change of
the 300 mb height level in the troposphere.

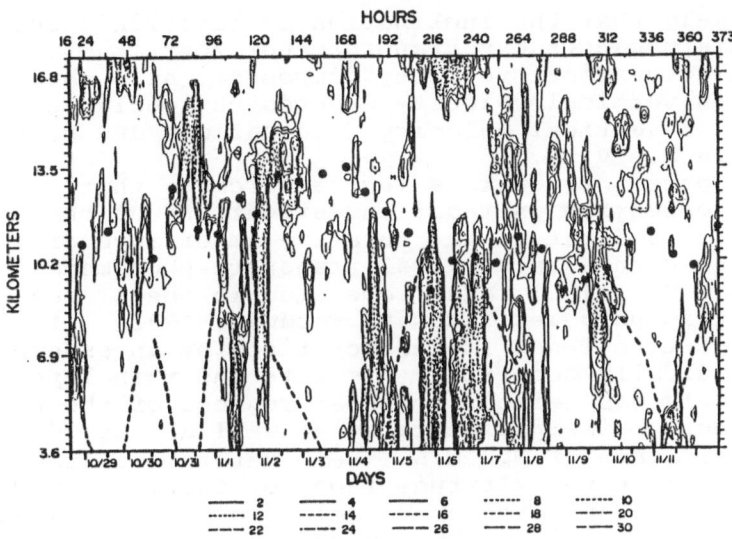

Fig.3 Upward velocity in cm s^{-1} with hourly profiles
and contours as indicated (from Dennis et al., 1986).

In Fig. 3 vertical velocity contours measured with VHF
radar are shown which clearly demonstrate penetrations from
the troposphere to the stratosphere. They penetrate the
tropopause, which is indicated by full circles. It appears
from these investigations that frontal passages (indicated
by the dashed lines in Fig. 3) may have caused these vertical velocity penetrations. The mean vertical velocity during
the 15 days of observations was positive at a few cm s^{-1}
between 9 and 12 km altitude, which is about the average
tropopause height. A net upward transport into the stratosphere thus can be concluded. The observations are confined
to a single location, however, which will not readily support a conclusion that the net transport over a wide area
was in the same direction. Since such synoptic-and meso-
scale disturbances are propagating horizontally and we took
the average over a time period (15 days), which was long as
compared to the coherence time of these disturbances, one
should assume that the spatial average is about similar to
the temporal average. A mean upward velocity through the
tropopause is conceivable, and it is thus expected that the
mean vertical velocity contributes to vertical transport
between the troposphere and stratosphere and also within
the lower stratosphere. Above the tropopause, however, the
mean vertical velocity was negative at a few cm s^{-1}, which
means a net downward transport in the lower stratosphere
during this time period. A caveat has to be raised here

again, namely that the inclination of baroclinic surfaces
of such disturbances may transform horizontal velocity into
the measured vertical velocity component. Also the accuracy
of the mean vertical velocity depends obviously on the
intensity of vertical velocity fluctuations due to short-
period wave activity.

It is apparent to quite a few researchers that the most
important objective in radar investigations of the stratosphere is to understand the relative importance of stratospheric turbulence in vertical transport phenomena. The
Arecibo 430 MHz (ISR) radar, operated in the ST mode, was
used for this purpose (Sato and Woodman, 1982). It was confirmed that turbulence is characterized by an assembly of
many sporadically occurring, discrete and thin layers, and
it apparently was related to shear regions of the wind. Turbulence energy dissipation rates as well as the eddy diffusion coefficients of 0.2 m s^{-1} were determined from the echo
power spectra in the altitude range of the lower stratosphere. Another approach by Woodman and Rastogi (1984),
using a probabilistic method to determine the turbulent diffusion from the occurrence of turbulence layers detected by
the same radar, yielded eddy diffusion coefficients between
0.2 and $0.3 \text{ m}^2 \text{ s}^{-1}$. These values are more than on order of
magnitude larger than those estimates determined from high-
altitude instrumented aircraft flights. It is therefore
likely that such large diffusion coefficients can account
for the stratospheric residence times of passive tracers.

It is reasonable that the turbulence is generated by
breaking gravity waves of a broad range of frequencies.
These waves are also observed with VHF radars. Particularly
observations of vertical velocity frequently show very distinct oscillations. The rms amplitudes of these velocity
oscillations do often not increase exponentially with
height, which points to a wave saturation or dissipation
mechanism. The frequency spectra of such velocity fluctuations, however, do not always fit to spectra expected for
gravity waves, rather than turbulence. It is, thus, claimed
that quasi two-dimensional turbulence related to quasi-geostrophic flow can coexist with a nearly universal spectrum
of gravity waves (see papers in Liu and Kato, 1985). This
idea is consistent with a red cascade of energy from small
to large scales in contrary to the commonly accepted cascade
of energy from large to small scales in a three-dimensional
inertial subrange of turbulence. It is presently quite
unclear if such a coexistence is real and how two-dimensional turbulence would impact on the vertical and horizontal
diffusion.

When gravity waves are dissipated, energy and momentum is
deposited to the atmosphere. Vincent and Reid (1983) used a
radar technique to measure gravity momentum fluxes. Its
divergence causes an acceleration of the mean wind. Such

accelerations are needed to damp the horizontal wind, obtained in numerical models, in order to satisfy the heat and momentum budget in the middle atmosphere. The results of Vincent and Reid (1983) suggest that the velocity fluctuations measured by radar are most likely due to gravity waves rather than due to meso-scale turbulence. However, these observations also show that there must be a substantial loss mechanism, either by breaking waves into turbulence and/or by the acceleration of the mean flow by momentum transfer. It is quite obvious, thus, that atmospheric waves and turbulence have not to be treated separately but as a coupled entity.

5. ATMOSPHERIC GRAVITY WAVES AND TURBULENT TRANSPORT WITHIN THE MESOSPHERE

It is accepted that there is an important influence on the dynamics as well as on the chemical composition of the middle atmosphere by the breaking of gravity waves into turbulence (e.g. Garcia and Solomon, 1985). Essential quantities to parametrize turbulent transport are the turbulent energy dissipation rate   and the eddy diffusion coefficient K. These can be determined from radar observations either through the turbulence refractive index structure constant, deduced from calibrated power measurements, or through the turbulent velocity fluctuations, deduced from the Doppler spectrum width. Besides of the radar-deduced parameters, power and spectrum width, the first approach needs knowledge the of profiles of temperature (and electron density in the mesosphere) and the fraction of the radar volume filled with turbulence, and the latter approach needs knowledge of the temperature profile, namely the Brunt-Väisälä frequency. According to Weinstock (1981) the energy dissipation rate in a turbulence layer embedded in a stable background atmosphere is

$$\varepsilon = 0.8 \cdot \langle w^2 \rangle \cdot N \qquad (1)$$

The mean squared velocity $\langle w^2 \rangle$ can directly be reduced from the width of the radar power spectrum, provided that the effects of wind shear and beam width broadening are negligible. It will be shown here that also the Brunt-Väisälä frequency N can be estimated from radar observations.

Recent work on the spectra of vertical velocity oscillations due to gravity waves in the troposphere, stratosphere and the mesosphere has revealed a typical feature which we call the "Brunt-Väisälä cut-off". Several observers noticed a spectral peak near the Brunt-Väisälä frequency. This peak is often characterized by a very steep slope at the high frequency section, but a fairly shallow slope towards lower

frequencies. This distinct spectral shape can be explained
by the fact that the vertical velocity amplitudes of atmospheric gravity waves increase with frequency up to their
natural cut-off at the Brunt-Väisälä frequency. It was shown
by VHF radar interferometer measurements that this spectral
peak is definitely due to gravity waves. It was suggested
(VanZandt, 1982) that the total spectra of vertical velocity
variations is a manifestation of a universal spectrum of
gravity waves, and detailed model computations tend to this
assumption.

Brunt-Väisälä cut-off compares reasonably well with the
profiles of the Brunt-Väisälä frequency deduced from radiosonde temperature profiles of the troposphere and stratosphere. The observed spectral shapes almost exactly resemble
the gravity wave (f < inertial frequency) model spectra (see
papers in Liu and Kato, 1985). Doppler shifts can substantially distort the spectra, and model computations showed
that the spectral energy is redistributed through the spectrum due to the Doppler shift (Liu, personal communication,
1986). This effect is most pronounced for just those waves
with frequencies very close to the Brunt-Väisälä frequency.

We assume that the distribution of gravity wave phase velocities is isotropic in azimuth with respect to the wind
velocity (within a suitable observation period). Then about
one quarter of the waves are shifted to higher frequencies,
one quarter to lower frequencies, and two quarters are not
or very little shifted because their phase velocities are
(exact or almost) perpendicular to the wind velocity. The
effect is that the spectrum is well smeared out, but the
peak at the Brunt-Väisälä frequency still remains unshifted
(due to the perpendicular waves) although it becomes less
distinguishable from the spread-out background spectrum.
Another effect, wave steepening due to amplitude growth of
gravity waves, can also have an influence on the spectral
shape (Weinstock, personal communication, 1985). A cut-off
above the Brunt-Väisälä frequency, however, is still apparent.

This effect can influence the velocity spectra during disturbed conditions (high wind), but also a spill-over from
the horizontal velocities through a wide antenna beam width
or sidelobes could have caused this effect. However, a peak
is still observable at the Brunt-Väisälä frequency, which
is particularly evident when one traces the spectra from
height to height.

We, thus, regard the measurement of the frequency of the
peak in a vertical velocity spectrum to yield most directly
the Brunt-Väisälä frequency from MST radar measurements.
Knowing the Brunt-Väisälä frequency profile, one can deduce
the potential temperature profile if one has a calibration
temperature at one height. It has to be noted, however, that
this method fails for superadiabatic lapse rates when the

Brunt-Väisälä frequency is imaginary. The application of
this method will also be difficult when the wind velocity
is too high, causing the Doppler effect to smear out the
total spectrum and blur the Brunt-Väisälä cut-off. A similar
deficiency will also appear if the gravity wave distribution
has a maximum in wind direction.

When the Brunt-Väisälä frequency is determined by the described method, the energy dissipation rate ε can only be
deduced from the radar data $\langle w^2 \rangle$ and N without need of any
supplementary observations or assumptions. The eddy diffusion coefficient K is similarly given by radar observations:

$$K = 0.8 \cdot \langle w^2 \rangle \cdot N^{-2} \qquad (2)$$

This formula is frequently used as a tool to obtain a most
appropriate estimate of K. The values of the constant 0.8,
however, vary quite a bit in the literature, and more work
is needed to reduce its uncertainties. In case of wave-
breaking events and the resulting strong turbulence, the

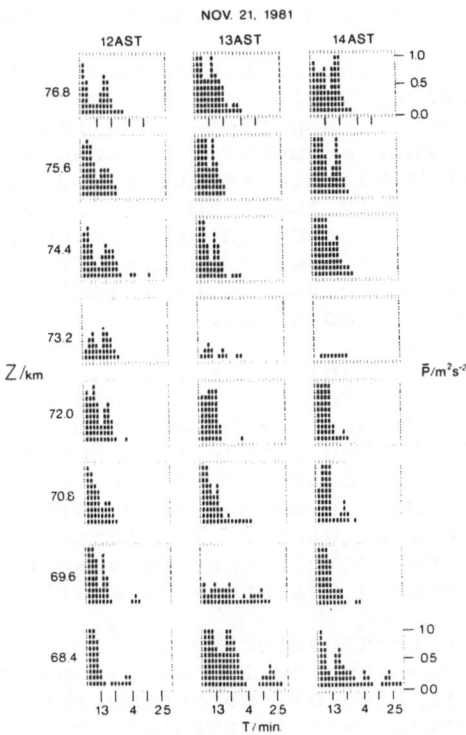

Fig.4 Power spectra of vertical velocity in the mesosphere, measured with the Arecibo VHF/MST radar.

Brunt-Väisälä frequency N is undefined. This leads to a
total break-down of the basic criterion for the referenced
formula (2). However, quite a few turbulence events seen by
VHF/MST radar may be regarded as some kind of remnant turbulence in stable stratifications or just resulting from dynamic instability of long-period waves. The criterion using
N^2 should then be valid. However, in addition to the uncertainty of the constant, we also have to know the thickness
of the turbulence layers, which frequently may not be
resolvable with the VHF radars. We, thus, can regard any
value of K only as the best estimate which presently is
achievable.

Mesospheric data were taken in November 1981 with the
SOUSY-VHF-Radar at the Arecibo Observatory using an average
power of 6 kW on 46.8 MHz and a height resolution of 1.2 km,
applying an 8-bit complementary code. The main dish of the
Arecibo Observatory was used as an antenna yielding a half-
power beam width of 1.7 degrees. The beam was pointed 2.3
degrees off the zenith such that a quasi-vertical velocity
was measured, allowing to investigate short-period gravity
waves. The beam was kept fixed at the E- or N- direction
for about one hour, such that also the mean horizontal wind
could be measured.

Fig. 4 shows the spectra of the quasi-vertical velocity
variations deduced from velocity time series. We notice
a clear cut-off of the spectra at periods of a few minutes.
Following the earlier outlined arguments, we assume that
this cut-off is at the Brunt-Väisälä frequency. Since the
mean wind velocities were fairly low (20 m s^{-1}), only a
small Doppler shift did result, and the spectra mostly indicate a clear cut-off. The cut-off deduced from Fig. 4, which
is assumed to be consistent with the Brunt-Väisälä frequency
N, is shown in Fig. 5. This is an average over 8 hours of
mesospheric observations from 08 to 16 local time.

We clearly see in Fig. 5 a significant increase of the
average N with height above 77 km. This increase is regarded
as an indication of the mesopause which we estimate from
these data to be around 80 km. We furthermore can deduce
the profiles of mean square velocity $\langle \hat{w}^2 \rangle$ due to waves from
the velocity time series, and the mean square turbulent
velocity $\langle w^2 \rangle$ from the spectral widths. The latter deduction
can be done without applying corrections, because the beam
width and wind shear broadening effects were small in this
experiment.

Since the mean square vertical velocity $\langle \hat{w}^2 \rangle$ is about
constant with height and does not follow an exponential increase (continuous curve in Fig. 5), we assume that the
waves causing these velocities were saturated (e.g. Fritts,
1984), which is consistent with other mesospheric observations. If the gravity waves are saturated due to dissipation

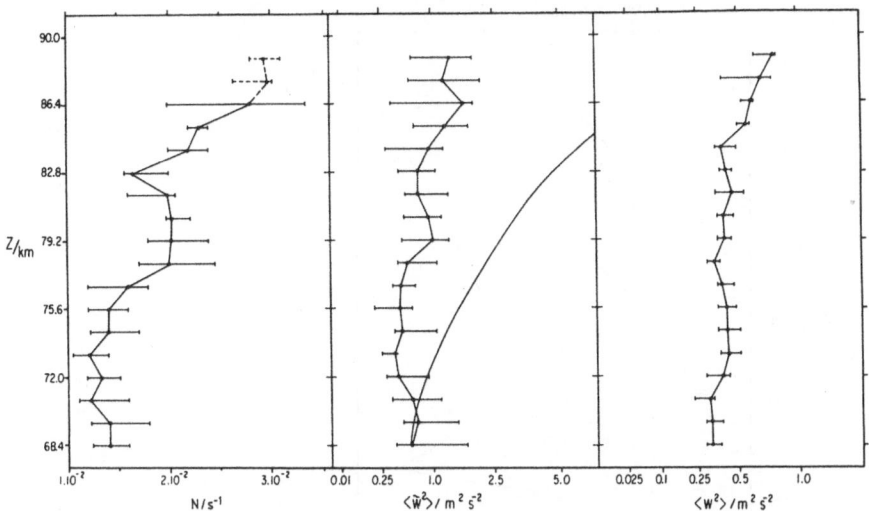

Fig.5   Profiles of estimate of average Brunt-Väisälä-
frequency N, mean square vertical velocity $\langle \hat{w}^2 \rangle$ due to
waves and mean square fluctuating velocity $\langle w^2 \rangle$ due to
turbulence, measured with the Arecibo VHF/MST radar.

into turbulence, we would assume an increase in the energy
dissipation rate. This quantity can be deduced from the mean
square turbulent velocity and the Brunt-Väisälä frequency
profiles (see eq. (1)). The energy dissipation rate is proportional to the eddy diffusion coefficient, which is shown
in Fig. 6. It apparently has a maximum between 70 and 78 km
where the wave velocity stays almost constant. Above 78 km,
the eddy diffusion coefficient is small and constant. This
is quite consistent with the observation that the mean vertical velocity due to waves again starts to increase with
height above 78 km.
     In Fig. 6 also the median eddy diffusion coefficient from
middle atmosphere models is included for heights above 80
km (Hocking, 1987). It is about a factor of 2 larger than
our values and it appears to be in the minimum of the mean
values. The radar-deduced K values of 20-50 $m^2 s^{-1}$ and the
decrease with altitude, however, are in accordance with in-
situ observations (Thrane et al., 1985). On the other hand,
we have also to regard that the K values in Fig. 6 are only
representative for the turbulence within a layer. Since the
turbulence is intermittent in time and altitude, the effective eddy diffusion coefficient (averaged over all time and
altitudes) has to be assumed to be even smaller. On the
average, these MST radar observations indicate that
turbulence layers occurred only 10-20% of the time and

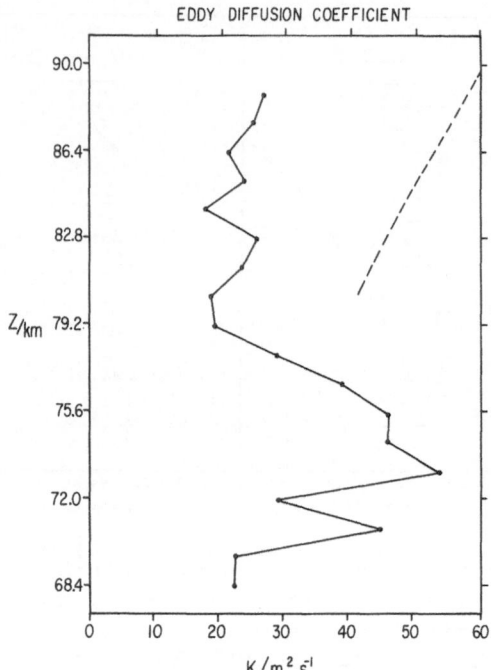

Fig.6   Profile of mean eddy diffusion coefficient K
within turbulence layers determined by VHF/MST radar
(solid line). The dashed line indicates median values
from model published by Hocking (1987).

altitude ranges (see for example Fig.1), which yields an
5-10 times smaller effective eddy diffusion coefficient than
that one displayed in Fig. 6. Since our data are averages
over one day only, they are, however, not representative for
average eddy diffusion coefficients and energy dissipation
rates. More data series have to be analyzed in this way to
yield climatological information.
    Since the effective eddy diffusion coefficient found in
the described experiments is remarkably small, one may ask
how the eddy transport in the mesosphere competes with the
vertical transport due to a mean vertical velocity. Mean
vertical velocities in the mesosphere of some 10 cm s^{-1} are
reported (e.g. Gage and Balsley, 1984). These values appear
reasonably large to yield an effective vertical transport
which is competitive to the turbulent transport. However,
since the reliability of mean vertical velocity measurements
with MST radars, particularly because of the intermittency
of echoes in the mesosphere, may not be sufficient, a conclusive comparison is still expected.

6. COUPLING PROCESSES BETWEEN THE MESOSPHERE AND THE LOWER THERMOSPHERE

Important influences are considered to take place in the mesosphere and lower thermosphere due to gravity wave and tidal variations of velocity and temperature in cases where chemical reaction rates are smaller than typical variations of vertical velocity and also due to the fact that reaction rates are temperature dependent. An appropriate complement to MST radars are incoherent scatter radars (ISRs) to observe such parameters in the upper mesosphere and lower thermosphere. The ISRs can thus observe the altitudes above the mesopause, which is generally less accessible by MST radars, but is still below the turbopause. The ISRs do not allow to measure turbulence but allow to measure horizontal as well as vertical velocities, electron density, temperature and neutral density, which are important parameters to study transport in the surrounding of the turbopause. Additionally, they can be used to measure composition, namely the number density ratio of negative cluster ions to electrons around the mesopause region. Further up in the thermosphere the ISRs are applied to measure energy and momentum input through magnetospheric electric fields and particles. These considerably influence the lower thermospheric composition. It is suggested that nitric oxide produced in polar regions is transported over large distances in the thermosphere and can diffuse or be transported downward by a mean motion into the mesosphere and stratosphere where it has a definite impact on the ozone concentration. Incoherent scatter radars at high latitudes, such as the EISCAT or the Sandestromfjord radar, can effectively contribute to study these processes, particularly the coupling between the lower thermosphere and the mesosphere.

7. CONCLUSION

Quite a few MST radar experiments have been carried out to measure quantities relevant to transport of energy, momentum and passive tracers, namely the vertical profiles of mean vertical velocity, power spectra of velocity fluctuations and their covariance as well as the distribution and intensity of turbulence layers and the resulting eddy diffusion coefficients. Although it appears that these results have led to important discoveries, the potential of the MST radars as well as the incoherent scatter radars to study transport processes in the atmosphere is not fully recognized and exploited yet. These potentials are expected to be considered efficiently in present and future investigations, campaigns and global projects.

REFERENCES

Balsley, B.B. and K.S. Gage (1980) 'The MST radar
technique: potential for middle atmospheric studies'.
Pure Appl. Geophys., 118, 452-493.

Bowhill, S.A. and B. Edwards (eds.) (1983,1984,1986)
Handbook for MAP, 9,14,23 (publ. by SCOSTEP Secretariat,
Univ. of Illinois, Urbana, IL).

Dennis, T.S., M.F. Larsen and J. Röttger (1986)
'Observations of mesoscale vertical velocities around
frontal zones'. Manuscript, Clemson University, SC.

Fritts, D.C. (1984) 'Gravity wave saturation in the
middle atmosphere: a review of theory and observations'.
Rev. Geophys. Space Phys., 22, 275-308.

Fritts, D.C., M.A. Geller, B.B. Balsley, M.L. Chanin, I.
Hirota, J.R. Holton, S. Kato, R.S. Lindzen, M.R. Schoeberl,
R.A. Vincent and R.F. Woodman (1984) 'Research status and
recommendations from the Alaska workshop on gravity waves
and turbulence in the middle atmosphere, Fairbanks, Alaska,
18-22 July 1983'. Bull. Amer. Meteor. Soc., 65, 149-159.

Gage, K.S. and B.B. Balsley (1984) 'MST radar studies of
wind and turbulence in the middle atmosphere'. J. Atmos.
Terr. Phys., 46, 739-753.

Garcia, R.R. and S. Solomon (1985) 'The effect of breaking
gravity waves on the dynamics and chemical composition of
the mesosphere and lower thermosphere'. J. Geophys. Res.,
90, 3850-3868.

Gossard, E.E. and K.C. Yeh (eds.) (1980) Radio Science
(Radar Investigations of the Clear Air), 15, No. 2.

Hocking, W.K. (1983) 'On the extraction of atmospheric turbulence parameters from radar backscatter Doppler spectra -
I. theory'. J. Atmos. Terr. Phys., 45, 89-102.

Hocking, W.K. (1986) 'Observation and measurement of turbulence in the middle atmosphere with a VHF radar'.
J. Atmos. Terr. Phys., 48, 655-670.

Hocking, W.K. (1987) 'Turbulence in the region 80-120 km'.
Adv. Space Res. (to appear).

Larsen, M.F. and J. Röttger (1982) 'VHF and UHF Doppler
radars as tools for synoptic research'. Bull. Amer. Meteor.
Soc., 63, 996-1008.

Liu, C.H. and S. Kato (eds.) (1985) Radio Science (Special Section: Technical and Scientific Aspects of MST Radars), 20, No. 6.

Reiter, E.R. (1975) 'Stratospheric-tropospheric exchange processes'. Rev. Geophys. Space Phys., 13, 459-474.

Röttger, J. (1980) 'Structure and dynamics of the stratosphere and mesosphere revealed by VHF radar investigations' Pure Appl. Geophys., 118, 494-527.

Sato, T. and R.F. Woodman (1982) 'Fine altitude resolution observations of stratospheric turbulent layers by the Arecibo 430 MHz radar'. J. Atmos. Sci., 39, 2546-2552.

Thrane, E.V., O. Andreasen, T. Blix, B. Grandal, A. Brekke, C.R. Philbrick, F.J. Schmidlin, H.U. Widdel, U. von Zahn and F.J. Luebken (1985) 'Neutral air turbulence in the upper atmosphere'. J. Atmos. Terr. Phys., 47, 243-265.

VanZandt, T.E. (1982) 'A universal spectrum of buoyancy waves in the atmosphere'. Geophys. Res. Lett., 9, 575-578.

Vincent, R.A. (ed.) (1984) Handbook for MAP (Ground-Based Techniques), 13 (publ. by SCOSTEP Secretariat, Univ. of Illinois, Urbana, IL).

Vincent, R.A. and I.M. Reid (1983) 'HF Doppler measurements of mesospheric gravity wave momentum fluxes'. J. Atmos. Sci., 40, 1321-1333.

Warnock, J.M., T.E. VanZandt and J.L. Green (1985) 'A statistical model to estimate mean values of parameters of turbulence in the free atmosphere'. Preprint Volume Seventh Symposium on Turbulence and Diffusion, (publ. by Amer. Met. Soc., Boston, MA).

Weinstock, J. (1981) 'Energy dissipation rates of turbulence in the stable free atmosphere'. J. Atmos. Sci., 38, 880-883.

Woodman, R.F. and A. Guillen (1974) 'Radar observations of winds and turbulence in the stratosphere and mesosphere'. J. Atmos. Sci., 31, 493-505.

Woodman, R.F. and P.K. Rastogi (1984) 'Evaluation of effective eddy diffusive coefficients using radar observations of turbulence in the stratosphere'. Geophys. Res. Lett., 11, 243-246.

LIDAR SOUNDING OF THE STRUCTURE AND DYNAMICS OF THE MIDDLE ATMOSPHERE.
A REVIEW OF RECENT RESULTS RELEVANT TO TRANSPORT PROCESSES˙

Marie-Lise CHANIN and Alain HAUCHECORNE
Service d'Aéronomie du C.N.R.S.
BP 3, 91371 Verrières le Buisson   CEDEX
France

ABSTRACT. Elastic backscattering of a laser beam by the atmosphere
has been used in the last few years to study with high temporal and
spatial resolutions some of the basic parameters of the middle atmosphere, density and temperature. The variability of their height
distribution has been interpreted in terms of gravity and planetary
waves. Some of the results relevant to dynamics are presented here.
They emphasize the major role played by dynamics in the temporal
variability of the middle atmosphere.

1.   INTRODUCTION

Using Rayleigh scattering of a laser beam, the lidar station set up at
the Observatory of Haute Provence (O.H.P., 44°N, 6°E) has carried out
since 1979 a long term survey of the middle atmosphere structure and
of its variability. The density and temperature profiles are measured
from that site with an average time-coverage of 3 hours per night, 100
nights per year, mostly since 1981 when the instrument started to be
run on a routine basis. The data bank which has been processed
consists on more than 600 nights of data representing several
thousands of individual profiles. Fluctuations of density and temperature have been observed from these data on time scales of the order of
hours as well as days and even on a long term basis. They have been
the object of three different types of studies related respectively to
gravity waves, planetary waves and long term trend, depending upon the
hourly, daily or yearly character of these fluctuations. This
classification implies a causal relationship between the period and
the origin of these changes which may not be as straight forward ; in
fact it does seem that variations extending to more than a day could
also be induced by the breaking of gravity waves ; furthermore our
results indicate that the long term variation can be related to
planetary waves ; this without taking into account the probable
coupling between planetary waves and gravity waves. Anyway for
459

G. Visconti and R. Garcia (eds.), Transport Processes in the Middle Atmosphere, 459–477.
© 1987 by D. Reidel Publishing Company.

convenience in this paper, the   results will be presented under   these
three topical titles, gravity waves, planetary waves and long term
trend, according to the time-scale of the observed variations.

2. THE METHOD : ITS PERFORMANCE AND ITS LIMITATIONS

The Rayleigh lidar technique has been   the subject of several
publications in the early phase of its development (Hauchecorne and
Chanin, 1980 ; Chanin and Hauchecorne, 1981). The more complete and
up-dated description of the method can be founded in Chanin and
Hauchecorne (1984). Let us just remind the reader than the parameter
directly derived from the Rayleigh backscattering of a laser beam is
the atmospheric molecular density. This assumes that the contribution
from Mie scattering by aerosols particles can be neglected ; it is
usually the case above 30 km except after a major volcanic eruption.
In the post El-Chichon period (1982-1983) the dust layer went up as
far as 35 km which then the lowest limit for the density
measurements. The upper limit is given by the signal to background
ratio and after successive improvements of the spectral and spatial
filtering of the sky contribution, this limit is now close to 100 km,
whereas it was limited to 70 km in the early phase of the
observations.
        The temperature is deduced from the density using the state
equation and assuming hydrostatic equilibrium. The initialisation of
the profile is done at the upper limit of the profile by using an
atmospheric model. Due to the difficulty in knowing the atmospheric
transmission, the density is only given in relative value. On the
other hand the temperature is obtained in absolute value, which
explains why we prefer to use it for studying the variability of the
atmosphere.
        The temporal analysis of the backscattered echo provides the
density and temperature profiles with a height resolution of 150
meters and the acquisition is done with a time resolution of 3
minutes. As the uncertainty is essentially due to statistical noise
(i.e. to the number of integrated photons) the final accuracy is the
result of a compromise between the time and space resolutions which
are choosen differently depending on the spatial and temporal scales
of the fluctuations to be studied. The upper range of the measurements
depends also upon that choice. Table I gives an idea of the accuracy
of the measurements for resolutions used when studying planetary waves
or long term fluctuations.

Altitude (km)	Density precision	Temperature accuracy
30 - 50	≤ 0.1 %	≤ 0.3 K
70	0.3 %	1 K
80	1 %	3 K
90	3 %	10 K

Table I. Precision and accuracy as a function of height for
          Δz = 3 km   Δt = 5 hours

As all measurements, this technique has its experimental limitations.
As already mentionned, the height range extends from 30 to 90 km at
present ; its extension up to 110 km is however possible with
to-day technology and depends only upon financial constraints. Even
with the current limits, the method covers a large part of the middle
atmosphere and mainly the altitude range where radars are blind. As
for the temporal coverage, we have been limited up-to-now to nighttime measurements. Recently, day-time data have been acquired up to 70
km but their accuracy is obviously degraded compared to night-time due
to the contribution of the sky background ; therefore they have not
been used in the studies reported here.

The main limit of the instrument is due to the fact that it is
groundbased, providing then only a survey at the vertical of the site.
We are trying to overcome that limit by setting up a network of such
lidars. A second one has been set up in France in February 1986 at
Biscarosse (44°N, 1°W) and has been operated on a routine basis since
mid-1986. The data are of a comparable quality, but it is premature to
include their analysis in this paper.

The set of data which was used for the study described hereafter
was obtained at OHP between January 1979 and September 1986 ; to
benefit from the successive improvements of the data quality, the
emphasis is put on the recent data for the short-term studies.

3. GRAVITY WAVES

Waves of short period have been detected in the data, even in the
early stage of the observations (Chanin and Hauchecorne, 1981), even
though the data accuracy left a lot to be desired compared to what have
been reached since. Recently better time and height resolutions can be
achieved : most of the data are now studied with Δz — 300 m and
Δt — 15 minutes ; it implies that the limits towards the high
frequency part of the spectra are 600 m in vertical wavelength and
half an hour in time. Though we know from radar and balloon studies
that the spectra (both in space and time) extend further, it appears
that most of the energy contribution comes from the lower frequency
part to which the lidar gives access. At the other end of the spectra,
the height range covered and the maximum duration of the sequences
place the limits around 20 km (limit due to the presence of the
stratopause) and 14 hours.

The wave characteristics which are directly obtained from the
profiles of either density or temperature are the vertical wavelength
λ_z the vertical phase speed C_z and the temperature gradient $\partial T/\partial z$,
which provides an indication on the stability. The major missing
quantity to describe fully the wave is the mean wind $\bar{u}$ which we can
only guess from statistics.

After a detailed study of the characteristics of the waves on
more than 60 cases, a statistical approach is now being performed ; a
summary of the main results is given here.

3.1. Occurrence and variability

Gravity waves seem to be always present between 30 and 80 km even though they are often at the limit of detection at the lowest altitude level ; Their amplitude can vary by one order of magnitude from day to day (as can be seen from the scattering of average daily power spectral density plotted in Fig. 3). Large variations are also observed with a time constant of one to a few hours and these sudden changes of behaviour seem to be related to meteorological disturbances. To understand this variability, a study of correlation between the wave activity and the presence of different meteorological situations (front, wind shear, convection...) is now being performed, but it is premature to report about its conclusions.

3.2. Change of amplitude with height

On a statistical basis the growth of amplitude with height is seen to be quite smaller than expected if energy would be conserved. When looking at each individual case, different behaviours have been observed. In about half of the cases the wave amplitude for long wavelengths ($\lambda_z \geq 5$ km) grows with height, following the $\rho^{-1/2}$ law as expected from theory for most of the altitude range from 30 to 70 km. However if the wave amplitude is already large at 30 km, the wave reaches saturation at lower height, often around 50-60 km as soon as the temperature gradient reaches the adiabatic value ($\partial T/\partial z \sim \Gamma$) ; then the wave amplitude drops and, a few kilometers above the first breaking level, the wave resumes growth until it breaks again at higher level (usually 70 km). This behaviour was not observed for small wavelengths ($\lambda_z < 5$ km) which are observed to keep a more constant amplitude indicating that they are closer to saturation.

3.3. Vertical wavelength spectrum

Fourier spectrum of individual profiles indicates usually the presence of one or more quasi-monochromatic waves. This statement stands for profiles obtained after an integration of one hour or more. Even though a slight change of wavelength may occur during consecutive hours of observation, the average of individual spectra for one night usually exhibits well defined wavelengths. This is no longer true on monthly average spectra which present a regular slope (between - 5/2 and 3), close to the saturated spectrum given by $N^2/2 \, m^3$. As expected from the theory of saturation spectrum (Van Zandt, 1982 ; Smith et al., 1987), the spectrum deviates from saturation and flattens for wavelengths larger than a characteristic wavelength λ^* which is shown to be around 5 km for the height range 45-75 km (Fig. 1). On the other hand the spectrum obtained for the upper stratosphere (30-45 km) appears to be below the saturation level even for small wavelengths.

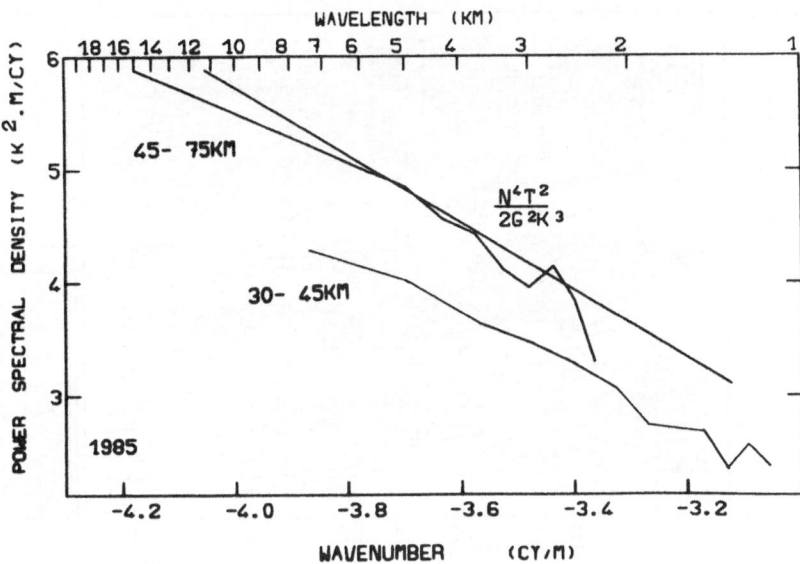

Figure 1.   Observed vertical wavelength spectra for the height   ranges
30-45 km (a)   and 45-75 km   (b) compared to   the theoretical saturated
spectrum for the mesosphere.

3.4. Vertical phase speed C_z

The height-time contour of the fluctuations provides the vertical
phase speed. When a wave with a well defined wavelength carries a
large part of the energy fluctuations, this contour map can be made
without filtering the data. This is the case in the map presented in
Fig. 2 which indicates a downward phase speed of 0.25 ms^{-1} between 30
and 55 km for the wavelength of 10 km. Otherwise one can use a spatial
filtering centered around one of the identified wavelength.   From the
large number of case studies it has been seen that the long wavelength
waves have phase speeds directed downwards around 0.2 $\pm$ 0.1 ms^{-1}. This
indicates that their energy is propagating upwards and may have their
origin in the troposphere. However waves of smaller wavelength (λ_z ~
5 km) are often seen to be stationary (C_z = 0) which indicates that
they may be of topographic origin. A statistical study is being
performed to estimate the relative contribution of stationary and
propagating waves to the energy spectrum. This will be done for our
two lidars sites which have quite different characteristics as far as
both orography and meteorology are concerned.

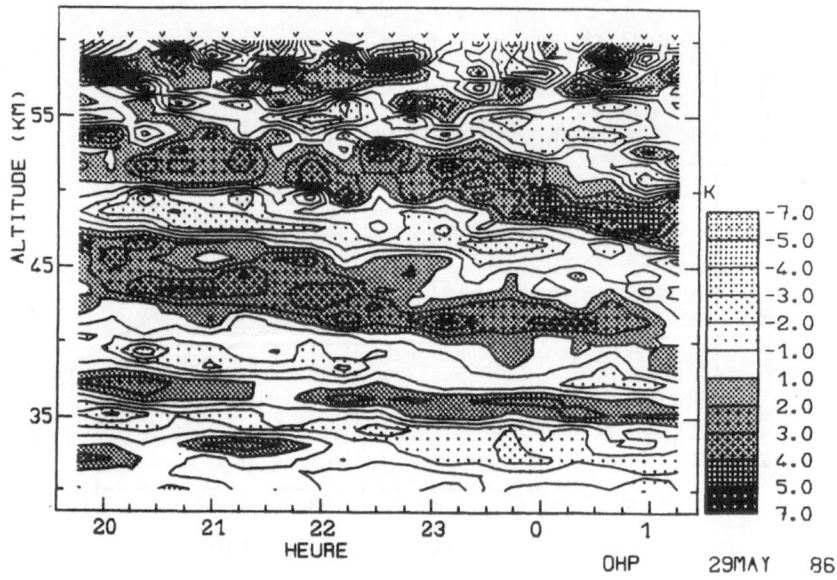

Figure 2. Contour map of the temperature deviation from the mean
night-time temperature from 30 to 60 km for May 29, 1985 at O.H.P.
(Δt - 15 minutes, Δz - 300 m).

3.5. Temporal spectrum

The length of the sequences of observation is often too short to
provide a significant Fourier spectrum in time, and the statistics
made from a few very long sequences is therefore not to be compared
with the spatial spectrum analysis. However the slope of the spectrum
is shown to be - 5/3 as expected. The periods appearing in the spectra
range from 1.5 to 10 hours (compatible to the values obtained from the
knowledge of λ_z and C_z) and it is found that the long wavelength (5 <
λ_z < 15 km) are usually associated with long periods (4 to 10 hours).

3.6. Seasonal variation of the wave activity

From the daily averaged vertical wavelength spectra, a statistical
analysis of the gravity wave activity has been performed for the year
1985. The extension of this work to other years is in progress. Fig. 3
presents the results of such an analysis for the wavelength range 6 to
10 km, respectively for the height range 30-60 km (Fig. 3a) and
45-75 km (Fig. 3b). Whereas a annual variation with a maximum in
winter is observed in the stratosphere and low mesosphere, the
mesosphere seems to present a semi-annual variation with two minima of
wave activity occurring at the equinoxes. This result for the

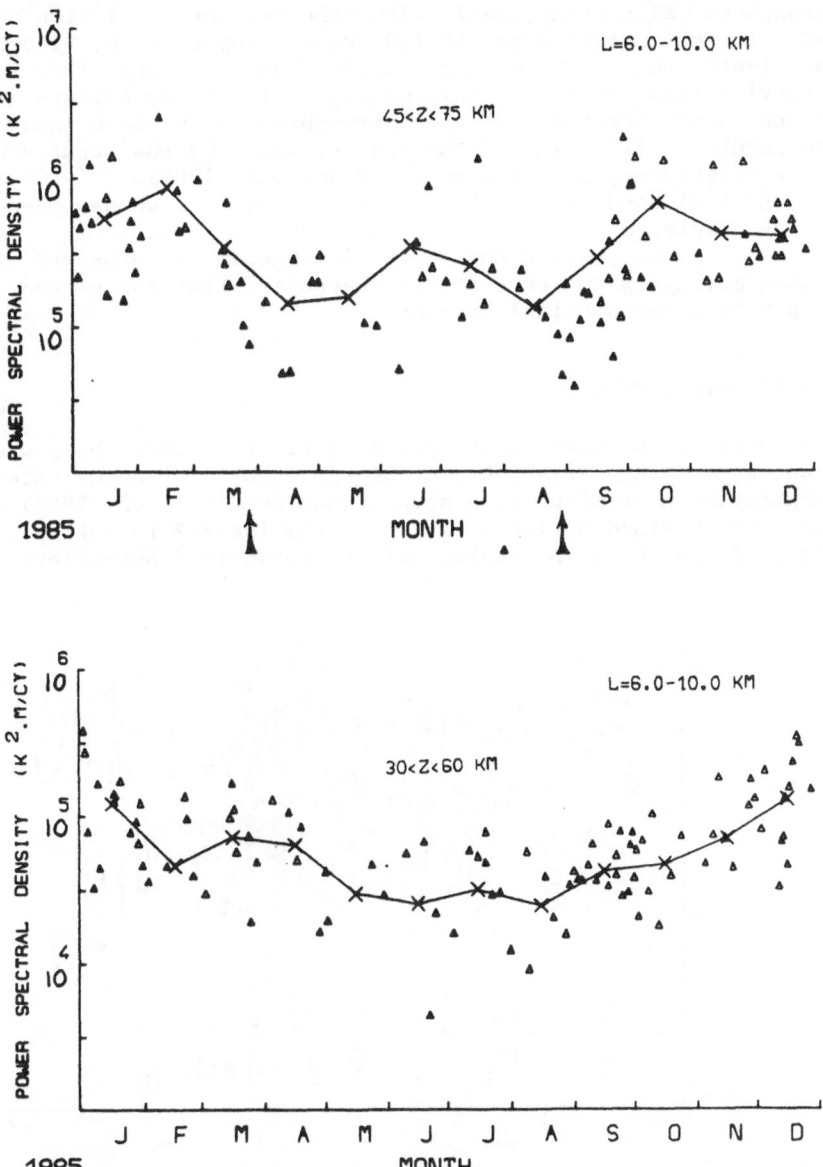

Figure 3. Seasonal variation of the gravity wave activity during the year 1985 at O.H.P., in the wavelength range 6-10 km and in the height ranges 30-60 km (a) and 45-75 km (b).
The crosses (×) correspond to monthly mean of the individual night power spectral density represented by triangles (Δ). The 2 arrows correspond to the time of turn over of the stratospheric wind.

mesosphere fits very well with the seasonal variation obtained by
radar in the height range 68-100 km and reported by Vincent (1987).
The facts that the minima occur when the zonal flow is changing
direction (indicated by arrows on Fig. 3b) and the difference observed
in the summer stratosphere and mesosphere can be both interpreted as
the result of filtering of the gravity waves by the stratospheric wind
systems (Lindzen, 1981 ; Garcia and Solomon, 1985).

As indicated earlier the wealth of data which has been acquired
by the Rayleigh lidar is far from being exhaustively studied in
terms of gravity-waves and more information is expected in a near
future mainly on the statistical characteristics and on the relationship with meteorological sources.

4. PLANETARY WAVES

The data have been used since 1981 to study the propagation of
planetary waves from 30 to 80 km and the occurrence of upper
stratospheric warmings (Hauchecorne and Chanin, 1982, 1983). Up to now
the data obtained during 5 winters, from 1981/82 to 1985/86, have been
analysed and the main results will be summarized hereafter.

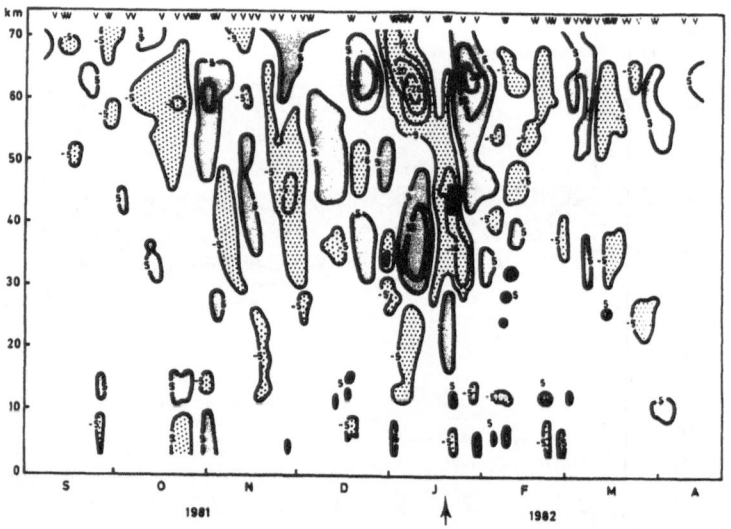

Figure 4. Time-height contour of the temperature deviation from a 45
days running mean, for the winter 1981-1982, obtained from radiosonde
data up to 30 km and by lidar above 30 km. Contour lines are plotted
by steps of 10 K. Gray areas correspond to warming whereas dotted
areas represent cooling from the mean value. The arrows above indicate
the days when lidar data were recorded. The large arrow in the bottom
indicate the time of a strong minor warming.

4.1. Temporal and vertical structure at OHP

During the period of westerly prevailing wind in the lower stratosphere, from October to March, the propagation of the planetary waves from the troposphere to the middle atmosphere is possible and a succession of warmings and coolings is observed with a quasi-systematic anticorrelation between the perturbations in the upper stratosphere and in the lower mesosphere, as it is shown in Figure 4 for the winter 1981/82. The periods of these perturbations are determined by a temporal spectral analysis of the data. This may be

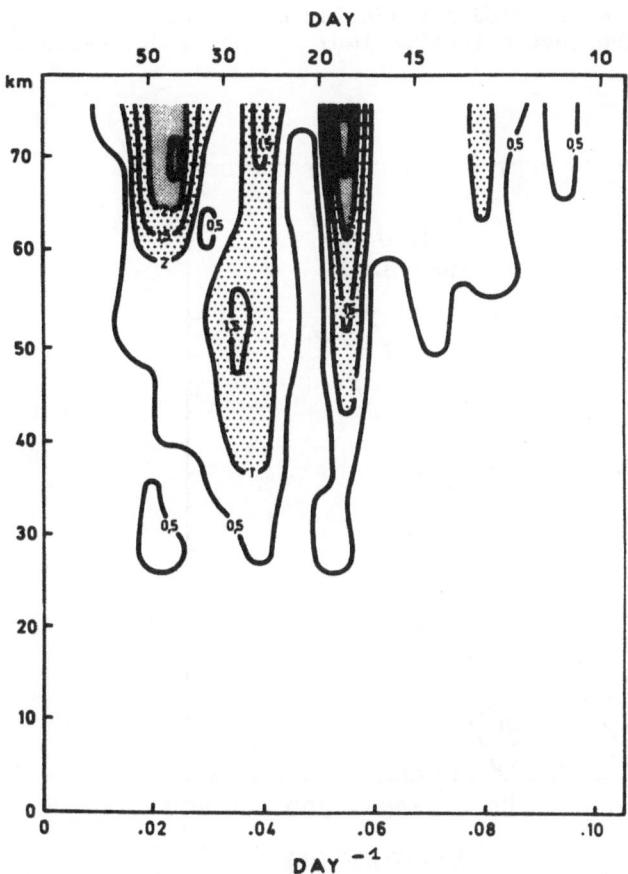

Figure 5. Spectral analysis of the pressure variations in percent by 0.005 day^{-1} for the winter 1981-1982. Contour lines are plotted in steps of 0.5 % / 0.005 day^{-1}.

done with the pressure data derived from the density profiles with an initialisation of the pressure profile at 30 km from radiosonde data. In general, the pressure spectra are easier to interpret than the temperature spectra in terms of planetary waves because they are less contaminated by the small scale fluctuations. The pressure spectrum for winter 1981/82 (Figure 5) shows a sharp peak at 18 days and broader components near 25 and 40 days. The 18 days oscillation has been identified as the well-known zonal wave 1 "16 days" Rossby wave with the help of the radiance maps of the SSU (Stratospheric Sounding Unit) near 45 km (Hauchecorne and Chanin, 1982). The nature of this wave is confirmed by the growth of its amplitude with altitude shown in Figure 6. The slope of growth is in very good agreement with the theoretical growth for Rossby waves with a scale height 7/2 H where H is the scale height of the atmosphere. The amplitude of this wave is half of that observed by Madden (1978) at 60°N, which may be explained by the fact that the theoretical maximum of the 16 days wave occurs at 60°N.

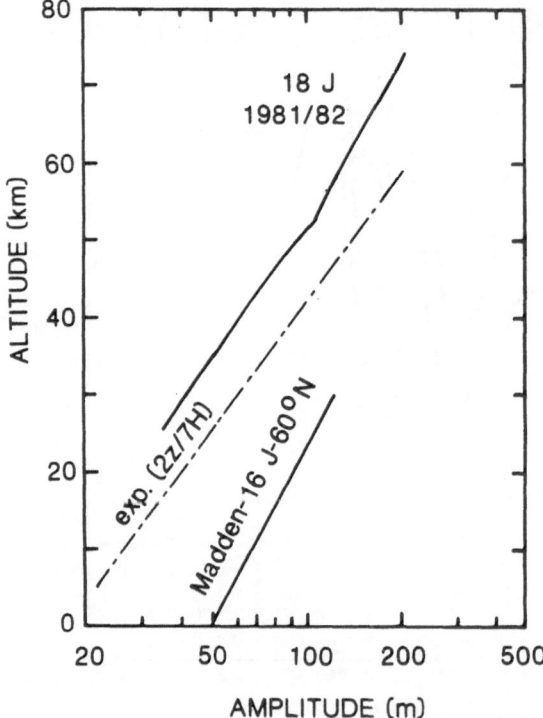

Figure 6. Rossby wave amplitude as a function of height for the winter 1981-1982 compared with the experimental result of Madden (1978) and the theoretical expectation.

In summary the main characteristics of the temperature perturbations observed by lidar in the period range 10 to 60 days are :
- Westward propagating Rossby waves of which two modes have been identified, a wave 1 mode with a period from 14 to 19 days and a wave 2 mode with a period from 11 to 13 days.
- Vacillations of the zonal circulation, with periods from 25 to 60 days, associated with the succession of upper stratospheric warmings in the polar upper stratosphere.

4.2. Horizontal structure during MAP/WINE

During the MAP/WINE campaign (Winter in Northern Europe) which took place during the winter 1983-84, the results obtained locally at OHP have been extended on the horizontal scale by using the rocketsonde temperature profiles obtained at Andoya (69°N, 18°E), Volgograd (48°N, 45°E) and Heiss Island (81°N, 58°E) (Hauchecorne et al., 1987). The time-height sections of the temperature above the 4 stations from December 1983 to March 1984 (Figure 7) show the horizontal extension of the observed perturbations. The warming that occurs near December 10 in the upper stratosphere is clearly visible at OHP, Andoya and Heiss Island, but not at Volgograd, whenever the one occurring near mid-February is observed at the 4 sites. It is followed by a warming in the middle stratosphere at the end of February, more developed at high latitude (Andoya and Heiss Island) than at middle latitude (OHP and Volgograd). The temporal spectral analysis of these data (Hauchecorne et al., 1987), confirms the results obtained at OHP. It shows that periods longer than 25 days are clearly related to the succession of upper stratospheric warmings, while periods from 10 to 20 days are (at least partially) due to the westward propagating Rossby waves which, with a 12.5 days period, are attributed to the second symetric mode of the wave 2.

4.3. Modeling of the wave 1- zonal flow interaction

A semi-spectral model of the wave 1 - zonal flow interaction, based on the primitive equation formulated by Holton (1976), has been used to interpret these results (Hauchecorne, 1985). The model takes into account the seasonal variation of the radiative equilibrium of the atmosphere. The forcing imposed at the lower boundary of the model at 10 km is either a stationary wave, or the sum of a stationary and a westward traveling wave. In the absence of forcing, the zonal wind follows a regular evolution (Fig. 8a) with a maximum of westerly wind at the winter solstice (day 180). In the presence of a stationary wave forcing with an amplitude below a critical value of about 200 m (Figure 8b) the regular evolution of the zonal wind is conserved, but as soon as the forcing reaches this value some vacillations of the zonal flow appear (Figure 8c). The time interval between two maxima of the wind is about 35 to 40 days, about equal to the time interval observed between two upper stratospheric warmings. The evolution of the temperature deviation above a fixed location in presence of a stationary wave surimposed on a 18 days traveling-wave at 45°N, 0°E

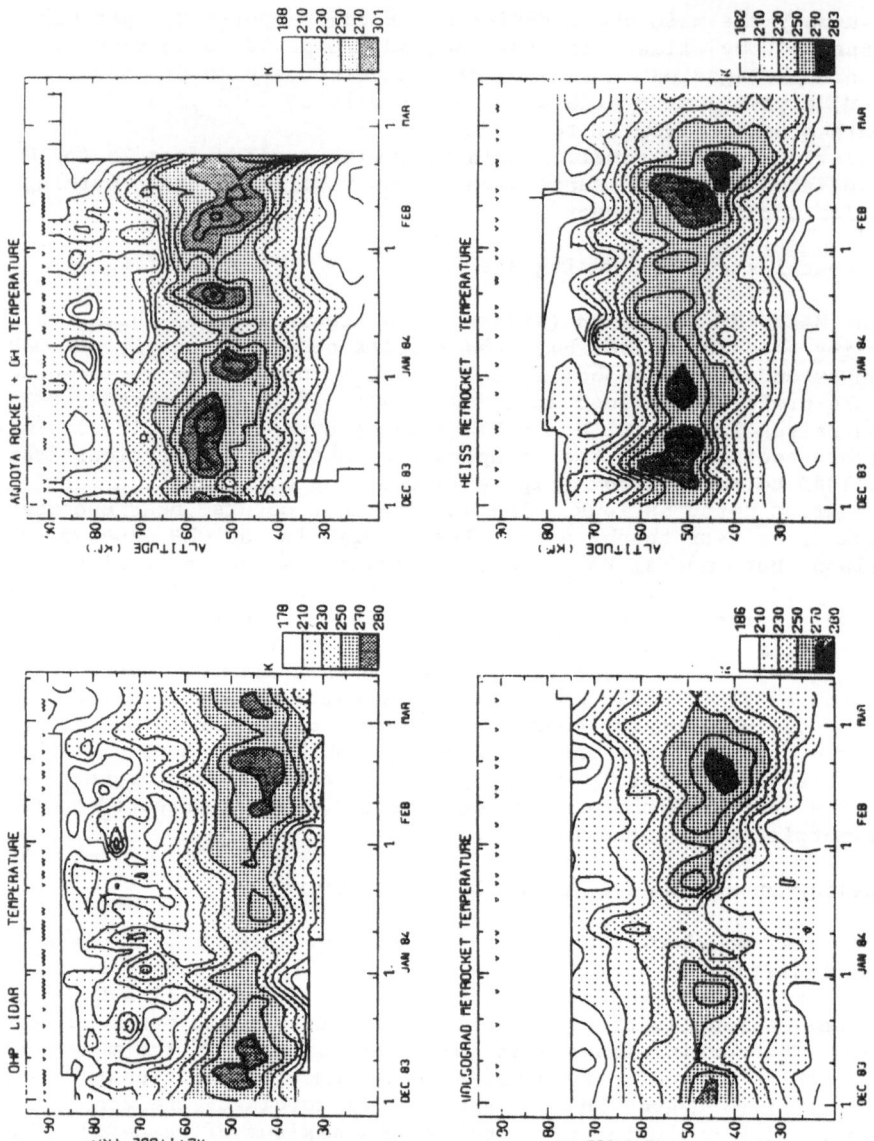

Figure 7. Temperature height contours obtained by lidar at O.H.P. (44°N, 6°E) and by rockets at Andoya (69°N, 18°E), Heiss Island (81°N, 58°E) and Volgograd (48°N, 45°E) during the MAP/WINE campaign.

(Figure 9) is qualitatively similar to the one observed in the lidar
data (Figure 4) with the same downward displacement of warm pulses
from the upper mesosphere to the stratosphere. However the model
underestimates by a factor of 2 the amplitude of the perturbations.

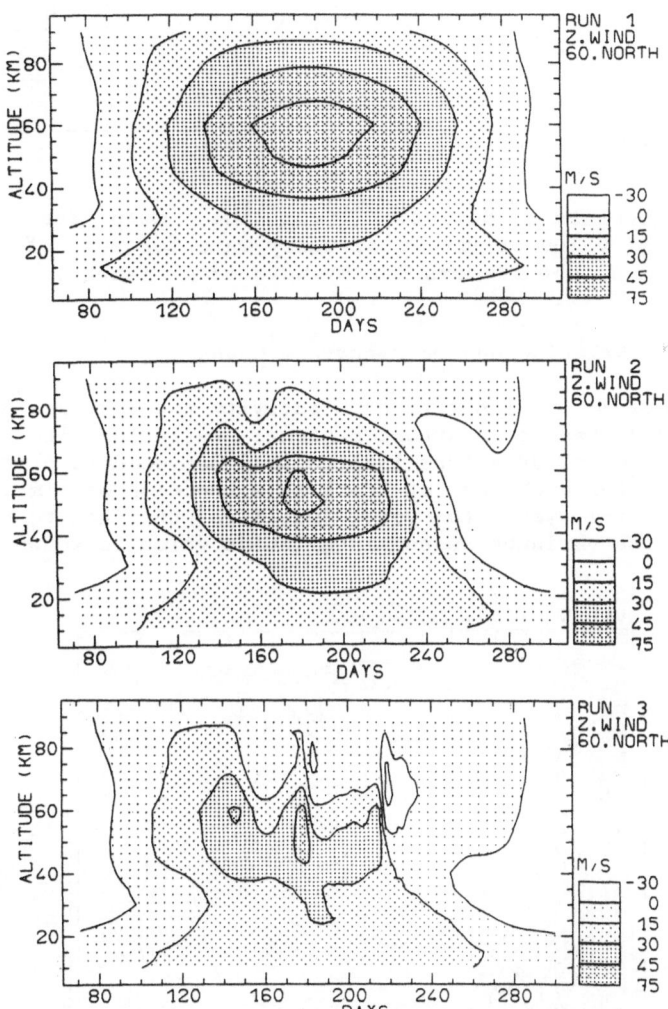

Figure 8. Wind contours calculated from the model for a latitude of
60°N when the amplitude of the stationary forcing is 0 (run 1), 182 m
(run 2), 212 m (run 3).

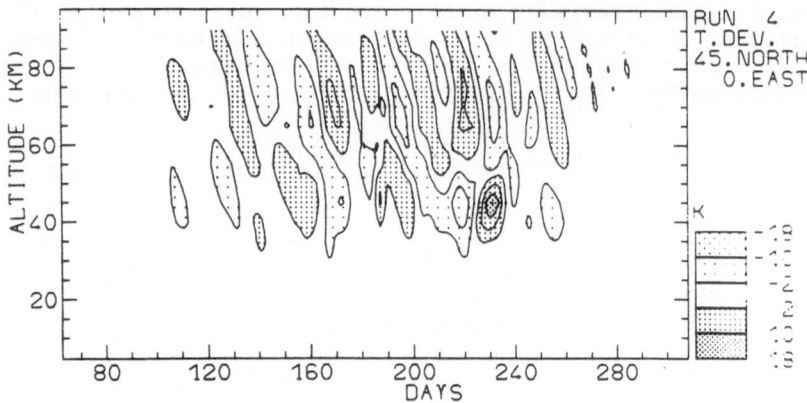

Figure 9. Temperature deviation from the mean for 45°N, 0°E calculated for a value of the stationary wave forcing of 212 meters and an amplitude of the 18-days Rossby wave of 42 meters.

4.4. Day-to-day variation due to planetary waves

As a consequence of planetary wave activity the variance of the temperature should present a strong seasonal variation with a maximum in winter. The set of data from 1981 to 1986 was used to plot contour of the variance for each year of data as a function of time. The strong similarity from year to year justifies the fact to represent the average of the variance seasonal variation which is shown in Fig. 10.

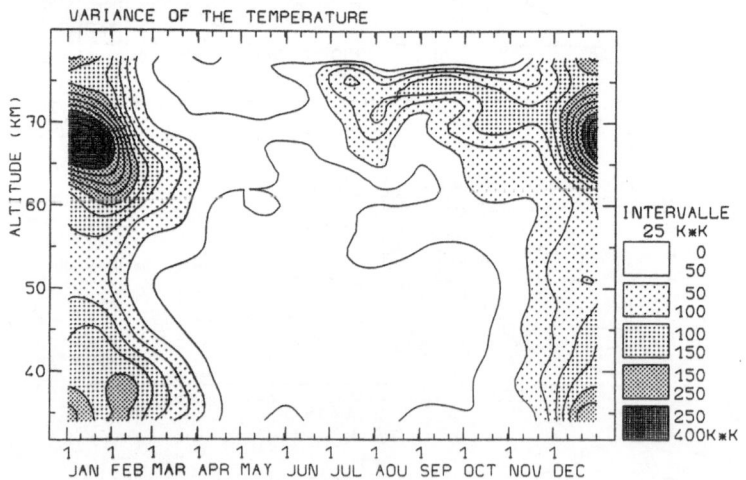

Figure 10. Contour map of the temperature variance for the period 1981-1986.

Note the sharp minimum of variance around 50-55 km, and the assymetry between Spring and Winter equinoxes in the mesosphere. The opposite behaviour of the stratosphere and mesosphere during Winter does not appear in this figure as the quantity represented is the square of the temperature fluctuations.

5. LONG TERM STUDY

The set of temperature data acquired since 1979 presents all the requirements to look at long term trends for the site of OHP. The large number of data, the regularity of their time coverage and the absolute accuracy without any external calibration make it a unique ensemble.

To carry out this study, taking into account the large seasonal variation of the temperature variance, we treated the data by separating winter, summer and equinoxes, and used the monthly average temperature. Furthermore to decrease the influence of gravity wave structures on the vertical profile, the night-time average profile was degraded in height resolution with a 3 km low frequency filter. The straight forward observation of the temperature monthly average as a function of time indicates clearly a decrease of the mean temperature of the mesosphere (- 20 K) and an increase (+ 20 K) in the strato-sphere since 1982. As the solar flux, after having been almost constant for the period 1979-1981, started also to decrease at that time, it seems obvious to relate both changes (Fig. 11). The correlation functions between the temperature and the 10.7 cm solar flux monthly means are presented for winter and summer in fig. 12. A strong and clear positive correlation appears in the mesosphere both in winter and summer with a maximum peaked around 65 km in winter and more largely spread in altitude in summer. In the stratosphere the correlation is clearly negative in winter whereas, it is barely significant in summer. A minimum of variability is observed around 50 km.

The similarity between the characteristics of the long term variation with the solar cycle and of the day to day variance is striking : i.e. different behaviours between summer and winter, maxima of variability with opposite phase at 65 and 35 km, minimum of variability at 50 km... It induces us to conclude that the solar induced effect on the mean temperature of the middle atmosphere is due to the indirect action of the solar flux on the planetary wave activity. If such an effect were to be explained by photochemistry, the maximum of variability should have been at 50 km where no variation is observed. Furthermore 2D models which take into account the direct solar flux influence on photochemistry predicted an amplitude of variation (< 2 K) one order of magnitude smaller than the one observed here (20 K). Those elements led us to conclude on the role of the solar cycle in the long term change of the planetary wave activity. A more detailed presentation and discussion of these results are given in Chanin et al. (1987).

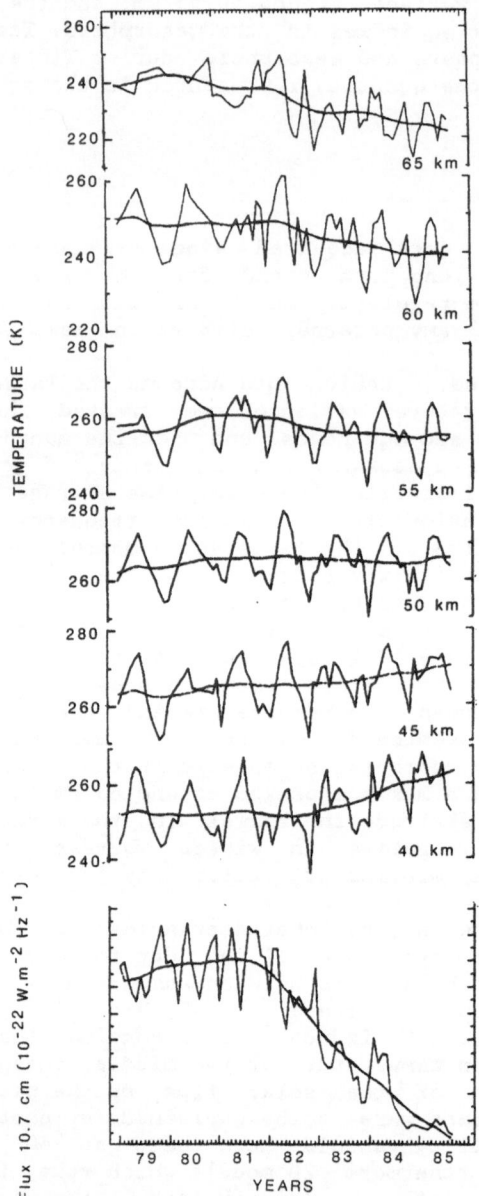

Figure 11. Monthly mean of the temperature measured by lidar between 40 and 65 km by step of 5 km and monthly mean of the 10.7 cm solar flux during the 1979-1985 period. A running mean over a year is superimposed on each set of data.

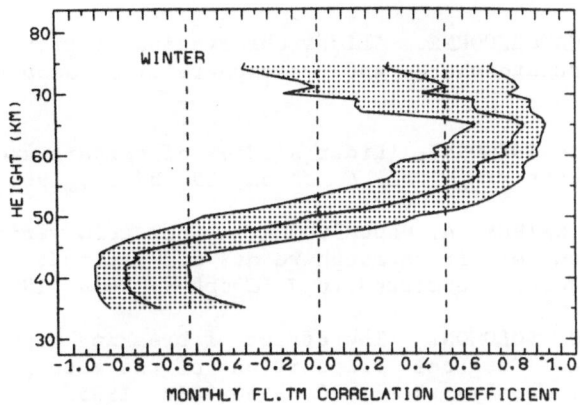

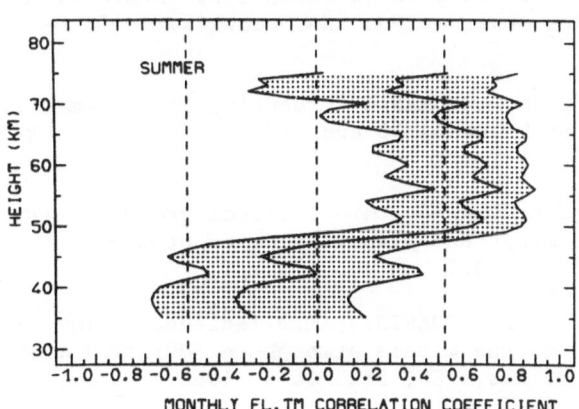

Figure 12. Correlation function (with ± 95 % confidence limits) between the monthly mean temperature and the solar flux at 10.7 cm for winter (N D J F) (a) and for summer (M J J A) (b) for the period 1979-1985. The significant level of 99 % is shown.

6. CONCLUSION

The selection of results obtained by the Rayleigh lidar technique at OHP emphasizes the large role played by dynamics in inducing variations of all time scales in the structure of the middle atmo- sphere. The interpretation of the results presented here are somehow limited by the local character of the measurements but the spreading of such instruments to form a world network should contribute largely to a better understanding of the dynamical processes and of their interaction.

REFERENCES

CHANIN M.L., A. HAUCHECORNE, 'Lidar observation of gravity and tidal waves in the stratosphere and mesosphere'. *J. Geophys. Res.*, 86, 9715-9721, 1981.

CHANIN M.L., A. HAUCHECORNE, 'Lidar studies of temperature and density using Rayleigh scattering'. *MAP Handbook*, 13, 87-99, 1984.

CHANIN M.L., N. SMIRES, A. HAUCHECORNE, 'Long term variation of the temperature of the middle atmosphere at mid-latitude : relationship with the solar cycle'. Submitted to *J. Geophys. Res.*, 1987.

GARCIA R.R. and S. SOLOMON, 'The effect of breaking gravity waves on the Dynamics and Chemical composition of the mesosphere and lower thermosphere'. *J. Geophys. Res.*, 90, 3850-3868, 1985.

HAUCHECORNE A., 'Planetary waves mean flow interaction in the middle atmosphere : Lidar observations and modelisation'. *MAP Handbook*, 18, 80-88, 1985.

HAUCHECORNE A., M.L. CHANIN, 'Density and temperature profiles obtained by lidar between 30 and 70 km'. *Geophys. Res. Letters*, 7, 564-568, 1980.

HAUCHECORNE A., M.L. CHANIN, 'Mid latitude ground-based lidar study of stratospheric warmings and planetary waves propagation'. *J. Atm. Terr. Phys.*, 44, 577-583, 1982.

HAUCHECORNE A., M.L. CHANIN, 'Mid-latitude lidar observations of planetary waves in the middle atmosphere during the winter of 1981-1982'. *J. Geophys. Res.*, 88, 3843-3849, 1983.

HAUCHECORNE A., T. BLIX, R. GERNDT, G.A. KOKIN, W. MEYER and N.N. SHEFOV, 'Large-scale coherence of the mesospheric and upper stratospheric temperature fluctuations'. To be published in *J. Atm. Terr. Phys.*, 1987.

HOLTON J.R., 'A semi-spectral numerical model for wave-mean flow interactions in the stratosphere : Application to sudden stratospheric warmings'. *J. Atmos. Sci.*, 33, 1639-1649, 1976.

LINDZEN R.S., 'Turbulence and Stress owing to gravity wave and tidal breakdown'. *J. Geophys. Res.*, 86, 9707-9714, 1981.

MADDEN R.A., 'Further evidence of travelling planetary waves'. *J. Atmos. Sci.*, 35, 1605-1618, 1978.

SMITH S.A., D.C. FRITTS and T.E. VANZANDT, 'Evidence for a saturation spectrum of atmospheric gravity waves'. *J. Atmos. Sci.*, 44, 1404–1410, 1987.

VANZANDT T.E., 'An universal spectrum of buoyancy waves in the atmosphere'. *Geophys. Res. Lett.*, 9, 575-578, 1982.

VINCENT R.A., 'Radar observations of gravity waves in the mesosphere'. *In this book*, 1987.

ALPHABETICAL INDEX OF THE AUTHORS

SUBJECT INDEX